PROGRESS IN COLLOID & POLYMER SCIENCE

Editors: H.-G. Kilian (Ulm) and G. Lagaly (Kiel)

Volume 93 (1993)

Trends in Colloid and Interface Science VII

Guest Editors:
P. Laggner (Graz) and O. Glatter (Graz)

Springer-Verlag Berlin Heidelberg GmbH

Die Deutsche Bibliothek — CIP-Einheitsaufnahme

Trends in colloid and interface science.

Früher begrenztes Werk in verschiedenen Ausg.

7 (1993)
(Progress in colloid & polymer science; Vol. 93)
ISBN 978-3-662-16052-7   ISBN 978-3-7985-1676-2 (eBook)
DOI 10.1007/978-3-7985-1676-2

NE: GT

ISSN 0340-255 X

Originally published by Dr. Dietrich Steinkopff Verlag GmbH & Co. KG, Darmstadt in 1993
Softcover reprint of the hardcover 1st edition 1993
Chemistry editor: Dr. Maria Magdalene Nabbe; English editor: James C. Willis; Production: Holger Frey, Bärbel Flauaus.

Type-Setting: Graphische Textverarbeitung, Hans Vilhard, D-64753 Brombachtal

Preface

From September 21 to 25, 1992, 284 active participants from 28 nations around the globe gathered for the VI. European Colloid and Interface Society (ECIS) conference in Graz. In addition to the classical physical and chemical disciplines normally represented at these events, a remarkable number of more biomedically or technologically oriented scientists gave contributions by lectures, oral communications or posters and thus demonstrated the fast-growing interdisciplinary character of today's research in colloid and interface science.

The event coincided, if not exactly by day and month, with the 90th birthday of Otto Kratky, one of the outstanding pioneers of x-ray small-angle scattering, a key method in the field of colloid science. The conference was honored by Kratky's presence and by the lively interest he still keeps in this field that owes much to him. Both of the institutions involved in the conference organization, the Institute of Biophysics and X-Ray Structure Research of the Austrian Academy of Sciences, and the Institute of Physical Chemistry of the Karl-Franzens University, have highly justifiable claims on his patronage.

This volume gives an account of the presentations. With the high number of registrants and the quality of their contributions it has, in the first place, been a delicate task to make the selection for the final program. For their invaluable help with this, we are most grateful to the members of the Program Advisory Committee. In shaping this volume, and to keep it within a tolerable size, we had to ask the authors to be rather brief in their written contributions. The plenary lectures, keynote lectures, and oral contributions appear as articles, alas with limited length, in the order as they were presented in the program. For the numerous posters, many of them absolutely current and outstanding in quality, we had to be strict in only allowing the presentation of an abstract. We apologize for this and ask the colleagues who fell into this category for their kind understanding.

We, as chairmen of the event, took great pleasure in hosting the VI. ECIS conference in Graz. This pleasure derives, to a great extent, from the tireless efforts of our collaborators in the scientific and organizational committees, who not only helped us in reaching decisions, but much more valuably, helped in the actual day-to-day running of the preparations and in all technical and logistic problem-solving during the days in Graz. In times like ours, when old political borders have fallen and are sometimes replaced by new ones, and where the mechanisms of transition are often not so clear, it was only due to the ingenuity of all concerned that the sudden appearance of administrative tasks did not affect the conference, where the climate was governed by stimulating and sometimes ardent scientific discussion. This compilation of papers cannot reflect that climate, but we hope nevertheless that it will serve as a reference base for further fruitful interaction among participants and also for those who, for one reason or the other, were unable to come to Graz.

On behalf of ECIS, we want also to acknowledge the financial support by corporate participants and sponsors (AVL-Laser Vertriebsges. GmbH, Bartelt GmbH, Bohlin Instruments GmbH, Brookhaven Instruments Corporation, Heraeus GmbH, Hoechst AG, Krüss GmbH, Lauda GmbH, MBraun Graz Optical Systems GmbH, Mütec GmbH Co KG, Malvern Instruments GmbH, A. Paar GmbH, Physica Meßtechnik GmbH, Suck-Wisent, Verder-Retsch GmbH). Their participation not only contributed valuably to interest in the event, but also made it possible through their contributions to award fellowships to young scientists and colleagues from the former East Bloc Countries. In the same sense, we want to express our gratitude to the Austrian Federal Ministry of Science and Technology, the Government of Styria, and the Mayor of Graz.

P. Laggner,
O. Glatter (Graz)

Contents

Poster session I

Poster session II

Synthesis "in situ" of nanoparticles in reverse micelles

M. P. Pileni[1,2]), I. Lisiecki[1,2]), L. Motte[2]), C. Petit[1,2]), J. Cizeron[1]), N. Moumen[1]) , and P. Lixon[2])

[1]) Université Pierre et Marie Curie, Laboratoire SRSI Paris, France
[2]) CEN Saclay, DRECAM-SCM, Gif sur Yvette, France

Abstract: Functionalized reverse micelles are used to control the size and the polydispersity of the metallic and semiconductor particles. The size of the particles can be controlled, either by the amount of water solubilized in the droplets, or by the solvent used to form reverse micelles, or by adding macrocyle molecules.

Key words: Reverse micelles — nanoparticles

Introduction

Nanosize particles constitute a wide class of catalysts, and because catalysis occurs on the surface, there is a substantial economic incentive to obtain small monodisperse particles. However, this raises two main problems. One is fundamental in nature and addresses the question as discussed below, regarding the particle size at which metallic or semiconductors properties are lost. The other is more practical and concerns the preparation of very small particles.

Surfactants dissolved in organic solvents form spheroidal aggregates called reverse micelles [1]. Water is readily solubilized in the polar core, forming a so-called "water pool," characterized by w, $w = [H_2O]/[AOT]$. With AOT as surfactant, the maximun amount of bound water in the micelle corresponds to a water-surfactant molar ratio $w = [H_2O]/[AOT]$ of about 6. Above $w = 15$, the water pool radius depends linearly on the water content $(R_w = 1.5\ w)^2$. Another property of reverse micelles is their dynamic character [1]. Upon collision, reverse micelles exchange their water content which makes possible co-precipitation or chemical reactions from reactants dissolved in separately droplets.

In aqueous solution the reducing of metallic ions or the coprecipitation reaction of semiconductor induce flocculation of the products. In reverse micellar solution, under various experimental conditions, particles clusters are formed. From these it can be concluded that AOT reverse micelles prevent flocculation.

Colloidal dispersions of metals exhibit either a well defined absorption band or broad regions of absorption in the ultraviolet-visible range. These are due to the excitation of plasma resonances or interband transitions and are characteristic property of the metallic nature of the particles. From Mies' theory [3], the absorption spectra have been simulated, for particles ranging in size from 1 nm to 10 nm and the resulting simulations are compared to the asorption spectrum in the bulk phase. For metallic particles with a diameter below 4 nm, the absorption band is broad. By increasing the size of the particles the absorption band is better resolved [4—6] with a peak centered at 570 nm. For silver clusters the plasmon peak is well resolved with a maximum centered at 400 nm. In both cases (copper and silver clusters) the shape of the absorption spectrum depends on the cluster sizes: the decrease in the size induces a strong decrease, either of the 570 nm peak or in the width of the 400 nm absorption band for copper and silver clusters, respectively.

Cadmium sulphide suspensions are characterized by an absorption spectrum in the visible range. In the case of small particles, a quantum size effect [8—11] is observed due to the perturbation of the electronic structure of the semiconductor with the change in the particle size. For CdS semiconductor, as the diameter of the particles approaches the excitonic diameter, its electronic properties start to change [11]. This gives a widening of the forbidden band and therefore a blue shift in the absorption threshold as the size decreases. This phenomena occurs as the crystallite size is comparable or below the excitonic diameter of 50—60 Å [11]. In first approximation a simple "electron-hole in a box" model can quantify this blue shift with the size variation [8, 10, 11]. Thus, the absorption threshold is directly related to the average size of the particles in solution.

In the present paper, we report the influence of the water content, on the solvent on the size and on the environment of metallic and semiconductors crystallites.

Experimental

Synthesis of nanoparticles [12, 13]

The synthesis is carried out by mixing two micellar solutions with the same ratio of water ($w =$ [H_2O]/[AOT]), one containing a solution in which either the reducing agent or sulphide ions (Na_2S) are solubilized and the other mixed sodium (AOT) and copper, silver or cadmium (AOT) reverse micelles in isooctane.

Results and Discussion

1. Change in the size of the crystallites with the water content

1.1. Metallic particles

Hydrazine as a reducing agent is added to a mixed micellar solution containing copper and sodium AOT. A change in the water content of the micelle, w, is obtained by adding H_2O to the micellar solution before the reaction takes place.

Figure 1 shows the absorption spectra of the colloidal particles, electronic microscopy pictures and the histograms observed 5 h after adding the reducing agent, for reverse micelles with various w values. At low water content a continuous absorption spectrum with a shoulder at 570 nm is observed. Upon increasing the water content, the 570 nm band characteristic to a copper cluster appears progressively. The electron micrographs show an increase in the size of the particles from 2 to 10 nm upon increasing the water content from 1 to 10. At water content up to 10, the size of the particles remains unchanged, but the polydispersity increases. Electron diffractograms compared to a simulated diffractogram of bulk metal copper are in good

agreement, indicating a crystalline face-centered cubic structure (f.c.c) with a lattice constant of 3.61.

The change in the particle size with water content can be explained in terms of interfacial water structure: at low water content copper ions are not totally hydrated and the effective number of ions participating in the chemical reduction is small. The increase in the water content induces an increase in the number of copper ions which react with hydrazine. This favors the growth of the average particles. At relatively high water content (up to 10), copper ions are totally hydrated and free water molecules are present. This favors diffusion of copper ions inside the droplet. Electrostatic interactions between the head polar groups of the surfactant and copper ions compete with hydration energy. The difference in these energies remains constant upon increasing the water content which keeps the particle size constant.

Similar behavior is observed by replacing copper by silver ions. At low water content the absorption band is broad. By increasing the water content the intensity of the absorption band increases with a decrease in the width. According to Mie's theory, this change in the aborption can be attributed to an increase in the size. This is confirmed by the electron microscopy pattern in which is observed an increase in the size from 5 to 20 nm with the water content.

1.2. Semiconductors particles

Figure 2 shows a red shift in the absorption spectra by increasing the water content. According to the data previously published using sodium di (ethyl-2-hexyl) sulfosuccinate (AOT) as a surfactant [13] this can be attributed to an increase in the average size of particles with the water content.

Below the absorption onset several shoulders are observed (Fig. 2) and can be clearly recognized in the second derivative (inserts Fig. 2). These weak absorption bands correspond to the excitonic transitions. This clearly shows a narrow size distribution [14]. At low water content the first excitonic peak is well resolved and is followed by a bump. The second derivative shows a very high intensity of this bump (insert Fig. 2 A). With small crystallite, according to the data previously published [14], several bumps due to several excitonic peaks are expected. Insert Fig. 2 A shows only one bump. This is due to the fact that the others are blue shifted and are not observable in our experimental conditions. By

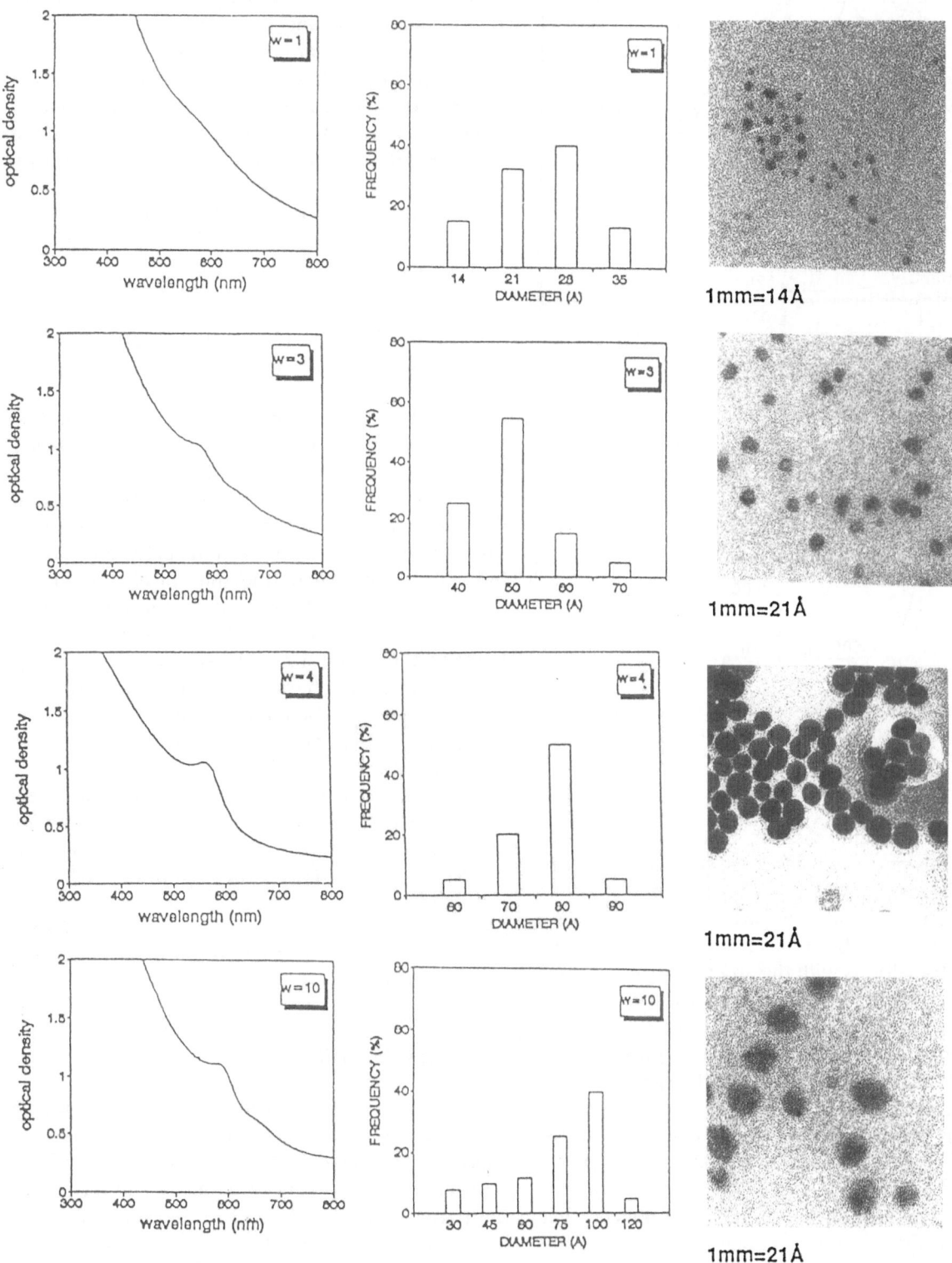

Fig. 1. Absorption spectra, electron microscopy, and histograms of copper metallic particles prepared in reverse micelles with various water contents

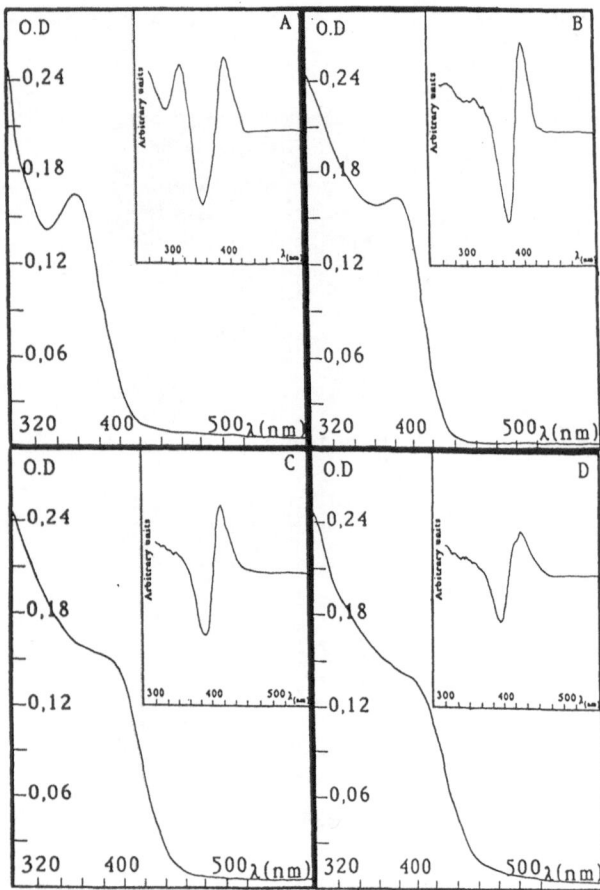

Fig. 2. Variation of the absorption spectrum of CdS in reverse micelles with the water content in isooctane: A (w = 5); B (w = 10); C (w = 20); D (w = 40)

increasing the water content, that is to say, by increasing the size of the particles, several bumps are observed (insert Fig. 2B). The intensity of these bumps decreases with the water content, w (inserts Fig. 2). This indicates a decrease in the number of excitonic transitions with the size of the particle and is in agreement to the theoretical calculations previously published for the Q-particles [14].

From the relation between [8, 11, 14] the absorption onset and the size of CdS particle, the average radius, r, is deduced. Figure 3A shows a strong change in the size of the particle with the relative ratio of cadmium and sulphide ions (x = $[Cd^{2+}]/[S^{2-}]$). The largest sizes are obtained for x = 1 and the smallest for x = 2. It can be noticed that the size of CdS is always smaller when one of the two reactants is in excess (x = 1/4, 1/2, 2). This confirms that the crystallization process is faster when one of the species is in excess [15].

The crystallites have been extracted from reverse micelles and the solution dried to obtaine powder. After adding solvent to this powder, nanosize particles remain unchanged. Electron microscopy has been performed from the powder semiconductor dispersed in pyridine. Figure 4 A and B shows an increase of the particle size with the water content. A good agreement between the size determination obtained from the absorption onset and from the extracted particles is obtained. The electron rays' diffraction pattern indicates the particles keep ZnS crystalline structure (f.c.c) with a lattice constant equal to 5.83 Å.

2. *Variation of the cluster's size by changing the composition of reverse micelles*

In AOT-water-solvent solution, the interaction between droplets strongly depends on the bulk solvent [18] and on the additives [19]: the attractive interactions between droplets increase by increasing the length of hydrocarbon chain of the bulk solvent. These changes in the interactions between droplets induce an increase of the intermicellar exchange kinetic rate constant by using as the bulk solvent cyclohexane, isooctane, and decane, respectively. Because of this, the size of the silver clusters observed by using cyclohexane as the bulk solvent is smaller than that obtained by using isooctane (Fig. 4 C and D). Similar data are obtained with CdS semiconductors (Fig. 3 B).

Similarly, it has been shown [19] that solubilization of macrocycles such as crown ether or kryptant induces an increase in the attractive interactions between droplets. This favors the intermicellar exchange process. From Fig. 3 D an increase in the size of CdS particles by crown ether and kryptant addition to the AOT-water-isooctane micellar solution can be observed.

3. *Changes of oxidation state of clusters*

By using sodium borohydride as the reducing agent, in the absence of oxygen, an absorption spectrum characteristic of copper metallic particles is observed. Figure 5 shows a decrease in the size of the particles from 28 to 3 nm upon increasing the water contents. At low water content (w = 3 and 5) the particles are homogeneously dispersed. Upon increasing the water content from w = 4 to w = 8, the particles become progressively associated. From

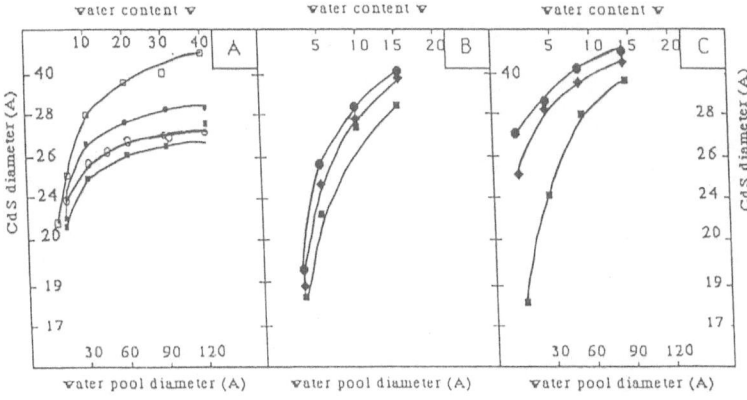

Fig. 3. Variation of the absorption onset and the CdS radius with the water content, w:
A) at various X values in isooctane ($X = 2$: ■, $X = 1/4$: ○, $X = 1/2$: ●, $X = 1$: □);
B) in varius solvent cyclohexane(■); isooctane (◆) and decane (●);
C) in isooctane, in the absence (■) and in the presence of crown ether (◆) and kryptant (●).

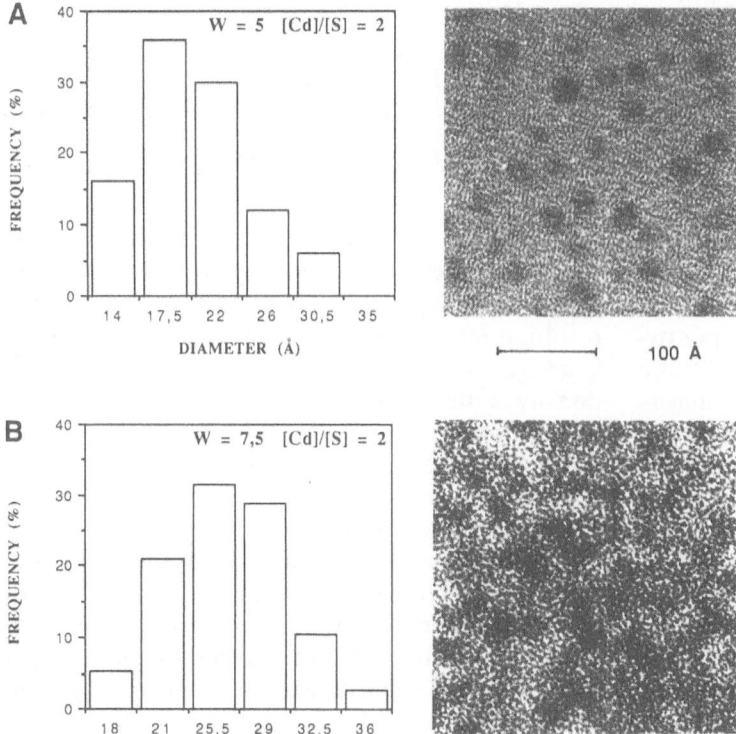

Fig. 4A,B. Electron microscopy and histrograms of CdS particles after extraction from a micellar solution in isooctane:
A) $w = 5$, $x = 2$, B) $w = 7.5$, $x = 2$

electron diffraction patterns for particles fromed in micelles with low water content ($< 3 < w < 5$), the clusters are face-centered cubic with a lattice dimension equal to 3.61, similar to what it is obtained in copper metallic bulk phase. At high water content ($w = 9$ and 10), a cubic phase with a lattice parameter equal to 4.27 characteristic to Cu_2O copper oxide is observed.

The shape of the absorption spectrum of the colloidal particles changes with water content: at $3 < w < 6$, the shapes of the absorption spectra are similar with changes in the intensity and a small residual absorption at 800 nm. Upon increasing the water content, an absorption around 800 nm appears. According to Yanase and Komiyama [7] the 800 nm absorption is attributed to an oxide monolayer surrounding the copper metallic cluster. By comparing the microscopy data and the absorption spectra, it can be deduced that at low water content metallic particles are formed. By increasing

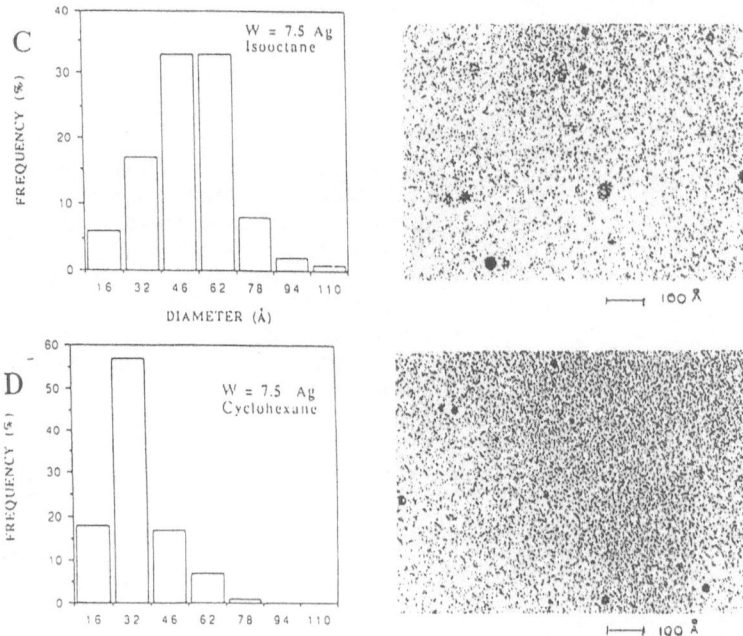

Fig. 4C,D. Electron microscopy and histograms of silver particles synthesised in micellar solution at $w = 7.5$, in isooctane (C) and in cyclohexane (D)

the water content, the copper metallic particles are surrounded by mono- and multi-layers of oxide which prevent further growth of the particles. At higher water content (above $w = 8$) pure copper oxide particles are obtained. This is confirmed by the fact that below $w = 7$, electron microscopy images show well-dispersed particles, where as at $w = 7$ and 8 they are associated. This is probably due to the high affinity of the surfactant for the oxide. At $w = 9$ and 10, the electron microscopy images do not show the presence of metallic particles. These data could again be related to the water structure in the water pool. At low water contents, water molecules are bound at the interface, where as at high water contents, free water is present and plays a role in oxide formation. Compared to the data presented above with hydrazine, the size of copper metallic particles formed with sodium borohydride is larger by one order of magnitude. This is due to a difference in the preparation mode.

4. Changes in the shape of the particles

By using pure copper AOT water-in-oil aggregates, it has been previously shown that a change in the shape of the aggregates is induced by increasing the water content. Droplets are observed at $w = 2$, while cylindrical aggregates are formed at $w = 4$[12]. The reduction of copper reverse micelles

was achieved with N_2H_4 as the reducing agent at various surfactant concentrations with a constant $w = 4$. The shape of the absorption spectrum of the resulting metal clusters is unchanged by increasing $Cu(AOT)_2$ concentration. The increase in optical density is due to an increase in the metallic particle concentration with the number of reverse micellar aggregates (Fig. 6). At low $Cu(AOT)_2$ concentration, the clusters are spherical. Upon increasing the $Cu(AOT)_2$ concentration, formation of cylindrical clusters is observed. The average number of cylindrical clusters remains constant (at about 10%) with 90% of the particles being spherical. The electron diffraction shows concentric circles characteristic of a face-centered cubic phase with a lattice dimension equal to 3.61 Å.

Conclusion

We report here the synthesis, in situ, of metallic and semiconductors clusters in reverse AOT micelles from functional surfactants. Reverse micelles are able to limit the size of growing metallic and semiconductor particles. The growth of the particles can be prevented by the local structure of the reverse micelle at low water content, by decreasing the intermicellar exchange process. The shape of copper metallic clusters is changed with the

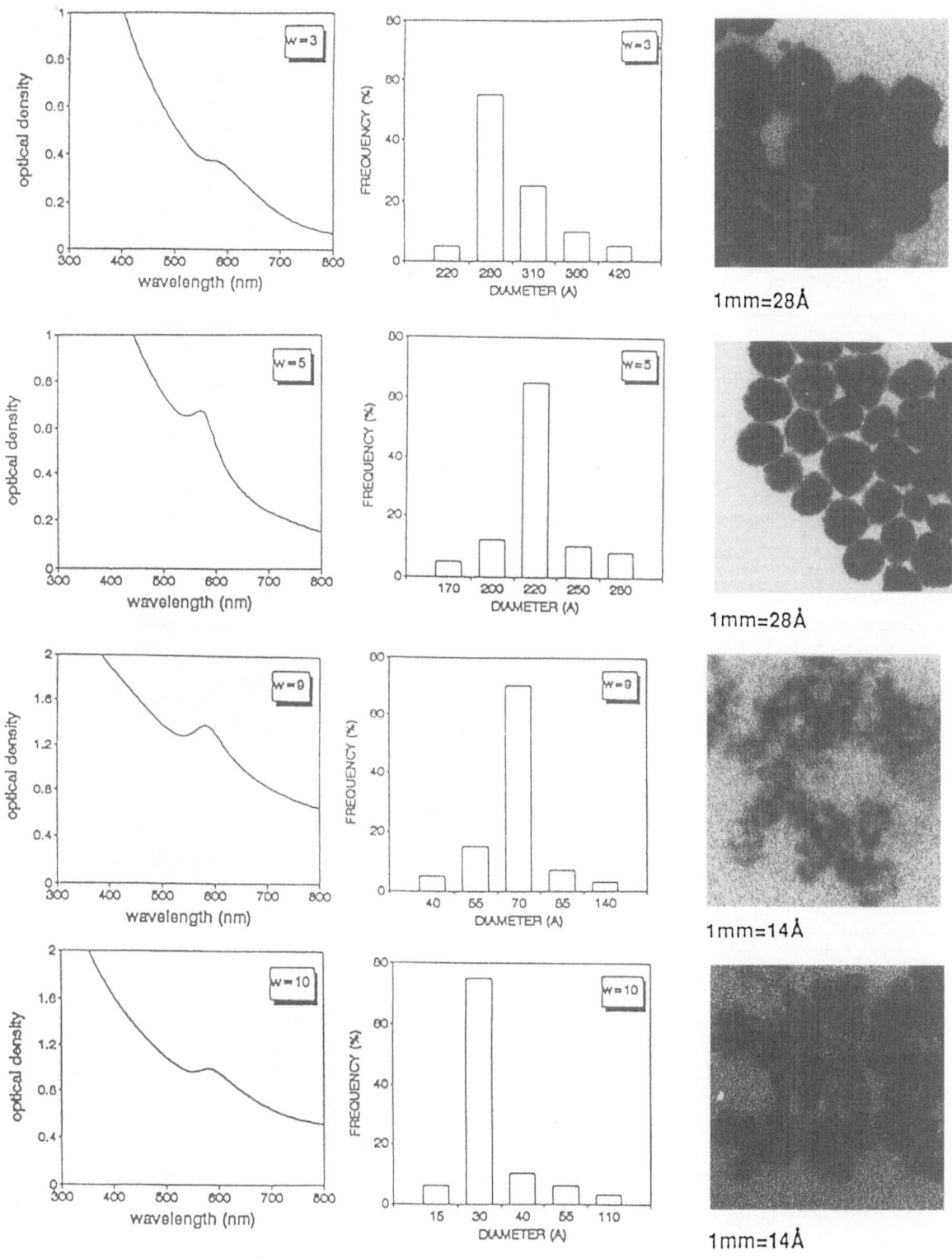

Fig. 5. Absorption spectra, electron microscopy, and histograms of metallic copper particles formed in reverse micelles at various water contents: $w = 3$; $w = 5$; $w = 9$; $w = 10$

Progress in Colloid & Polymer Science, Vol. 93 (1993)

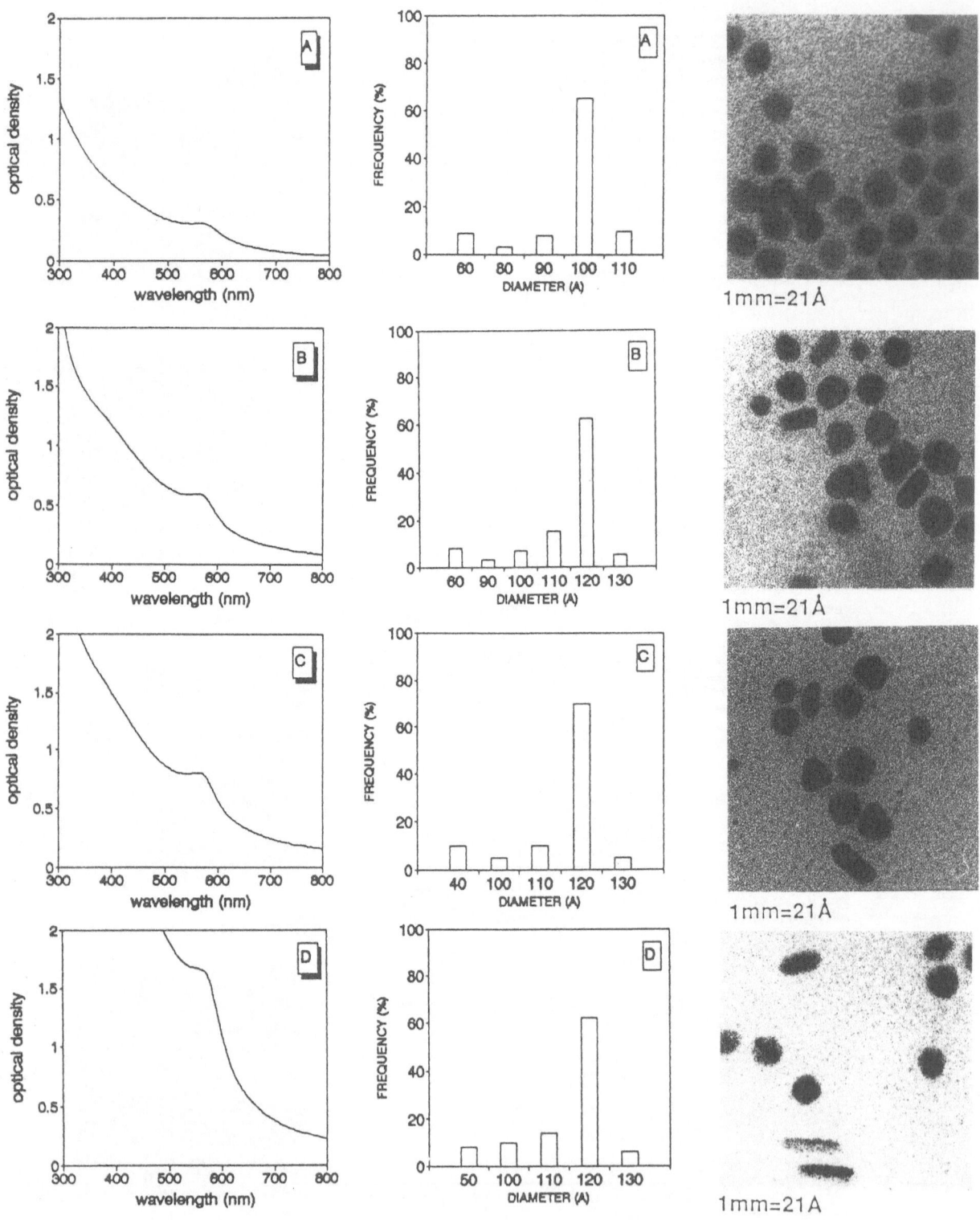

Fig. 6. Absorption spectra, electron microscopy, and histograms of metallic copper particles formed in reverse micelles at different copper concentrations:
A) $[Cu(AOT)_2] = 10^{-2}$ M; B) $[Cu(AOT)_2] = 2 \cdot 10^{-2}$ M, C) $[Cu(AOT)_2] = 4 \cdot 10^{-2}$ M; D) $[Cu(AOT)_2] = 10^{-1}$ M.

shape of the microaggregates containing the reactants. The microenvironment of the copper clusters can be changed, either to accomodate an oxide monolayer surrounding the metallic particles, or to the formation of borohydride, a copper oxide clusters.

References

1. Structure and reactivity in reverse micelles (1989) Pileni MP (ed) Elsevier: Amsterdam
2. Pileni MP, Zemb T, Petit C (1985) Chem Phys Lett 118:414
3. Mie G (1908) Ann d Physik 25:377
4. Creighton JA, Desmond GE (1991) J Chem Soc Faraday Trans 87(24):3881—3891
5. Truong VV, Scott GD (1977) J Opt Soc Am 67:502
6. Anno E, Tanimoto M, Yamaguchi T (1988) Phys Rev B 38:3521
7. Yanase A, Komiyama H (1991) Surf Scien 248:11—19
8. Brus LE (1983) J Chem Phys 79:5566
9. Bawendi MG, Steigerwald ML, Brus LE (1990) Annu Rev Phys Chem 41:477
10. Henglein A (1989) Chem Rev 89:1861
11. 11a. Wang Y, Herron N (1990) Phys Rev B 41:6079 — 11b. Wang Y, Herron N (1991) J Phys Chem 95:525
12. Petit C, Lixon P, Pileni MP (1991) J Phys Chem 7:2620
13. Petit C, Pileni MP (1988) J Phys Chem 92:2282
14. Katsikas L, Eychmüller A, Giersig M, Weller H (1990) Chem Phys Lett 172:201
15. Fisher CH, Weller H, Lume-Periera C, Janata E, Heinglein A (1986) Ber Bunsenges Phys Chem 90:46
16. Binks BP, Meunier J, Abillon O, Langevin D (1989) Langmuir Vol 5:(No)2
17. Aveyard R, Binks BP, Clark S, Mead JJ (1986) Chem Soc Faraday Trans I 82:125
18. Towey TF, Khon-Lodhi A, Robinson BH (1990) J Chem Soc Faraday Trans 86:(22)3757
19. Leser M, Kooijman M, Pollitte J, Magid LJ (1991) J Phys Chem 95:9013

Authors' address:

M. P. Pileni
Université P. et M. Curie
Laboratoire SRSI
Bâtiment F(74) BP 52
4 place Jussieu
75005 Paris, France

Preparation and properties of dispersions of colloidal boehmite rods

P. A. Buining, A. P. Philipse, C. Pathmamanoharan, and H. N. W. Lekkerkerker

Van't Hoff Laboratory for Physical and Colloid Chemistry, University of Utrecht, The Netherlands

Abstract: Dispersions of colloidal boehmite (γ-A1OOH) rods (of controllable length) are prepared from alkoxide precursors. A polyisobutene grafting procedure leads to sterically stabilized dispersions. In the organophilic system ($L \approx$ 200 nm $\pm$ 50%; $D \approx$ 10 nm $\pm$ 25%) an isotropic-nematic phase separation is observed for boehmite volume fractions between 3.8% and 14.6%. The effect of polydispersity on the phase diagram is well predicted qualitatively by Onsager's theory extended to bidisperse mixtures.

Key words: Boehmite rods — organophilic — isotropic-nematic — phase separation

Introduction

Concentrated dispersions of colloidal particles of rodlike shape may show interesting physical-chemical behavior, like the formation of an ordered nematic phase. The phenomenon of rod alignment has been observed in several aqueous dispersions, like V_2O_5 [1], β-FeOOH [2], γ-AlOOH [3], cellulose microcrystals [4] and tobacco mosaic virus [5]. These species show a hard-core interaction with an additional soft repulsion due to the surface charge.

Systematic phase studies of rodlike colloids were, until now, hindered by the lack of suitable systems. Further, it is desirable to study a system of rods showing a hard repulsion. Recently, we were able to prepare aqueous dispersions of boehmite (γ-AlOOH) particles of controllable rodlength in the range 100-500 nm and a width of around 15 nm [6]. These particles were sterically stabilized in cyclohexane by a polyisobutene grafting procedure [7].

This study concerns the isotropic-nematic phase separation as observed in the organophilic boehmite system of rods which are assumed to model a hard interparticle potential.

Isotropic-nematic phase separation

Concentrated dispersions of the organophilic boehmite rods ($L \approx$ 200 nm $\pm$ 50%, $D \approx$ 10 nm $\pm$ 25%) show an isotropic-nematic phase separation for boehmite volume fractions between

ϕ = 3.8% and ϕ = 14.6%. The nematic phase nucleates as a tactosol of spindle-shaped droplets (tactoids) in the isotropic phase which settle under gravity to form the continuous nematic phase. The experimental phase diagram of the measured volume fractions of the coexisting isotropic and nematic phases is shown in Fig. 1. Clearly, the phase behavior deviates from that expected for the monodisperse system, where the boehmite volume fractions of the coexisting isotropic and nematic phases would be constant with increasing overall volume fraction.

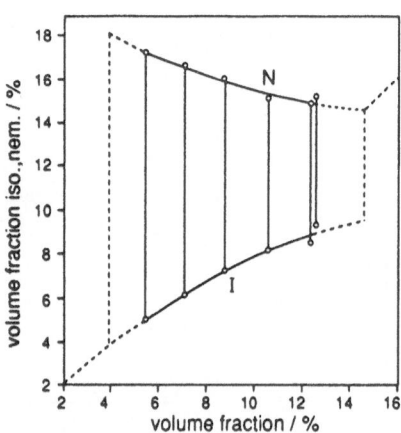

Fig. 1. Experimentally determined boehmite volume fractions of coexisting isotropic (I) and nematic (N) phases as a function of the overall boehmite volume fraction. From these values, the phase diagram is constructed.

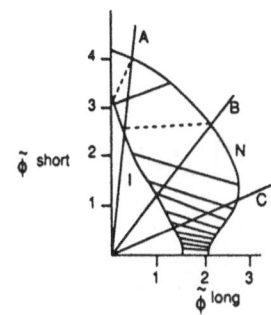

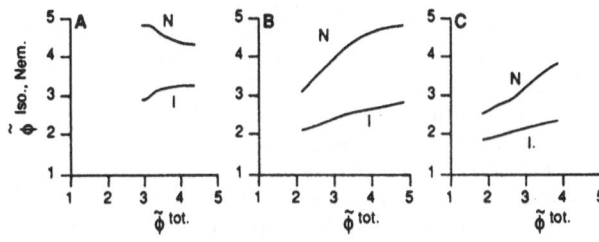

Fig. 2. Upper plot: Theoretical phase diagram for the isotropic (I) to nematic (N) phase separation of the bidisperse system of short (L_1) and long (L_2) rods with $L_2/L_1 = 2$, based on calculations of Lekkerkerker et al. [8]. The tielines imply a fractionation effect: the long rods go preferentially into the nematic phase. The presented volume fractions are scaled $(\tilde{\phi} = (L_1/D)\phi)$. Lower plots: Theoretical scaled volume fractions of coexisting isotropic and nematic phases as a function of the overall volume fraction, constructed from the tielines in the upper plot

To investigate the effect of polydispersity on the phase diagram, we took the model of the bidisperse mixture. Figure 2 gives the theoretical phase diagrams for a bidisperse mixture of short (L_1) and long (L_2) particles having the same diameter D. The diagrams are constructed for $L_2/L_1 = 2$, using the calculations of Lekkerkerker et al. [8], who extended the Onsager [9] approach to the bidisperse case. The qualitative trends in the experimental phase diagram can be well explained by a large ϕ^{short}/ϕ^{long} ratio. The dispersions of the boehmite rods indeed contain a minor amount of very long particles (see [6]).

Concluding remarks

This is the first time, to our knowledge, that the I-N phase transition is observed and studied ac-

curately in a system of rodlike particles where the interparticle potential is assumed to be hard. The effect of polydisperity on the phase diagram is qualitatively predicted by the theory for bidisperse hard rods.

References

1. Zocher H (1925) Z Anorg Chem 147:91
2. Maeda Y, Hachisu S (1983) Colloids Surfaces 6:1—16
3. Zocher H, Török C (1960) Kolloid Z 170:140—144
4. Marchessault RH, Morehead FF, Walter NM (1959) Nature 184:632—633
5. Fraden S, Hurd AJ, Meyer RB, Cahoon M, Caspar DLD (1985) J Physique 46:C3-85 — C3-113
6. Buining PA, Pathmamanoharan C, Jansen JBH, Lekkerkerker HNW (1991) J Am Ceram Soc 74:1303—1307
7. Buining PA, Veldhuizen YSJ, Pathmamanoharan C, Lekkerkerker HNW (1992) Colloids Surfaces 64:47—55
8. Lekkerkerker HNW, Coulon P, van der Haegen R, Deblieck R (1984) J Chem Phys 80:3427—3433
9. Onsager L (1949) Ann NY Acad Sci 51:627—657

Authors' address:

P. A. Buining
Van't Hoff Laboratory
University of Utrecht
Padualaan 8
3584 CH Utrecht, The Netherlands

Progress in Colloid & Polymer Science

Progr Colloid Polym Sci 93:12—14 (1993)

Dynamic light scattering from concentrated particle dispersions

P. Štěpánek

Institute of Macromolecular Chemistry, Academy of Sciences of the Czech Republik, Prague, Czech Republic

Abstract: Using dynamic light scattering, we studied the properties of light scattered from latex dispersions over a very broad range of volume fraction of latex, ϕ, between very dense ($\phi \sim 0.35$) and very dilute ($\phi \sim 10^{-6}$) dispersion. We show that, at intermediate ϕ, the spectrum of decay times obtained by Laplace inversion of the measured autocorrelation function is bimodal. The slow component, corresponding to single scattering, can be determined up to volume fraction of latex of $\phi \sim 0.01$. In these dense dispersions the second virial coefficient of diffusion, k_D, is about two orders of magnitude larger than the value expected for hard spheres. For volume fraction of spheres above $\phi \sim 0.05$, the limit of strong multiple light scattering is attained, where diffusive transport of light is observed.

Key words: Multiple light scattering — dynamic light scattering — Laplace inversion — particle diffusion

Introduction

Multiple light scattering (MLS) is a feature that is often present in light scattering experiments (both static and dynamic), but of which it is difficult to evaluate the effect on the measured quantities. We generally try to suppress multiple light scattering by sufficiently diluting the system under consideration, or by index matching the scatterers to their surrounding. In a dynamic light-scattering experiment, a cross-correlation arrangement can also be used to eliminate MLS [1]. This technique is, however, experimentally difficult regarding aligning both scattered beams to intersect in the scattering volume at complementary angles.

Cases where MLS cannot be avoided include concentrated particle dispersions, phase separating systems, critical phenomena, and samples of industrial interest. Properties of multiply scattered light have been theoretically analyzed by several authors, both for the static [2] and dynamic [3, 4, 5] case; in particular, corrections for double scattering have been evaluated [4, 6]. It has also been found that the intensity of light multiply scattered from concentrated suspensions is enhanced in the backscattering direction [7].

In this contribution we extend our previous work [8] on MLS to include depolarized light scattering and to focus on the properties of single scattering as extracted from MLS data.

Experimental

The multiple scattering medium was formed by various latex dispersions prepared by diluting with distilled water a stock dispersion of polystyrene latex with diameter $d = 120$ nm and volume fraction of latex particles $\phi = 0.35$. The dispersions were placed in square spectroscopic cells with side length 10 mm.

The incident light had a wavelength $\lambda = 632.8$ nm (Spectra Physics 125 A HeNe laser), the correlation function was generated by an ALV 5000 correlator. Correlation functions were measured at an angle of $90°$, both for polarized and depolarized scattering. The incident light beam and the observation axis intersected in the center of the scattering cell.

The measured autocorrelation functions were analyzed, using the technique of inverse Laplace transformation, described by Eq. (1) below. For a homodyne experimental setup, we have

$$g^2(t) - 1 = \alpha [\, \textstyle\int A(\tau) e^{-t/\tau} d\tau]^2 \,, \tag{1}$$

where $g^2(t)$ is the normalized intensity correlation function which is measured, and $A(\tau)$ is the distribution of decay times τ characterizing the scattered field, which is the quantity of physical interest. The Laplace inversion was performed using a non-linear fitting routine REPES [9].

Results and discussion

Figure 1 shows typical correlation functions and spectra of decay times for polarized scattering for low, intermediate, and high volume fractions of latex, $\phi = 1.1 \times 10^{-5}$, 6×10^{-2}, and 1×10^{-2}, respectively. At low volume fraction, only single scattering is present, as seen in Fig. 1d, where the single component corresponds to the normal Brownian diffusion of latex particles in the dispersion. At intermediate volume fractions, Fig. 1e, a second, faster component appears, which represents the multiple scattering contribution. The single scattering process is still visible, but with a shorter decay time and a smaller amplitude. At high volume fractions, Fig. 1f, only the multiple scattering component is present in the decay times spectrum.

The decay rates $\Gamma = 1/\tau$ for both components, determined from the moments of the peaks as in Figs. 1d—f, are collected as a function of volume fraction ϕ in Fig. 2. It is seen that the single scattering contribution can be determined from the data in an interval of 3.5 decades in volume fraction, up to $\phi = 0.011$ (which is already a very turbid sample), where the amplitude of the single scattering only represents a few percent of the total intensity scattered by the dispersion. From this decay rate the diffusion coefficient D can be evaluated for each particular volume fraction, using the standard relation [10] $\Gamma = DK^2$; here, K is the scattering vector. For the lowest volume fraction, $\phi = 3 \times 10^{-6}$, we get $D = 3.58 \times 10^{-8}$ cm²/s, from which we obtain the hydrodynamic radius of the latex spheres, $R_h = 62$ nm. This value was obtained using the Stokes-Einstein relation

$$D = \frac{kT}{6\pi\eta R_h} , \qquad (2)$$

where k is the Boltzmann constant, T the absolute temperature, and η the viscosity of the solvent.

It is of interest to evaluate the second virial coefficient of diffusion, k_D from the relation

$$D = D_0(1 + k_D\phi) . \qquad (3)$$

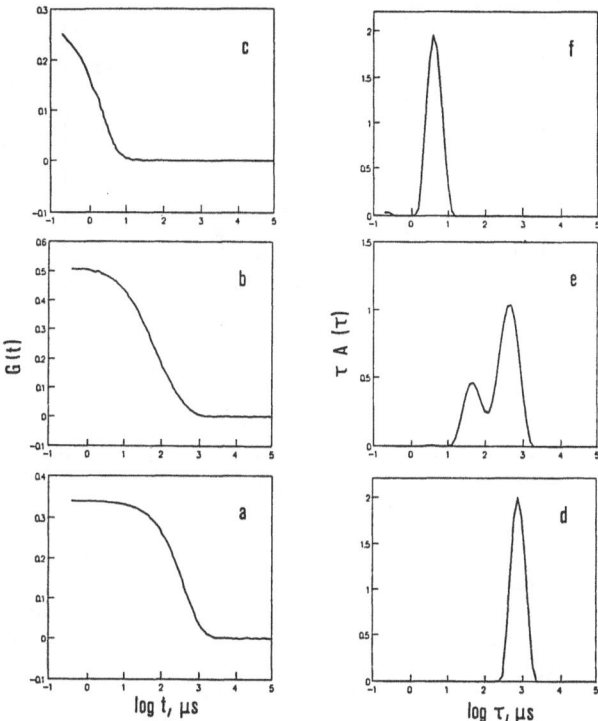

Fig. 1. Normalized intensity correlation functions obtained at an angle of 90° on a latex dispersion of volume fraction a. 1.1×10^{-5}, b. 6×10^{-2}, c. 1×10^{-2}. Figures 1d—f are the corresponding spectra of decay ties as obtained by REPES

Using the D values corresponding to the single scattering process at all the concentrations where this component could be determined, we obtain $k_D \sim 100$. This value is two orders of magnitude higher than the values of k_D reported for a dilute dispersion of hard spheres, $K_D = 1.0$ to 1.3 [11 — 13]. This discrepancy is probably due to extra interactions between particles present in this latex system and to the influence of the third virial coefficient of diffusion in these nondilute systems. Further experiments are in progress to elucidate this point.

The multiple scattering contribution can be detected starting from $\phi = 3.5 \times 10^{-5}$ in depolarized scattering and from $\phi = 1.7 \times 10^{-5}$ in polarized scattering, and then its decay rate coincides with that determined for the depolarized scattering (see Fig. 2). The decay rate of the muliple scattering component is at low volume fractions about twice that for single scattering, which approximately corresponds to the decay rate of double scattering [3, 6]. The multiple scattering rate increases with increasing volume fraction to reach, above $\phi = 3.5 \times 10^{-2}$, the limit for strong multiple scattering.

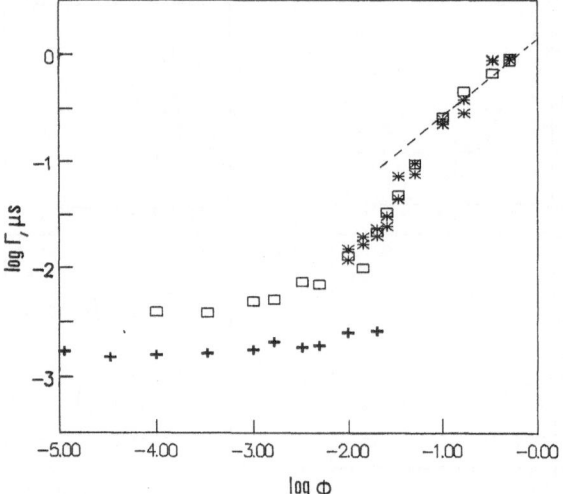

Fig. 2. Decay rates Γ of the fast and slow component of the peaks in the spectra of decay times; + polarized single scattering, □ depolarized double, oligo and multiple scattering, * polarized multiple scattering. The straight line shows the expected exponent of 2/3 valid in the limit of strong multiple scattering

This limit is characterized by a diffusive transport of light [13] through the dispersion; the scattering is isotropic and fully randomized, i.e., the intensities of the polarized and depolarized scattering are equal. It has been shown [13] that, for the diffusive transport of light, the following relation is valid

$$\Gamma \sim L^2, \tag{4}$$

where L is the path length of the primary laser beam in the sample, i.e., the geometrical thickness of the dispersion. Relation (4) is usable for a given dispersion placed in cells of variable thickness. It can be readily transformed for the case of dispersions of a variable volume fraction of latex placed in a cell of a fixed thickness, to yield

$$\Gamma \sim \phi^{2/3} . \tag{5}$$

The expected exponent is represented by a straight line in Fig. 2. It is seen that the last four samples with ϕ above 3.5×10^{-2} satisfy relation (4).

Conclusion

We have shown that, as volume fraction ϕ of latex particles in a dispersion is increased, five different regimes of scattering can be observed:
— region I: only polarized single scattering;
— region II: in addition to I, depolarized double or oligo scattering is observed;
— region III: both single and multiple scattering components can be resolved in the DLS data;
— region IV: only multiple scattering is seen, both polarized and depolarized, but the dispersion is not yet dense enough to produce diffusive transport of light;
— region V: the strong multiple light scattering limit, leading to diffusive transport of light.

References

1. Phillies GDJ (1981) J Chem Phys 74:260—262
2. Dhont JKG (1983) Physica 120A:238—262
3. Van Rijsvijk, FC Smith UL (1976) Physica 83A: 121—129
4. Dhont JKG (1985) Physica 129A:374—380
5. Maret G, Wolf PE (1987) Z Phys B65:409—414
6. Ivanov AP, Chairulina AY, Chaikovski AP (1973) Opt Spectrosk 25:1153—1160
7. Wolf PE, Maret G (1985) Phys Rev Lett 55:2696—2699
8. Konak C, Jakes J, Stepanek P, Petras F, Karska M, Krepelka J, Perina J (1991) Applied Optics 30:4865—4871, Konak C, Jakes J, Petras F, Karska M, Perina J (1990) Coll Czech Chem Commun 55:1022
9. Jakes J (1988) Czech J Phys B38:1305—1316
10. Berne BJ, Pecora R (1976) Dynamic Light Scattering, Wiley, New York
11. Konak C, Tuzar Z, Štěpánek P, Sedlacek B, Kratochvil P (1985) Progr Colloid Polym Sci 71:15—19
12. Kops-Werkhoven MM, Fijnaut HM (1981) J Chem Phys 74:1618—1622
13. Hess W, Klein R (1976) Physica 85:509—513
14. Weitz DA, Pine DJ, Pusey PN, Tough RJA (1989) Phys Rev Lett 63:1747—1750

Author's address:

Dr. Petr Štěpánek
Institute of Macromolecular Chemistry
of the Czech Republic Academy of Sciences
Heyrovskeho nam.2
162 06 Praha 6, Czech Republic

Progress in Colloid & Polymer Science   Progr Colloid Polym Sci 93:15—21 (1993)

Reaction-controlled self-assembly of amphiphilic membranes

H. J. Watzke

Institut für Polymere, ETH-Zürich, Switzerland

Abstract: Reaction-controlled vesiculation of single-chain surfacant mixtures is proposed as a versatile model system to study self-assembly processes coupled to chemical reactions. In particular, mixtures of single-chain cationic and anionic amphiphiles have proven to be useful self-vesiculating systems. The strong electrostatic interactions of the headgroups induce strong non-ideal behavior which leads eventually to the formation of closed-shell bilayer structures (vesicles). The use of amphoteric zwitterionic amphiphiles as components in the mixtures enabled the design of switchable headgroup interactions by protonation/deprotonation equilibria (switch-box systems). Reaction-controlled self-assembly of amphiphilic membranes could be established in mixtures of N,N-dimethyl-N-dodecylamine oxide (DDAO) [1643-20-5] and sodium N-lauroyl-N-methyltaurinate (SLT) [4337-75-1]. Emphasis was directed to the in situ study of the micellar-vesicular reorganization. The amphiphilic mixtures were characterized by various physico-chemical methods (electron microscopy, light scattering, pH titrations, etc.).

Key words: Amphiphilic membranes — self-organization — reaction-controlled — self-assembly — N,N-dimethyl-N-dodecylamine oxide — sodium N-lauroyl-N-methyltaurinate — headgroup interactions

Introduction

Most reactions are performed in solutions. Optimization of the reaction medium is an important part of chemistry because chemical processes are sensitive to changes in their reaction environment. In recent years a new concept was introduced in designing non-homogeneous reaction media based on amphiphilic aggregates which locally change the distribution of reacting species due to accumulation and depletion or solubilization of otherwise insoluble substances [1].

For this purpose, amphiphilic solutions provide a variety of different aggregational structures, both in aqueous and non-polar organic solutions. An intriguing aggregate structure is found in vesicles which are formed by a closed-shell bilayer isolating a water volume from the bulk. However, the drawback of closed-shell bilayer structures is their impermeability to ionic or hydrophilic molecules after formation. Furthermore, vesicles exhibit an intrinsic instability which leads to aggregation and fu-

sion ending in extended lamellar liquid crystalline bodies.

A reversible reorganization from micellar state to vesicles would allow the incorporation and the release of hydrophilic guest molecules in a controlled way. It also provides a route to the self-formation of stable vesicular aggregates because there is no need to apply external energy for their formation.

Vesicles are usually produced by forced dispersion of double-chain amphiphiles in aqueous solutions. Over the last 20 years reports sporadically appeared which indicated that for certain amphiphilic systems spontaneous formation of vesicles (or liposomes) occurs [2]. Recently, it could be shown that mixtures of cationic-anionic single-chain amphiphiles form unilamellar vesicles under proper conditions [3—8]. The headgroup interactions in cationic-anionic mixtures are known for their non-ideal behavior leading to synergistic effects like strong reduction in surface tension and lower critical micellar concentrations (CMC) [9]. The headgroup interactions also determine the self-vesiculation process of cationic-anionic amphiphile mixtures [10].

We report on the use of a single-chain amphiphilic system which is able to switch, under chemical control, between non-ionic-anionic and cationic-anionic interactions.

Description of the system

Our switch-box system is depicted in scheme 1. N,N-dimethy-N-dodecylamine oxide (DDAO) can be transformed by protonation from the zwitterionic to the cationic from, which eventually produces with N-lauroyl-N-methyltaurinate (SLT) an ion-pair.

protonation   ion-pair formation

DDAO   DDAOH   SLT

DDAO itself forms micelles above a CMC of 0.002 moles dm^{-3} [11, 12]. The zwitterionic form (DDAO) is maintained at high pH where the oxygen is deprotonated. The cationic form (DDAOH) appears under protonation. The pK_a of DDAO was found to be 4.95 in the monomeric solution [11]. Both DDAOH and DDAO have the same CMC of 0.002 moles dm^{-3} [11]. DDAOH and DDAO show attractive interactions in micellar aggregates because of the hydrogen bonding between the OH group of DDAOH and the negatively charged oxygen of DDAO [11]. DDAO was also found to form vesicles in a three-component system containing DDAO, n-hexanol, and water at certain compositions [13].

Sodium N-lauroyl-N-methyltaurinate (SLT) forms micelles above a CMC of 0.0087 moles dm^{-3} [14]. SLT was chosen as anionic component because it

contains a tertiary amid bond which was found to enhance the self-vesiculation properties of single-chain anionic amphiphiles in mixtures [7, 8]; secondly, it has a sulfonate headgroup which will not interfere with the DDAO protonation equilibrium because of its low *pKa*.

The DDAO-SLT system has two parameters controlling its vesiculation. It was found that the self-vesiculation in single-chain mixed micelles depends on the overall composition [3—8]. Therefore, the total composition of DDAO and SLT will determine the occurrence of the vesiculation. On the other hand, the protonation degree of DDAO in the micelles will control the interacting composition of cationic and anionic species and, therefore, the realization of the vesiculation.

Additionally, the DDAO-SLT system provides us with the opportunities to study chemical reactions under the conditions of mixed micelles or closed-shell bilayer structures using the same amphiphilic molecules and to control the net charge of the aggregates both in number and sign.

Experimental Methods

N,N-dimethyl-N-dodecylamie oxide (DDAO) and sodium N-lauroyl-N-methyltaurinate (SLT) were purchased from Fluka, Switzerland, and used as received. Mixtures of both amphiphiles were prepared in different aqueous solutions of the same concentration. Mixing the respective volumina of DDAO and SLT stock solutions gave samples with the desired overall composition. The overall composition was expressed by Ψ, the compositional parameter. Ψ is defined by $\Psi = [SLT]/([SLT] + [DDAO])$.

Buffer reactions

The concentration of the buffer solutions was 0.05 moles dm^{-3}. Phosphate buffers were made from mixtures of NaH_2PO_4/Na_2HPO_4 of different ratios, producing two solutions of different starting pH (e. g., pH 7.05 and the other of 5.00). The buffer solution of the higher pH contained the surfactant mixture. Appropriate volumes of these buffer solutions were rapidly mixed to produce a solution with the desired final pH. Turbidity and pH were measured simultaneously over an extended period of time (typically around 40 min). The turbidity measurements were performed on a Hewlett-Packard Diode Array spectrometer 8452 A equipped with a Hewlett-Packard Vectra 05/65 computer.

Titration procedure

Boiled deionized millipore water was stored under CO_2 exclusion and used as titration solvent. The background electrolyte NaCl [0.01 moles dm^{-3}] was added prior to titration. The NaCl solution was purged by water-vapor-saturated nitrogen gas. Stock solutions of different compositions were neutralized by addition of 1 M HCl, giving an overall neutralization degree of r, which is defined by $r = [HCl]/[C_t]$. The concentrated stock solutions of the respective surfactant mixtures were introduced into the sodium chloride solution by a peristaltic pump (MS-Reglo, Ismatec SA) with a constant pump velocity of 0.0772 mL min^{-1}. During titration the solution was kept under nitrogen gas. Changes of the pH were measured with Ingold integrated pH electrodes (Ingold, Switzerland) and recorded by a Metrohm Potentiograph E 336 A from Metrohm, Switzerland.

Quasi-elastic light scattering

Hydrodynamic radii of vesicles were determined by quasielastic light-scattering at 25°C. The light-scattering was performed on a Malvern Spectrometer 4700, under an angle of 90°, using a Coherent INOVA 200 argon ion laser ($\lambda = 488$ nm) as light source. The data were analyzed by an exponential sampling inverse Laplace transform program as described by Ostrowsky and coworkers [15]. The samples were centrifuged prior to the measurement to remove larger dust particles.

Freeze-fracture electronmicroscopy

Freeze-fracture electronmicroscopy was perfomed following established procedures of Müller and coworkers [16]. Samples for freeze-fracture were prepared by propane jet vitrification. The platinum-carbon replicas were inspected in a Philips EM 301 electron microscope.

Results and Discussions

Sodium N-lauroyl-N-methyltaurinate (SLT) easily forms vesicles in mixtures with permanent charged cationic partners like N-dodecylpyridinium chloride at appropriate compositions (Fig. 1). The vesicles appear to be similar to those observed in mixtures of N-acylated amino acid anionic amphiphiles with cationic surfactants [7, 8].

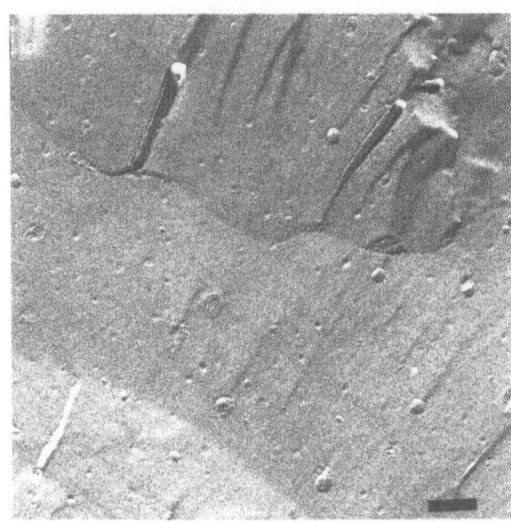

Fig. 1. Freeze-fracture electron micrograph of a mixture of sodium N-lauroyl-N-methyltaurinate (SLT) with N-dodecylpyridinium chloride (DPC). Total surfactant concentration was 0.050 moles dm^{-3} ($\Psi = 0.6$). Bar length indicates 100 nm. Samples were taken from the surfactant solution at 25°C

Mixing DDAO with SLT leads to an increase of the solution pH. Figure 2 summarizes this behavior for three different total concentrations of 20 mM, 2 mM and 0.2 mM. Even at the low concentration of 0.2 mM a slight increase is observed around the composition of $\Psi = 0.5$. The equilibria involved in the formation of the ion-pair are [17]

$$DDAO + H_2O \rightleftharpoons DD\overset{\oplus}{A}OH + OH^{\ominus}$$

$$DD\overset{\oplus}{A}OH + LT^{\ominus} \rightleftharpoons DDAOH \cdot LT .$$

The ion-pair formation drives the equilibria to the right side and increases the solution pH. This behavior was also reported for other mixed systems, e. g., DDAO with sodium dodecylsulfate (SDS) [18], with sodium perfluorooctanoate [19], or with sodium dodecylbenzenesulfonate [20]. Rosen and coworker investigated the interaction of DDAO with potassium dodecylsulfonate (PDS) below the CMC by surface-tension measurements [13]. They found a strong interaction between DDAO and PDS leading to ion-pair formation already in the monomeric solution, which solely determined the surface tension reduction [13].

The CMC values of the different mixtures were determined by the pH response methods introduced by Engberts and Brackmann for alkyl phospha-

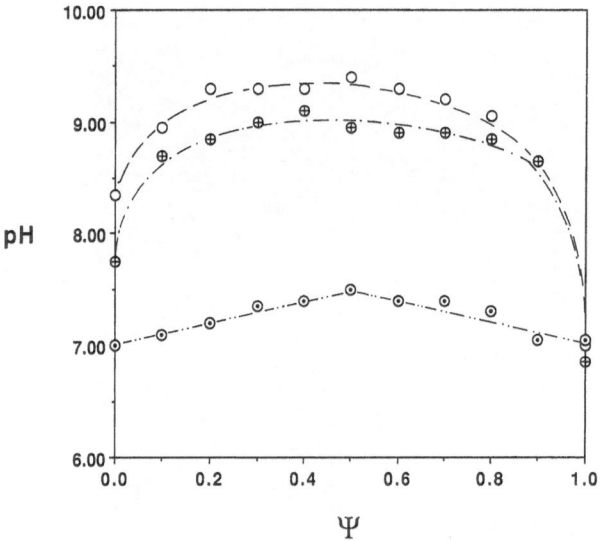

Fig. 2. pH changes in solutions of various compositions of sodium N-lauroyl-N-methyltaurinate (SLT) with N,N-dimethyl-N-dodecylamine oxide (DDAO) at different total concentrations. O $[C_t]$ = 0.020 moltes dm^{-3}, ⊕ $[C_t]$ = 0.002 moles dm^{-3} and ⊙ $C_t]$ = 2.0 × 10^{-4} moles dm^{-3}. Sample temperature was 25°C

Table 1. Values of CMC$_{mix}$ of SLT/DDAO mixtures of different overall compositions (Ψ)

Ψ	cmc/M[a]
0.00[b])	2.00 × 10^{-3}
0.25	3.58 × 10^{-4}
0.40	1.60 × 10^{-4}
0.50	2.31 × 10^{-4}
0.60	1.90 × 10^{-4}
0.75	1.85 × 10^{-4}
1.00[c])	8.70 × 10^{-3}

[a]) Determined at r = 0; [b]) from ref. [11]; [c]) from ref. [14].

tes and DDAO-polymer interactions [21]. Table 1 collects the CMC value of the DDAO-SLT mixtures. The determined CMC concentrations lie within the range of the reported values for DDAO-SDS and DDAO-sodium dodecylbenzenesulfonate mixtures [17, 19].

In comparison with the CMC of the parent micellar solution, the CMCs of the mixtures are significantly lower, which confirms the strong interaction of the two surfactants within the mixed aggregates.

Titration of the mixtures DDAO-SLT by hydrochloric acid results in turbid solutions below pH = 6.0. Using quasi-elastic light scattering, we determined the hydrodynamic radii of the aggregates at the onset of turbidity. An average hydrodynamic radius of 103.8 ± 1.5 nm was obtained for Ψ = 0.4 (at pH$_{onset}$ = 6.4) and 103.0 ± 2.1 nm for Ψ = 0.6 (at pH$_{onset}$ = 5.4), respectively. Video light microscopy reveals that the formed aggregates are single-walled hollow spheres (vesicles). Solutions with Ψ < 0.15 and Ψ > 0.8 do not form vesicles under protonation, but turn into oily-turbid droplet emulsions (coacervates) at low pH. Coacervation is a common phenomena observed in mixed micellar systems [22].

Solutions above pH 7 contain mixed micellar aggregate with hydrodynamic radii of 13.8 ± 5 nm (Ψ = 0.4) and 19.0 ± 9 nm (Ψ = 0.6), respectively. These results indicate rather non-spherical micelles considering molecular geometry of the DDAO and SLT molecules. The existence of a reversible reorganization from mixed micellar solutions into vesicles and vice versa could be proven by forward and backward titration with HCl and NaOH (Ψ = 0.4 to Ψ = 0.6). The pH for the onset of vesicle formation did not change upon these repeated titrations.

The results of these first experiments confirmed the possibility for controlled and reversible changes in the organizational states of the DDAO-SLT system. The next step included the coupling of the DDAO-SLT self-vesiculation to a chemical reaction system containing species which could interact with the amphiphilic aggregates. The simplest chemical systems fullfilling these conditions are buffer equilibria. The experiment was performed by changing the ratio of buffer components upon mixing. The simultaneous pH changes led to adaptation of the DDAO protonation equilibrium. The result was eventually the vesiculation of the DDAO-SLT micellar solutions.

The solutions exhibit a large sensitivity to the nature of the buffer components (valency of the ions, ion-binding properties, etc.). For example, buffer containing citrate or cacodyl ions led to droplet emulsions. In the range of DDAO excess (Ψ = 0.20 - 0.25), phosphate buffer also produced coacervation at higher total concentrations and showed liquid-liquid phase separations under salt addition. Electron microscopic inspection of the start phosphate buffer solution (pH 7.05) revealed that vesicles had been already formed at a total concentration of 0.001 moles dm^{-3} and a overall composition of Ψ = 0.6 (Fig. 3).

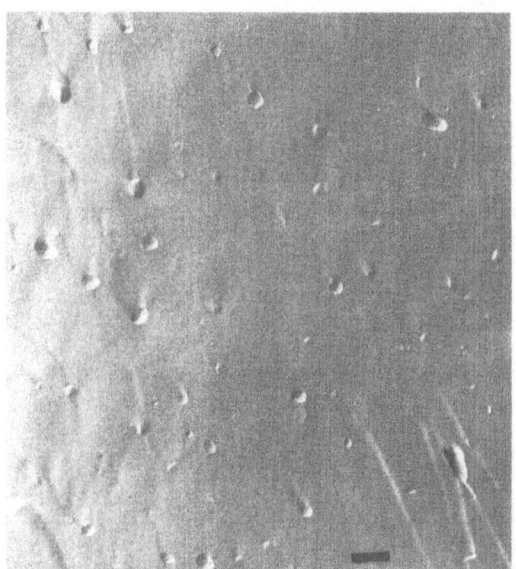

Fig. 3. Freeze-fracture electron micrograph of a mixture of sodium N-lauroyl-N-methyltaurinate (SLT) with N,N-dimethyl-N-dodecylamine oxide (DDAO) in a NaH$_2$PO$_4$/Na$_2$HPO$_4$ buffer (0.050 moles dm^{-3}) at pH 7.05. Total surfactant concentration was 0.001 moles dm^{-3} and $\Psi = 0.6$. Bar length indicates 100 nm. Samples were taken from the solution at 25 °C

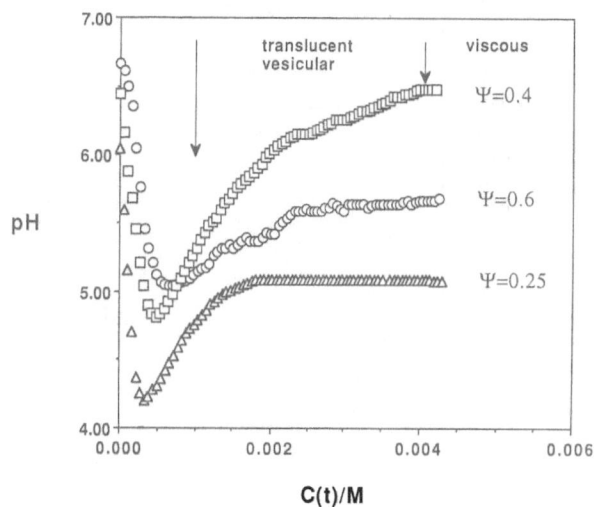

Fig. 4. Titration curves of SLT/DDAO mixtures of different compositons. The concentrations of the stock solution were 0.1 moles dm^{-3}. NaCl (0.01 moles dm^{-3}) was employed as background electrolyte. Ratios of added HCl to total surfactant concentration $r = 0.5$ ($\Psi = 0.25$), $r = 0.66$ ($\Psi = 0.4$), and $r = 0.4$ ($\Psi = 0.6$). Samples were titrated at 25 °C

Further decrease of the pH induced vesicle aggregation and, eventually coacervation and phase separation. Tests indicated that the DDAO-SLT vesicles are responding to salt in the same way as other association colloids. It seems that the screening effect of divalent ions foster the bilayer formation already at low overall protonation degrees (e. g., at pH 7.05).

Titration of 10 mM NaCl solutions with concentrated and partial neutralized DDAO-SLT mixtures revealed the dependence of the vesiculation on the protonation degree and the total amphiphile concentration. Figure 4 shows three titration curves of the compositions $\Psi = 0.25$, $\Psi = 0.4$, and $\Psi = 0.6$ with different neutralization degrees ($r = 0.5$, $r = 0.66$, and $r = 0.4$, respectively). (For the definition of r see experimental methods). The onset of the aggregation can be seen clearly by the steep rise of the pH. The optical appearance during the titration changes shortly after the onset of aggregation from transparent to translucent solution. Samples drawn from this part of the titration contained vesicular aggregates.

Further increase of the total concentration changes the appearance from a translucent back to a transparent but now viscous solution. The DDAO-SLT system adopts different aggregational forms in the progress of the titration depending on the neutralization degree and total concentration.

To illustrate the behavior of the DDAO-SLT system, Fig. 5 summarizes the titrations for a single composition ($\Psi = 0.25$) at different degrees of neutralization (from above: $r = 0.0$, $r = 0.25$, $r = 0.5$, and $r > 1.0$). The titrations progress through five distinct areas of different appearance and with different aggregational states. Area A gives the range below the CMC$_{mix}$ where a monomolecular solution of DDAO-SLT can be found. Above the CMC$_{mix}$ two main areas exist. Area B represents the area of mixed micellar solution. Below the line of pH 6.7 (area C) vesicles appear and change to coacervates (area D) or to viscous solutions (area E) depending on neutralization degree and total concentration. The occurrence of the coacervate phase can be related to the approach of the point of zero charge (PZC) in the aggregates due to neutralization between r = 0.5 and r > 1.00 in the composition area of DDAO excess ($\Psi < 0.5$). The viscous solutions most likely contain wormlike micelles, as were found for other DDAO anionic mixtures [23].

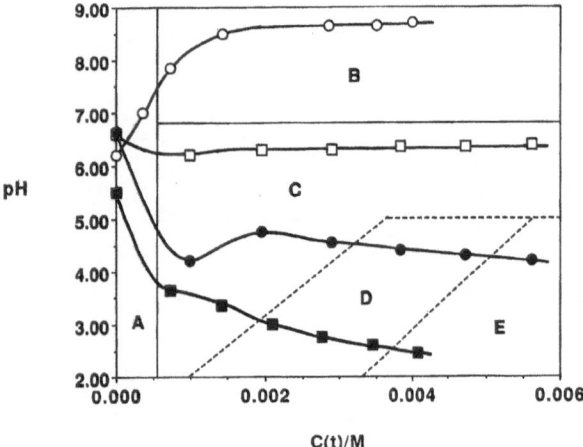

Fig. 5. Titration curves of SLT/DDAO with a composition of $\Psi = 0.25$ at different HCl ratios (from above: $r = 0.0$ (○), $r = 0.25$ (●), $r = 0.5$ (□), and $r > 1.0$ (■)). Letters indicate areas of different optical appearance and aggregational form: **A** transparent molecular solution ($C_t < CMC_{mix}$); **B** transparent mixed micellar range; **C** translucent vesicular range; **D** oily turbid coacervation area, and **E** transparent viscous solutions. C_t of stock solutions were 0.1 moles dm^{-3}; NaCl was added (0.01 moles dm^{-3}). Samples were titrated at 25°C

Conclusions

We have introduced a novel amphiphilic system (DDAO-SLT) as a versatile model to study the reorganization from the micellar state to closed-shell bilayer structures like vesicles. We are currently extending our studies to multicomponent systems containing permanent charged cationic and anionic amphiphiles. The DDAO/DDAOH couple figures in these systems as a molecular switch which shifts the cationic composition from the mixed micellar range into that of vesicles (results to be published). We have also extended the investigation to other chemical reaction systems containing interactive species to study the mutual feedback between the self-vesiculation and the progress of the chemical transformations.

Acknowlegdement

The author gratefully acknowledges the collaborations with Peter Schurtenberger, Peter Skrabal, Ernst Wehrli, Michael Ambühl and Stephane Szönyi.

References

1. Fendler JH (1982) Membrane Mimetic Chemistry. John Wiley and Sons, New York
2. Vesiculation of fatty soaps by half neutralization: a) Gebicki JM, Hicks M (1973) Nature (London) 243:232; b) Gebicki JM, Hicks, M (1976) Chem. Phys. Lipids 16:142; c) Hargreaves WR, Deamer DW (1978) Biochemistry 17:3759; d) Haines TA (1983) Proc Natl Acad Sci (USA) 80:160
   Vesiculation of single and double chain synthetic surfactants: a) Talmon Y, Evans DF, Ninham DW (1983) Science 221:1047; b) Hundscheid FJA, Engberts JBFN (1984) J Org Chem 49:3088; c) Wakita M, Edwards KA, Regen SL, Turner D, Gruner SM (1988) J Am Chem Soc 110:5221; d) Miller DD, Bellare JR, Kaneko T, Evans DF (1988) Langmuir 4:1363; e) Fukuda H, Kawata K, Okuda H, Regen SL (1990) J Am Chem Soc 112:1635
   Vesiculation of phopholipids: a) Carnie S, Israelachvili JN, Pailthorpe BA (1979) Biochim Biophys Acta 554:340; b) Jain MK, van Echteld CJA, Ramirez F, de Gier J, de Haas GH, van Deenen LLM (1980) Nature (London) 284:486; c) Jain MK, de Haas GH (1981) Biochim Biophys Acta 642:203; e) Gabriel NE, Roberts MF (1984) Biochemistry 23:4011; f) Hauser H, (1987) Chem Phys Lett 43:283; g) Tilcock CPS, Cullis PR, Hope MJ (1986) Biochemistry 25:816; h) Cantu L, Corti M, Musolino M, Salina P (1990) Europhys. Lett. 13:561
3. Kaler EW, Kamalakara Murthy A, Rodriguez BE, Zasadzinski JAN (1989) Science 245:1371
4. Kaler EW, Kamalakara Murthy A, Rodriguez BE, Zasadzinski JAN (1992) J Phys Chem 96:6698—6707
5. Szönyi S, Cambon A, Watzke HJ, Schurtenberger P, Wehrli E (1992) In: Lipowsky R, Richter D, Kremer K (eds.) Structure and Conformation of Amphiphilic Membranes. Springer Proceedings in Physics, Vol. 66, Springer, Berlin, pp. 198
6. Szönyi S, Watzke HJ (1992) Prog Colloid Polym Sci (this volume)
7. Ambühl M, Bangerter F, Luisi PL, Skrabal P, Watzke HJ (1992) Prog Colloid Polym Sci (this volume)
8. Ambühl M, Bangerter F, Luisi PL, Skrabal P, Watzke HJ (1993) Langmuir 9:36—38
9. Rosen MJ (1989) Surfactants and Interfacial Phenomena. 2nd edition, John Wiley and Sons, New York
10. a) Safran SA, Pincus P, Andelman D (1990) Science 248:354; b) Safran SA, Pincus P, Andelman D, MacKintosh FC (1991) Phys Rev 43:1071
11. Rathman JF, Christian S (1990) Langmuir 6:391—395 and references therein
12. Faucompre B, Lindman B (1987) J Phys Chem 91:383—389
13. Hoffmann H, Thunig C, Munkert U (1992) Langmuir, (1992) 8:2629, The author is grateful to Prof . H. Hoffmann for the information. Simons BD, Cates ME (1992) J Phys II (France) 2:1439—1451
14. Bistline RG, Rothman ES, Serota S, Stirton AJ, Wrigley AN (1971) J Am Oil Chem Soc 48:657—660

15. Ostrowsky N, Sornett D, Parker P, Pike R (1981) Opt Acta 28:1059
16. Müller M, Meister N, Moor H (1980) Mikroskopie (Wien) 36:129
17. Rosen MJ, Friedman D, Gross M, (1964) J Phys Chem 68:3219
18. Myzakawa K, Ogawa M, Mitsui T (1984) Int J Cosm Sci 6:33—46
19. a) Hoffmann H, Ebert G (1988) Angew Chem 100:933—944; b) Haegel FH, Hoffmann H (1988) Prog Coll Polym Sci 76:132—139
20. Kolp DG, Laughlin RG, Krause FP, Zimmerer RE (1963) J Phys Chem 67:51
21. a) Brackmann JC, Engerts JBFN (1992) Langmuir 8:424—428; b) Brackmann JC, Engberts JBFN (1989) J Coll Interf Sci 132:250—255; c) Arakawa J, Pethica BA (1980) J Coll Interf Sci 75:441—450
22. for a review on coacervation of mixed micellar systems see: Tomlinson E, Davis SS, Mukhayer GI (1979) In: Mittal KL (ed.) Solution Chemisty of Surfactants. Vol. 1, Plenum Press, New York, pp 3
23. Hoffmann H, Ebert G (1988) Angew Chem 100: 933—944

Author's address:

Dr. Heribert J. Watzke
ETH-Zürich
Institut für Polymere
Universitätsstrasse 6
8092 Zürich, Switzerland

Progress in Colloid & Polymer Science　　　　　　Progr Colloid Polym Sci 93:22—24 (1993)

Spinodal fractals

M. Carpineti and M. Giglio

Physics Department, University of Milan, Italy

Abstract: We report for the first time the observation of a finite wave vector peak in the scattering pattern from a dense solution of aggregating colloids. Measurements have been conducted under isopycnic conditions and at various monomer concentrations. The scattered intensity $S(q)$ scales according to a universal law, $S(q/q_m) = q_m^{-d} F(q/q_m)$, in close analogy with spinodal decomposition dynamics (q_m is the position of the scattered peak). Although for spinodal decomposition, d is equal to 3, here, we measured $d = d_f$, the fractal dimension of the colloidal clusters. We also observed that $S(q)$ attains a terminal shape. The position of the peak in this terminal phase scales linearly with the initial monomer concentration, as predicted by a simple model.

Key words: Colloids — aggregation — spinodal decomposition — static light scattering — fractals

Introduction

Substantial interest has been devoted in the past to both spinodal decomposition and colloidal aggregation, in particular, because of the universal features exhibited by these phenomena.

Spinodal decomposition, whose modality have been demonstrated to be common to very different systems, ist the process of phase separation of thermodynamically unstable fluids [1—4]. The phase separation takes place via a periodic density modulation whose length and amplitude grow in time. In fact, the thermodynamically unstable states under the coexistence curve are characterized by a negative diffusion coefficient, so that fluctuations grow instead of decaying. In particular, it can be demonstrated [1] that, at each instant, a typical lenght exists at which fluctuations grow in the fastest possible way. Accordingly, a system which phase separates via spinodal decomposition produces a ring-shaped scattering pattern with increasing maximum intensity and a collapsing radius.

As it is well known [5], colloidal particles in solution are prevented from aggregating by their surface charges and, by adding some electrolyte to the solution, the coulombic repulsion can be reduced so as to induce the aggregation. Also, colloidal aggregation has been demonstrated to exhibit a universal behavior independent of the system used [6, 7]. In particular, the morphology of the aggregates is fractal and the fractal dimension d_f is related to the aggregation dynamic followed by the process. As in the case of spinodal decomposition, a good probe for studying colloidal aggregation is static light scattering. The scattering pattern produced by a dilute solution of fractal aggregates has a maximum at zero angle proportional to the weight average cluster mass and exhibits a power law behavior at high q values $I(q) \propto q^{-d_f}$, the exponent d_f being the fractal dimension.

In this work, we present some experimental results showing an interesting analogy between spinodal decomposition and colloidal aggregation [8]. We have studied colloidal aggregation in dense solutions with static light scattering and we have found that the intensity distributions $I(q)$ present a maximum at $q > 0$. Furthermore, the peak position moves to lower q values as time elapses, while the peak intensity grows until both reach some terminal value. In close analogy with spinodal decomposition, the intensity distributions $I(q)$ after an initial period can be scaled onto a unique master curve with a simple scaling law

$$I(q/q_m) = q_m^{-d_f} F(q/q_m) \, , \tag{1}$$

where q_m is the peak position and $F(q/q_m)$ is a time-independent function. It is interesting to point out that the same scaling law holds for spinodal decomposition, but with $d = 3$ instead of d_f.

Experimental details

We have performed measurements with polystyrene spheres of 95 Å in radius in dense solutions (volume fractions ranging from 3×10^{-5} and 3×10^{-3}) and in isodensity condition by matching the density of the polystyrene with a mixture of water and deuterated water. Aggregation has been induced by a divalent salt ($MgCl_2$) and the molarity has been chosen so that the evolution of the aggregation was adequately slow to be followed at ease.

We have studied the aggregation processes with low-angle static light scattering. The peculiarity of our scattering instrument, that has already been described [9], is that we can collect signals over two decades of wave vectors ranging from $q \simeq 4 \times 10^2$ cm^{-1} and $q \simeq 3 \times 10^4$ cm^{-1}. A collimated beam impinges onto the sample cuvette and both the scattered and the transmitted light are collected by a lens in which focal plane a multielement sensor is located. We use a custom-made sensor composed of 31 sensing elements shaped as concentric quarters of an annulus. The lens realizes a correspondence between the scattering angles and the radii of the elements. The transmitted beam is focused in a tiny hole placed in the center of the sensor and behind which a photodiode is located.

The 31 sensors, together with the monitor and the transmitted beam, are read in sequence with an interface with a personal computer. With this instrument, the time required for collecting the light scattered at 31 different angles is quite small so that many intensity distributions can be evaluated at different times after the start of the reactions.

Experimental results and conclusions

In Fig. 1, we present some intensity distributions taken at different times for an aggregation process at monomer concentration $c_0 = 8.25 \times 10^{13}$ cm^{-3}. As can be noticed, there is a peak at $q_m > 0$ moving in time toward smaller and smaller values. Furthermore, the peak height increases while its width shrinks. Actually, this behavior is quite reminiscent of that of spinodal decomposition [10—12] and it has never been observed before during an aggregation process. In the inset a plot of the curves in a log-log scale is presented. Particularly interesting is the asymptotic behavior for high q— values. In fact,

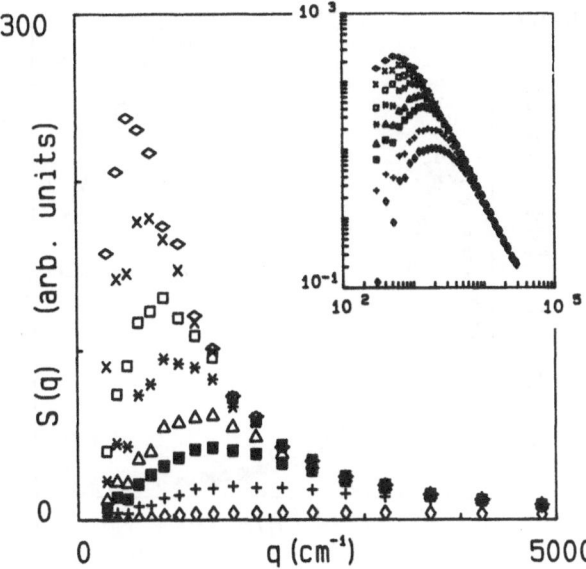

Fig. 1. Intensity distributions taken at different times after the start of an aggregation performed at monomer concentration $c_0 = 8.25 \times 10^{13}$ cm^{-3}. In the inset the curves are plotted on a log-log scale and the power law behavior at high q-values can be noticed

while the presence of a peak in the scattering pattern indicates the existence of a spatial ordering over a length scale of about $1/q_m$, the power law behavior at high q shows that, on a short length scale, the system has a fractal morphology, with fractal dimension d_f given by the slope of the asymptote. Furthermore, the fact that all the curves fall onto the same asymptotic straight line indicates that no loss of mass from the scattering volume has occurred [9].

We found that all the curves taken after an initial period can be scaled onto the same master curve with a simple scaling law (Eq. (1)), very similar to that working in the case of spinodal decomposition. In Fig. 2, we present the scaled curves corresponding to the same aggregation of Fig. 1. Again, in the inset the data in log-log scale are presented from which it can be noticed that the scaling is good over almost two decades in q—vectors. As in the case of spinodal decomposition, the scaling of the curves indicates that only one typical length is enough for characterizing the system.

Finally, we observed another typical aspect of these high-density aggregation processes. For the higher monomer concentrations used the reactions stop as soon as the peak position has reached a terminal value, and we found that this final position of q_m depends on the monomer concentration c_0. We performed measurements at different c_0 varying

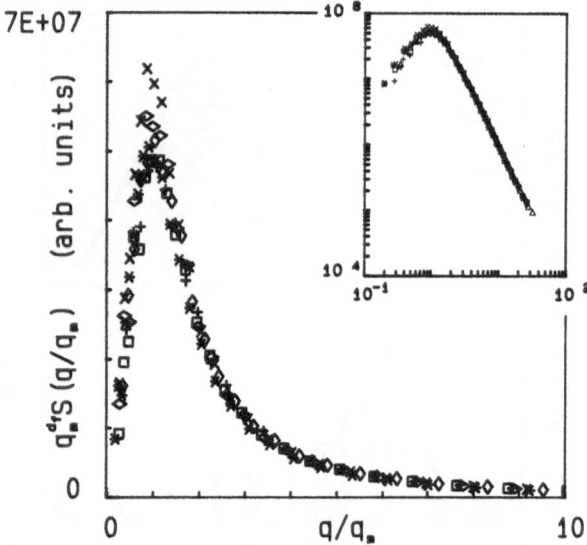

Fig. 2. Plots of the scaled functions $q_m^{d_f} S(q/q_m)$ as a function of q/q_m. Again, in the inset the curves are plotted on a log-log scale for showing that the scaling is good over the entire q-range

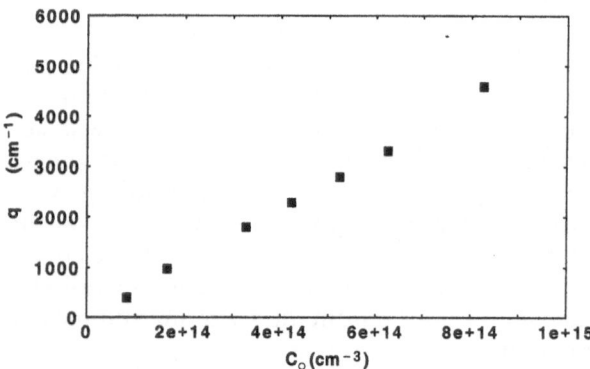

Fig. 3. Terminal peak position as a function of the monomer concentration. As can be noticed, there is a linear dependence as predicted by a simple model

from $c_0 = 8.25 \times 10^{13}$ cm^{-3} and $c_0 = 8.25 \times 10^{14}$ cm^{-3}. In Fig. 3 the terminal peak position as a function of the initial concentration is shown and, as can be clearly noticed, there is a linear dependence between these quantities.

This final behavior can be explained simply. As the reaction proceeds the volume occupied by the clusters increases until all the available space is completely filled. This can happen because the average density of fractal objects decreases when their dimensions grow, the volume being proportional to M^{3/d_f}, where M is the total mass of an aggregate. Using a simple model, it can be easily calculated that the maximum radius that can be

obtained starting with a particular value of c_0. In fact, assuming a monodisperse distribution and requiring that the terminal situation is reached when the cluster volume fraction is equal to one, we obtain $R_{max} = 1/(4c_0 R_0^2)$, where R_0 is the monomer radius. As can be noticed, this results is in good agreement with the data shown in Fig. 3, assuming $q_{max} \propto 1/R_{max}$.

In conclusion, our measurements have shown that during the cluster growth at high monomer concentrations, a spatial ordering inside the fluid is generated over the length scales probed by our instrument. Furthermore, after an initial period, the system is characterized by a unique lenght and the intensity distributions can be scaled onto the same master curve with a scaling law very similar to that holding for spinodal decomposition. In spite of the regularity generated over large length scales, the aggregates retain on a finer scale a typical fractal morophology as indicated by the asymptotic behavior of the intensity distributions. It is the very fractal nature of the aggregates which is responsible for the existence of the upper limit of the linear dimensions of the clusters and, therefore, of the termination of the aggregation reactions.

References

1. Furukawa H (1986) Adv Phys 34:703—750
2. Langer JS, Baron M (1973) Ann Phys (New York) 78:421
3. Langer JS (1971) Ann Phys (New York) 65:53
4. Binder K, Billotet C, Mirold P (1978) Z Phys B 30:183—195
5. See, for example, Kinetics of Aggregation and Gelation (1984) Family F, Landau DP (eds) North Holland, Amsterdam
6. Ball RC, Weitz DA, Witten TA, Leyvraz F (1987) Phys Rev Lett 58:274—277
7. Lin MY, Lindsay HM, Weitz DA, Klein R, Ball RC, Meaking P (1990) J Phys (Condens Matter) 2:3093—3113
8. Carpineti M, Giglio M (1992) Phys Rev Lett 68:3327—3330
9. Carpineti M, Ferri F, Giglio M, Paganini E, Perini U (1990) Phys Rev A 42:7347—7354
10. Huang JS, Goldburg WI, Bjuerkaas AW (1974) Phys Rev Lett 32:921—923
11. Chou Y, Goldburg WI (1981) Phys Rev A 36:858—864
12. Wiltzius P, Bates FS, Heffner WR (1987) Phys Rev Lett, 36:1538—1541

Authors' address:

Prof. M. Giglio
Dipartimento di Fisica
Università degli studi di Milano
Via Celoria 16
20133 Milano, Italia

Progress in Colloid & Polymer Science   Progr Colloid Polym Sci 93:25—29 (1993)

Time-resolved simultaneous small- and wide-angle x-ray diffraction on dipalmitoylphosphatidylcholine by laser temperature-jump

G. Rapp[1]), M. Rappolt[1]), and P. Laggner[2])

[1]) European Molecular Biology Laboratory (EMBL) Hamburg Outstation at DESY
[2]) Institut für Biophysik und Röntgenstrukturforschung, Austrian Academy of Sciences, Graz, Austria

Abstract: The kinetics of the structural phase transitions of dipalmitoylphosphatidylcholine (DPPC) in excess water have been investigated by simultaneous small- and wide-angle detection in millisecond time-resolved synchrotron x-ray diffraction. The transition from the rippled to liquid-crystalline phase ($P_{\beta}'(\text{mst}) - L_a$) has been induced by an approx. 5°C temperature-jump within 2 ms using an erbium glass laser. The structural parameters, i.e., the long-range order of the bilayer stacking and the short-range order of the lateral lipid packing have been detected with 5 ms time-resolution. The experiments show that the intensities of both the small- and the wide-angle reflections indicative of the $P_{\beta}'(\text{mst})$-structure decay in parallel and on the same time-scale as the first- and second-order reflections of the lamellar lattice of the L_a-phase grow in intensity. For the pretransition, we have made the important observation that the rearrangement of the hydrocarbon chains is slower than the initial changes in the long-range lamellar lattice.

Key words: Small- and wide-angle x-ray diffraction — synchrotron radiation — IR-laser — T-jump — phospholipids — phase transitions

Introduction

Phospholipids, the major constituents of biological membranes, occur in a variety of thermotropic and lyotropic mesophases. Their structures and thermodynamic properties under near equilibrium conditions have been studied in great detail (cf. Small, 1986; Cevc and Marsh, 1987). One of the most widely studied compounds in this area is dipalmitoylphosphatidylcholine (DPPC), for which a sound base of reference data exists with respect to structure and thermodynamics of its various phases and transitions in aqueous media (Chapman et al., 1967; Tardieu et al., 1973; Janiak et al., 1976; Ruocco and Shipley, 1982; Cevc, 1991; Tenchov, 1991; Yao et al., 1991; Katsaras et al., 1992). Comparatively little is known, however, about the structural mechanisms and dynamics of the transitions. Recent experiments on DPPC/water dispersions under non-equilibrium conditions, i.e., upon laser induced T-jumps in the order of 5—10°C, have, in some cases, already shown qualitative dif-ferences in comparison to near-equilibrium investigations (Laggner and Kriechbaum, 1991; Laggner, Kriechbaum and Rapp, 1991). From these results we have argued that distinct differences may exist between the mechanisms of transition close to and far from equilibrium, respectively. This might be of particular interest in the context of the biological role of phospholipids in membranes, since important membrane functions, as for example, signal amplification or membrane fusion are clearly non-equilibrium processes. To better understand the dynamic properties of membranes, such investigations under non-equilibrium conditions are indispensable.

Already the early attempts to follow the phospholipid transition kinetics with real-time x-ray diffraction (for reviews, see Laggner, 1986; 1988; Gruner, 1987; Caffrey, 1989; Lis and Quinn, 1991) have shown that this method bears unique potential to study structural transitions cinematographically. More recently, important progress was made through the development of an IR-laser T-jump method, which fully exploits the millisecond time-resolution potential of diffraction experiments with synchrotron radiation, as could be shown by some exemplary studies on lipid phase transitions (Laggner et al., 1989). One very interesting example is the "pretransition" of DPPC in excess water, where the system changes from the gel-phase with tilted chains (L_{β}') to the rippled P_{β}'-phase at 35°C, followed by the main transition at 42°C to the lamellar bilayer, liquid-crystalline (L_a) phase. The pretransition is associated with a symmetry change from a one-dimensional lamellar to a two-dimensional monoclinic lattice (Fig. 1). Here, short-lived ordered intermediate structures have been detected by the IR-laser T-jump technique, which are not seen in transitions under near-equilibrium conditions (Laggner, Kriechbaum and Rapp, 1991). In these experiments, however, only the small-angle reflections have been recorded, so that the details of rearrangement in the hydrocarbon chain packing remained elusive.

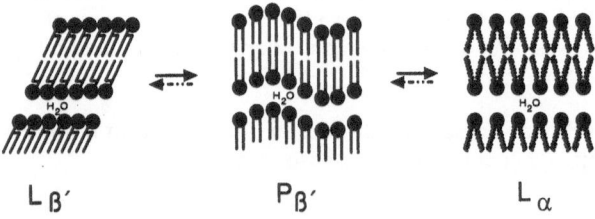

Fig. 1. Scheme of the types of phases and transitions investigated in this study

In studies on lipid phase transitions, where hysteresis loops can be considerably broad and strongly rate-dependent, it is essential to simultaneously measure the structural parameters in the wide- and small-angle range with the same sample. A camera designed for this purpose for conventional x-ray tubes has been described recently (Laggner and Mio, 1992). In the present study, we describe the performance of a related approach with two detectors for time-resolved work at synchrotron sources. We have been able to simultaneously monitor the small- and wide-angle reflections in the diffraction pattern and with millisecond time-resolution. The reflections in the wide-angle region around $(0.4 \text{ nm})^{-1}$ give additional insight not only into the structure (e.g., the gel phase with tilted hydrocarbon chains can be distinguished from the one with untilted chains),

but also in the coordination and arrangement of the lipids during the transition.

Materials and Methods

Sample preparation

DPPC (1,2-dipalmitoyl-*sn*3-glycero-phosphocholine, purity 99%) was purchased from SIGMA, Munich, and used without further purification. Weighted amounts of lipids (20% *w/w*) were dispersed in bidistilled and degassed water (pH 5.3) and incubated for several hours at 52°C. Then the samples were vortexed and incubated at the above-mentioned temperature for another hour. The dispersions were then stored for a few days at 4°C. For measurements the samples were transferred into a 0.7-mm-thick stainless-steel holder with 10 μm-thick mylar windows.

Laser T-jump

For rapid heating of the sample an erbium glass laser was used. The system allows T-jumps up to 12°C within 2 ms. The unfocused beam diameter is about 6 mm. The wavelenght of the laser is 1.54 μm, at which the absorption length of water is 1.5 mm. (For details see [Rapp and Goody, 1991]). To reduce the temperature gradient in the sample the transmitted laser beam was reflected by an aluminum-coated kapton foil at the beam-exit of the cell. The temperature of the samples before the T-jump was measured with a 200 μm-thick Ni—CrNi thermocouple positioned close to the intersection point of the x-ray beam with the sample. It was controlled by a Peltier-element.

X-ray diffraction

The experiments have been performed on the beamline X33 of the EMBL at DESY in Hamburg (Koch and Bordas, 1983; Boulin et al., 1988). Sample-to-detector distances up to 5 m are possible and, except for an air-gap of a few centimeters long around the sample, the whole x-ray pathway is in vacuum. Details of the camera combining small- and wide-angle x-ray diffraction will be described elsewhere (Rapp, Rappolt & Koch, submitted). Briefly, two linear detectors with delay-line readout (Gabriel, 1977) are connected electronically in series. One of the detectors covers the wide-angle part, $s = (0.37—0.47 \text{ nm})^{-1}$, the other the small-angle part $s = (2—15 \text{ nm})^{-1}$ of the diffraction pattern. The small-angle detector was calibrated with a dried collagen sample from rat tail tendon giving the following spacings: L_β' = 6.43 nm at 20 °C, L_a = 6.65 nm at 43 °C. In this arrangement, the two detectors appear to the data-acquisition system as one delay-line with approximately twice the length of a single detector. The time-resolution of this system is better than 20 μs. The number of exposures is limited only by the size of the memory. The signal-to-noise ratio in the wide-angle part of the diffraction pattern is critically dependent on the air path in the x-ray beam. This has been minized to about 1 cm behind the sample.

A typical T-jump experiment with this arrangement consists of one or more cycles of 64 sequential exposures of variable length freely programmable by the data acquistion system. Usually, the first exposure time is chosen with 1 s, followed by 60 exposures of 5 ms each, and three more exposures of 1 s. The laser was fired regularly at the 10th frame. After one cycle of 64 exposures the sample was allowed to passively cool to the starting temperature. Under these conditions, starting from about 38 °C, the DPPC-system is jumped through the main transition and relaxes reversibly to the metastable ripple phase (Tenchov et al., 1989; Yao et al., 1991) within 20 s.

Results and Discussion

To characterize the dispersions of DPPC (20% w/w), static exposures of the gel (L_β'), rippled (P_β') and liquid-crystalline phase (L_a) were simultaneously taken in the small- and wide-angle region. Figure 2 shows such a representative x-ray diffraction pattern of DPPC in the gel phase at 20 °C. Between 0.1 and 0.4 nm^{-1} the first- and second-order reflections of a one-dimensional lattice are seen, and at about 2.4 nm^{-1} there are two clearly resolved wide-angle reflections originating from the chain packing in an orthogonal sub-cell (Ruocco and Shipley, 1982).

A demonstration of the information that can be resolved by such a dual-detector experiment is

shown in Fig. 3, where a moderately slow temperature scan with 14 °C/min (by Peltier heating) is performed with DPPC between 28 and 48 °C. While the small-angle patterns show the loss of long-range order between the pre- and the main transition in a similar way as reported earlier (Laggner, 1988), the wide-angle pattern remains virtually unchanged until the main transition. This indicates that, at

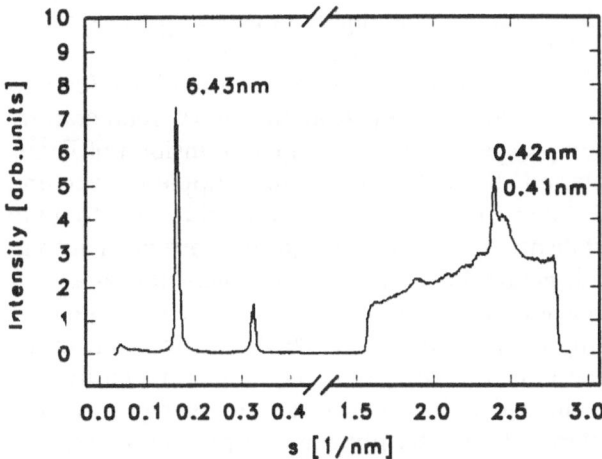

Fig. 2. Small- and wide-angle x-ray pattern of DPPC in the gel phase at 20 °C: The raw data set of a 50 s exposure is plotted vs. the scattering vector $s = (2 \cdot \sin \Theta)/\lambda$, where 2Θ is the scattering angle and λ = 0.15 nm. The peak at about 1.9 nm^{-1} arises from the mylar windows

the time-scale of this experiment, the chains have insufficient time to relax into the hexagonal subcell-arrangement typical for the P_β'-phase, and suggests that the chain-tilt with respect to the bilayer plane remains intact. It is therefore likely that the first stage of the transition involves an accumulation of defects in the regular lamellar packing of bilayers (as seen from the broader small-angle patterns), in which the chains are still tilted, while the formation of the monoclinic zig-zag-lattice is a slow process, longer than the time needed in this experiment to heat from the pre- to the main transition tempera-

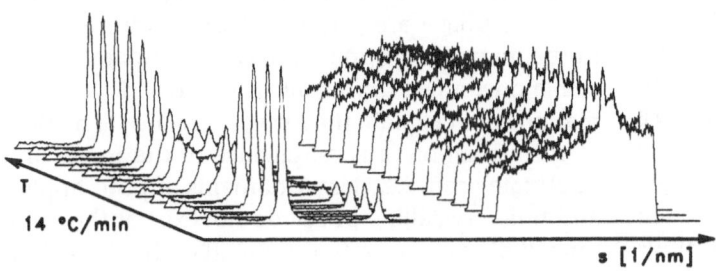

Fig. 3. Series of small- and wide-angle diffraction patterns during a heat scan of DPPC from 28 °C to 53 °C with a scan rate of 14 °C/min: Exposure time was 2 s followed by a 3-s pause. The pre- and subsequent main transition ($L_\beta' - P_\beta' - L_a$) is seen between the fifth and 15th frame

ture, which was about 30 s. The formation of un-tilted chains in a hexagonal subcell is therefore thought to coincide with the relaxation of the L_β'-phase to a well-annealed zig-zag lattice of the bilayers. It is also to be noted that under these conditions, i.e. relatively slow scan rate, no intermediate, thinner lamellar structures can be resolved, as are seen under laser T-jump conditions (Laggner, Kriechbaum and Rapp, 1991). This again demonstrates that lipid transition pathways and mechanisms can be rate-dependent.

The second type of experiment has been focused on the kinetics of the main transition. Temperature-jump experiments on the main transition of DPPC started from the metastable rippled structure, $P_\beta'(mst)$. To attain 5 ms time resolution with good statistics the data of 100 T-jumps have been averaged. Figure 4a shows the first small-angle x-ray diffraction patterns of a T-jump experiment. The starting temperature was at 39.5°C, in the 10th frame the laser was triggered. The final temperature was estimated to be about 44.5°C by energy measurements of the laser pulse and in parallel by comparing the final peak position in the L_a-phase with the corresponding one of a slow heat scan.

As shown in Fig. 4b, the changes of the normalized intensities of the small- and wide-angle reflections coincide. The transition is almost completed in the 11th frame, indicating that the transition is only slightly slower than the time for heating. In the determination of the values in Fig. 4b only those parts of the pattern have been used which have no intensity above background in the angular range of the L_a signal. At a time-resolution of 5 ms no additional effect or further intermediate structure could be detected.

As summarized in Cevc and Marsh (1987) the events occurring during the main transition are trans-gauche isomerization of the hydrocarbon chains as a nucleation process, migration of these distortions causing density fluctuations in the bilayer followed by relaxation of the high-density regions via lateral expansion. The latter is coupled to a redistribution of water in the interbilayer space, which is thought to be the rate-limiting step for this nucleation and growth mechanism. On the other hand, as discussed in Laggner and Kriechbaum (1991), any transformation between symmetry related phases can be made to a diffusionless martensitic type, provided that the driving force for the transition is big enough. This is certainly the case in the experiments described where the non-

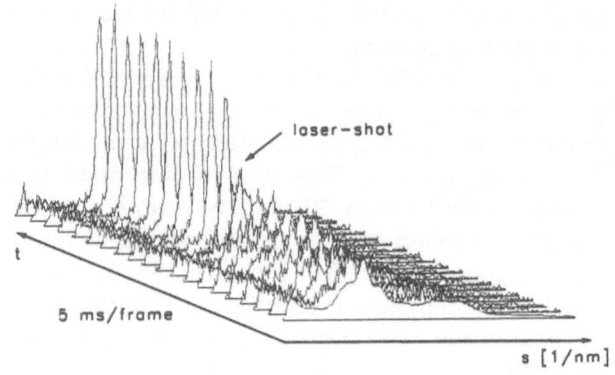

a)

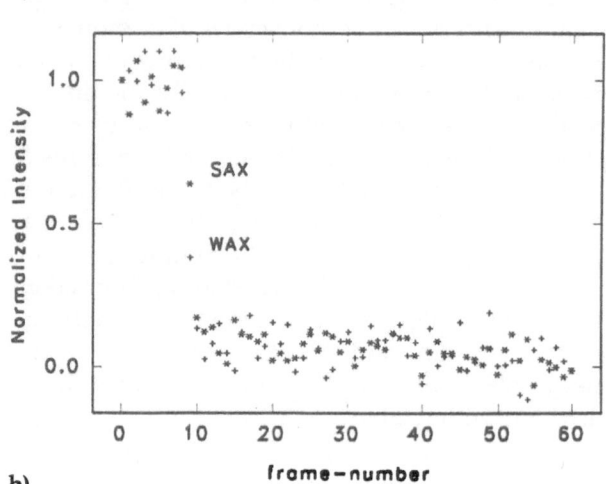

b)

Fig. 4. Laser-induced T-jump from the $P_\beta'(mst) - L_a$ phase in DPPC. a) First 20 frames with the laser triggered at frame 10. b) Time-course of the normalized intensities of the SAX- and WAX-reflections of the $P_\beta'(mst)$ with 5 ms per frame

equilibrium situation in the system is created both rapidly and far enough from the equilibrium transition temperature.

Interpreting the rippled phase as a periodically modulated gel phase, it is evident that the modulation does not kinetically hinder the main transition, although it has a dramatic influence on the pretransition. Suggesting the additional reflections of the $P_\beta'(mst)$ phase are due to a higher water content in the metastable rippled phase as compared to the stable one, a fast diffusionless martensitic transformation could explain the results. Second, at no time of the transition is a coexistence of the initial and final state seen, i.e., larger domains of pure rippled and fluid crystalline phase do not exist at the same time. It displays a continuous shift of the rippled into the liquid crystalline structure in analogy to the stretching of an accordion. This is in contrast to the main transition in PE's which show a continuous two state decay and growth mechanism (Laggner et al., 1989)

Conclusion

The combination of laser T-jump induced phase transitions with time-resolved small- and wide-angle x-ray diffraction at synchrotron radiation sources allows accumulation of valuable information on the millisecond kinetics and mechanisms of lipid phase transitions. It opens a new approach to characterize the structural dynamics with simultaneous monitoring of long- and short-distance regularities and serves as a promising approach to solve parts of the remaining ambiguities about the kinetics and mechanisms of phase transitions in phospholipids. Detailed studies on the kinetics of the transitions as a function of the degree of superheating and of the heating rate should give a better insight into the mechanism of the transition.

Acknowledgements

We gratefully acknowledge the help of Drs. A. Gabriel and M.H.J. Koch with the double detector setup, of Mr. R. Kläring with the construction of the mechanical components, and A. Schuster with the preparation of the manuscript. This work has been supported in part by the Österreichischer Fonds zur Förderung der Wissenschaftlichen Forschung under grant no. S4614 (P.L.).

References

1. Boulin C, Kempf R, Gabriel A, Koch MHJ (1988) Nucl Instr Methods Phys Res A 269:312
2. Caffrey M (1989) Annu Rev Biophys Chem 18:159—186
3. Cevc G, Marsh D (1987) Phospholipid Bilayers, Physical Principles and Models Wiley, New York
4. Cevc G (1991) Chem Phys Lipids, Special Issue on Phospholipid Phase Transitions (Kinnunen P, Laggner P eds) 57:293—307
5. Chapman D, Williams RM, Ladbrooke BD (1967) Chem Phys Lipids 1:445—475
6. Gabriel A (1977) Rev Sci Instr 48:1303
7. Janiak M, Small DM, Shipley GG (1976) Biochemistry 15:4575—4580
8. Gruner SM (1987) Science 238:305—312
9. Katsaras J, Yang DSC, Epand R (1992) Biophys J 63:1170—1175
10. Koch MHJ, Bordas J (1983) Nucl Instr Methods Phys Res A 208:461
11. Laggner P (1986) New Methods in Absorption, Scattering and Diffraction for Applications in Structural Biology (Chance B, Bartunik HD eds) Academic Press London pp 171:182
12. Laggner P (1988) Top Curr Chem 145:173—202
13. Laggner P, Kriechbaum M (1991) Chem Phys Lipids, Special Issue on Phospholipid Phase Transitions (Kinnunen P, Laggner P eds) 57:121—145
14. Laggner P, Kriechbaum M, Rapp G (1991) J Appl Cryst 24:836
15. Laggner P, Kriechbaum M, Hermetter A, Paltauf F, Hendrix J, Rapp G (1989) Progr Colloid Polym Sci 79:33—37
16. Laggner P, Mio H (1992) Nucl Instr Methods Phys Res A 323:86—90
17. Lis LJ, Quinn PJ (1991) J Appl Cryst 24:48—60
18. Rapp G, Goody RS (1991) J Appl Cryst 24:857
19. Ruocco MJ, Shipley GG (1982) Biochim Biophys Acta 691:309—320
20. Small DM (1986) Handbook of Lipid Research, Vol IV: The Physical Chemistry of Lipids, from Alkanes to Phospholipids, Plenum Press, New York
21. Tardieu A, Luzatti V, Reman FC (1973) J Mol Biol 75:711—733
22. Tenchov BG, Yao H, Hatta I (1989) Biophys J 56:757
23. Tenchov BG (1991) Chem Phys Lipids, Special Issue on Phospholipid Phase Transitions (Kinnunen P, Laggner P eds) 57:165—177
24. Yao H, Matuoka S, Tenchov BG, Hatta I (1991) Biophys J 59:252

Authors' address:

Dr. G. Rapp
M. Rappolt
EMBL c/o DESY
Notkestr. 85
20253 Hamburg, FRG

Prof. Dr. P. Laggner
Institut f. Biophysik
und Röntgenstrukturforschung
Steyrergasse 17
8010 Graz, Austria

Static and dynamic depolarized laser light scattering from large aggregates in equilibrium with small vesicles

L. Cantù[1]), M. Mauri[1]), M. Musolino[2]), S. Tomatis[1]), and M. Corti[3])

[1]) Study Center for the Functional Biochemistry of Brain Lipids, Dept. of Chemistry and Biochemistry, Medical School, University of Milan, Italy
[2]) Dept. of Physics, Politecnico of Milan, Italy
[3]) Dept. of Electronics, University of Pavia, Italy

Abstract: Depolarized laser light-scattering measurements have been performed on dilute solutions of the ganglioside GM3 in order to confirm the existence of a small amount of axisymmetric particles in thermodynamic equilibrium with a main population of unilamellar vesicles (average radius 250 Å). Both static and dynamic data are consistent with a disklike form with an average radius of 3500 Å. The use of depolarized light scattering has been proved to be useful to optically separate the contributions coming from different coexisting aggregate shapes in solution.

Key words: Depolarized light scattering — vesicles — GM3 ganglioside

Introduction

Spontaneous vesicle formation is an interesting problem which has been discussed mainly in connection with multicomponent systems [1, 2]. Recently, the occurrence of spontaneous vesiculation in water solution has been reported also for a single-component system [3—5], a glucosidic surfactant of biological origin, namely, the ganglioside GM3. GM3 is an amphiphilic molecule with two hydrocarbon chains, 18 and 20 hydrocarbons, respectively, and a headgroup made up of three sugars, one of which is a sialic acid residue. Laser light-scattering experiments with polarized light show that GM3 vesicles, with an average radius of about 250 Å, are in thermodynamic equilibrium with a small amount of large aggregates with nonspherical symmetry [3]. The coexistence of the two types of aggregates constitutes an important question which justifies the effort to obtain an independent experimental verification.

In this paper, we show that depolarized light-scattering measurements provide the necessary discrimination between the two coexisting forms of aggregation, and that the large aggregates can be represented as discs of approximately 3500 Å in radius.

Depolarized light from anisotropic discs

Since GM3 aggregates are made up of GM3 molecules which can be optically anisotropic, the light scattered by a GM3-water solution can have a depolarized component. Depolarized light-scattering theory predicts that for thin spherical vesicles, even if made up of anisotropic molecules, no depolarization can occur, due to internal cancellation effects [6]. On the other hand, if the shape of the aggregate is axisymmetric the optical anisotropy of the molecule gives rise to a depolarization effect. This is the case with an amphiphilic cylindrically symmetric molecule embedded into a cylindrically symmetric aggregate like a disk.

Theoretical calculations [7] predict the angular dependence of the scattered light intensity by anisotropic thin disks in the three different polarization configurations, in terms of their optical anisotropy δ and their radius R:

$$I_{VV} = A\,\frac{2}{h^2}\left\{I_1 + \delta(3I_2 - 2I_1)\right.$$

$$\left. + \delta^2\left(I_1 - 3I_2 + \frac{27}{8}\,I_3\right)\right\}$$

$$I_{HV} = I_{VH} = A \frac{9\delta^2}{h^2} \left\{ I_2 - I_3 + \sin^2 \frac{\vartheta}{2} \right.$$

$$\left. \cdot \left(\frac{5}{4} I_3 - I_2 \right) \right\}$$

$$I_{HH} = A \frac{2}{h^2} \left\{ I_1 \cos^2 \vartheta - \delta \cos \vartheta \left(2 - \sin^2 \frac{\vartheta}{2} \right) \right.$$

$$\cdot \left(3I_2 - 2I_1 \right) + \delta^2 \left[\left(2 - \sin^2 \frac{\vartheta}{2} \right)^2 \right.$$

$$\cdot \left(I_1 - 3I_2 \right) + 9I_3$$

$$\left. \left. \cdot \left(1 - \sin^2 \frac{\vartheta}{2} + \frac{3}{8} \sin^4 \frac{\vartheta}{2} \right) \right] \right\},$$

where A is a normalization constant [7] and:

$$I_n = \int_0^\pi J_1^2 (h \sin \gamma)(\sin \gamma)^{2n-3} d\gamma$$

with $h = \dfrac{4\pi R}{\lambda} \sin \dfrac{\vartheta}{2}$.

In our case, interesting observations can be made in HV configuration, where contribution to the scattered intensity comes only from axisymmetric particles all through the angular range, and in HH configuration at scattering angles around 90°, where any shift of the minimum of the scattered intensity from the exact 90° position is due only to the presence of depolarizing particles, which is connected to their optical anisotropy.

Moreover, also dynamic light-scattering measurements can be made in the HV configuration to obtain information on the translational and rotational diffusion coefficients of the nonspherical particles, which, as already stated, are the only ones which contribute to the scattered intensity. The translational, D, and rotational, Θ, diffusion coefficients of an oblate ellipsoid of rotation were calculated by Perrin [8] in terms of its minor, b, and major, a, dimensions and axial ratio $\rho = b/a$ as:

$$D = \frac{k_B T}{6\pi\eta b} \rho G(\rho)$$

$$\Theta = \frac{3 k_B T}{16\pi\eta b^3} \rho^3 \frac{(2 - \rho^2) G(\rho) - 1}{1 - \rho^4}$$

with $G(\rho) = (\rho^2 - 1)^{-1/2} \tan^{-1}((\rho^2 - 1)^{1/2})$.

The time-dependent part of the field correlation function, g_1, is expressed in terms of the translational and rotational diffusion coefficients as [9]:

$$g_1 \div \beta^4 e^{-(Dk^2 + 6\Theta)t} \quad \text{with} \quad k = \frac{4\pi}{\lambda} \sin \frac{\vartheta}{2},$$

and β, the out-of-plane polarizability.

Results and discussion

Depolarized HV static and dynamic laser light-scattering measurements were performed as a function of the scattering angle for a 1 mM GM3-water solution at a temperature of 25°C. The static data are shown in Fig. 1. Dots represent the measured intensity values. By means of a dilute solution of a nonionic-surfactant small spherical micelles, leakage of the vertically polarized component of the scattered intensity has been tested to be less then a few percent of the measured depolarized intensity at low angles and has been subtracted. This is consistent with a rejection ratio of the apparatus to be of the order of 10^{-5}. The full line represents the calculated distribution for thin disks with a radius of 3500 Å. Fitting with the predicted behavior for the intensity scattered by rods [10] is not acceptable.

From the ratio between the horizontal and vertical components of the scattered light the value of the intrinsic anisotropy δ can be found to be equal to 0.028, a very low value which is an index of the optical anisotropy of the single GM3 monomer.

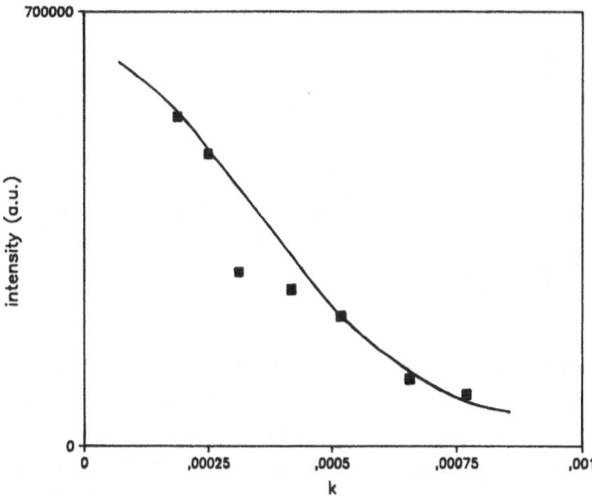

Fig. 1. k-dependence (Å^{-1}) of the depolarized HV intensity scattered by a 1 mM GM3-water solution

In Fig. 2 the angular distribution of the scattered intensity in the HH configuration is shown. The behavior in a narrow range around 90° is plotted in the insert. Besides showing the high degree of confidence of the collected data, the slight shift of the minimum from the 90° position confirms the presence in solution of depolarizing objects with low anisotropy.

Figure 3 shows the k-square dependence of the decay constant Γ_h of the field correlation function of the depolarized scattered intensity, HV configuration. Dots are the experimental values. Full lines show the expected behaviors for thin disks with radius 3000 Å (top), 3250 Å (intermediate) and 3500 Å (bottom). Low-angle data are better fitted by assuming larger dimension, while high-angle values are likely to be due to smaller discs. This indicates that a polidispersity is present in the dimensions of the ellipsoids: in fact, larger objects give a higher contribution to the intensity scattered in the forward direction with respect to smaller particles, which, instead, can be better appreciated at higher scattering angles.

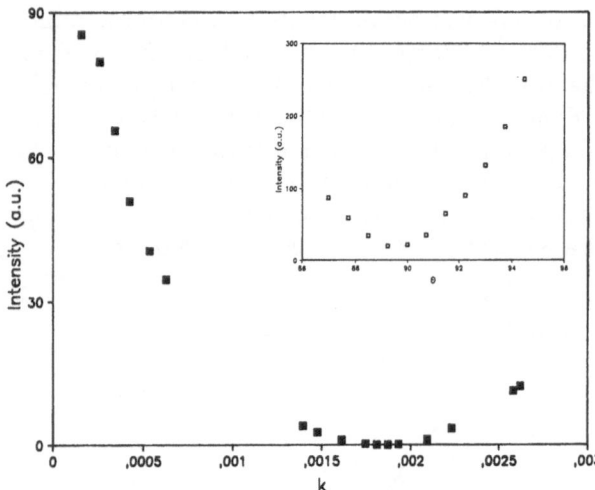

Fig. 2. k-dependence (Å^{-1}) of the HH intensity scattered by a 1 mM GM3-water solution. The 90° region is expanded in the insert

In conclusion, the internal consistency of independent static and dynamic laser light-scattering measurements, both in polarized and in depolarized configurations, performed on dilute GM3-water solutions, confirm the existence of axisymmetric particles which can be schematized as thin polydisperse disks with average dimension of 3500 Å.

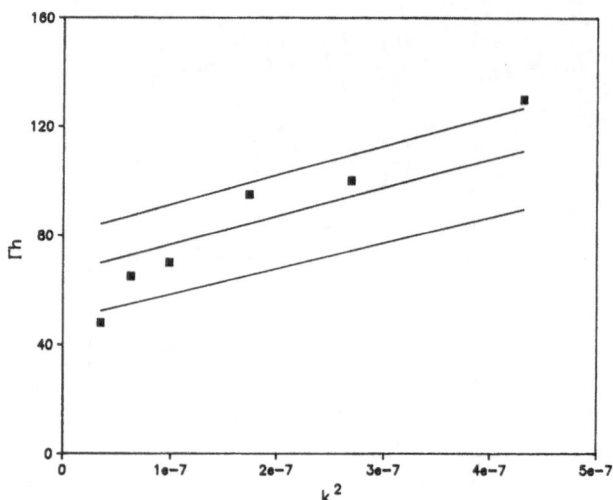

Fig. 3. k-square dependence (Å^{-2}) of the decay constant Γ_h of the field correlation function of the depolarized scattered intensity

The coexisting feature of this family of disks with the vesicle population is a stimulating thermodynamic problem.

Acknowledgements

Work partially supported by CNR Progetto Finalizzato CHIMICA FINE II.

References

1. Kaler EW, Murthy AK, Rodriguez BE, Zasadzinsky JAN (1989) Science 245:1371
2. Safran S, Pincus P, Andelman D (1990) Science 248:354
3. Cantù L, Corti M, Musolino M, Salina P (1990) Europhys Lett 13(6):561
4. Cantù L, Corti M, Lago P. Musolino M (1991) SPIE, Photon Correlation Spectroscopy Multicomponent Systems 1430:144
5. Cantù L, Corti M, Musolino M (1992) Springer Proc in Physics, Springer-Verlag Berlin Heidelberg 66:185
6. Aragon SR, Elwenspoek M (1982) J Chem Phys 77:3406
7. Picot C, Weill G, Benoit H (1968) J Coll Int Sci 27:360
8. Perrin F (1934) J Phys Rad 10:33; (1936) J Phys Rad 1:1
9. Berne BJ, Pecora R (1975) Dynamic Light Scattering, Wiley NY
10. Horn P (1955) Ann Phys 12(10):386

Authors' address:

Prof. M. Corti
Dept. of Electronics
University of Pavia
Via Abbiategrasso 209
27100 PAVIA, Italy

Progress in Colloid & Polymer Science

Progr Colloid Polym Sci 93:33—36 (1993)

Structure of clathrin-coated vesicles from contrast-variation small-angle neutron scattering data

J. Skov Pedersen

Department of Solid State Physics, Risø National Laboratory Roskilde, Denmark

Abstract: Previously published small-angle neutron scattering data from clathrin-coated vesicles have been analyzed in terms of a structural model. The data consist of contrast variation measurements at three different D_2O solvent concentrations: 0%, 42%, and 75%. The model used for interpreting the data has spherical symmetry and explicitly takes into account polydispersity, which is described by a Gaussian distribution. A constant thickness of the clathrin coats is assumed. The fitting of the model shows that the coated vesicles consist of a low-density outer protein shell (clathrin) and a central protein shell (accessory polypeptides and receptors) of approximately six times higher density. The polydispersity of the samples is about 90 Å (full-width-at-half-maximum value) and the average outer radius is approximately 400 Å. The inner high-density shell has an inner and outer radius of 115 and 190 Å, respectively. A simultaneous fit to the three neutron contrast variation data sets identifies the lipid membrane with a thickness of 40 Å and an outer radius of 196 Å. The molecular mass of the average particle is 27×10^6 Da. The coated vesicles consist, on average, of approximately 85% protein and 15% lipids. About 40% of the protein mass is situated in the central high-density shell which gives a large amount of protein in the lipid membrane. The densities of the central shell and the lipid membrane show that the hydration is small in the central region.

Key words: Small-angle neutron scattering — contrast variation — clathrin — vesicles — molecular structure

The protein clathrin plays an important role in connection with endocytosis in living cells [1]. The clathrin forms a shell of a polygonal network on the cytoplasmaic side of the membrane, stabilizing the structure of the coated vesicle. Clathrin (the heavy chain) is a 180 kDa protein which forms trimers, termed triskelions (see, e.g. [1] and references therein). The triskelion is a star-shaped molecule of about 800 Å in diameter with kinks at the three legs at about 200 Å from the center of the molecule. The outer lattice network is formed by packing of the triskelions. The second important component in the coats is the 100 kDa accessory polypeptide (AP), which together with an associated 50 kDa polypeptide, is believed to form the contact between the clathrin shell and the internal vesicle [2]. Most of the structural information quoted above was obtain-ed from electron microscopy [1, 3, 4].

The present work is a structural analysis of the small-angle scattering data previously published by Bauer et al. [6]. A more detailed description of the analysis and results is given in [8], which also include the analysis of the x-ray and neutron scattering data for coated vesicles, reassembled coats (without membrane), and stripped vesicles published by Bauer et al. [5]. The advantage of the contrast variation technique for neutron scattering is that it allows the constituents of the particles to be identified. This is done by using the known linear variation of the scattering density of the constituents with D_2O % of the solvent. The aim of the structural analysis is thus to determine the protein distribution in the particles and the position of the lipid membrane.

The samples were prepared by G. Jones, M. Behan, and D. Clark in the Biological Support Laboratory at Daresbury Laboratory, U. K.. The details of the sample preparations are given in [5, 6]. The clathrin-coated vesicles were purified from bovine brains by successive centrifuging in sucrose gradients and by gel exclusion chromatography. The small-angle neutron scattering experiments were performed at the instrument at the reactor DR 3 at Risø National Laboratory in Denmark. The measured data were corrected for background by the conventional procedure and put on absolute scale by dividing with the incoherent scattering from water [9]. In the data analysis the instrumental smearing effects were described by the analytical expressions given in [10].

The neutron scattering data [6] for the three contrasts of 0, 42, and 75% D_2O are displayed in Fig. 1. A secondary maximum is clearly present at $q = 0.027$ Å^{-1} in the 0% data, and a weaker maximum is observed in the 75% data. For the 42% data, the scattering length density of the protein is matched, and the secondary maximum is nearly absent.

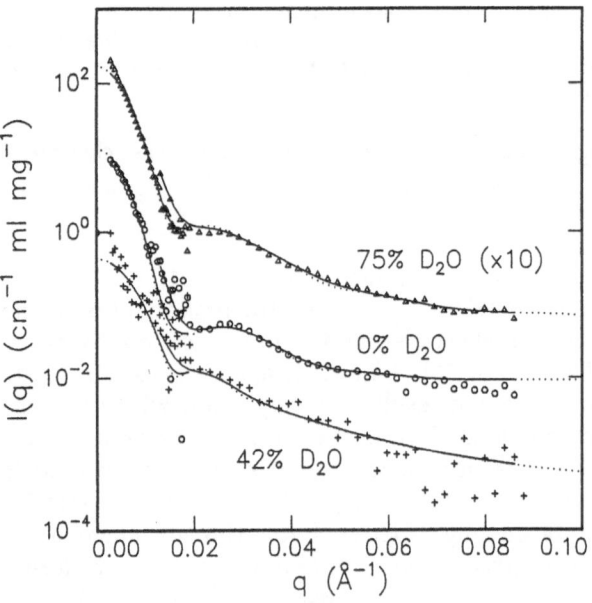

Fig. 1. Small-angle neutron scattering data for coated vesicles. The lower, middle, and upper data sets are for 42, 0, and 75% D_2O, respectively. The data for 75% are multiplied by 10. The solid lines are the simultaneous fit of the three-shell model, smeared by instrumental resolution. The dotted lines are the corresponding ideal fits

In the present work the data have been analyzed using spherical models consisting of several concentric shells, as have previously been used in the analysis of small-angle scattering from virus particles [11—14]. The application of models with spherical symmetry means that the finer structural features of the clathrin lattice are neglected. The parameters in the models are the radii of the shells R_i and the scattering length density ρ_i of the shells. These parameters are optimized by means of nonlinear least-squares methods. However, light-scattering experiments and electron microscopy [5, 6] show that the particles have a significant polydispersity of about 20—30%. This was included in the analysis by assuming a Gaussian size distribution. Furthermore, a constant coat thickness independent of the radius of the particles has been observed by electron microscopy [3], and this was also incorporated in the model.

Vigers et al. [3] calculated the radial electron density for the particles from their electron micrographs. The density is only due to protein in the coats as the lipids have an electron density which is close to that of the water. The radial density distributions show a shell structure with two main shells: a low density outer shell and a higher density central shell with an empty core. These two protein shells were included in the model. Finally, the lipid membrane was included in the model as a shell of thickness (~ 40 Å). This value was estimated from the membrane thicknesses determined in [11, 14].

The data have previously been analyzed [6] in terms of model-independent information. By assuming monodispersity of the samples a composition of 25 ± 5% lipids and 75 ± 5% proteins was found. From the scattering length densities of lipids and protein and their variation with D_2O concentration [9, 14], one finds that the lipids contribute only 12% of the total scattering amplitude at 0% D_2O. At 75% D_2O, the lipids and protein both contribute about 50% of the scattering. Due to these considerations, we have used the following procedure. First the 0% D_2O data was fitted by including only two shells in the structure factor, that is, we neglected the smaller contribution to the scattering from the lipids. Then, the 75% D_2O data were fitted, including a shell to describe the membrane. The scattering length densities in the two shells, also used for the 0% data, were multiplied by a factor of 0.76 in order to take into account the reduction in effective scattering density between 0 and 75%, as expected for protein. Only the parameters describing the membrane were optimized in the second fit. The fit to the 0% D$_2$O data is excellent. The central high density shell has an outer radius of 194 Å and a thickness of 76 Å. The polydispersity is 70 Å and the outer radius of the average particle is 391 Å, giving a relative polydispersity of 18%. When fitting the 75% data, the polydispersity was fixed at the value found for the 0% data. The fit gave a membrane with an outer radius of 195 Å, which shows that the membrane and the inner high density protein shell overlap.

Finally, the three contrasts were fitted simultaneously taking into account the variation of the scattering length density of lipids and protein. The match points are 12% D$_2$O for the lipids and 42% D$_2$O for the protein [9, 14]. These values were fixed in the fits, so that the prefactors in the linear expression for the scattering length density directly relate to the amount of lipids and protein. A common size distribution and polydispersity were assumed. The model has nine fitting parameters, including two parameters describing the residual background in the 0 and 75% D$_2$O data.

The average density profile from the fit is shown in Fig. 2. The continuous curves in Fig. 1 are the fits of the model, smeared by instrumental resolution, and the dotted curves are the ideal fits. The fit to the data is very good, and the variation in the shape of the secondary maxima is well reproduced by the model. The polydispersity is 23% (90 Å). The overall match point for the average particle is 36% D$_2$O. The composition of the particles can be calculated from this by using the known values for the scattering length densities [6, 14]. Using this, we find a protein mass fraction of 86% for the average particle. This is in reasonable agreement with the values determined previously [6, 7]. The molecular mass of the average particle is 27 × 10^6 Da. For the average particle, 38% of the protein mass is in the central high-density shell.

The radial structure of the coated vesicles is shown in Fig. 2. The outer radius of average particles is close to 400 Å, and the central high-density shell has an outer and inner radius of 190 and 115 Å, respectively. We associate the outer shell with the clathrin protein and the inner shell with the 100—50 kDa accessory polypeptides and receptor proteins. The lipid membrane has an outer radius of 196 Å and a thickness of 40 Å.

By simply dividing the masses of the various components by the volume of the corresponding

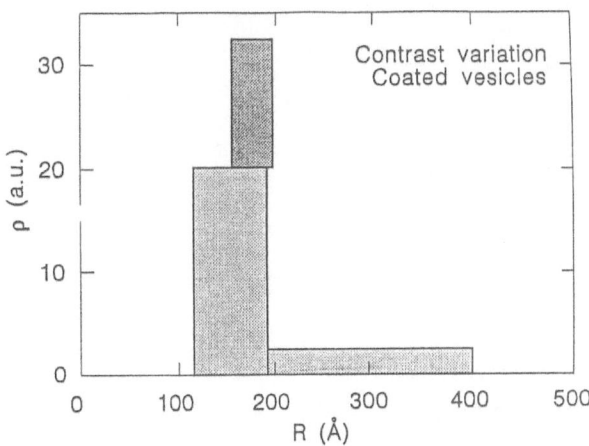

Fig. 2. Average particle radial density profiles. The profile is from the simultaneous fit of the neutron data. The membrane component is dark shaded

shells the densities can be calculated. The outer protein shell has a density of 0.10 g/cm^3. This is quite low as compared to the density of close-packed globular proteins in simple vira of ~1.0 g/cm^3 [12]. As already mentioned, we associate the outer protein shell with the clathrin polygonal lattice and the low density in the outer shell is obviously connected with the open structure of the lattice. The inner shell has a density of 0.66 g/cm^3, which is much closer to the value expected for close-packed globular proteins [12]. The membrane shell which overlaps with the inner protein shell has a density of 0.41 g/cm^3. This value can be compared to, for example, the one found for the membrane in influenza virus [14], which is 0.8—0.9 g/cm^3. From the determined values of the density in the central protein shell and the lipid membrane, one sees that the hydration is very low in the central region.

The main new conclusions from the present work are: i) the smaller particles from preparations of coated vesicles contain also a lipid membrane; ii) the position of the membrane coincides with the central high-density protein shell and, consequently, the membrane contains a large amount of protein. This leads us to suggest the schematic models shown in Fig. 3, which also incorporate the structural information from cryoelectron microscopy [3, 4]. The structures are shown as cuts through the center of a barrel-shaped particle [3, 4]. The present results show that the accessory polypeptides are situated in the cell membrane. These molecules probably work as receptors for the clathrin molecules when the clathrin lattice is formed.

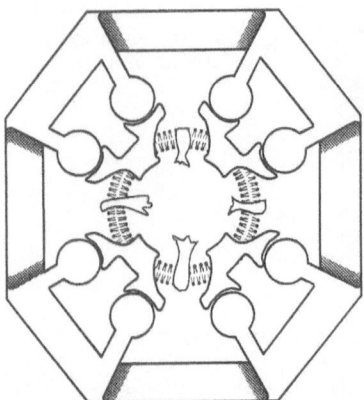

Fig. 3. Schematic drawings of the coated vesicles. A cut through the center of barrel-shaped particles is shown

Acknowledgements

This work was supported by the Danish Natural Science Council. The collaboration with Rogert Bauer, Steen Hansen, Kell Mortensen, Gareth Jones, Moira Behan, and David Clark during the work described in [6] is gratefully acknowledged.

References

1. Pearse BMF, Crowther RA (1987) Ann Rev Biophys Chem 16:49—68
2. Unanue ER, Ungewickell E, Branton D (1981) Cell 26:439—446
3. Vigers GPA, Crowther RA, Pearse BMF (1986) The EMBO Journal 5:529—534
4. Vigers GPA, Crowther RA, Pearse BMF (1986) The EMBO Journal 5:2079—2085
5. Bauer R, Behan M, Hansen S, Jones G, Mortensen K, Særmark T, Øgendal L (1991) J Appl Cryst 24:815—821
6. Bauer R, Behan M, Clark D, Hansen S, Jones G, Mortensen K, Pedersen J Skov (1992) Eur Bioph J 21:129—136
7. Pearse BMF (1975) J Mol Biol 97:93—98
8. Pedersen J Skov (1992) Eur Bioph J in the press
9. Jacrot B (1976) Rep Prog Physics 39:911—953
10. Pedersen J Skov, Posselt D, Mortensen K (1990) J Appl Cryst 23:321—333
11. Schneider D, Zulauf M, Schäfer R, Franklin RM (1978) J Mol Biol 124:97—122
12. Chauvin C, Witz J, Jacrot B (1978) J Mol Biol 124:641—651
13. Sjöberg B (1978) J Appl Cryst 11:73—79
14. Cusack S, Ruigrok RWH, Krygsman PCJ, Mellema JE (1985) J Mol Biol 186:565-582

Author's address:

Dr. Jan Skov Pedersen
Department of Solid State Physics
Risø National Laboratory
4000 Roskilde, Denmark

Progress in Colloid & Polymer Science   Progr Colloid Polym Sci 93:37—44 (1993)

On the zeta-potential of sulfonated polystyrene model colloids

D. Bastos and F. J. de las Nieves

Biocolloids and Fluid Physics Group Department of Applied Physics, University of Granada, Spain

Abstract: Highly sulfonated polystyrene latex particles were prepared by a two-stage "shot-growth" emulsion polymerization process in the absence of emulsifier. Sodium styrene sulfonate was used as an ionic comonomer to produce latex particles with the same particle size and different surface charge densities. The conversion of electrophoretic mobility measurements into zeta-potential (ζ) data was carried out according to several theoretical approaches: Smoluchowski, O'Brien and White, and Dukhin and Semenikhin. The current electrophoretic theories give rise to a maximum in ζ-potential. This behavior contradicts the classical electric double layer (e.d.l.) models which predict a continuous decrease in the electrokinetic potential. These theoretical approaches assume the absence of ionic conduction inside the shear plane. Dukhin and Semenikhin have developed an attempt to account for this phenomenon theoretically. In this work, we have calculated the ζ-potential of sulfonated polystyrene latex particles in the presence of symmetrical 1:1 and 2:2 electrolytes. The zeta-potentials estimated by the O'Brien and White (ζ_{O-W}) theory display a maximum and their values were lower than those calculated by the Dukhin and Semenikhin (ζ_{D-S}) theory, with which the maximum disappears. The considerations made by both theories about the contribution of all ions of the e.d.l. to the ionic conduction within this layer, is the reason for that differences.

Key words: Sulfonated polystyrene latexes — zeta-potential

Introduction

Monodisperse spherical polystyrene latexes have proven to be model systems which are widely used in many practical applications (calibration standards, support for biomolecules, etc.) and are also suitable for studying fundamental colloidal phenomena. The synthesis of polystyrene latexes prepared in the absence of emulsifier to produce particles with sulfate end groups have been widely described [1—6]. However, the surface of these particles may change with the time because of the hydrolysis of the sulfate end groups to carboxyl groups. Recently, papers have appeared [7—10] describing the preparation of sulfonated polystyrene latexes with independent control of the particle size and surface charge density. The sulfonate functionality was selected because of its stability against hydrolysis.

The electrokinetic characterization of highly sulfonated polystyrene particles with similar particle size and different surface charge densities was started in previous papers [11, 12]. In the first article [11], the effect of an extensive and systematic cleaning process on the electrokinetic behavior was studied. In the second [12], the effect of electrolyte type on the electrophoretic mobility (μ_e) was also studied.

The objective of this work is to continue the electrokinetic study of sulfonated polystyrene latexes by estimating the zeta-potential (ζ), because this parameter play an important role in the characterization of the electrical double layer (e.d.l.) at solid-liquid interfaces. The current electrophoretic theories used in the conversion of mobility into ζ-potential give rise to a maximum in ζ-potential as well. This behavior contradicts the classical double layer models which predict a continuous decrease in potential. Various explanations for this maxima have been proposed [13—16], and some authors [17] have even pointed out that a maximum mobility value does not necessarily imply a maximum in ζ-potential, indicating that the conversion of mobility into ζ-potential of polystyrene microspheres/electrolyte solution interface should be done by means of a theoretical approach which takes into account all possible mechanisms of double-layer polarization. Overbeek [18] and Booth [19] were the first to incorporate polarization of e.d.l. into the theory. Also, O'Brien and White [20], starting with the same set of equations as Wiersema [21], have more recently published a theoretical approach of electrophoresis which takes into account any combination of ions in solution and the possibility of very high ζ-potential, far enough from the values to be expected in most experimental conditions. Nevertheless, the method of O'Brien and White assumes the absence of ionic conduction inside the slipping plane, and Midmore and Hunter [13] have recently shown that this condition is not obeyed by the negatively charged latex/electrolyte system below electrolyte concentrations of about 0.001 M. If the influence of surface conductance on electrophoresis is substantial, this invalidates the above theoretical approaches. The anomalous surface conduction is related to the tangential charge transfer between the slipping plane and the particle surface [22]. The Dukhin and Semenikhin theory [23] considers only a particular case of anomalous conduction associated with the presence of a boundary layer.

The main purpose of this work is to check whether the e.d.l. theory, which includes the anomalous conduction, explains the electrophoretic behavior of sulfonated polystyrene microspheres. Special attention has been paid to the conversion of mobility into ζ-potential at different ionic strength and with several electrolytes of different valences. To make this conversion, two sulfonated latexes with the same particle size and different surface charge densities were used.

Materials and methods

Two sulfonated polystyrene latexes with the same particle size and different surface charge densities were used throughout this work. The preparation method and cleaning process were described in previous papers [11, 12]. The particle size of the latexes was obtained by transmission electron microscopy and the polydispersity index estimated

by the method described in [12]. Surface charge densities (σ) of the latexes were determined by conductometric and potentiometric titrations of the cleaned samples and by means of an automatic set up. The experimental details were also shown in [12].

Table 1. Particle size, polydispersity index, and surface charge density of sulfonated polystyrene latexes

Latex	Nass (g) 2^a injec.	DN (nm)	Stand. Deviat.	Dw (nm)	PDI	σ_0 (μC/m^2)
SN-8	1.2	178.5	5.5	179.0	1.003	12.3
SN-13	1.6	177.8	8.3	179.0	1.007	17.0

As a summary, Table 1 shows the number-average (D_n), weight-average (D_w) diameters, the polydispersity index (PDI) and the surface charge densities (σ) for the two sulfonated latex samples.

The different electrolytes used throughout this work were analytical grade reagents (all from Merck) and they were used without further purification. Double-distilled and deionized (DDI) water was used throughout.

The electrophoretic mobilities of the cleaned latexes were measured at 25 °C under different electrolyte types and concentrations by using a Zetasizer IIc (Malvern Inst., England). The electrophoretic mobility values were obtained by taking the average of (at least) six measurements at the stationary level in a cylindrical cell, changing the suspension sample twice. The experimental error was taken as the standard deviation in these measurements. The suspensions were prepared by adding approximately 0.1 ml of the latex to 40 ml of each electrolyte solution.

Results and discussion

Figures 1 and 2 show the ζ-potentials of latexes SN-8 and SN-13, respectively, versus the electrokinetic radius κa (κ reciprocal of the e.d.l. thickness and a radius of the spherical particles) for three different monovalent electrolytes (KBr, NaCl and NaNO$_3$). The ζ values were calculated by the simple Smoluchowski equation, which only considers the effect of the liquid phase by the dielectric constant and the viscosity. Both figures show a maximum in the ζ-potential curves when κa is around 65, which corresponds to an electrolyte concentration of $5 \cdot 10^{-2}$ M. This behavior is in marked contrast with the classical theory of the e.d.l. which predicts a continuous decrease in the electrokinetic potential with increasing, salt concentration or electrokinetic radius. Various explanations for this maximum have been proposed [13—17, 22, 24] which consider preferential adsorption of co-ions onto the surface or physical changes in particle surface properties (including anomalous surface conductance).

The conversion of mobility values into ζ-potential can be made by different theoretical treatments. We have chosen those developed by O'Brien and White [20] and Dukhin and Semenikhin [23].

O'Brien and White have shown that the interpretation of electrophoresis can be broken into two simpler problems: the force required to move the particle at a certain velocity with no applied electric field and the force required to hold the particle fixed in the presence of the applied electric field. This approach assumes the absence of ionic conduction inside the shear plane. Using the computer program developed by O'Brien and White [20], we have calculated the ζ-potential of the sulfonated polystyrene latexes. The curves again showed a maximum, as did those in Figs. 1 and 2, although with higher ζ-values at low electrokinetic radius. The results are not shown in order to simplify the figures (for comparison, the result with KBr and latex SN-13 will be shown in Fig. 6). At these conditions, there were some problems in the calculation of ζ-potential with KBr and NaCl electrolytes. The low particle size of our latexes and its high surface charge density yielded high electrophoretic mobility values at low electrolyte concentration. Under this conditions the O'Brien and White program failed to give acceptable ζ-potential values when the electrokinetic radii were lower than 8. Even consulting the tables derived by Ottewill and Shaw [25], the calculation was impossible in some cases; the low electrokinetic radius together with the high μ_e gave ζ values beyond the tables.

Dukhin and Semenikhin [23] have studied the influence of polarization on the electrophoretic mobility of spherical particles with a thin double layer ($\kappa a > 25$). Thus, we will have some difficulties with our latexes at low electrolyte concentration, where the κa values are lower than 20. If we consider only relatively large values of ζ and ψ_d (diffuse potential), the equations derived by Dukhin-Semenikhin can be simplified to [22, 27]:

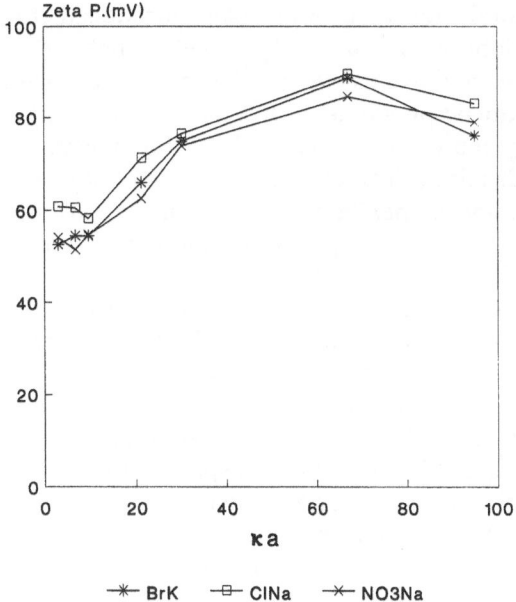

Fig. 1. Smoluchowski's ζ-potential of latex SN-8 versus κa for 1:1 electrolytes: *, KBr, □, NaCl, ×, NaNO₃

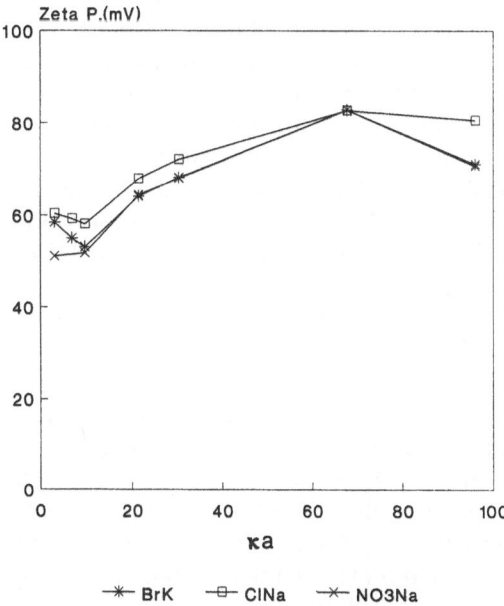

Fig. 2. Smoluchowski's ζ-potential of latex SN-13 versus κa for 1:1 electrolytes: *, KBr; □, NaCl; ×, NaNO₃

$$\mu_e = (3/2) \, [(1 + \text{Rel})/(1 + 2\,\text{Rel})] \, \zeta \tag{1}$$

$$\text{Rel} = (\lambda_s/\lambda a) = [\exp(\psi_d/2)$$
$$+ \, 3 \, m \, \exp(\zeta/2)] \, / \, \kappa a \,, \tag{2}$$

where m is the dimensionless ionic drag coefficient, λ_s is the surface conductance, and λ the solution bulk conductivity. Dukhin [26] introduced the dimensionless relaxation parameter Rel as a measure of the effect of surface conductance on electrokinetic phenomena. In Eqs. (1) and (2) μ_e, ζ and ψ_d represent dimensionless parameters. A more detailed view of O'Brien and White, and Dukhin and Semenikhin theories can be found in an interesting review [27] about the conversion of electrokinetic data into ζ-potential values.

We have calculated the ζ-potential values by the Dukhin and Semenikhin theory [23] (ζ_{D-S}) for both sulfonated latexes and several 1:1 electrolytes. The programs and calculus methods were the same used in [22, 28]. The results are shown in Figs. 3 and 4 (latexes SN-8 and SN-13, respectively). The ζ-potential values are substantially greater than those calculated by Smoluchowski or O'Brien and White theories. These differences are gradually smoothed out as the e.d.l. becomes thinner, as would be expected. The greater values of ζ_{D-S} in comparison with ζ_{O-W} are readily explained on the basis that, in the first theory, the contribution to polarization from all ions of the diffuse layer is taken into account, whereas O'Brien and White account for only the ions of the hydrodynamically mobile part of the e.d.l. Consequently, the passage of ζ through a maximum is presumably due to the influence of the surface conductance on electrophoresis.

The calculation of ζ within the framework of the Dukhin and Semenikhin theory requires the value of the diffuse potential (ψ_d). This potential has been calculated by using the Gouy-Chapman model of the e.d.l., and the surface charge density (σ) of the sulfonated latexes determined by conductimetric titration. The ψ_d values are shown in Fig. 5 for both latexes. As would be expected, the latex SN-13 with a high σ displays higher ψ_d values than those of latex SN-8. As the e.d.l. is compressed (higher κa values) the diffuse potential decreases (as can be seen in Fig. 5) and the same trend should be expected for the electrokinetic potential (as can be seen in Figs. 3 and 4).

For comparison, Fig. 6 shows ψ_d and ζ potentials for latex SN-13, as calculated by Smoluchowski, O'Brien-White and Dukhin-Semenikhin theories. Similar results were found for latex SN-8. As κa increases, the contribution of surface conductance drops off, and beyond $\kappa a \approx 90$ it becomes negligibly small; as a consequence, the difference among the ζ-potential values disappears. But, at low κa, the ζ-potential obtained from Dukhin-Semenikhin theory shows a different trend in com-

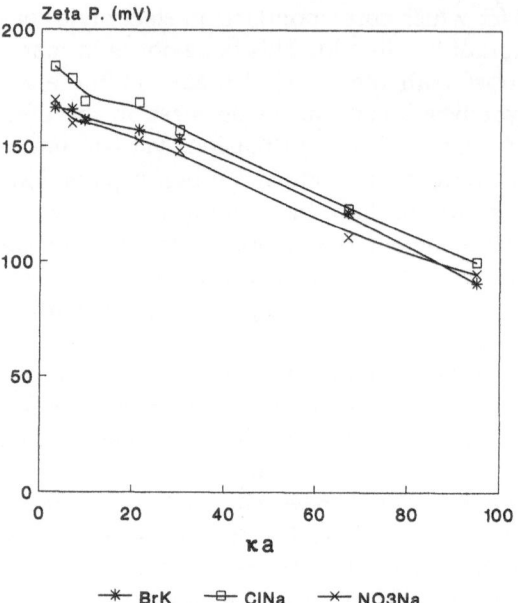

Fig. 3. Dukhin-Semenikhin's ζ-potential of latex SN-8 versus κa for 1:1 electrolytes: *, KBr; □, NaCl; ×, NaNO$_3$

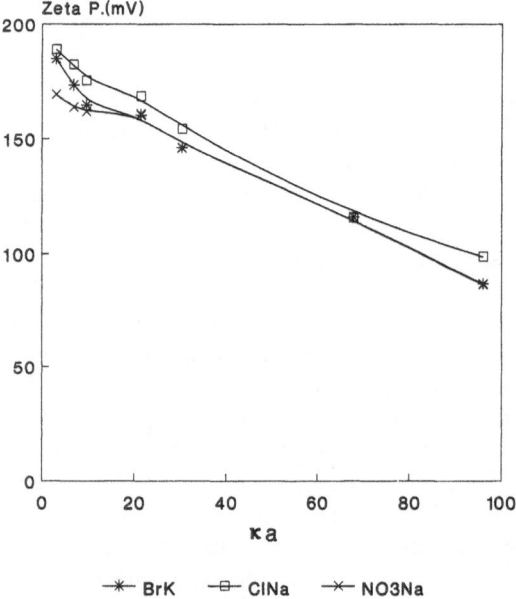

Fig. 4. Dukhin-Semenikhin's ζ-potential of latex SN-8 versus κa for 1:1 electrolytes: *, KBr; □, NaCl; ×, NaNO$_3$

parison with the others, and its variation is more similar to that obtained for the diffuse potential, as would be expected.

We have also studied the ζ-potential of sulfonated latexes in the presence of a 2:2 electrolyte such as MgSO$_4$. Figures 7 and 8 show the ζ-potential

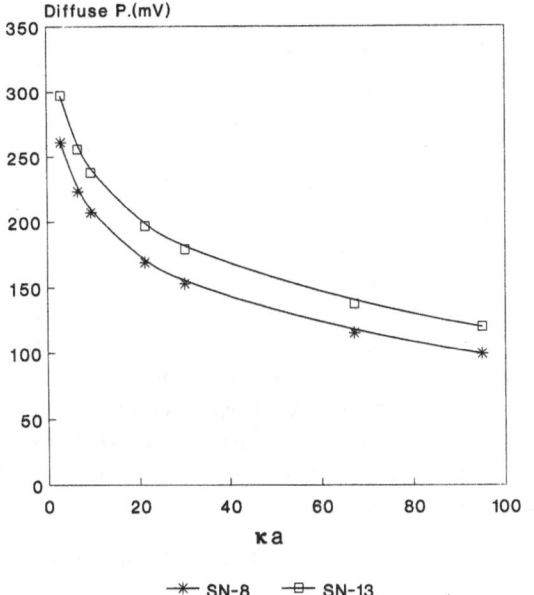

Fig. 5. Diffuse potential of latexes SN-8 (*) and SN-13 (□) versus κa

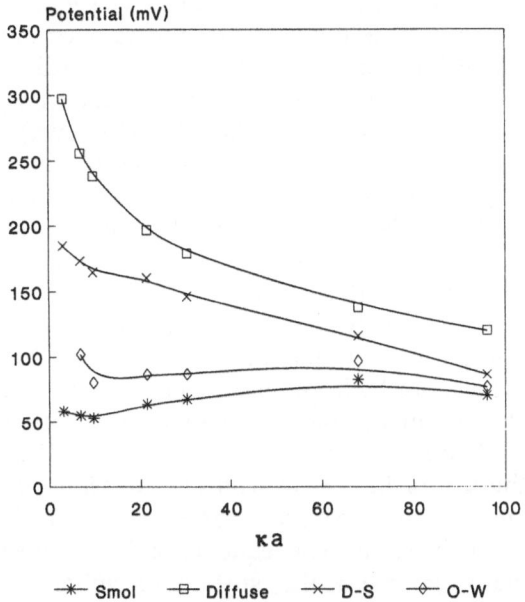

Fig. 6. Smoluchowski (*), diffuse (□), Dukhin-Semenikhin (×), and O'Brien-White (◇) potentials versus κa, for latex SN-13 and a 1:1 electrolyte (KBr)

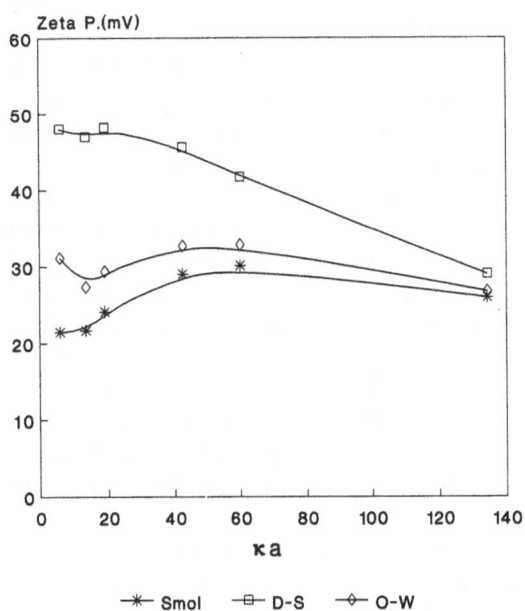

Fig. 7. Smoluchowski (*), Dukhin-Semenikhin (□) and O'Brien-White (◇) potentials versus κa, for latex SN-8 and a 2:2 electrolyte (MgSO$_4$)

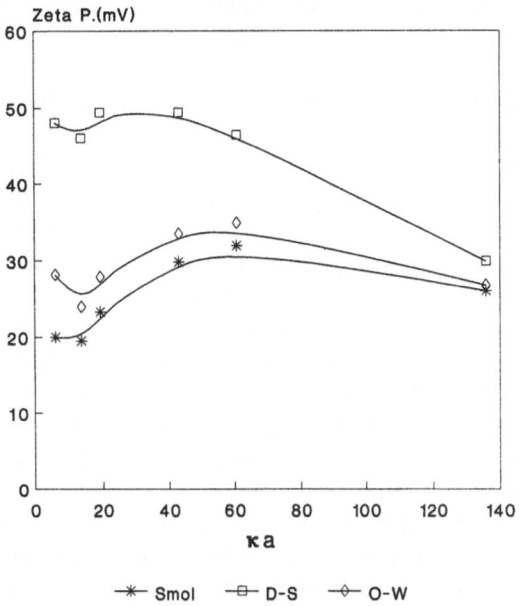

Fig. 8. Smoluchowski (*), Dukhin-Semenikhin (□) and O'Brien-White (◇) potentials versus κa, for latex SN-13 and a 2:2 electrolyte (MgSO$_4$)

values as calculated by Smoluchowski, O'Brien and White, and Dukhin and Semenikhin theories. As was found with 1:1 electrolytes, the $\zeta_{O\text{-}W}$ values maintain the maximum, although they are higher than those calculated by Smoluchowski's equation.

The $\zeta_{D\text{-}S}$ values are again greater, in comparison with $\zeta_{O\text{-}W}$, and both tend to the same values as $\kappa a > 120$ and the surface conductance becomes negligible. In both Figs. 7 and 8 the maximum in $\zeta_{D\text{-}S}$ disappears, although at low κa (lower than 20) the ζ-potential values show a plateau, when a continuous decrease would be expected. This is a consequence of the difficulties of the application of the Dukhin and Semenikhin theory when the electrokinetic radius is extremely low [22, 27, 28].

An important result of the combined measurements we have made is our finding that even when the ζ-potential values are corrected for the e.d.l. polarization, the values for the sulfonated latex/electrolyte solution interface are less than ψ_d over a wide range of electrokinetic radius. These results have been found for 1:1 and 2:2 electrolytes (Figs. 6—8). As pointed out by Baran et al. [29], this may be due to the formation of a liquid layer of the latex particle surface with low hydrodynamic mobility, in which the ions retain high mobility.

It is also possible that the relationship $\psi_d > \zeta$ that we have found is a consequence of surface roughness of the latex particles. Actually, Chow and Takamura [16] found that the maximum in mobility may be attributed to the surface roughness of the polystyrene particles, which results in a larger value for the location of the shear plane, and then a smaller value for the ζ-potential. Particularly with sulfonated polystyrene latexes prepared by shot growth [11, 12] or seed [7] emulsion polymerization, it has been found that the surfaces of these particles are covered by a layer of oligomers or polymer chains chemically bound on the surface. These chains also shift the shear plane toward the bulk solution, yielding a decrease in the ζ-potential at low ionic strength and producing a maximum in the ζ-potential/κa curves. Thus, the surface roughness or oligomers layer (hairy layer) could produce the same effect: the shifting of the shear plane toward the bulk solution. This displacement would increase with the increase in the amount of comonomer added at the second injection [11, 12], or during the seed [7] and, therefore, when the surface charge density of the latex particle increases. But, also, the displacement of the shear plane could produce a thicker e.d.l. and, therefore, a higher surface conductance, with the maximum appearing at higher ionic strengths. This is the result that we have found with sulfonated latexes: the maximum in the ζ-potential appears at a concentration of 5.10^{-2} M with 1:1 electrolytes, or 10^{-2} M with 2:2, 2:1 and 1:2 electrolytes [12]. However, with cationic [14, 22] or sulfate [15] latexes prepared by conventional emulsion polymerization, the maxima appeared at a concentration of around 5.10^{-3} M for 1:1 electrolytes, or 10^{-3} M for 2:2 or 2:1 electrolytes

[30]. Thus, the appearance of a hairy layer on the surface of latex particles seems to be related with the preparation method of the latexes. When no-conventional emulsion polymerization is used a thicker layer appears on the particle surface, shifting the shear plane toward the bulk solution and increasing the anomalous surface conduction. As a result, for the sulfonated latexes the maximum in the ζ-potential appears at higher electrolyte concentration (5.10^{-2} M for monovalent electrolytes) and the colloidal stability of the sulfonated latexes is greater in comparison with sulfate ones [31].

In order to estimate the thickness of the layer which coated the particle surface, Δ, and shifts away the shear plane, we can use the Eversole and Boardman equation [32]:

$$ln\ \tanh\ (ze\zeta/4kT) = ln\ \tanh\ (ze\psi_d/4kT) - \kappa\Delta\ , (3)$$

where z is the valence of the ions, e is the unit charge of an electron, k is the Boltzmann constant, and T is the absolute temperature. Thus, the position (Δ) can be estimated from the slope of the straight line obtained when plotting ζ_{D-S} versus κ. Figure 9 shows the experimental points used and the straight lines fixed for both latexes and KBr as electrolyte. The thickness of this layer (water layer with low hydrodynamic mobility or hairy layer) is 3.5 Å for latex SN-8 and 4.7 Å for SN-13. Other authors [33, 34] have found values between 10 and 15 Å for PS/PHEMA copolymer, which indicates a thicker layer of hydrated poly-HEMA on the surface of this particles. With cationic latex the thickness of this layer varied between 2 and 30 nm [22], which indicates a thicker water layer around latex particles. The more hydrophobic character of the cationic latexes in comparison with anionic latexes [35] could be the reason for that differences. Thus, from the use of Eq. (3) we can only consider the position, Δ, of the shear plane to be shifted away by approximately 5 Å as a consequence of the existence of a hairy layer on the surface of sulfonated polystyrene latexes.

In conclusion, the conversion of mobility data of sulfonated latex particles into ζ potentials for electrokinetic radii between 7 and 100, should be made by the Dukhin and Semenikhin theory, since it takes into account the inherent anomalous surface conductance of the polystyrene microsphere/electrolyte solution interface. Futhermore, the high surface charge density and low particle size of these sulfonated polystyrene latexes introduces some difficulties for the calculation of ζ at very low elec-

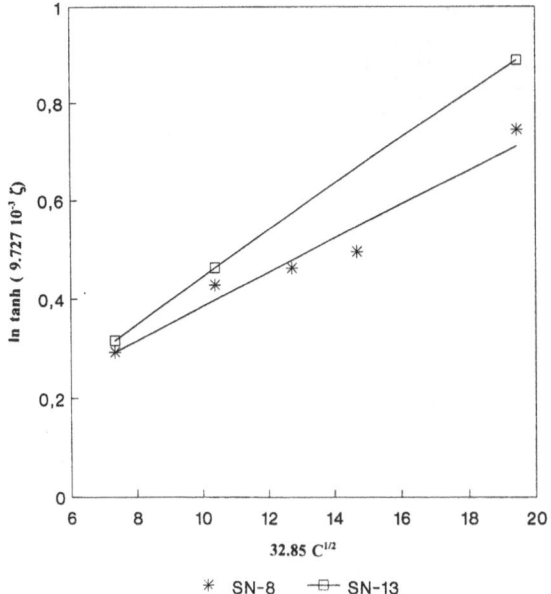

Fig. 9. Experimental values and fixed straight lines of the function $\ln[\tan h\,(ze\zeta/4kT)]$ versus κ (= 32, 85 × $C^{1/2}$), for latexes SN-8 (*) and SN-13 (□) (where ζ is in mV and C in mol/l)

trokinetic radius, when the O'Brien and White theory is used. Under these conditions the Dukhin and Semenikhin theory can be used, although it yields a plateau at low κa (lower than 20). Also for sulfonated latexes, the anomalous conductance seems to be higher at low electrolyte concentrations because of the thicker e.d.l. which sorround the particles. This is a consequence of a layer of oligomers or polymer chains chemically bound on the particle surface.

Acknowledgements

This work is supported by the Comisión Interministerial de Ciencia y Tecnología (CICYT), projects n° MAT 90-0695-C02-01 and MAT 93-0530-CO2—CO1, by Centro para el Desarrollo Technológico Industrial (CDTI), project n° 89-IE-0044-C02-00, and by Consejería de Educación y Ciencia de la Junta de Andalucía (Ayuda Consolidación de Grupos).

References

1. Ottewill RH, Shaw JN, (1967) Kolloid Z Z Polym 218:34
2. Kotera A, Furusawa K, Takeda Y (1970) Kolloid Z Z Polym 239:677
3. Furusawa K, Norde W, Lyklema J (1972) Kolloid Z Z Polym 250:908
4. Bijsterbosch BH (1978) Colloid Polym Sci 256:343
5. Goodwin JW, Hearn J, Ho CC, Ottewill RH (1973) Br Polym J 5:347
6. Brouwer WM, Zsom RLJ (1987) Colloids Surfaces 24:195
7. Tamai H, Niino K, Suzawa T (1989) J Colloid Interface Sci 131:1
8. Kim JH, Chainey M, El-Aasser MS, Vanderhoff JW (1989) J Polym Sci, Polym Chem Ed 27:3187
9. Kim JH, Chainey M, El-Aasser MS, Vanderhoff JW (1992) J Polym Sci, Polym Chem Ed 30:171
10. Tsaur SL, Fitch RM (1987) J Colloid Interface Sci 115:450
11. de las Nieves FJ, Daniels ES, El-Aasser MS (1991) Colloids Surfaces 60:107
12. Bastos D, de las Nieves FJ, Colloid Polym Sci, in press
13. Midmore BR, Hunter RJ (1988) J Colloid Interface Sci 122:521
14. Hidalgo-Alvarez R, de las Nieves FJ, Van der Lind de AJ, Bijsterbosch BH (1986) Colloids Surfaces 21:259
15. Elimelech M, O'Melia Ch (1990) Colloids Surfaces 44:165
16. Chow RS, Takamura K (1988) J Colloid Interface Sci 125:226
17. Van der Linde AJ, Bijsterbosch BH (1990) Croatica Chim Acta 63:455
18. Overbeek JThG (1943) Kolloid Chem Beih 54:287
19. Booth F (1950) Proc Roy Soc (London) Ser A 203:514
20. O'Brien RW, White LR (1978) J Chem Soc, Faraday Trans II 77:1607
21. Wiersema PH (1964) Rijkuniversiteit, Utrecht
22. Moleón-Baca JA, Rubio-Hernández FJ, de las Nieves FJ, Hidalgo-Alvarez R (1991) J Non-Equilib Thermodyn 16:187
23. Semenikhin NM, Dukhin SS (1975) Kolloidn Zh 37:1127
24. Zukoski CF, Saville DA (1985) J Colloid Interface Sci 107:322
25. Ottewill RH, Shaw JN (1972) J Electroanal Chem 37:133
26. Dukhin SS, Derjaguin BV (1974) in "Surface and Colloid Science", Ed. Matijevic E, vol 7, Wiley, New York
27. Hidalgo-Alvarez R (1991) Adv Colloid Interface Sci 34:217
28. Hidalgo-Alvarez R, Moleon JA, de las Nieves FJ, Bijsterbosch BH (1992) J Colloid Interface Sci 149:23
29. Baran AA, Dukhina LM, Soboleva NM, Chechik OS (1981) Kolloidn Zh 43:211
30. Galisteo F, de las Nieves FJ, Cabrerizo M, Hidalgo-Alvarez R (1990) Progr Colloid Polym Sci 82:313
31. Bastos D (1992) M S These, University of Granada, Spain
32. Eversole WG, Boardman WW (1941) J Chem Phys 9:798
33. Shirahama H, Suzawa T (1984) J Appl Polym Sci 27:3651

34. Shirahama H, Suzawa T (1985) J Colloid Interface Sci 104:416
35. Rubio-Hernández FJ, de las Nieves FJ, Bijsterbosch BH, Hidalgo-Alvarez R (1989) In: The Plastic and Rubber Institute (ed) "Polymer Latex III", p 15/1, London

Authors' address:

Dr. F. Javier de las Nieves
Biocolloids and Fluid Physics Group
Department of Applied Physics
University of Granada
18071 Granada, Spain

Progress in Colloid & Polymer Science   Progr Colloid Polym Sci 93:45—50 (1993)

Electrical properties of polystyrene particles from negative electrolyte adsorption measurements

T. Gilányi

Department of Colloid Chemistry, Loránd Eötvös University Budapest, Hungary

Abstract: The surface charge density of the polystyrene particles was determined from the concentration and activity of a binary electrolyte added to the latex dispersion. The distribution of ions was calculated by means of the non-linearized Poisson-Boltzmann equation and cell model. The effective charge of the latex particles was found to be smaller than the analytical charge. The particle charge was constant when the macro ions were compressed into smaller volumes and, unexpectedly, it was independent of the ionic strength in the range investigated. It was concluded that the small effective charge cannot be interpreted by specific binding of counter ions to the particles.

Key words: Electric double layer — Poisson — Boltzmann — polystyrene latex — negative electrolyte adsorption

Introduction

The theories of macro ionic systems involve at least one electrical — charge or potential — parameter to be experimentally determined. The difficulties with the electrical characterization of the macro ions are well known; the electrical parameters derived from different equilibrium and transport methods are model dependent.

The straightforward electrical parameter might be the total (analytical) charge of the particles, which is well defined in many systems, e.g., in the case of ionic micelles, latex particles, etc. However, it has long been recognized that macro ions behave in solution with an effective charge quite different from their analytical one. The so-called counter ion association phenomenon has been the subject of extensive theoretical and experimental study. The generally accepted idea is that the counter ions bind specifically (by non-coulombic forces) to the particle surface, i.e, the macro-ions are partially dissociated, or, the binding is a consequence of the high surface charge density of the particles. Obviously, in certain systems the counter-ions certainly bind specifically, but in the cases when the surface ionic groups are strong electrolytes, this assumption has not been justified. How can some 70—80 percent of the

sodium counter ions be bound onto the alkyl sulfate micelles [1] with only a small fraction of the micelle charge remaining "effective"? It has been shown that counter ion binding to micelles is apparent; it can be interpreted by the long-range electrostatic interactions between micelles and small ions [2].

In this work an attempt was made to describe the macro ionic system without measured electrical parameter of the macro ions, from the influence of macro ions on the small ion distribution in the system. The distribution of small ions was calculated by means of the non-linearized Poisson-Boltzmann equation. For finite macro ion concentration the cell model was applied [3]. The input parameters to the numerical solution of the PB-cells were the size and volume fraction of the macro particles, the concentration and mean activity of a binary supporting electrolyte.

PB cell model

Let us consider the simplest case of a system consisting of uniform spherical macro ions immersed in a 1:1 electrolyte solution. In a sound description of such systems, beyond the interactions between macro ions and small ions, the macro ion — macro ion and small ion — small ion correlations should be treated together, which has been a formidable task until recently. The effect of the electrostatic interactions among simple ions and macro ions is estimated via the solution of the non-linearized Poisson-Boltzmann equation, subject to appropriate boundary conditions. The macro ion — macro ion correlation is taken into account by the application of the cell model [3], where the system is treated as a collection of independent cells, each containing a macro-ion and the corresponding amount of solvent and simple ions. The main criticism against the use of the simple PB model is the neglect of the correlations between small ions. Recent statistical mechanical investigations predict that the PB theory gives reasonably good results for 1:1 electrolytes over a wide range of electrolyte concentration, particle size and charge [4, 5].

The PB equation for spherical symmetry is

$$\varepsilon \frac{1}{r^{-2}} \frac{d}{dr} \left(r^2 \frac{d}{dr} \Psi \right)$$
$$= - F c_e [\exp(-e\Psi/kT) - \exp(e\Psi/kT)] \,, \tag{1}$$

where Ψ is the electrostatic potential, r is the distance from the center of macro ion, ε is the permittivity of the medium, F is the Faraday constant, and c_e is the electrolyte concentration at $\Psi = 0$. The boundary conditions are

$$\frac{d\Psi}{dr}\bigg|_R = 0 \tag{2}$$

at the cell boundary (R) and

$$\Psi = \Psi_a \tag{3}$$

at the particle surface (a). The supporting electrolyte concentration is determined through the relation

$$c_o = \frac{3c_e}{R^3} \int_a^R \exp(-e\Psi/kT) r^2 dr \,. \tag{4}$$

The cell size $R = a/\phi^{1/3}$, where ϕ is the volume fraction of the macro-ions.

The cell model as applied here is similar to that used by Gunnarson et al. [3] for ionic micelles, except that the input experimental parameter is c_o/c_e instead of the surface charge density of the particles. Furthermore, to avoid computational difficulties the reference potential ($\Phi = O$) was chosen

at c_e. It is not necessary to have such a point in the system, because $c_+(r)c_-(r) = c_e^2$ is independent of r and c_e is a measurable quantity.

Equation (1) was solved by a fifth order Runge-Kutta method. First at fixed c_e, a and ϕ values, Eq. (1) was solved for an arbitrary Φ_a, and c_o/c_e was calculated from Eq. (4). The calculated c_o/c_e value was compared to the measured one and the computation was repeated with new Φ_a values until the experimental and calculated c_o/c_e ratio were the same within 0.01%. The solution of Eq. (1) at the four experimental parameters gives the surface potential, the surface charge density $\left(\sigma = -\varepsilon \frac{d\phi}{dr}\bigg|_a \right)$, and the potential Φ_R at the cell boundary from which the osmotic pressure can be calculated [6].

The important relation concerned in this work is the dependence of c_o/c_e on the surface charge density. In Fig. 1 the c_o/c_e values are plotted against σ. At low charge densities c_o/c_e strongly decreases with σ and with increasing surface charge density c_o/c_e becomes almost independent of σ. The range where σ can be determined from c_o/c_e measurements also depends on the particle size and ionic strength of the system.

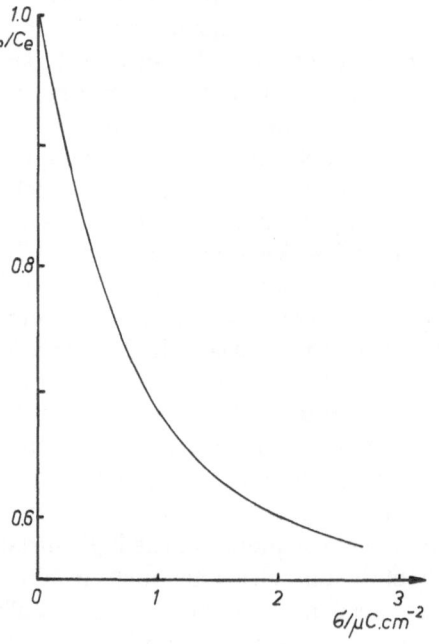

Fig. 1. The theoretical $c_o/c_e(\sigma)$ function at $c_e = 2.5 \ 10^{-4}$ mol $\cdot$ dm^{-3} $\phi = 8 \cdot 10^{-3}$ and 2a = 20 nm)

It is important to note here that in this method the electrical parameters of the macro ions are derived from the total potential distribution function. The contribution of the potential function to the integral in Eq. (4) is relatively little in the vicinity of the particles. Thus, the inner, or Stern layer does not have a direct importance since it is only several molecules thick, however, it has a significant indirect role in determining the value of the potential of the diffuse portion of the double layer. The advantage of the method is that it can be applied for concentrated systems and it is an equilibrium method. The applicability is restricted for monodisperse dispersions and for not too high surface charge densities.

Experimental

Polystyrene latex particles were prepared by aqueous emulsion polymerization with sodium dodecyl sulfate emulsifier and $K_2S_2O_8$ initiator at 70°C [7]. The samples were steam stripped at 100°C for 3—4 h. The latex dispersions were dialyzed against deionized water, followed by ion exchange using Vanderhoff's method [8]. The particle diameter was determined by light-scattering measurements and controlled by electron microscope micrographs. Conductometric titration of the ionexchanged sample with NaOH solution showed the presence of strong acid groups only. The diameter and surface charge density of the sample were 20 nm and 1.25 μC cm^2, respectively.

Potentiometric measurements were performed without liquid junction by a pair of pH-glass and Ag/AgCl electrode at 25.0 ± 0.1°C. The latex stock solution was titrated together with water and HCl stock solution, keeping a pre-selected e.m.f. value constant. The latex volume fraction (ϕ) and the HCl concentration (c_o) were calculated from the volume change of the system. The titration curve gives the $c_o(\phi)$ function at constant mean chemical potential of the supporting electrolyte. The e.m.f. value was converted into c_e value by means of the e.m.f. vs. lgc_e calibration curve. c_e called equilibrium electrolyte concentration, is the electrolyte concentration of a latex-free electrolyte solution in equilibrium with the latex at constant pressure and temperature.

In each case c_e was also determined by linear extrapolation of the c_o vs. ϕ curves to $\phi = 0$, when $c_o = c_e$. The measurements were accepted if the c_e

values determined in the two ways were the same within 1%.

The chemicals used in the experiments were reagent grade and twice freshly distilled water was used.

Results and discussion

In Fig. 2 the electrolyte concentration is plotted against the latex volume fraction at different constant c_e values. The amount of the supporting electrolyte which ensures the constant mean chemical potential of the electrolyte in the system decreases with the latex volume fraction.

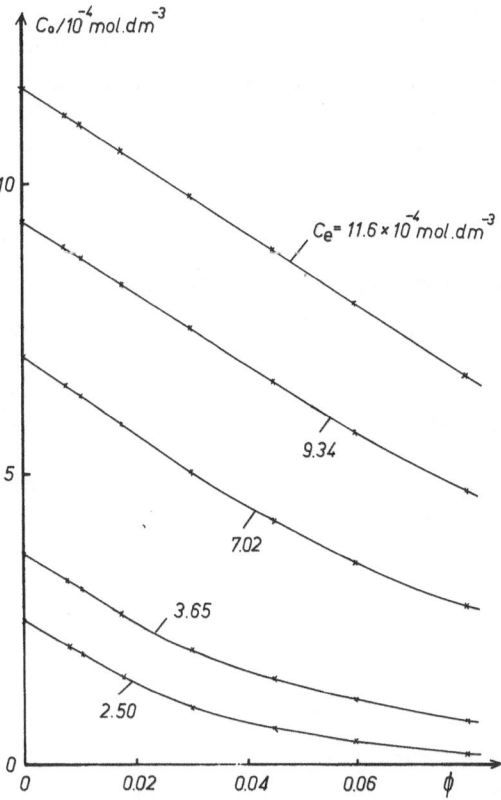

Fig. 2. Change of the electrolyte concentration (c_0) with the latex volume fraction (ϕ) at different mean electrolyte activities

The thermodynamic explanation of the effect of the macro-ions on the small ion activity (Donnan equilibrium) is based on the equality of the mean chemical potential in the macro-ionic and macroion free reference system, from which it follows for 1:1 electrolyte (neglecting the small ion activity coefficients) that

$$c_o (c_o + aZc_M) = c_e^2 , \tag{5}$$

where Z and c_M is the valence and concentration of the macro ions, respectively, and a is the degree of dissociation. Equation (5) explains the influence of macro ions on the mean small ion activity; the counter ions dissociated from the macro ions (aZc_M) contributes to the small ion concentration. Equation (5) can be found in almost every textbook in the same or different forms to explain the Donnan equilibrium. It is based on the underlying supposition that the entropy of mixing of small ions can be expressed simply by means of the volume average concentration of the small ions. This would be the case if the small ions were homogeneously distributed in the system, which is an erroneous supposition for macro ionic solutions.

Another explanation of the phenomenon is based on the electrostatic interactions between the macro ion and small ions. These interactions can be formally expressed as an excluded volume effect (V^*) of macro ions on the supporting electrolyte

$$c_o = c_e (1 - V^*/V) \tag{6}$$

or as a negative electrolyte adsorption [9]

$$\Gamma_-^{el} = (c_e - c_o)/\phi - c_e , \tag{7}$$

where Γ_-^{el} is the deficiency of the coions around the macro ions. Both V^* and Γ_-^{el} can be calculated from Eq. (4) if $\Phi(r)$ is known.

Figure 3 shows how the relative excluded volume (V^*/V) varies. The points are joined by lines at constant volume fraction and at constant c_e. The excluded volume increases with the latex volume fraction and decreases with the concentration of the supporting electrolyte. At very low electrolyte concentrations all curves tend to 1, i.e, practically the total volume of the system is excluded for the electrolyte.

The presentation of data as negative electrolyte adsorption is shown in Fig. 4. The negative electrolyte adsorption decreases with the latex volume fraction. This dependence is significant at small ionic strengths where the range of electrostatic interactions between macro ions extends to larger distances. The important conclusion drawn from Fig. 4 is that in a system of strongly interacting colloid particles the concentration of the particles together with the composition of the fluid phase is also a variable parameter in the excess adsorption.

The surface charge density and surface potential calculated from the electrolyte concentration and activity measurements are plotted in Fig. 5. σ was found to be independent of the latex volume fraction at constant c_e, i.e., the charge of the particles is constant when they are compressed to each other. σ is also constant with increasing ionic strength, while the potential monotonously decreases. These results fit to the picture that the electrical state of the strong electrolyte type macro ions is determined by their charge. It is unexpected, however, that the surface charge density (0.60 μC cm^{-2}) is only 48% of the total charge density (1.25 μC cm^{-2}) determined by acid-base titration.

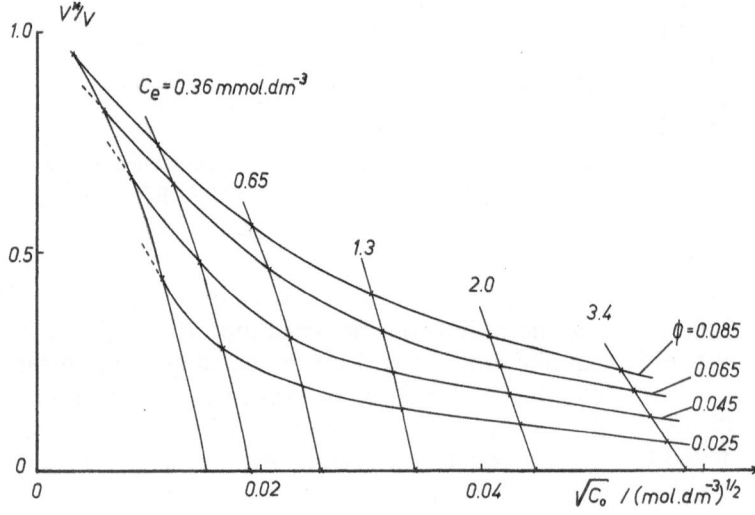

Fig. 3. Plots of the relative excluded volume ($V^*/V = 1 - c_0/c_e$)

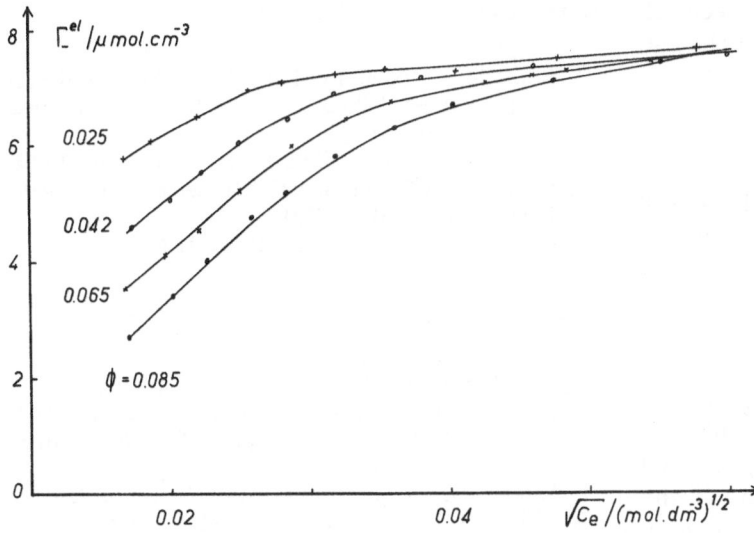

Fig. 4. The negative electrolyte adsorption against $\sqrt{c_e}$ at different latex volume fractions

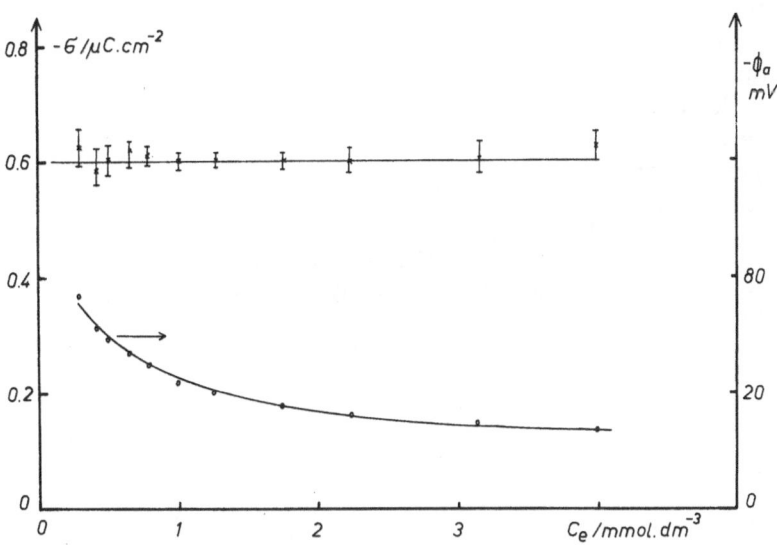

Fig. 5. Surface charge density ($\phi = 0.25 \ldots 0.085$) and surface potential ($\phi = 0.085$) of the latex particles from c_0/c_e measurements

Accepting that the measurements are correct, we are faced with the problem of how to interpret the small "effective" charge of the latex particles:

a) the counter ions bind specifically to the latex particles, especially, because they are hydroxonium ions;

b) the discrepancy is due to the approximations and neglects in the PB theory;

c) we use unrealistic model assumptions.

It is a strong argument against the specific counter ion binding that σ does not decrease with the ionic strength as shown in Fig. 5. It is noted here that, prior to computations, we made potentiometric measurements on latex dispersions (135 nm, 7.1 μC cm^{-2}) both in H$^+$-latex form with HBr and in Na$^+$-latex form with NaCl. Within the experimental error the results were identical, both in the cases of Na$^+$ and H$^+$ counter ions, from which we concluded that the counter ions do not bind specifically. However, the latex sample was not well chosen; the solution of the PB equation showed that the surface charge density is too high to draw a decisive conclusion.

The possible error due to the cell model can be eliminated by extrapolating to infinite large cell when we get the separated macro ion — small ions situation. Since σ is the same at $\phi \to 0$ as it is at finite concentrations, the small effective charge cannot be attributed to an error in the cell model or to particle — particle interactions.

The assumption that the potential in the diffuse double layer can be reasonably well-represented by the solution of the PB equation is generally believed to be true below eletrolyte concentrations of the order 10^{-2} M for 1:1 electrolytes [10]. The range of validity of the simple PB theory has not been settled yet. In spite of this, we believe that the explanation of the low surface charge density lies in the physical picture used for the particle surface. If the surface charges are distributed in an aqueous layer, into which the counter ions may penetrate, the effective charge of the particles may drastically decrease without specific interactions between the ions. In the case of such surface structures the concept of the Helmholtz and Stern layer should be revised.

Acknowledgement

The work was supported by the Hungarian Research Fund (No. 2162).

References

1. Shinoda K (1963) In: Hutchinson E, Van Rysselberghe P (eds) Colloidal surfactants. Academic Press, New York, pp 20—25
2. Gilányi T (1987) J Colloid Interface Sci 125:641—648
3. Gunnarsson G, Jönsson B, Wennerström H (1980) J Phys Chem 84:3114—3121
4. Outhwaite CW, Bhuiyan LB (1991) Molecular Phys 74:367—381
5. Akesson T, Jönsson B (1991) Electrochim Acta 36:1723—1727
6. Loeb AL, Overbeek JThG, Wiersema PH (1961) In: The electrical double layer around a spherical colloid particle. The MIT Press, Cambridge, Massachusetts
7. Ottewill RH, Richardson RA (1982) Colloid Polym Sci 260:708
8. Vanderhoff JW (1980) Pure Apple Them 52:1263
9. Gilányi T, Horváth-Szabó G (1991) Colloids Surfaces 57:273—281
10. Napper DH, Hunter RJ (1972) In: Keizer M (ed) Surface chemistry and colloids. Butterworths, London

Author's address:

Tibor Gilanyi
Department of Colloid Chemistry
Lorand Eötvös University
P.O. Box 32
Budapest 112, Hungary 1518

Progress in Colloid & Polymer Science   Progr Colloid Polym Sci 93:51—52 (1993)

Coalescence mechanisms of polymer colloids

F. Dobler, M. Lambla and Y. Holl

Institut Charles Sadron (EAHP-CRM), Université Louis Pasteur et Centre National de la Recherche Scientifique, Strasbourg, France

Key words: Latex — film formation mechanisms — coalescence in water

Coalescence of polymer colloids or latices is the process by which a dispersion of submicron polymer particles in water is transformed into a continuous, transparent and non-porous polymeric film. One can distinguish three steps in the process [1, 2]. In the first step, water evaporates at a constant rate until a stage is reached where particles form a dense packing of spheres. At the beginning of the second step, top particles become exposed to the air. They are brought into close contact and progressively lose their shape in such a way that polymeric material fills the entire space. In the film, interfaces between particles still exist. The third step corresponds to the evolution of these interfaces. It is sometimes called maturation. Film properties like mechanical strength and permeability are altered during this step [3]. Our present work dealt only with phenomena taking place during the second step [4, 5]. We did not tackle the problem of maturation.

The nature of the forces ensuring particle deformation in the second step of the latex film formation process has been much discussed in the literature. The three main theories are the following. For Brown [6], coalescence is due to capillary forces which appear at the surface of the drying latex when particles become partially exposed to the air. Capillaries full of water are formed in the spaces between the close packed latex spheres and capillary forces compress the packing of particles and make them coalesce. Vanderhoff et al. [7, 8] consider that the driving force for coalescence is the particle-water interfacial tension. Laplace's equation is used to demonstrate that a pressure gradient exists which pushes matter from the central part of the particle to the interparticle contact area. The third important theory was proposed by Sheetz [9]. According to this author, a thin layer of coalesced particles closes the surface of the drying latex. The remaining water evaporates after diffusion through this polymer layer and the packing of particles is compressed, as by a piston. The coalescence of the surface layer is ensured by capillary forces.

Some attempts have been made to verify these theories. These experimental verifications were not very convicing. In order to further investigate this important problem of deformation mechanisms, we synthesized model core-shell latices with particles having same radius, same bulk properties but different surface compositions [10]. This allowed us to study the influence of the particle-water or particle-air interfacial tensions on the deformation mechanisms. In the first part of our work [4], coalescence was studied in water under the influence of the singular particle-water interfacial tension. It was established that particle-water interfacial tension was indeed able to provoke coalescence of the particles, even at temperatures less than the glass transition temperature. It was also shown that the kinetics of the coalescence strongly depend on the particle-water interfacial tension. When the tension decreases, the rate of coalescence decreases. Activation energies of coalescence in water were measured. They are equal to the activation energy of the motions of the polymer at the glass transition as determined by dynamic mechanical spectrometry. However, coalescence in water is different from coalescence in standard conditions, i.e., with evaporation of water. The questions at the end of the first part were the following. Is coalescence in standard conditions ensured by particle-water interfacial tension or do other forces contribute to the process? And if other forces contribute, what is their relative importance?

In the second part [5], we compared coalescence kinetics in water to those when water simultaneously evaporates and established the limit conditions for film formation in terms of temperature and relative humidity. It was demonstrated that, for limit conditions, particle-water interfacial tension has a negligible contribution to the mechanism of coalescence of polymer colloids when films are formed at the same time as water evaporates. Evaporation of water is at the origin of forces which ensure deformation of particles. It was shown that the origin of particle deformation is not the capillary forces which can develop at the surface of the latex when a solid volume fraction around 75% is reached. This conclusion was drawn after preliminary research and it would be important to further investigate this point in order to establish it more firmly. Our results support Sheetz's theory of coalescence. In our case, iridescence of the latex surface indicates that the continuous polymeric film is formed at a very early stage of the process, at solid volume fractions in the range of 30 to 40%. This implies that the actual coalescence process takes place in very similar conditions for all latices, whatever the surface composition of the particles. The consequence is that all latices have the same limit temperature and relative humidity for film formation. Light diffraction as well as other techniques could be used to confirm the formation of a closed surface layer.

Coalescence of polymer colloids cannot be described by a unique mechanism. In water, coalescence is due to the particle-water interfacial tension. In standard conditions, evaporation of water is the cause of forces which ensure particle deformation. However, interfacial tensions indirectly contribute to the process via their effect on the coalescence of the surface particles.

When the coalescence is conducted above the limit conditions, the relative contribution of the acting forces on coalescence can be altered. For instance, in standard conditions, when temperature and relative humidity are well above the limit, coalescence under the influence of the particlewater interfacial tension can be faster than evaporation of water and thus become dominant in the coalescence process. Mechanisms of coalescence are complex and still deserve further study.

References

1. Vanderhoff JW, Bradford EB, Carrington WK (1973) J Polym Sci Symp 41:155
2. Chainey M, Wilkinson MC, Hearn J (1985) J Polym Sci Polym Chem Ed 23:2947
3. Voyutskii SS, Ustinova ZM (1977) J Adhes 9:39
4. Dobler F, Pith T, Lambla M, Holl Y (1992) J Colloid Interface Sci 152:1
5. Dobler F, Pith T, Lambla M, Holl Y (1992) J Colloid Interface Sci 152:12
6. Brown GL (1956) J Polym Sci 22:423
7. Vanderhoff JW, Tarkowski HL, Jenkins MC, Bradford EB (1966) J Macromol Chem 1:131
8. Vanderhoff JW (1970) Br Polym J 2:161
9. Sheetz DP (1965) J Appl Polym Sci 9:3759
10. Dobler F, Pith T, Holl Y, Lambla M (1992) J Appl Polym Sci 44:1075

Authors' address:

Dr. Y. Holl
EAHP
4, rue Boussingault
67000 Strasbourg, France

Progress in Colloid & Polymer Science   Progr Colloid Polym Sci 93:53—56 (1993)

The specific interactions between HEUR associative polymers and surfactants

A. P. Mast

Department of Interfaces, Colloids & Applied Physics, DSM Research, The Netherlands

Abstract: The influence of the surfactant structure on the thickening mechanism of low molecular weight thickeners is examined. Experimental results from rheological experiments are evaluated. The surfactants used are Akyposal BA 56, Akyposal RLM 56, Akyposoft 100 NV, Surfagene S 30, Akypo RLM 25, Avanel S 90 and Avanel S 30. The thickening mechanism in concentrated surfactant systems of low molecular weight thickeners is based on the increase of the hydrodynamic volume of the micelles by the building of EO-shells around the micellar core. The presence of long EO-chains in the surfactant molecules makes it impossible to build the additional shells. The ionic headgroup present in the surfactant molecule can, to some extent, either reinforce of diminish this effect. Sulphate surfactants have a stronger specific interaction with the EO-groups of the rheology modifier than sulphonates, and these are again stronger than carboxylic acid salts.

Key words: Low molecular weight HEUR polymer — surfactant head group — surfactant-polymer interaction — surfactant structure

Introduction

Although associative thickeners are today used increasingly more, the specific interactions of these polymers with surfactants is still not understood completely. Bridging, either between particles or micelles, is thought to be the way in which associative polymers act as a rheology modifier [1—4]. In a preceding paper [5] the influence of the structure and composition of the *H*ydrophobically modified *E*thoxylate *UR*ethanes (HEUR) was examined. The question was raised of whether these polymers are really "associative". The HEUR polymers with a low molecular weight performed best with respect to thickening properties, but these molecules are not capable of bridging between micelles for the simple reason that they are too small. An alternative thickening mechanism for low molecular weight HEUR polymers was discussed. This mechanism can be summarized as follows. The hydrophobic tails of these molecules are incorporated into the micelles and the hydrophillic part builds a structured EO-shell around the micellar core, giving rise to an increase in hydrodynamic volume. Besides the fact that the rheology modifier molecules have to have at least one hydrophillic and one hydrophobic part, there also has to be some interaction between the surfactant head groups and the EO-groups of the rheology modifier. This specific interaction between the surfactant head group and the rheology modifier is examined here and the available experimental results discussed.

Methods, materials and samples

Both the experimental methods and the polymeric materials are described extensively in the preceding paper [5]. Here, a range of different surfactants is used to evaluate the importance of the surfactant structure. An overview of all materials is given in Table 1. All surfactants are commercially available and used without further purification. The overall concentration of surfactant in each sample is, in all cases, 9.4 wt% active matter, and the ionic strength is adjusted to 0.43 M using NaCl

Table 1. The rheology modifiers and surfactants used in this study

Specification	Material	Producer	A.M.[1]	Chemical Information
Modifier A	PEO(8K)C18	NDSU[2]	—[3]	dioctadecylmodified PEO(8000)[4]
Modifier B	PEG6000distearate	AKZO	—[3]	distereatemodified PEO(6000)[4]
Modifier C	Aminol A15	CHEM-Y	97	$R_1\text{-}(OCH_2CH_2)_{1.5}OCH_2CONHCH_2CH_2OH$
Surfactant 1	Akyposal BA56	CHEM-Y	56	$R_1\text{-}(OCH_2CH_2)_3OSO_3^-Na^+$
Surfactant 2	Akypo RLM25	CHEM-Y	90	$R_2\text{-}(OCH_2CH_2)_{2.5}CH_2COO^-Na^+$
Surfactant 3	Avanel S30	PPG	35	$R_3\text{-}(OCH_2CH_2)_4SO_3^-Na^+$
Surfactant 4	Akyposal RLM56	CHEM-Y	56	$R_2\text{-}(OCH_2CH_2)_2OSO_3^-Na^+$
Surfactant 5	Avanel S90	PPG	35	$R_3\text{-}(OCH_2CH_2)_{10}SO_3^-Na^+$
Surfactant 6	Akyposoft 1000NV	CHEM-Y	22	$R_2\text{-}(OCH_2CH_2)_{10}CH_2COO^-Na^+$
Surfactant 7	Surfagene S30	CHEM-Y	39	$R_2\text{-}(OCH_2CH_2)_3COCH_2CHSO_3^-COO^-(Na^+)_2$

[1] active matter surfactant (wt%)
[2] North Dakota State University
[3] unknown, assumed to be 100%
[4] not fully substituted (see ref. [5])

R_1: pareth
R_2: laureth
R_3: decyl

(Aldrich, 99+%). The rheology modifier A is used at a concentration of 0.83 wt%, while B and C are used at 2.0 wt%.

Results and discussion

The rheology modifiers used here all have a molecular weight less than 20 000, which means the thickening mechanism is based on an increase in hydrodynamic volume of the micelles [5]. In Figs. 1 to 3 experimental results of a limited number of samples are presented. The results represent the trends observed for all samples very well.

Surfactants 1 to 3 all have approximately the same number of EO-groups but different head groups. What we see here is that the sulphate surfactant (1) always has the highest low shear viscosity, while the sulphonate (3) comes next, and the carboxylic acid salt (2) last. The influence of the number of EO-groups is deduced from the comparison of surfactants 1 with 4 and 3 with 5. Going from surfactant 1 to 4, we lose a CH_2- and a EO-group (on average). Thus, the chainlength of surfactant 4 is less and this will result in a decrease of the hydrodynamic volume of the micellar core. Assuming that these two groups together have a length of 3.5 Å, the reduction in volume is about 30%. A

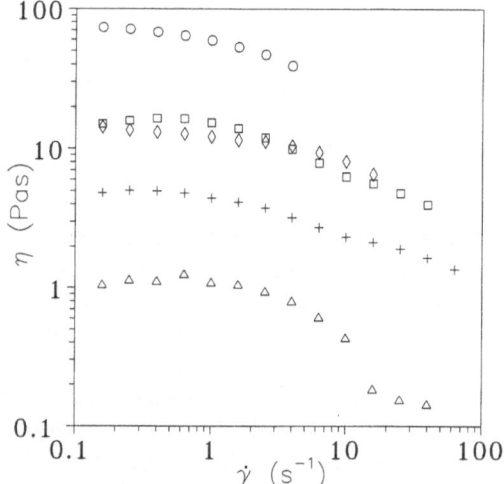

Fig. 1. Steady shear viscosity as a function of shear rate for dispersions containing modifier A at 0.83 wt%. ○) surfactant 1 only; ◇) surfactant 4 only; □) surfactant 1 and 6 conc. ratio 9:1; △) surfactant 1 and 6 conc. ratio 8:2; +) surfactant 1 and 7 conc. ratio 8:2

decrease in thickening property is therefore certainly not surprising (see Fig. 1). Comparing surfactants 3 and 5 the number of EO-groups increases from 4 to 10 while the low shear viscosity decreases. The gain in hydrodynamic volume of the micellar core is destroyed by some other, negative effect on the viscosity. Either the hydrophobic groups of the rheology modifiers are not incorporated into the micellar core or the interaction with the head group is diminished because they already interact with their "own" EO-groups. By using combinations of surfactants it is possible to get an idea about the relative importance of these two effects.

Using surfactants 1 and 6 together in the concentration ratios of 9:1, 8:2, and 7:3, respectively, gives a gradual decay of low shear viscosity with increasing concentration of surfactant 6 (see Figs. 1 to 3). Surfactant 6 has a COO^- instead of a SO_4^- head group, a CH_2-group less, and seven EO-groups more (on average) than surfactant 1. The specific interaction of EO-groups with a COO^- is known to be less than with a SO_4^- [6, 7]. Thus, increasing the relative concentration of surfactant 6 decreases the influence of the head group and increases the effects of more EO-groups. But we should not forget that two different ionic groups are present. Surfactant 7 possesses two kind of ionic groups, namely, a COO^- and a SO_3^-. Although we already saw that the sulphate is not totally comparable with the sulphonate, it at least gives us a clue towards the effect of combined ionic head groups. A combination of surfactant 1 with 7 always has a higher low shear viscosity than the same combination of 1 with 6 (see Figs. 1–3), but also is always less than a pure sulphate. We can therefore conclude that, in the case of the combination of surfactant 1 with 6, the effect of the increased number of EO-groups is the major reason for the decrease in viscosity.

As we have seen up to now, the viscosity of the dispersion decreases with increasing amount of surfactant 6. Comparing the samples with the same surfactant combinations but different rheology modifiers, we can see that rheology modifier C performs relatively better. The loss in thickening properties is less than with rheology modifiers A and B. In fact, we have a system of mixed micelles when rheology modifier C is used. The length of the EO-chain does not have such a great influence on the micellar aggregation.

A few remarks should be made before we can proceed to the conclusions. Both the surfactants as the rheology modifiers used, expect for modifier A, are all commercially available materials. These materials are, in general, polydispers with respect to the number of EO-groups. In the preceding paper [5] it was already mentioned that the impurities and the polydispersity in the modifiers can

play a major role [8]. Varadaraj et. al. [9, 10] showed that a polydispersity in the number of EO-groups is not that important for the aggregation behavior of surfactants.

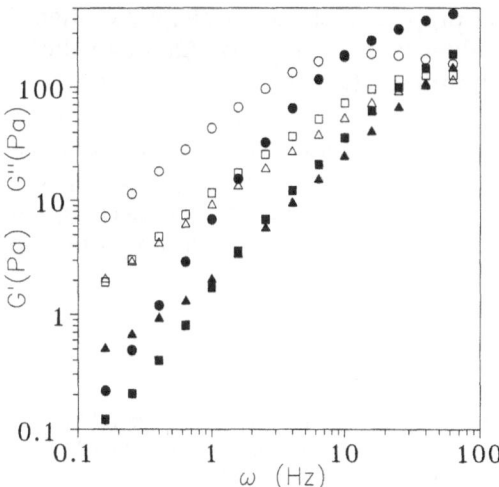

Fig. 2. Storage (G') and loss (G'') modulus as a function of frequency for dispersions containing modifier B at 2.0 wt%. ●) G' surfactant 1 only; ○) G'' surfactant 1 only; ■) G' surfactant 1 and 6 conc. ratio 9:1; □) G'' surfactant 1 and 6 conc. ratio 9:1; ▲) G' surfactant 1 and 7 conc. ratio 8:2; △) G'' surfactant 1 and 7 conc. ratio 8:2

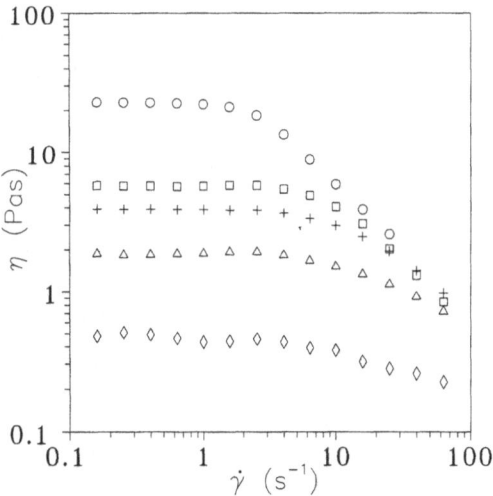

Fig. 3. Steady shear viscosity as a function of shear rate for dispersions containing modifier C at 2.0 wt%. ○) surfactnat 1 only; □) surfactant 1 and 6 conc. ratio 9:1; △) surfactant 1 and 6 conc. ratio 8:2; ◇) surfactant 1 and 6 conc. ratio 7:3; +) surfactant 1 and 7 conc. ratio 8:2

Concluding remarks

When the surfactant structure changes this has an impact on the interactions these surfactant molecules have with the rheology modifiers. Because this interaction is very important for the thickening properties of the rheology modifiers it directly influences this property. Although the effects of impurities should not be forgotten, all experimental results presented indicate that the effect of a longer EO-chain within the surfactant can, in effect destroy any thickening behavior. The specific interactions of the head group can make up for some of this destruction, but it certainly does not cancel this effect.

Acknowledgements

I want to thank Robert Prud'homme (Princeton University, Dept. Chem. Eng) for offering the possibility to work at his lab and Ed Glass (North Dakota State University, dept. Polym. Coatings) for providing the HEUR polymers. Furthermore, I want to thank CHEM-Y, especially Hamke Meijer, for making this research project possible.

References

1. Bieleman JH, Riesthuis FJJ, van der Velden PM (1986) Polym Paint Col J 176:450—460
2. Karunasena A, Glass JE (1989) Polym Mater Sci Eng 61:544—549
3. Char K, Frank CW, Gast AP (1989) Langmuir 5:1335—1340
4. Siano DB, Bock J, Myer P, Valint Jr PL (1987) Polym Mater Sci Eng 57:609—612
5. Mast AP, Prud'homme RK, Glass JE: submitted for publication in Langmuir
6. Brackman J (1991) PhD-thesis, Groningen, The Netherlands
7. Aalbers JG (1964) PhD-thesis, Amsterdam, The Netherlands
8. Glass JE. Kaczmarski JP (1991) PMSE-proc. ACS 65:175—176
9. Varadaraj R, Bock J, Geissler P, Zushma S, Brons N, Coletti T (1991) J Coll Interf Sci 147:387—395
10. Varadaraj R, Bock J, Geissler P, Zushma S, Brons N, Coletti T (1991) J Coll Interf Sci 147:396—402

Author's address:

Drs. A. P. Mast
DSM RESINS B.V., dept. DPCRA
P.O. box 615
8000 AP Zwolle, The Netherlands

Progress in Colloid & Polymer Science

Progr Colloid Polym Sci 93:57—58 (1993)

Diffusion and sizes of micelles in polymer solutions and gels *)

J.-E. Löfroth, P. Hedberg, and L. Johansson

Drug Delivery Research, Astra Hässle AB, Mölndal, Sweden

Abstract: The sizes and self-diffusion coefficients of the non-ionic micellar systems octaethylene- and hexaethylene glycol mono n-dodecylether, $C_{12}E_8$ and $C_{12}E_6$, respectively, were studied in gels and solutions of the polyelectrolyte κ-carrageenan in water as functions of polymer and surfactant concentration at 25°C. A recently presented tracer technique [1] was used to determine the self-diffusion coefficient of the surfactant monomer, free or in the aggregated state. The time-resolved fluorescence quenching technique was used to determine aggregation numbers of the micellar aggregates giving the number average aggregation numbers and the standard deviations [2]. The polydispersity was analyzed assuming a Schultz distribution [3] of the aggregation numbers.

Key words: Diffusion — micelle — polymer — gel

It was found that the combination of the results from the self-diffusion and fluorescence measurements was well suited to study the behavior of the surfactants, both in water and in the polymer systems. Thus, the hydrodynamic radius of $C_{12}E_8$ in water was estimated to be 30 Å from the diffusion studies. The same radius was determined from the fluorescence results when using two water molecules per ethylene group in the surfactant and the estimated number average aggregation number $\langle S \rangle_N = 79$. Good agreement was found between the calculated radius distribution and reported small-angle neutron scattering results [3]. Moreover, the diffusion quotient D/D_0, where D_0 is the micellar self-diffusion coefficient extrapolated to infinite dilution, decreased as $1-2\phi$, where ϕ is the surfactant volume fraction. This hard sphere behavior [4], which also was earlier reported for the $C_{12}E_8$ micelle in [5], was also reflected in the diffusion results in Na^+-κ-carrageenan (solution) and K^+ — κ-carrageenan (gel above 0.5% v/v of polymer) from experiments at varying polymer concentration. It was found that the obstruction from the polymer chains could be described by the hard sphere

theory recently presented [6, 7] predicting a faster diffusion in the gel than in the solution. The size of the $C_{12}E_8$ in these systems was marginally effected. Also, the ratio D/D_{aq}, where D_{aq} refers to a system without polymer, was independent of the surfactant concentration.

The diffusion and fluorescence results for $C_{12}E_6$ in water showed, on the other hand, that these micelles were larger, $\langle S \rangle_N = 192$, than the $C_{12}E_8$ micelles and were best described as prolates. From the diffusion experiments in water the short axis was estimated to be 28.3 Å, similar to the radius of $C_{12}E_8$. Assuming the short axis to be 28.3 Å, the diffusion experiments at different concentrations of surfactant showed that the axial ratio of the assumed prolate increased as $1 + 0.0385 \, {}^*C$, where C is the total surfactant concentration in mM. Also it was concluded that the aggregate concentration reached an upper level when increasing the total surfactant concentration. This showed that initially both the number of the aggregates and the growth of the aggregates increased, while at higher surfactant concentration mainly the size of already formed aggregates increased. The calculations were in good agreement with the fluorescence results that gave $\langle S \rangle_N = 192$ at 50 mM total $C_{12}E_6$ concentration in water.

---

*) J Phys Chem 79:747 (1993) Johansson L, Hedberg P, Löfroth J-E.

Contrary to the other surfactant, no differences could be detected between the diffusion of $C_{12}E_6$ in Na^+ — and K^+-κ-carrageenan. Also, the diffusion ratio D/D_{aq} of the surfactant in 3% w/w K^+-κ-carrageenan first decreased with increasing surfactant concentration, again contrary to what was found for $C_{12}E_8$. However, above a surfactant volume fraction of typically 0.02, the ratio was independent of the surfactant concentration. Also, a slight increase in $\langle S \rangle_N$ to about 250 in the presence of the polymers was found for the $C_{12}E_6$ micelle at 50 mM surfactant concentration. One possible explanation would be that aggregate diffusion was not the dominating transport mechanism. Instead, the intra- and intermicellar diffusion of the monomer substantially contributed to the observed diffusion coefficient. This was concluded by considering two effects; first, a strongly reduced aggregate diffusion and, second, the micelles being pushed closer together by the polymer chains due to repulsive interactions. The latter effect was also shown by calculations of the radial pair distribution function for hard spheres in polymer networks. By considering a water phase in equilibrium with a polymer phase, both containing hard spheres, a measure of the local micelle concentration in the polymer networks was obtained by simulation. Also, the simulation results compared well with an approximative theory. This formed the basis for quantitative predictions with the prolate cell mode [8] of the diffusion of the $C_{12}E_6$ surfactant in the polymer systems. The results showed the role of polymer structure on polymer-induced phase transitions in a surfactant/polymer system with repulsive interactions.

Acknowledgements

LJ was financially supported by grants to Chalmers University of Technology from Astra Hässle AB.

References

1. Johansson L, Löfroth JE (1991) J Coll Interface Science 142:116
2. Almgren M, Löfroth JE (1982) J Chem Phys 76:2734
3. Magid LJ (1987) In: Schick MJ (ed) Nonionic Surfactants, Physical Chemistry, Marcel Dekker, NY; Surfactant Science Series vol 23:677
4. Cichocki B, Hinsen K (1990) Physica A 166:473
5. Jonströmmer M, Jönsson B, Lindman B (1991) J Phys Chem 95:3293
6. Johansson L, Skantze U, Löfroth JE (1991) Macromolecules 24:6019
7. Johansson L, Elvingson C, Löfroth JE (1991) Macromolecules 24:6024
8. Jönsson B, Wennerström H, Nilsson PG, Linse P (1986) Colloid Polymer Sci 264:77

Authors' address:

Jan-Erik Löfroth, Associate Professor
Drug Delivery Research
Astra Hässle AB
43183 Mölndal, Sweden

Progress in Colloid & Polymer Science   Progr Colloid Polym Sci 93:59—62 (1993)

Influence of solvents on the formation of lyotropic mesophases in binary systems

H.-D. Dörfler

Lehrstuhl für Kolloidchemie, TU Dresden, FRG

Abstract: Comparing the phase diagrams of three binary soap systems KC_{18}/ethylene glycol, KC_{18}/buthylene glycol, and KC_{18}/glycerol, we found systematic alterations in the types of phase diagrams that depend on the chemical structure of the solvents. The following mesophases were observed: lamellar phases, hexagonal phases isotropic phases, and gel phases. Differences in number, position, extension, and the sequence of the occurring mesophases were found as a function of concentration of the mixtures and temperature. We came to the conclusion that the influence which the solution medium exerts on the formation of structures in the binary systems is very distinct and pronounced.

Key words: Phase diagrams — lyotropic mesophases — polymorphism — microscopic textures — nonaqueous polar solvents — K-stearate

1. Introduction

Up to now, lyotropic mesophases have been tested mostly in binary and ternary surfactant systems or in other aqueous multicomponent systems [1—11]. Therefore, investigations on influences of non-aqueous solvents on the formation of lyotropic mesophases presents a new aspect in this field. In the literature, we find only the very first attempts to get some information about this topic [12—27]. The aim of our studies was to prove the influence of ethylene glycol, buthylene glycol, and glycerol on the phase formation in the three binary K-stearate/ethylene glycol-, K-stearate/buthylene glycol and K-stearate/glycerol systems.

2. Methods and sample preparation

The texture was observed by applying a polarizing microscope (Jenapol 30-0060, Carl Zeiss, Jena) equipped with a heating stage. On average, we tested 15 singular concentrations per binary system. The mixtures were investigated in so-called flat capillaries. The mixtures obtained by a special pretreament procedures were put into the flat capillary. Then, the capillary was hermetically closed by melting it at its open end.

Texture examination was performed at temperatures between about $T \approx 303 - 623$ K. Heating was carried out by means of a Mettler-Thermosystem. The temperatures for the phase regions, given in the phase diagrams, represented average values resulting from three observations. The heating rate was 0.5 K $\min^{-1}$. The concentration range from $c = 5$ to 80 mol —% was referred to potassium stearate KC_{18}.

3. Phase diagram of the system potassium stearate (KC_{18})/ethylene glycol (E)

The results from texture observation on singular concentrations of the system KC_{18}/E are summarized by the preliminary phase diagram in Fig. 1. In all, six different phases have been observed and the following concentration dependent phase sequences (phase notation see in the legend) have been obtained:

from $c = 5$ to 24 mol —%: C ↔ G ↔ H ↔ L;
from $c = 24$ to 36 mol —%: C ↔ G ↔ I ↔ L ↔ S;
from $c = 36$ to 80 mol —%: C ↔ G ↔ L.

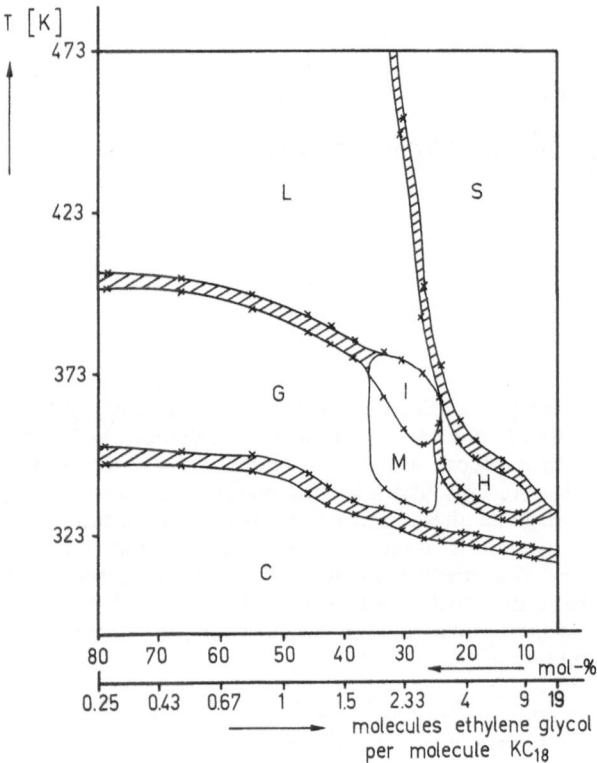

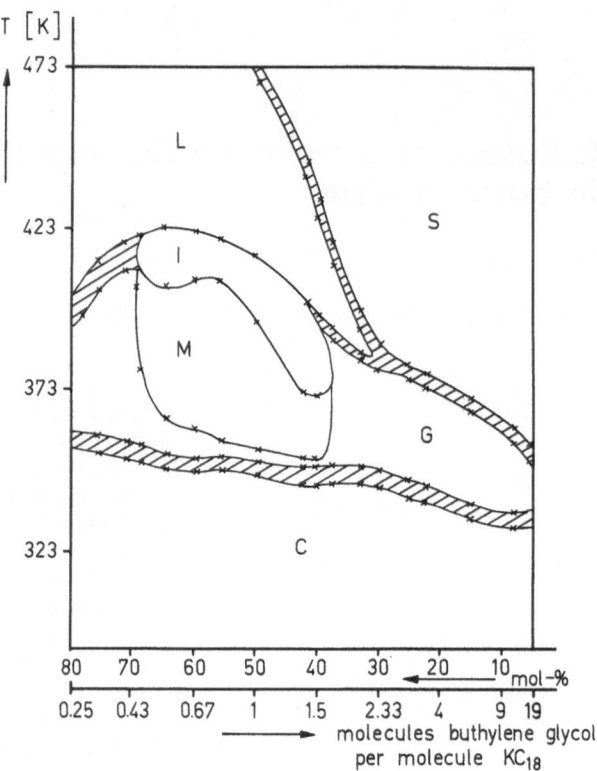

Fig. 1. Preliminary phase diagram of the binary system potassium stearate (KC_{18})/ethylene glycol (E). Concentrations are given in mol-% KC_{18} or molecules ethylene glycol per molecule KC_{18}. Phase notation: C = crystalline phase; G = gel phase; L = lamellar phase; H = hexagonal phase; S = isotropic solution; M = "melting region" (two-phase region); I = isotropic phase

Fig. 2. Preliminary phase diagram of the binary system potassium stearate (KC_{18})/buthylene glycol (B). Concentrations are given in mol-% KC_{18} or molecules buthylene glycol per molecule KC_{18}. Phase notation: C = crystalline phase; I = isotropic phase; L = lamellar phase; S = isotropic solution; M = "melting region" (two-phase region)

4. Phase diagram of the system potassium stearate (KC_{18})/buthylene glycol (B)

The results obtained by the texture observations on singular mixtures of the KC_{18}/B system is summarized in the preliminary phase diagram in Fig. 2. Up to five phases were observed. When temperature rose within the various range of concentration in the binary system the following polymorphism were found:

from c = 5 to 32 mol-%: C ↔ G ↔ S;
from c = 40 to 70 mol-%: C ↔ G ↔ I ↔ L.

A noteworthy fact concerning the KC_{18}/B system is that an extended optically isotropic phase I and a distinct "melting range" existed. The crystalline phase C extended over the whole range of concentration. With increasing temperatures the gel phase G was formed. The temperature of the region of transition from the gel phase G to the lamellar phase L were followed by the temperatures at which the lamellar texture developed from the optically isotropic phase I. The lamellar mesophase L was observed as a multicolored lamellar texture within the concentration range rich in soap.

5. Phase diagram of the system potassium stearate (KC_{18})/glycerol (G)

The results of the texture examinations of singular mixtures are summarized by the preliminary phase diagram of Fig. 3.

According to the phase diagram, no lyotropic mesophases have been observed within the concentration range of c = 0 — 14 mol-%. Toward higher temperatures the crystalline phase melts and turns into the isotropic solution S. The hexagonal phase

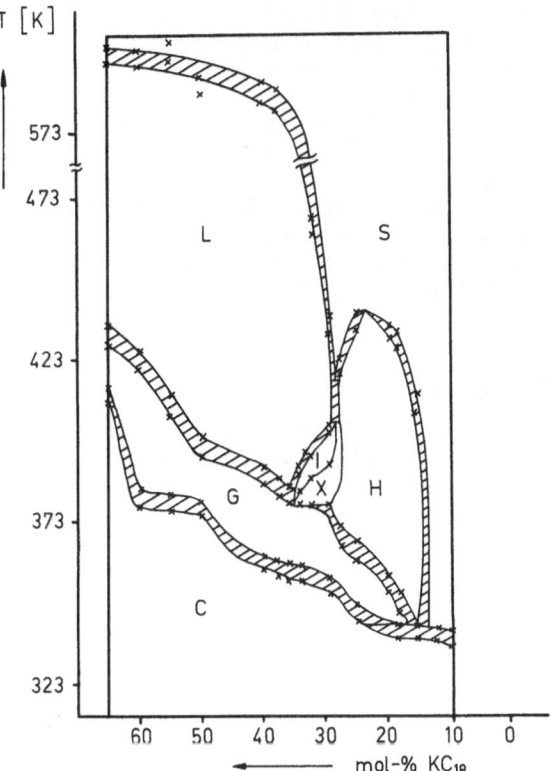

Fig. 3. Preliminary phase diagram of the binary system potassium stearate (KC_{18})/glycerol (G). Concentrations are given in mol-% KC_{18}. Phase notation: C = crystalline phase; G = gel phase; I = isotropic phase; X = heterogeneous gel phase region; H = hexagonal phase; L = lamellar phase; S = isotropic solution

around $\Delta T \approx 20$ K. When the temperature is continuously elevated, this gel phase melts at $T = 378$ K and the formation of the so-called heterogeneous two-phase region X starts. The phase now obtained probably represents a region of a gel phase which coexists with an optically isotropic phase. These observations result only from test performed with a polarizing microscope. The application of other methods such as calorimetry and X-ray diffraction are important, too.

The flocculent gel structure melts with increasing temperature. The photograph shows dark areas other than regions of remaining gel structure until the gel has completely changed into the optically isotropic phase. The optically isotropic phase exists between $T = 378 - 400$ K and has an average temperature width of $\Delta T \approx 6$ K. At temperatures beyond $T = 400$ K the optically isotropic I-phase passes in the L-phase.

The lamellar phase L exists within the concentration range of $c = 29 - 65$ mol-%. The temperature of phase formation depends again on the soaps concentration and runs through a maximum. A steep increase of the temperature of the phase transition G ↔ L occurs within the concentration range at $c = 34 - 65$ mol-%, where it finally reaches a value of $T = 430$ K. At higher temperatures the L-phase exists over a large temperature range, before it turns into the isotropic solution at $T = 598$ K.

is built up at concentrations between $c = 15 - 28$ mol-% and covers the temperature range of $T = 340 - 435$ K. The temperature of the phase transition H ↔ S runs through a maximum value. Beyond $T = 435$ K the hexagonal phase H is succeeded by the isotropic solution S. Between the C- and H-phase region a gel phase G with a temperature extension of around $\Delta T \approx 20$ K is inserted. But the region of the gel phase is not restricted to the concentration range at the H-phase. It is also found at concentrations between $c = 17 - 65$ mol-%. The phase diagram further demonstrates that the temperature of the phase transition C ↔ G slowly increases.

The concentration range of the optically isotropic phase is situated between $c = 29 - 35$ mol-%. In this part of the phase diagram the C-phase melts with rising temperature by forming a gel. The mean width of the temperature range of this gel phase is

References

1. Ekwall P (1975) Advances in Liquid Crystals, Vol. 1, Academic Press, New York — San Francisco — London
2. Fontell K (1990) Colloid Polym Sci 268:264
3. Vold RD (1939) J Phys Chem 43:1213
4. McBain JW, Vold RD, Frick M (1940) J Phys Chem 44:1013
5. McBain JW, Lee WW (1943) Oil and Soap 17:
6. Luzzati V, Mustacchi H, Skoulios A (1958) The Faraday Society's Discussion 25:43
7. Leigh JD, McDonald MP, Wood RM, Tiddy GIT, Trevethan MA (1981) J Chem Soc, Faraday Trans I, 77:2867
8. Mourachafieh N, Friberg SE (1979) Mol Cryst Liq Cryst 49:231
9. Hoffmann H, Ebert G (1988) Angew Chem 100:933
10. Schambil F, Schwuger MJ (1990) Tenside Surf Det 27:380
11. Heusch R (1991) Tenside Surf Det 28:38
12. Schwandner B (1988) Theses, Universität Bayreuth

13. El-Nokaly MA, Ford LD, Friberg SE, Larsen DW (1981) J Colloid Interface Sci 84:228
14. El-Nokaly MA, Friberg SE, Gan-Zuo L (1982) Mol Cryst Liq Cryst 72:183
15. Friberg SE, Liang P (1984) J Phys Chem 88:1045
16. Friberg SE, Podzimek M (1984) Colloid Polym Sci 262:252
17. Friberg SE, Wohn CS (1985) Colloid Polym Sci 263:156
18. Friberg SE, Liang P (1986) Colloid Polym Sci 264:449
19. Friberg SE, Liang P, Liang YC (1986) Colloids and Surfaces 19:249
20. Rananavare SB, Ward AJI, Friberg SE, Larsen DW (1986) Mol Cryst Liq Cryst 133:207
21. Friberg SE, Ward JI (1987) Langmuir 3:735
22. Friberg SE, Chang SW, Uang YJ, Lockwood FE (1987) J Dispersion Sci Technology 8:429
23. Friberg SE, Liang YC, Heuser J, Benton W (1990) Colloid Polym Sci 268:749
24. Friberg SE, Sun WM (1990) In: Bloor M, Wyn-Jones E (eds) The Structure, Dynamics and Equilibrium Properties of Colloidal Systems, Kluwer Academic Publishers p. 529
25. Friberg SE (1990) In: Tribology and the Liquid Crystalline State, ACS Symposium Series 441, Girma Biresaw.
26. Belmajdoub A, Marchal JP, Canet D, Rico I, Lattes A (1987) New J Chem 11:415
27. Auvray X, Petipas C, Anthore R, Rico I, Lattes A (1989) J Phys Chem 93:7458

Author's address:

Prof. Dr. habil. Hans-Dieter Dörfler
TU Dresden, Lehrstuhl für Kolloidchemie
Mommsenstr. 13
01062 Dresden

Progress in Colloid & Polymer Science   Progr Colloid Polym Sci 93:63—65 (1993)

The study of a colloidal crystal using ultra small-angle x-ray scattering (USAXS)

J. S. Ridgen[1], A. N. North[1] and A. R. Mackie[2]

[1]) Physics Department, University of Kent, Canterbury, United Kingdom
[2]) AFRC Institute of Food Research, Norwich Research Park, Colney, Norwich, United Kingdom

Abstract: USAXS, using the Bonse-Hart twin crystal diffractometer system, has been developed at a synchrotron source for the study of colloidal dispersions and other heterogeneous systems [1]. Many improvements have been made, and the full potential of this newly enhanced scattering system in both scientific and industrial fields is only just beginning to be realised [2]. One well-characterised commercial colloidal dispersion, a highly monodisperse colloidal latex, has shown some very interesting results using the improved system. At high concentrations some sharp Bragg Peaks have been observed in the scattering profile. The eperimental structure factor for the colloidal latex clearly shows at least five scattering peaks, the positions of which may be used to identify the structure of the ordered colloid. This structure is thought to be randomly close packed.

Key words: USAXS — twin crystal diffractometer — colloidal crystal

Introduction

The Bonse-Hart twin crystal diffractometer has recently been developed [1, 2] at the USAX station 2.2 at the SERC Daresbury Laboratory, Warrington, UK. The use of well-characterised colloids has helped to highlight the many problems which have restricted the functionality of this system in the past. Refinement of the instrument has made the precise examination of scattering from samples showing larger feature sizes (500 Å to 1.0 μm) possible. These studies have previously been restricted due to experimental resolution (e.g. SAXS) or turbidity effects (e.g. light scattering).

A schematic diagram of the USAXS system is presented in Fig. 1. Four Bragg reflections from the 111 plane of a channel cut silicon monolith produce a highly collimated, monochromatic beam from the synchrotron source. These incident x-rays are scattered at the sample, and rotation of a second, analyser crystal allows scattering at any angle to be collected; the inherently low scattered intensities produced by this method mean that high intensity synchrotron radiation must be used in order to fully utilise the system for many weak scatterers, such as colloidal latices or very dilute systems.

Standard and more advanced corrections to the collected data are necessary; corrections for dead time losses, smearing effects, sample absorption and multiple scattering. The improved USAXS instrument shows a very high angular resolution, $\sim 5 \times 10^{-7}$ Å^{-1} in scattering vector Q, where $Q = 4\pi/\lambda \sin(\Theta/2)$ for a scattering angle Θ. The system yields useful data at Q-values as low as 5×10^{-5} Å^{-1}.

Colloids

Scattering from a low concentration colloidal dispersion of spherical particles is given by the particle form factor $P(Q)$ [2], where

$$P(Q) \propto \left(\frac{\sin QR - QR\cos QR}{(QR)^3} \right)^2 .$$

For higher concentration dispersions, the scattering is modulated by an interference term, the structure factor $S(Q)$, and hence for N scatterers the scattered intensity is given by $I(Q) = N(Q)\, S(Q)$. It is therefore possible to obtain an experimental structure factor for a sample by dividing the corrected scattering distribution from a high concentration solution by that from a low concentration.

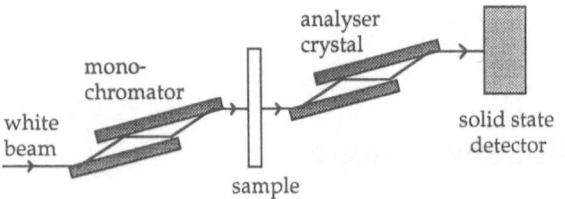

Fig. 1. A schematic diagram of the USAXS arrangement

A highly monodisperse, chlorinated colloidal latex ('haloflex', ICI plc.) with radius ~1000 Å has been made available for USAXS studies. In order to remove the surfactant present in this dispersion due to manufacturing processes, the stock solution (50% w/w) was dialysed against distilled deionised water for 2 weeks; this resulted in an increase in stock concentration to 57%.

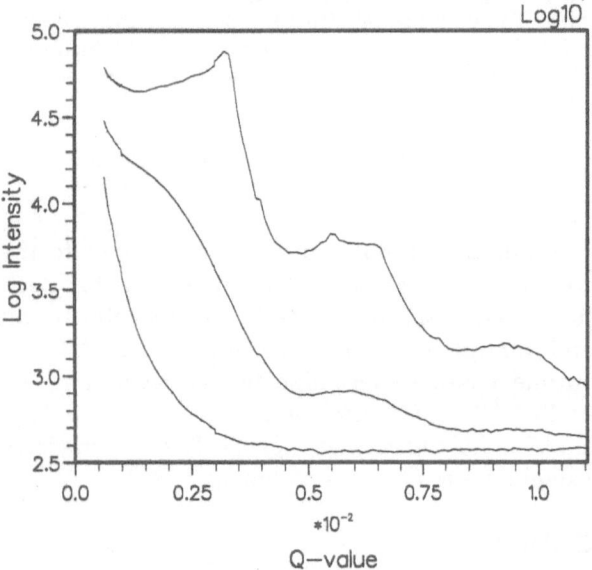

Fig. 2. Scattering distributions for the dialysed haloflex, concentrations 57% (top curve) and 1% (middle curve), and the water background (bottom)

Scattering from low concentration haloflex showed the expected form, however, scattering from the dialysed stock solution showed up to six sharp peaks, superimposed on the expected distribution (see Fig. 2). Peak positions could be found by examination of the experimental structure factor (Fig. 3); these are tabulated in Table 1, together with the expected positions if the Q-value of the first peak was assumed to be the 100 reflection of a simple

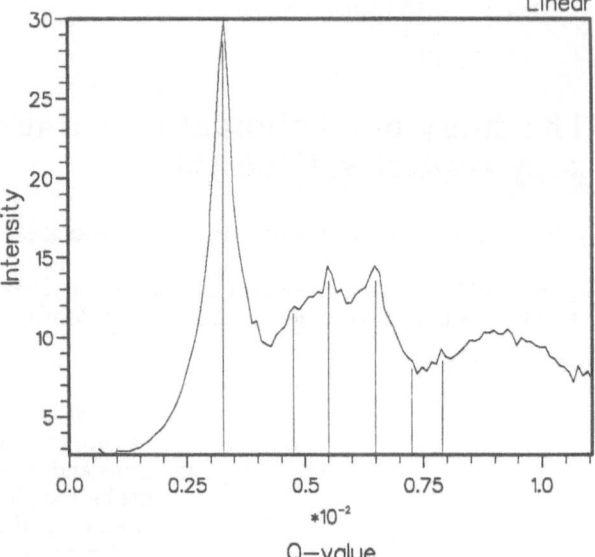

Fig. 3. Experimental Structure Factor for the high concentration colloidal latex, showing indicated Bragg peaks

cubic structure. The discrepancy between calculated and measured peak positions is well within experimental error.

Interpretation

The presence of the 100 peak in the scattering distribution indicates the presence of either a simple cubic (sc) or hexagonal closed packed (hcp) structure, the presence of the 111 peak indicates either a sc or face centred cubic (fcc) structure. From these two observations, it is possible to conclude that the observed structure is simple cubic. However, there are a number of reasons to suggest that this may not be the case: the presence of the experimentally observed 'liquid-like' structure factor background to the Bragg peaks — a true ordered crystal would show only the Bragg peaks; and most real systems show much denser packing than simple cubic will allow — bcc, fcc and hcp structures have been found experimentally in real systems but not sc. For these reasons, it is necessary to consider the possibility of the coexistance of both hcp and fcc states.

Work by Pusey and van Meegan et al. [3, 4] has been carried out into the three states of a hard sphere colloid. It was found that at certain ranges of particulate concentrations the colloid could be observed to be in either a liquid, solid or glassy

Table 1. Comparison of calculated peak positions (assuming a 100 reflection at 0.33×10^{-3} Å^{-1} (measured)) with experimentally determined values

Reflection {hkl}	100	110	111	200	210	211
$\sqrt{h^2 + k^2 + l^2}$	1	$\sqrt{2}$	$\sqrt{3}$	2	$\sqrt{5}$	$\sqrt{6}$
Measured Q-value for peaks $\times 10^{-3}$ Å^{-1}	3.3	4.7	5.5	6.5	7.3	7.9
Calculated peak position $\times 10^{-3}$ Å^{-1}	3.3	4.6	5.7	6.5	7.3	8.0

phase. In a certain range of concentrations, between the freezing and melting concentrations (~ 0.494 and ~ 0.545 [4]), a quotes round 'polycrystalline state' was defined, in which gradual crystallisation of the colloid into many crystallites occurred. Several samples within this concentration range showed an unusual structure factor — a sharp Bragg peak was evident, superimposed on a wider diffuse band of scattering.

This unusual scattering was postulated to be due to the "randomly close packed" structure [5]. In this structure, a single closest packed layer A is formed, and the position of successive similar layers over this basis determines the final structure of the crystal. For layers packing in a formation ABABAB... the final crystal will be hcp, for layers packing in a formation ABCABC... the final crystal formation will be fcc. The randomly closed packed structure originates from a faulted stacking of these layers, for instance ABACBC..., resulting in the presence of both fcc and hcp Bragg peaks, plus a diffuse background.

Conclusion

The experimental structure factor for the high concentration colloidal latex clearly shows the presence of up to six Bragg peaks with an underlying trend which may be interpreted as a diffuse background. These observations seem to indicate that the structure present in the high concentration colloidal latex is randomly closed packed, and not simple cubic as might be suggested by initial inspection. These sharp Bragg peaks could not have been observed in a colloid of this size without the high resolution available with USAXS; and the success of this newly improved system.

References

1. North AN, Dore JC, Mackie AR, Howe AM, Harries J (1990) Nucl Instr Meth B47:283—290
2. North AN, Rigden JS, Mackie AR (1992) Rev Sci Instr 63:1741—1745
3. Pusey PN, van Meegan W, Bartlett P, Ackerson BJ, Rarity J, Underwood SM (1989) Phys Rev Lett 63:2753—2756
4. Pusey PN, van Meegan W, Underwood SM, Bartlett P, Ottewill RH (1991) Physica A176:16—27
5. Guinier A (1963) X-Ray Diffraction, Freeman, New York.

Authors' address:

Jane S Rigden
Physics Department
University of Kent, Canterbury
Kent, CT2 7NR, United Kingdom

Progress in Colloid & Polymer Science

Progr Colloid Polym Sci 93:66—71 (1993)

Characterization of a PEO-PPO-PEO block copolymer system

K. Schillén[1,2]), O. Glatter[1]), and W. Brown[2])

[1]) Institute of Physical Chemistry University of Graz, Austria
[2]) Department of Physical Chemistry University of Uppsala, Sweden

Abstract: The investigated EO_{27}-PO_{39}-EO_{27} triblock copolymer, Pluronics P-85, is a nonionic surface-active agent which has a monomeric molecular weight of 4600, of which 50% corresponds to the PEO component. P-85 exist in complex states of aggregation in aqueous solution. Monomers, micelles and larger aggregates (clusters) coexist at lower concentrations in relative proportions which depend on temperature. The micelles are composed of a core of the water-insoluble PPO-block with a swollen shell of PEO-endblocks, having a hydrodynamic radius of 75 Å. The monomers have a hydrodynamic radius of 18 Å. At higher temperatures and concentrations, the micelles are close-packed in a body-cente-red-cubic (b.c.c.) liquid crystalline phase which is solid-like and glass-clear. At even higher temperatures this gel of micelles "dissolves" again. For structural (internal) information on the monomer, the micelle, and to study the interacting micelles and the gel, small-angle x-ray-scattering measurements and dynamic light-scattering measurements have been made in the concentration range 1 to 25% (w/w) with the temperature range 5 to 70°C. The micellization was detected clearly by ultra-sound velocimetry measurements. To characterize the viscosity of the gel (cubic phase) advanced densimetry was performed.

Key words: Triblock copolymer — micelle — gel — small-angle x-ray scattering — ultra-sound velocimetry — densimetry

Introduction

Nonionic triblock copolymers of poly(ethylene oxide)/poly(propylene oxide)/poly(ethylene oxide) (Pluronics) have attracted much interest in the literature [1—3]. The complex aggregation behavior of these copolymers in water solution has been described in some detail by Zhou and Chu [4], Wanka, Hoffmann and Ulbricht [5], and Almgren et al. [6].

PEO-PPO-PEO block copolymer systems exist in complex states of aggregation in aqueous solution which depends on the relative block sizes [7]. Monomers, micelles, and clusters (larger aggregates of copolymer molecules) coexist at lower concentrations in relative proportions which depend on the temperature. The micelles are composed of a core of the water in-soluble PPO-blocks with a swollen shell of PEO endblocks [4, 5, 10]. At higher temperatures and concentrations, a solid-like glass-clear "gel" is formed [5, 10] by close-packing of micelles in a body-centered-cubic (b.c.c) crystalline structure [8, 9]. At higher temperatures this gel "dissolves" again. Both formation and breakdown of the gel are thermally reversible. At even higher temperatures a phase of hexagonally packed rods appears prior to clouding [9].

To be able to understand the behavior of these systems the results from different kind of methods have been combined, e.g., scattering methods (light, neutron, x-ray), rheology together with advanced densimetry and ultra-sound velocimetry. In a recent communication [10] the triblock copolymer EO_{27}-PO_{39}-EO_{27}, Pluronics P-85 (with a monomeric molecular weight of 4600, of which 50% corresponds to the PEO component) was investigated using light-scattering and oscillatory shear viscosity measurements together with other measurements. The main purpose was to study the micelle and gel formation in the system. A neutron-scattering study on the cubic phase (gel) of P-85 has also been reported [9].

In order to obtain information on the micellization and the gel of P-85 and to study the interactions between micelles, small-angle x-ray scattering and dynamic light-scattering measurements have been performed together with ultra-sound velocimetry and advanced desimetry. Detailed results will be published elsewhere.

Micelle formation

Dynamic light scattering

Dynamic light scattering demonstrates the complex forms of aggregation in these triblock copolymer systems. The relaxation time distributions, obtained by inverse Laplace transformation of the autocorrelation functions from dynamic light scattering measurements, for P-85 at low concentra-

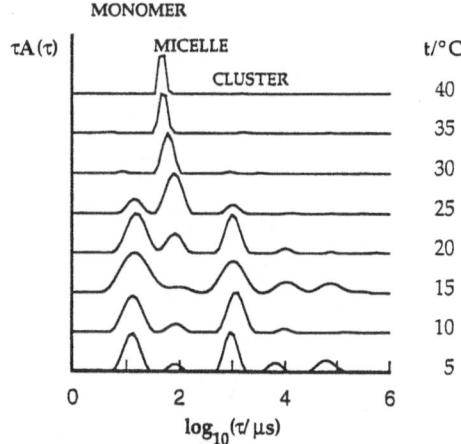

Fig. 2. Micelle formation. Relaxation time distributions from DLS measurements for P-85 at $c = 5\%$ (w/w) as a function of temperature

tions (below 10% (w/w)) and at temperatures between 5 and 50°C are shown in Fig. 1, where τ is the relaxation time (decay time). Peaks attributed to the coexisting monomers (fast mode), micelles, and clusters are indicated. All modes are diffusive. At 25°C the distribution are bimodal, showing monomers and micelles. At 40° and 50°C only micelles are present in the concentration range 1% < c < 10%. The critical micelle concentration (CMC) is much lower at 40° and 50°C than at a lower temperatures.

The hydrodynamic radius of the monomer and the micelle of P-85 is found to be 18 Å and 75 Å, respectively. The relaxation time distributions, in Fig. 2, at different temperatures for P-85 at a concentration of 5% (w/w) show micelle formation. The micellization occurs gradually over a range of temperature in the vicinity of 25°C.

Ultra-sound velocimetry

An alternative method for detecting the micellization is the use of ultra-sound velocimetry [11]. Measurement of the time required for a short pulse to run through a well-defined sample cell is used to calculate the speed of the ultrasonic wave. The speed of sound is a function of the adiabatic compressibility and the density of the solution. It can be shown that the speed of sound is dependent on the number of particles. It decreases as the number density of particles decreases as is the case when micelles form in the solution. The specific speed is the relative change in the solution compared to the

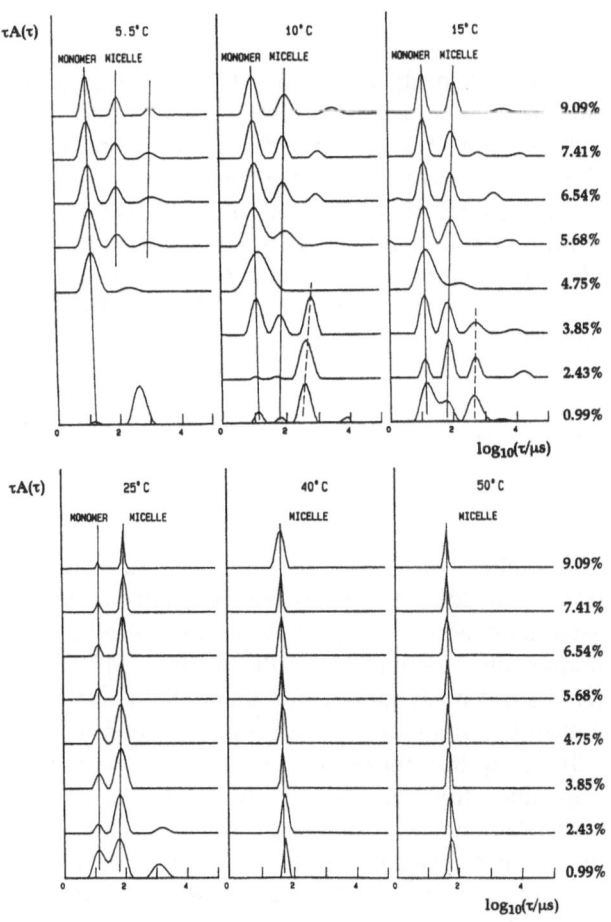

Fig. 1. Relaxation time distributions obtained by Laplace inversion of dynamic light-scattering (DLS) data. For P-85 at low concentrations ($c > 10\%$ (w/w)) and at temperatures between 5° and 50°C

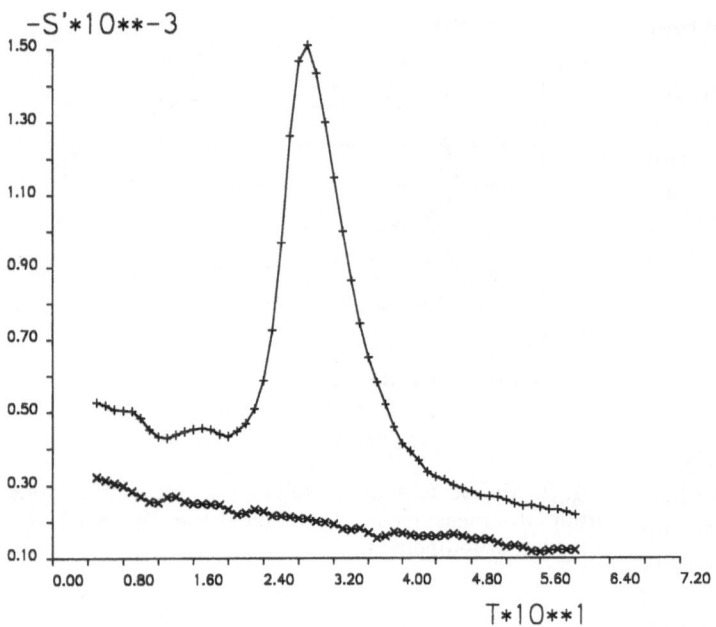

Fig. 3. Micelle formation. The first negative derivative of the specific speed obtained by measurement of the speed of sound as a function of temperature in a P-85 solution of 5% (w/w) (+++) and a 10% (w/w) sucrose solution (× × ×)

speed in water. The negative first derivative of the specific speed shows the changes very clearly.

In Figure 3 the negative first derivative of the specific speed is given as a function of temperature for a 5% P-85 solution and a sucrose solution. The sucrose curve shows a monotonically decreasing behavior. The P-85 curve matches the sucrose curve at low and high temperatures. Deviations from the temperature behavior of sucrose start at about 17°C with a maximum at 29°C, where micelle formation is most pronounced, and vanishes at about 49°C, so micelle formation starts at about 17°C and is complete at 49°C. It is a gradual process, as indicated in the dynamic light-scattering results in Fig 2.

Small-angle x-ray scattering

The small-angle x-ray scattering measurements (SAXS) done on the P-85 system also give information on micelle formation. Figure 4 shows the smeared experimental scattering curves at different temperatures for a 5% (w/w) P-85 solution. At 10° and 20°C the curves are flat, indicating small particles corresponding to the monomers of P-85. The radius of gyration of the monomer at 10°C is 14 Å. At 25 and 30°C the increase in the innermost part of the scattering curves is due to the beginning aggregation in the system. The scattering curve at 40°C corresponds to the particle form factor of the micelle. The radius of gyration of the micelle at 40°C is 46 Å. These results clearly show the transition from monomers to micelles.

From SAXS measurements on a 1% (w/w) P-85 solution a micellar maximum dimension of 155 Å is obtained and it is also seen that the electron density of the PPO curve is lower than that of the PEO shell.

Gel formation

Small-angle x-ray scattering

At higher concentrations there is a sharp transition to a solidlike glass-clear gel [10] which is associated with close-packing of the micelles in a body-centered-cubic crystalline phase [9]. The results from dynamic and static light-scattering measurements point in the same direction.

To study the interacting micelles, the gel formation and the ordering within the gel, SAXS measurements were made on different concentrations of P-85 at a constant temperature of 40°C. The desmeared experimental scattering curves for the different concentrations are shown in Fig. 5.

For the concentration of 5% (w/w) the particle form factor of the micelle with a secondary side maximum and with some interparticle interaction

lg(Intens.)

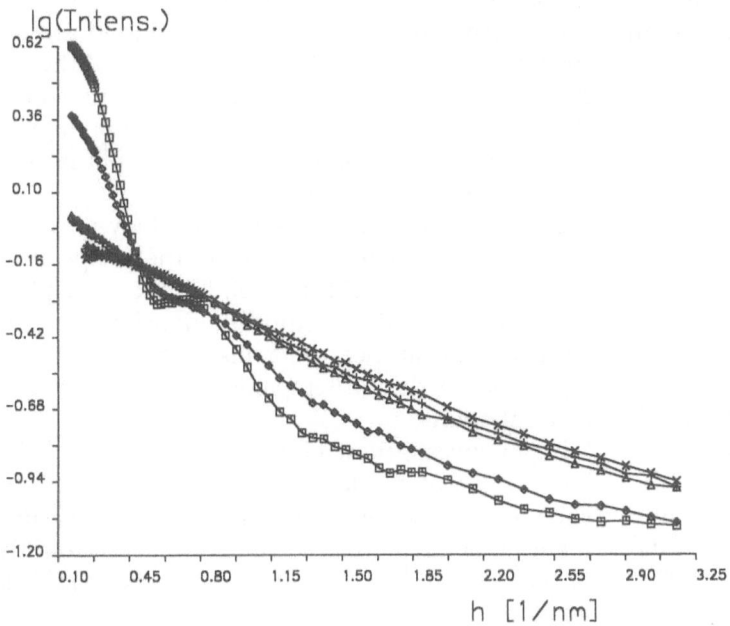

Fig. 4. Micelle formation. Smeared scattering curves from SAXS measurements at different temperatures for a P-85 solution of 5% (w/w) (10°C (× × ×), 20°C (+++), 25°C (ΔΔΔ), 30°C (◇◇◇), 40°C (□□□))

lg(Intens.)

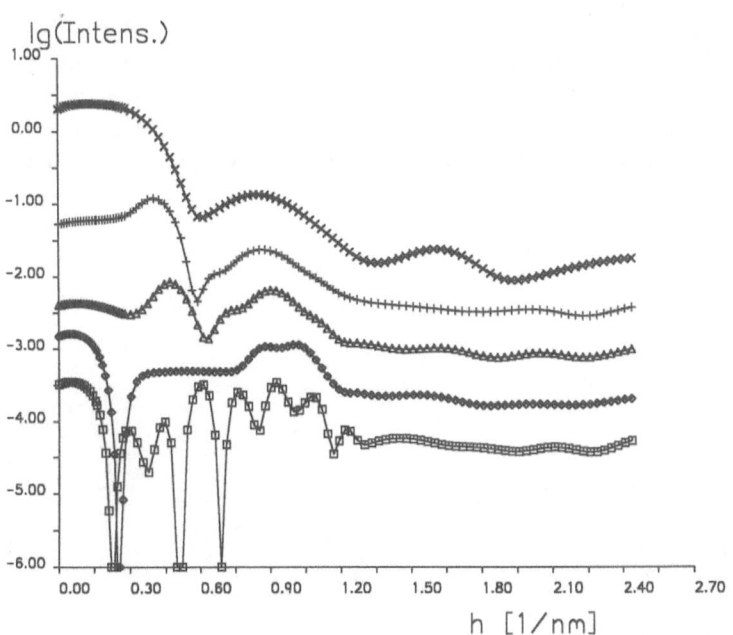

Fig. 5. Gel formation. Desmeared scattering curves from SAXS measurements at 40°C for different concentrations of P-85 (× × ×) 5% w/w); 15% (w/w) (+++), 25% (w/w) (ΔΔΔ), 30% (w/w) (◇◇◇), 35% (w/w) (□□□)). The curves are multiplied by arbitrary factors to prevent strong overlap in this figure

can be seen. At 15% (w/w) an interaction peak is evident in the form factor which moves to higher h-values (shorter distance in real space) with higher concentrations. This is understandable since the micelles approach each other at higher concentrations. The form factor of the micelle can still be seen at 25% (w/w) and 40°C even though the gel is already formed. The situation is changed with the concentration of 30% (w/w). The interaction peak disappears and the system is more dense. The scattering curve of 35% (w/w) is flat with a lot of side maxima. This is the scattering curve of a highly ordered system and can be compared with neutron-scattering measurements done on the P-85 system [9].

Densimetry

Due to the very high viscosity of the gel (cubic phase) precision densimetry can, like oscillatory shear viscosity mesurements [7, 10], give information about the gel formation. Precision densimetry is a mechanical oscillator method where the sample is filled into a U-shaped glass tube. The instrument measures the damping ratio (DR), i.e., the ratio between the damping force (viscosity) and the spring force, which is related to the oscillation period.

The damping ratio for five different concentrations of P-85 as a function of temperature is given in Fig. 6. As mentioned earlier, the P-85 system exist in a gel-like state (cubic phase) with sharp boundaries in rheology at concentrations of about 25% (w/w) and more. The 25% (w/w) solution shows a slight increase in DR (i.e., viscosity) between 20° and 30°C. At 32°C the solution is a gel with such a high viscosity that the DR is smaller than that of water (and therefore has negative values). The stiff gel is "dissolved" at 48°C. The width of the gel regime increases with concentration. This can be shown by comparing with the 30% (w/w) data

(25°C — 60°C). The oscillations in the curve at higher temperatures at this concentration are not readily understood. Note that at 20% (w/w) there is no gel formation, but at all concentrations shown there is an increase in the DR at the highest temperatures, indicating that another phase with rather high viscosity appears.

Depending on which kind of rheology measurement is made, there is a shift in the temperature regime of the gel (cubic phase), but the principle is the same. Both methods, densimetry and oscillatory shear measurements, show that the gel region is broader (in temperature range) at higher concentrations.

In an investigation of the influence of relative block sizes of block copolymers [7] it has been shown, by oscillatory shear viscosity measurements, that the thermal stability range and elasticity of the gel increases with increasing PEO endblock length.

Conclusions

Triblock copolymers of PEO-PPO-PEO exist in complex forms of aggregation in aqueous solution. At lower concentrations monomers, micelles, and clusters (larger aggregates) coexist in proportions that depend sensitively on temperature and con-

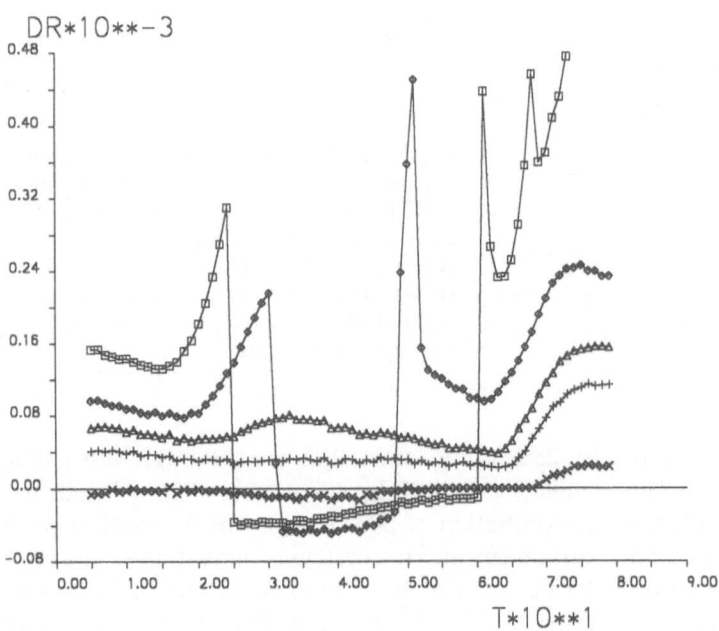

Fig. 6. Gel formation. The damping ratio obtained from densimetry measurements for different concentrations of P-85 as a function of temperature. (× × ×) 5% (w/w); 15% (w/w) (+++), 20% (w/w) (△△△), 25% (w/w) (◇◇◇), 30% (w/w) (□□□))

centration. The micelle formation can be followed using dynamic (and static) light scattering, SAXS and ultra-sound velocimetry. The monomers have hydrodynamic radii in the size range 15 to 30 Å, the micelles 75 to 130 Å. The clusters are 800 Å and larger.

At higher concentrations, solid-like gels are formed. The different scattering methods indicate that the gel consists of close-packed micelles in a cubic crystalline pattern. The viscoelastic properties of the gel can be studied using densimetry and oscillatory shear viscosity measurements.

References

1. Schick MJ (1987) In: Non-ionic surfactants, Physical Chemistry. Marcel Dekker, New York
2. Tuzar Z, Kratochvil P (1976) Adv Colloid Interface Sci 6:201—232
3. Al-Saden AA, Whatley TL, Florence AT (1982) J Colloid Interface Sci 90:303—309
4. Zhou Z, Chu B (1987) Macromolecules 20:3089—3091
   Zhou Z, Chu B (1988) Macromolecules 21:2548—2554
   Zhou Z, Chu B (1988) J Colloid Interface Sci 126:171—180
5. Wanka G, Hoffmann H, Ulbricht W (1990) Colloid Polym Sci 268:101—117
6. Almgren M, Alsins J, Bahadur P (1991) Langmuir 7:446—450
7. Brown W, Schillén K, Hvidt S (1992) J Phys Chem 96:6038—6044
8. Mortensen K, Brown W, Nordén B (1992) Phys Rev Lett 13:2340—2343
9. Mortensen K (1992) Europhys Lett 19(7):599—604
10. Brown W, Schillén K, Almgren M, Hvidt S, Bahadur P (1991)
11. Rassing J, Attwod D (1983) Int J Pharm 13:47—55

Authors' address:

Otto Glatter
Institute of Physical Chemistry
University of Graz
Heinrichstraße 28
8010 Graz, Austria

Progress in Colloid & Polymer Science   Progr Colloid Polym Sci 93:72—75 (1993)

Block copolymer in aqueous solution:
Micelle formation and hard-sphere crystallization

K. Mortensen

Department of Solid State Physics, Risø National Laboratory Roskilde, Denmark

Abstract: The phase behavior of triblock-copolymers of poly(ethylene oxide) and poly(propylene oxide) dissolved in water shows unique characterisctics: dissolved polymers at low temperature and low polymer concentration, spherical micellar aggregates at intermediate temperatures with a range in temperatures where both dissolved polymers and micelles are present, and prolate ellipsoidal or rod-like micelles at high temperatues. In the high polymer concentration regime, an apparent "inverse crystallization" characteristic is observed as the micellar volume fraction approaches the critical value $\phi = 0.53$ of hard-sphere crystallization. The crystalline phase is cubic, characterized by true long-range correlation in the bond-angle correlation, but with absence of long-range order in the bond-length correlation. At higher temperatures a crystalline phase of hexagonally packed rods is observed.

Key words: Block copolymer micelles — hard sphere crystallization — cubatic phase — small-angle neutron scattering

I. Introduction

Block copolymers of poly(ethylene oxide) and poly(propylene oxide), $EO_{m1}PO_nEO_{m2}$, are well known non-ionic polymer surfactants, which in recent years have attracted great interests. Due to the marked change in water solubility of the center poly(propylene oxide) block, the copolymers form, at elevated temperatures, aggregates and, possibly, "thermo reversible" sol-gel transitions. A number of tri-block copolymers has been shown to aggregate in the form of micelles [1—7], with a core presumably dominated by PPO and a corona dominated by hydrated PEO-blocks [8, 9].

Using small-angle neutron-scattering technique, we have studied the P85 system, $EO_{25}PO_{40}EO_{25}$. The polymer was dissolved in deuterated water, D_2O, at 5°C, forming a transparent, homogeneous solution with good scattering contrast between solvent and polymers. The scattering contrast between PPO and PEO is negligible. The neutron-scattering experiment was performed using the Risø-SANS facility. More details on the experiment and data analysis have been published elsewhere [5, 6, 8].

II. Micelle formation

In Fig. 1 is shown the azimuthally averaged scattering function, $I(q)$ vs q, obtained for 25% polymer solution (q is the scattering vector, $q = 4\pi \sin(\Theta/2)/\lambda$, corresponding to scattering angle Θ and neutron wavelength λ). At least four temperature regimes appear from this scattering function. At low temperature, $I(q)$ is small and shows only weak q-dependence. Above approximately 10°C, the scattering intensity increases markedly, and develops into a well resolved correlation peak close to 15°C. Above 25°C a small, but significant change appear in the peak width, and above approximately 60°C, the form of the correlation peak changes.

The low temperature scattering function agrees well with the Debye-function

$$I(q) \sim x^{-2}(1 + x + e^x), \qquad (1)$$

$x = (qR_g)^2$, R_g being the polymer radius of gyration, thus showing that the suspension in the low

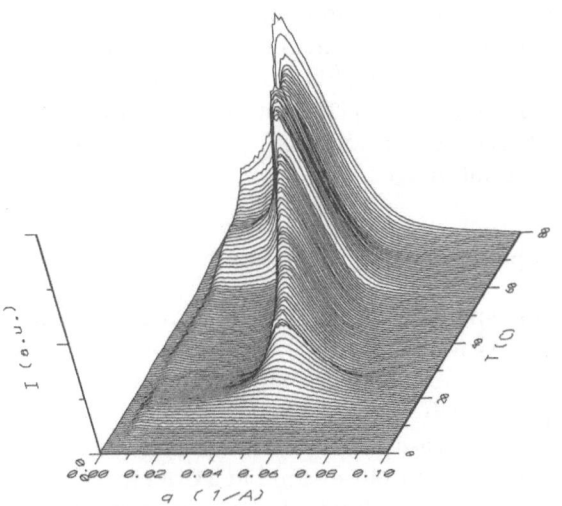

Fig. 1. Scattering function of 25% aqueous suspension of $EO_{25}PO_{40}EO_{25}$, as obtained in the $T = 5-80\,°C$ range

temperature regime is a solution of fully dissolved individual polymer coils. Fitting Eq. (1) to the data gives $R_g = 17$ Å.

The marked increase in scattering intensity above approximately 10°C reveals formation of micelles. As the temperature is increased further, the scattering function becomes dominated by a pronounced correlation peak reflecting important intermicellar correlations. Assuming mono-dispersed particles, the scattering function can be written as a product of the single-particle form factor $P(q)$ and a structure factor, $S(q)$, describing the inter-particle interference:

$$I(q) = \Delta p^2 \cdot N \cdot V \cdot P(q) \cdot S(q) , \qquad (2)$$

where Δp^2 is the contrast factor N is number density of scatterers, and V the volume.

As a first approach to analyze the data, we use the form factor of a dense sphere

$$P(q) = \left[\frac{3}{(qR_c)^3} (\sin(qR_c) - qR_c\cos(qR_c)) \right]^2 \qquad (3)$$

characterized by the sphere-radius R_c. The observed scattering functiondoes, however, not show the characteristic ripples and the q^{-4}-approach at large q-values as expected from Eq. (3). These deviations are attributed to the PEO sub-chains dispersed into the water-phase, in agreement with a simplified model of the micellar structure with a central, dense core of dominantly poly(propylene oxide)

and an outer corona of hydrated poly(ethylene oxide) effectively grafted on to the surface of the core, as schematically illustrated in Fig. 3.

The structure factor is given by

$$S(q) = 1 + 4\pi N \int (g(R) - 1) \frac{\sin(qR)}{qR} R^2 dR, \qquad (4)$$

where $g(R)$ is the distribution function describing the arrangement of the micelles. Using the Ornstein-Zernike approximation to express the correlation fluctuations $(g(R) - 1)$ in terms of a short-range direct interaction term, and using the Percus-Yevick expression for this direct term with the hard-sphere interaction potential characterized by the interaction radius R_{hs}, the structure factore has the form [10]

$$S(q) = 1 / [1 + 24\phi G (2q R_{hs}, \phi)/(2qR_{hs})] , \qquad (5)$$

where G is a trigonometric function of $2qR_{hs}$ and the micellar volume fraction ϕ. Fits to the experimental data thereby result in three parameters: micelle core-radius, R_c, hard-sphere interaction radius, R_{hs}, and volume fraction of hard spheres micelles, ϕ. In Fig. 2 is shown the resulting micellar volume fraction as obtained for samples with polymer concentrations in the range 1—35%.

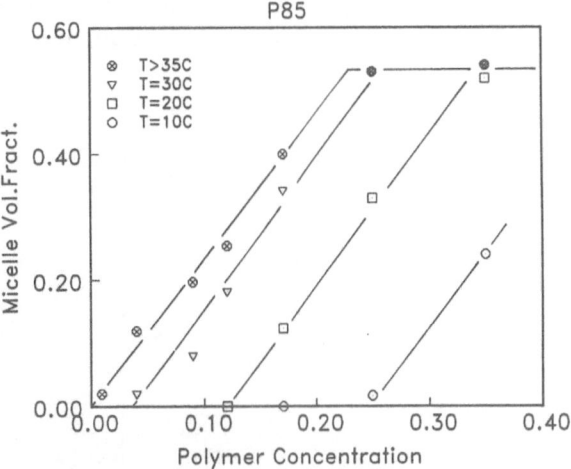

Fig. 2. Micelle volume fraction versus polymer concentration and temperature, as obtained by fitting the hard-sphere Percus Yevick model to the experimental scattering function

In opposition to many classical surfactants, a wide regime appears in the phase diagram where both micelles and dissolved polymers are present. Only above approximately 35°C does the simple relation between micellar volume fraction ϕ and polymer concentration c appear:

$$\phi = V_m/(NV_p) \cdot c \, , \tag{6}$$

reflecting that "all" polymers have aggregated into micelles. The pre-factor is determined by the aggregation number N, the dry polymer volume V_p, and the hard-sphere micelle volume V_m.

The micellar sizes, as given by R_c and R_{hs}, appear (from the Percus-Yevick fits) to be independent of polymer concentration, but have a temperature dependence reflecting significant changes in aggregation number [8].

III. Hard-sphere crystallization

When the $\phi_c = 0.53$ line is crossed the micellar liquid undergoes a first-order phase transition to a cubic crystal [5, 6], in agreement with hard-sphere crystallization. It appears from Fig. 2 that this critical volume fraction can be reached only for polymer-concentrations above approximately 20%. On the other hand, it is possibly to cross the $\phi_c = 0.53$, not only by increasing the polymer concentration, but also by increasing the temperature, thus explaining the apparent "inverse melting" characteristic.

The ability to cross the critical volume-fraction by changing the temperature gives unique opportunity to study thermodynamic details of concentration-induced hard-sphere crystallization. Vadnere et al.

already made an extended study on the thermodynamics involved in the phase-transition of a number of $EO_mPO_nEO_m$ systems dissolved in water, including P85 [11]. However, their discussion was based on traditional gelation-process for polymer solutions.

IV. Cubatic phase

The suspension is, in the cubic phase, a transparent, paste-like material. If the poly-crystalline material is exposed to shear, the poly-crystal abruptly transforms into a single crystal. In Fig. 3 is shown the two-dimensional scattering pattern as obtained on a sample sheared in a Couette cell [12]. The figure clearly shows the formation of six-fold symmetric Bragg-peaks corresponding to the [111]-plane of the BCC lattice with (110)-type of reflections. The q_{110}-value of 0.05 Å^{-1} gives the lattice constant 170 Å. The single-crystal is characterized by true long-range order in the "bond-angle" correlation length ($\xi_A \sim$ sample size), as evidenced by the well resolved scattering pattern, but liquid-like "bond-length" correlation ($\xi_l \sim$ 400 Å, i.e., 2—3 lattice periodicities). These findings lead us to the conclusion that this micellar phase is "cubatic", in which the bond-angle correlation is the relevant order parameter [5].

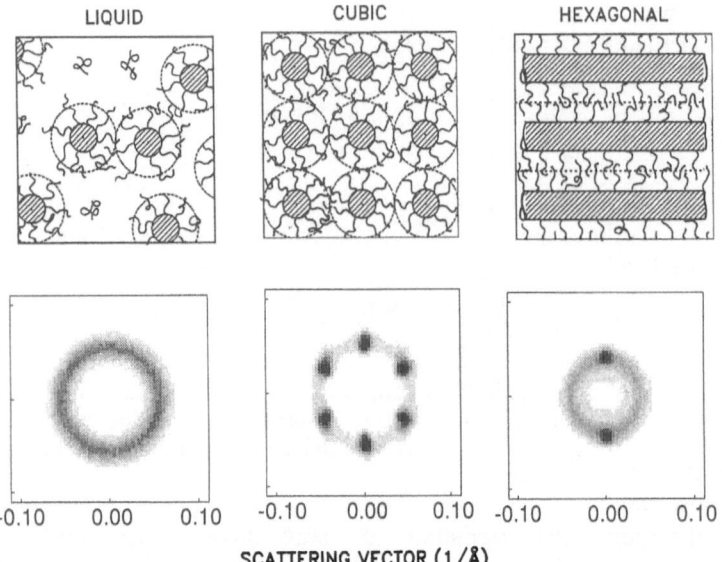

Fig. 3. Top row: Schematic illustration of interacting micelles, spherical at low temperature, characterized by the core-radius R_c and hard-sphere interacting radius R_{hs}; and rod-like at high temperature. Bottom row: Two-dimensional scattering funtions as obtained perpendicular to the shear plane of 25% solution $EO_{25}PO_{40}EO_{25}$. The three columns represent P85 at $T = 25°C$, $T = 27°C$, and $T = 68°C$

V. Rodlike micelles

At $T \sim 60\,^{\circ}\mathrm{C}$ structural changes appear, as seen in the scattering pattern shown in Fig. 1, and in melting of the cubic structure [6].

From the form factor derived from the scattering function, it appear that the spherical micelles transform into larger, prolate ellipsoidal form [8].

The liquid phase of prolate aggregates appears as a channel between two crystalline phases. Increasing the temperature further leads to formation of another crystalline mesophase of hexagonally ordered rod-like micelles, giving the scattering pattern shown in Fig. 3. Close to the transition temperature of $T = 63\,^{\circ}\mathrm{C}$ a clear coexistence of the hexagonal phase and the liquid phase appear, revealing first-order nature of the transition. At higher temperatures, the form of the Bragg-peaks reflects presence of dislocations in the crystalline structure, before it at $T \sim 79\,^{\circ}\mathrm{C}$ melts into a new liquid phase.

VI. Conlusions

In conclusion, we have shown that poloxamers, as represented by the $EO_{25}PO_{40}EO_{25}$ P85 block copolymer, constitute a unique class of materials which in water aggregate into micelles forming crystalline mesophases with unusual characteristics. At temperatures below approximately $60\,^{\circ}\mathrm{C}$, the form of the micelles is spherical, whereas, close to $60\,^{\circ}\mathrm{C}$, it undergoes a transition into ellipsoidal, or rod-like structure.

Acknowledgement

I gratefully acknowledge illuminating discussions with J. Skov Pedersen (Risø National Laboratory) and W. Brown (Uppsala University). This work was supported by the Danish Natural Science Research Council.

References

1. Rassing J, and Attwood D, (1983) Int J Pharm 13:47—55
2. Zhou Z, and Chu B (1987), Macromolecules 20:3089—3091; Zhou Z, and Chu B (1988) J Colloid Interface Sci 126:171—180
3. Wanka G, Hoffmann H, and Ulbricht W (1990) Colloid Polym Sci 268:101—117
4. Brown W, Schillén K, Almgren M, Hvidt S, and Bahadur P (1991) J Phys Chem (1991) 95:1850—1858
5. Mortensen K, Brown W, and Nordén B (1992) Physical Review Letters 68:2340—2343
6. Mortensen K (1992) Europhys Lett 17:599—604
7. Mortensen K (1992) Supp Colloid Polym Sci 270:
8. Mortensen K, and Pedersen JS (1993) Macromolecules 26:805—812
9. Linse P, and Malmsten M (1992) Macromolecules 25:5434—5439
10. Ashcroft NW, and Lekner J (1966) Phys Rev 145:83—90
11. Vadnere M, Amidon GL, Lindenbaum S, and Haslam JL (1984) Int J Pharm 22:207—218
12. Nordén B, Elvingson C, Eriksson T, Kubista M, Sjöberg B, Takahashi M, and Mortensen K (1992) J Mol Biol 216:223—228

Author's address:

Kell Mortensen
Dept. Solid State Physics
Risø National Laboratory
4000 Roskilde, Denmark

Progress in Colloid & Polymer Science　　　　　Progr Colloid Polym Sci 93:76—80 (1993)

Phase behavior of surfactant-alcohol-oil-water cubic liquid crystals*

A. de Geyer

Institut Laue-Langevin, Grenoble, France**

Abstract: X-ray small-angle diffraction experiments have been performed on the mesophases of a system of single-chain surfactant, short-chain alcohol, and comparable amounts of oil and salt water. We focus in this report on two cubic liquid crystalline phases of probable space-groups Pm3n and Fd3m. It is suggested that the structures of both cubic mesophases correspond to an aqueous film delimiting networks of polyhedral cells (with oil interior and hydrophilic walls).

Key words: Lyotropic liquid crystals — cubic mesophases — microemulsions — ordering of surfactant films — polyhedral networks — x-ray small-angle diffraction

I. Introduction

Cubic liquid crystalline phases have been studied widely in water/surfactant or surfactant-like-lipid systems (see [1—4] for recent reviews). During recent years, much attention has been devoted to cubic mesophases in some ternary water/oil/surfactant systems containing a low proportion of oil [5—8]. It is now well recognized that a large structural diversity arises from variation of the volume fractions, but also from modifications in the interactions between neighboring surfactant molecules. Various cubic structures have been reported with ternary systems containing soap molecules, single-chain surfactants, double-chain surfactants or mixture of single- and double-chain surfactants. The addition of short-chain alcohol molecules to single-chain surfactants can favor the formation of other cubic mesophases which have the particularity to contain comparable amounts of oil and water and to exist at much lower surfactant concentration [9]. We report here preliminary results on the phase behavior for one such surfactant/alcohol/oil/brine multicomponent system. Special attention is devoted to the formation of two Pm3n and Fd3m cubic mesophases.

---

*) A full account of this work will be published elsewhere.

**) Present address: Centre d'Etudes Nucléaires de Grenoble.

II. Materials and methods

Investigating the mesophases formed in a system of comparable amounts of toluene and salt water (2.8 w/w% NaCl added) and about 20 v/v% sodium dodecylsulfate and various proportions of short-chain alcohol (butanol), we have found a rich variety of different phases upon the variation of the butanol concentration and of the temperature. The components were put together in glass tubes with different concentrations of alcohol and gently mixed. The samples are fluid at room temperature. They have been left for several days to equilibrate. The change in phases when increasing the temperature (modification of the viscosity and/or of the isotropic or birefringent aspect) were observed visually. In addition to fluid isotropic phase at low temperature and high alcohol/surfactant ratio and birefringent phases at high temperature and high alcohol/surfactant ratio, the phase diagram contains two further phases which are perfectly clear and transparent, optically isotropic and of stiff consistency (Fig. 1). X-ray investigations into these viscous isotropic phases reveal the existence of two cubic mono-phases of probable space-groups Pm3n and Fd3m.

X-ray diffraction experiments were performed on a point focusing camera with a rotating anode (EMBL, Grenoble) at a wavelength of 1.54 Å and a sample-to-film distance of 741 mm. The samples were introduced as fluid phases at room

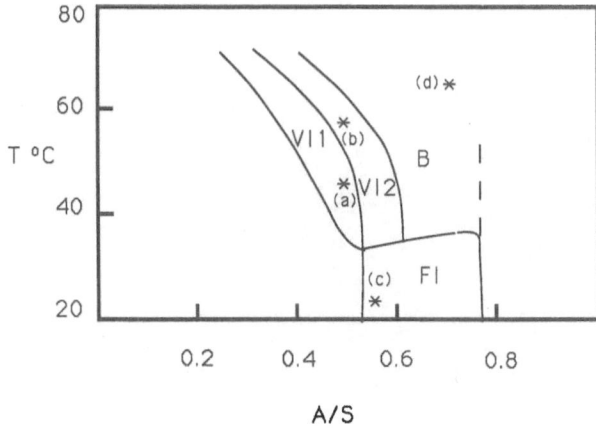

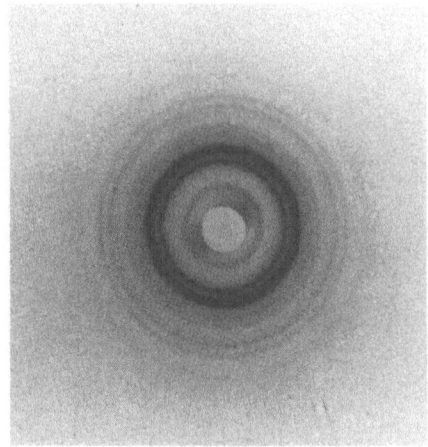

Fig. 1. Schematic phase diagram for a system composed of 40 v/v% toluene/40 v/v% salt water (2.8 w/w% NaCl)/20 v/v% SDS/butanol. The butanol concentration is given in butanol/SDS mole ratio (A/S). FI: fluid isotropic phase. VI1 and VI2: viscous isotropic phases. B: viscous birefingent phase. The other parts of this phase diagram refer to non-optically transparent mixtures

0.1 Å^{-1}

Fig. 2. X-ray diffraction pattern for the VI1 phase ((a) in Fig. 1) at A/S = 0.5 and T = 45°C. Due to the short time of exposure, only the first diffraction rings are visible on this film. The scattering around the beam stop is due to imperfect focalization. The reflections are in agreement with the space-group Fd3m (a = 386 Å). Q (Å^{-1}) represents the momentum transfer $Q = (4\pi \sin\theta)/\lambda$

temperature into 1 mm glass capillaries which were sealed, mounted on a thermostated block and maintained in slow rotation into the beam. All cubic samples were found to have a good stability. A long exposure time was required (from 1 to several days) to obtain sufficient intensity in the higher diffraction orders. To allow a quantitative interpretation of the diffraction patterns, the photographic films were scanned. The profiles of diffraction were obtained after radial averaging.

III. Results

We report (Figs. 2 and 3) powder diffraction patterns of samples with a butanol/SDS mole ratio of 0.5 at two different temperatures (45°C and 57°C) in those neighboring regions VI1 and VI2 of viscous isotropic phases. The diffraction patterns are characterized by a large number of sharp Debye-Scherrer rings centered around the direct beam which reveal the existence of very well ordered structures with relatively large unit cells.

At 45°C (Fig. 2), 13 diffraction rings were observed which can be indexed as the 111, 220, 311, 222, 400, 331, 422, 511 or 333, 440, (531, 442, 620, 533), 622, 444, 711 or 551, 642 reflections of a cubic phase of probable space-group Fd3m (Q^{227}) with a lattice parameter of 386 Å. Indices hkl that satisfying general reflections of Fd3m but not observed here, either because of sets of special positions or because of their weakness, are placed in brackets.

At higher temperature this system undergoes a transition to another cubic phase of different symmetry. The diffraction pattern obtained at a temperature of 57°C (Fig. 3) contains 18 diffraction rings which can be indexed as the 110, 200, 210, 211, 220, 310, 222, 320, 321, 400, (410), 330 or 411, 420, 421, (332), 422, (430), 510 or 431, 520 or 432, (521), 440, (530 or 433, 531, 600 or 442), 610 reflections of a cubic phase of probable space-group Pm3n (Q^{223}) with a lattice parameter of 268 Å. Similar Pm3n cubic phases were also observed at little higher surfactant concentration. Figure 4 shows a diffraction pattern (Pm3n phase: a-parameter 239 Å) obtained when increasing the surfactant concentration from 20 v/v% SDS to 24 v/v% SDS.

We can notice for these Pm3n and Fd3m cubic mesophases that the intensity of the first reflection is relatively weak. This can indicate crystallographic structures based on distinct sets of special positions as will be detailed elsewhere. The Pm3n phase

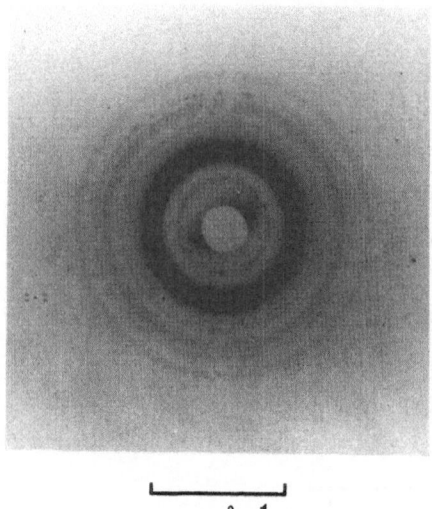

0.1 Å⁻¹

Fig. 3. X-ray diffraction pattern for the VI2 phase ((b) in Fig. 1) at A/S = 0.5 and T = 57°C. The reflections are in agreement with the space-group Pm3n (a = 268 Å)

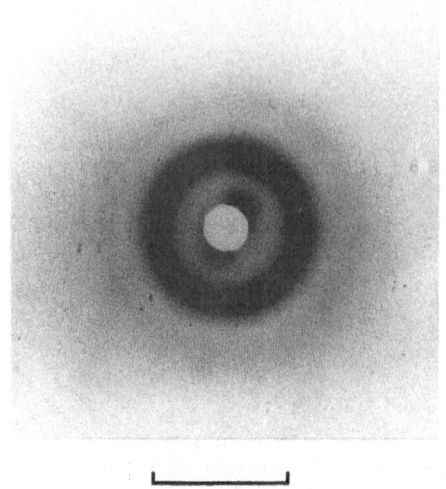

0.1 Å⁻¹

Fig. 5. X-ray scattering for the fluid isotropic phase ((c) in Fig. 1) below the Pm3n cubic crystalline phase. A/S = 0.57 and T = 23°C

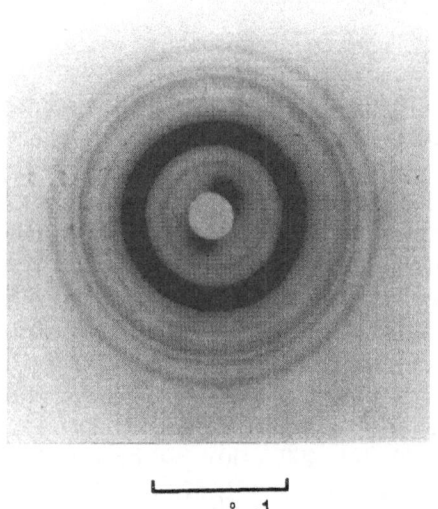

0.1 Å⁻¹

Fig. 4. X-ray diffraction pattern for a Pm3n cubic phase (a = 239 Å) at a composition of 38 v/v% toluene/38 v/v% salt water (2.8 w/w% NaCl)/24 v/v% SDS/butanol at A/S = 0.34 and T = 35°C

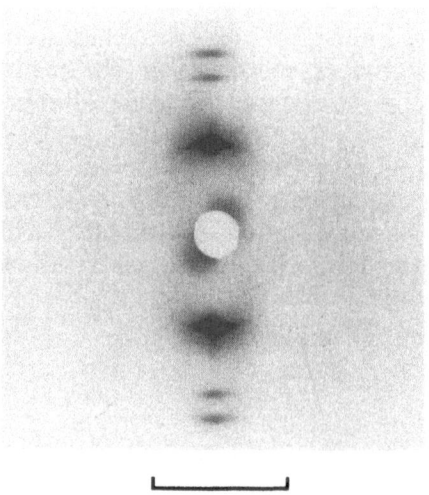

0.1 Å⁻¹

Fig. 6. X-ray diffraction pattern for an oriented sample of the viscous birefringent phase ((d) in Fig. 1) at A/S = 0.71 and T = 65°C. The alignment occurs in the direction of the capillary axis (horizontal axis on the figure). The diffraction spots correspond to wave vectors exploring equatorial directions

shows a certain tendency to grow large crystallites (granular aspect of the diffraction patterns if the capillary is not rotated) whereas the Fd3m phase produces almost perfect powder-like samples.

In a part of the phase diagram (Fig. 1), cooling the Pm3n cubic phase produces melting of the crystalline structure into a fluid isotropic phase. The x-ray scattering pattern for this fluid phase (Fig.5) shows a broad correlation peak at scattering angles corresponding to the most intense 200, 210 and 211 reflections observed with the cubic crystal. Other correlation peaks of weak intensity are also observed at larger scattering angles. At higher temperature, the Pm3n cubic phase undergoes a transition into a birefringent phase whose structure is not very well understood. We think that different intermediate mesophases probably intervene in the domain of the birefringent phases and further studies are in progress. Some of these phases have strong tendency to orient spontaneously along the capillary axis. We notice in Fig. 6. 3 intense diffraction lines which can be indexed as the $\sqrt{1}$, $\sqrt{3}$, $\sqrt{4}$ reflections of an hexagonal phase (a parameter 111 Å).

IV. Discussion

Pm3n and Fd3m symmetries have already been reported in a variety of binary water/surfactant or water/lipid systems (or eventually also in ternary systems containing low proportion of hydrocarbon) [10–14]. In some cases, the structures have been described in terms of networks of identical surfactant aggregates (spheres) occupying one set of special positions: as positions (k) [see International tables of crystallography] of Pm3n [12]. In other cases, indentical prolate surfactant aggregates occupy two different sets of special positions: as positions (a) and (d) of Pm3n. Prolate micelles on positions (a) are isotropically disordered while those on positions (d) undergo anisotropic rotations on their sites [14]. Finally, in other cases, it is two types of surfactant aggregates (cylinders and spheres) occupying two sets of special positions (as positions (a) and (j) of Pm3n [10]) which form cubic networks.

With respect to similar previous works on binary or ternary surfactant systems, the Pm3n (a-parameter: 268 Å) and Fd3m (a-parameter: 386 Å) mesophases found in this sudy have the particularity to being formed by more diluted (typically 20 v/v% surfactant) systems (which explain large unit cell dimensions) and to containing short-chain alcohol molecules. Furthermore, this system has the hitherto unknown property to present a transition between the cubic phases of Fd3m and Pm3n symmetries when increasing the alcohol content or the temperature. All these observations suggest that we have to deal with a new type of cubic phase.

The system composition is sufficient to have a dense packing of objects as, for instance, oil droplets. The coalescence of the particles is avoided due to the strong repulsions existing between the polar-heads forming the outer part of the particles. But these hypothetical droplets (oil interior surrounded by a thin surfactant layer) must instead behave as soft spheres, especially because as the surfactant layer is composed of single-chain surfactant and short-chain alcohol molecules (components which are well known to forming less rigid interfaces [15]). For these two cubic mesophases, it is likely that such soft particles transform into polyhedral cells with a water film between them.

Different sets of special positions are compatible with the reflections that we observe for these Pm3n and Fd3m mesophases. From the volume fractions of the various components and the surfactant concentration, it has been possible to calculate the dimensions of the objects and the surface area per polar head in different structural models. From comparison of these values to known molecular dimensions and comparison of the calculated intensity profiles to the experimental data, we have found that the remaining models were those involving positions (a, d) for the Pm3n cubic phase and (d, a) for the Fd3m cubic phase. Such positions can be those of space-filling polyhedra such as dodecahedra and tetrakaidecahedra in a Pm3n lattice, and dodecahedra and hexakaidecahedra in an Fd3m lattice. Similar polyhedral frameworks have been found for the hydrogen-bonded water molecules in the two Pm3n (12 Å) and Fd3m (17 Å) forms of cubic crystalline clathrate hydrates [16, 17]. For the surfactant system described in this study, the faces of the polyhedra would support a continuous layer of one medium (water) delimiting closed polyhedral cells of the other medium (oil), with alcohol and surfactant molecules lying at the oil/water interfaces. The choice of the "oil-in-water" type of structure is consistent with the strong head group-head group electrostatic repulsions which are not screened at low salt concentration. It can be noticed that the Pm3n (268 Å) and Fd3m (386 Å) cubic mesophases found in this surfactant system have the same axis-ratio as those Pm3n (12 Å) and Fd3m (17 Å) cubic phases, with polyhedral structures, in clathrates. The validity of these polyhedral models for this surfactant system has been confirmed by simulating the corresponding powder diffraction profiles and fitting the experimental curves [18].

Similar Pm3n and Fd3m polyhedral structures have been predicted from considerations on the geometrical frustration arising in dense systems of amphiphilic layers with curved interfaces [19]. These polyhedral models have been proposed to account for, respectively, the type I (hydrocarbon chains "inside") and type II (hydrocarbon chain "outside") forms of Pm3n and Fd3m cubic mesophases observed in different surfactant-rich binary (water/surfactant) systems. In the case of more diluted surfactant systems, as encountered with these cubic phases, this interpretation in terms of frustrated surfactant films is not suitable, however. By the thickness of the water layers (typically 20 Å) compared with the dimensions of the polyhedral cells (typically 120 Å), this system instead presents some analogy with biliquid foams as encountered in some oil/water/soap systems [20]. However, the ordering of the water layers to form polyhedral crystalline structures, as observed here, remains to be understood. It is this issue and the crucial role of the alcohol in the formation of such crystalline networks which will be addressed in a later report in more detail.

Acknowledgement

I would like to thank the EMBL (Grenoble) for use of the x-ray small angle camera and C. Berthet for her help during the experiments.

References

1. Mariani P, Luzzati V, Delacroix H (1988) J Mol Biol 204:165—189
2. Lindblom G, Rilfors L (1989) Biochimica et Biophysica Acta 988:221—256
3. Fontell K (1990) Colloid Polym Sci 268:264—285
4. Mariani P (1991) Current Opinion in Structural Biology 1:501—505
5. Rädler JO, Radiman S, de Vallera A, Toprackcioglu C (1989) Physica B 156:398—401
6. Radiman S, Toprakcioglu C, Faruqi AR (1990) J Phys (Paris) 51:1501—1508
7. Barois P, Hyde ST, Ninham B, Dowling T (1990) Langmuir 6:1136—1140
8. Ström P, Anderson DM (1992) Langmuir 8:691—709
9. Tabony J (1986) Nature 319:400
10. Tardieu A, Luzzati V (1970) Biochim Biophys Acta 219:11—17
11. Seddon JM, Bartle EA, Mingins J (1990) J Phys Condens Matter 2:285—290
12. Burns JL, Cohen Y, Talmon Y (1990) J Phys Chem 94:5308—5312
13. Luzzati V, Vargas R, Gulik A, Mariani P, Seddon JM, Rivas E (1992) Biochemistry 31:279—285
14. Fontell K, Fox K, Hansson E (1985) Mol Cryst Liq Cryst Lett 1:9
15. Szleifer I, Kramer D, Ben-Shaul A, Roux D, Gelbart WM (1988) Phys Rev Lett 60:1966—1969
16. Mc Mullan RK, Jeffrey GA (1965) J Chem Phys 42:2725—2732
17. Mak TCW, Mc Mullan RK (1965) J Chem Phys 42:2732—2737
18. de Geyer A (to be published)
19. Charvolin J, Sadoc JF (1988) J Phys (Paris) 49:521—526
20. Ebert G, Platz G, Rehage H (1988) Ber Bunsenges Phys Chem 92:1158—1164

Author's address:

Dr. A. de Geyer
Groupe Physico-Chimie Moléculaire
Service d'Etude des Systèmes et Architectures Moléculaires
Département de Recherche Fondamentale sur la Matière Condensée
Centre d'Etudes Nucléaires de Grenoble
85 X — 38041 Grenoble Cedex, France

Progress in Colloid & Polymer Science   Progr Colloid Polym Sci 93:81—84 (1993)

Laser scattering studies of structural and dynamic colloidal properties of protoplasm and blood

N. N. Firsov[1]), N. B. Lapteva, B. A. Levenko, A. V. Priezzhev, S. G. Proskurin, and O. M. Riaboshapka

M. V. Lomonosov Moscow State University, Moscow, Russia
[1]) Russian State Medical University, Moscow, Russia

Abstract: We have investigated some structural and dynamic properties of such complicated biological colloids as protoplasm and blood. These properties are manifested by phase transitions and aggregation phenomena which can only be studied *in situ* inside a living cell or an embryo or, in the case of blood, shortly after the preparation of the sample. We have shown that laser Doppler microscopy and backscattering nephelometry can give new information on basic intracellular mobility mechanisms, dynamic patterns of the developing fish embryo, and aggregational properties of red blood cell, sensitive to different pathologies.

Key words: Protoplasm — red blood cells — light scattering — laser doppler microscopy — laser nephelometry

Introduction

Historically, the recognition of colloid and surface science as a particular branch of physics and chemistry originated with Thomas Graham's papers in 1861 and 1864, stemming from his investigations of biological products [1]. Cell protoplasm and blood, being complex fluids of major importance for living species from cellular up to mammalian organism level, exhibit many properties common with colloidal systems. They comprise an integrity of macromolecular and larger structures: organelles or forming elements (in the case of blood). Their components can reversibly aggregate and/or undergo reversible phase transitions. When illuminated with a beam of light they form typical to colloids scattering patterns, and can be studied efficiently with elastic and quasielastic light-scattering techniques.

We have applied the technique of laser Doppler microscopy (LDM) [2] to study the dynamic properties of protoplasm and blood *in situ* while streaming along living cells and blood capillaries correspondingly and, also, the technique of backscattering nephelometry (BN) [3] to study the structural (aggregational) properties of red blood cells (RBCs) in whole blood in *in vitro* conditions of a Couette sample cell with controllable shear stress.

The investigation of intracellular dynamics is focused to get more knowledge on how the cell protoplasm, as an active colloid, is involved in the force generating mechanism driving the protoplasm into fast streaming. Measuring the pulsatile blood flows inside a living fish embryo is aimed to quantitatively distinguish different stages of its cardiovascular system development. The nephelometric study of aggregational phenomena in the whole blood was performed in order to see whether the aggregational properties of RBCs are sensitive to different pathologies.

Materials and methods

The velocity patterns of the intracellular mobility of protoplasm were registered in protoplasmic strands of migrating plasmodia of *Physarum polycephalum* slime mold [4]. The plasmodia were obtained from a mass of protoplasm placed onto a 0.5-mm-thick layer of agar-agar substrate in a Petri dish. The measurements were carried out at room temperature.

The velocities of directed blood flows were measured in arteria and vena of fish *(Danio rerio)* embryos at early stages (2 or 3 days) of their development.

To perform the above in vivo measurements in an noninvasive way, we used a direction-sensitive, dual beam, fringe-mode laser Doppler microscope based on a He-Ne laser [2]. The probe volume during these measurements was around 300 μm^3. The output beat signal of the photomultiplier was fed to a real-time spectrum analyzer and then to a PC which yielded the time dependencies of the measured velocities.

The aggregation properties of RBCs were measured from the whole blood stabilized with the EDTA solution (0.2 ml of 5% solution per 10 ml of blood). Under such conditions the properties of the sample stay constant during 6 h after the blood had been taken from a healthy donor or patient. Our experiments were regularly performed not later than 2 h after the preparation of the sample. The state of RBCs was controlled visually with a microscope.

The measurements were carried out with the aggregometer, which is basically a backscattering nephelometer [3], whose sample cell is a 1-mm-thick gap between two vertical coaxial cylinders with a total volume of 5 ml. The outer cylinder is transparent and stationary while the inner one is darkened to eliminate stray light and it can be rotated so that shear rates can be induced stepwise-ly in the blood sample in the range from 1.5 up to 837 s^{-1}. The probing radiation from a He-Ne laser has the intensity density of 3 mW/mm^2, and the probe volume is around 2 mm^3. The measurements with one sample were performed during 10 to 15 min.

The successive stages of the experiment are as follows. First, the backscattering intensity is measured at the maximum shear rate, which induces the total disaggregation of RBCs, and their partial deformation take place. Then the rotating cylinder is stopped, which leads to rather quick restoration of the RBCs shape, causing a peakwise increase of scattering, and to spontaneous aggregation of the cells manifested by gradual decrease of the measured intensity. The latter process undergoes several stages, each reflecting the appearance of the aggregates of different dimensions.

Results and discussion

1) Figure 1 presents an example of the time-course of the velocity of the endoplamic flow inside a pro-

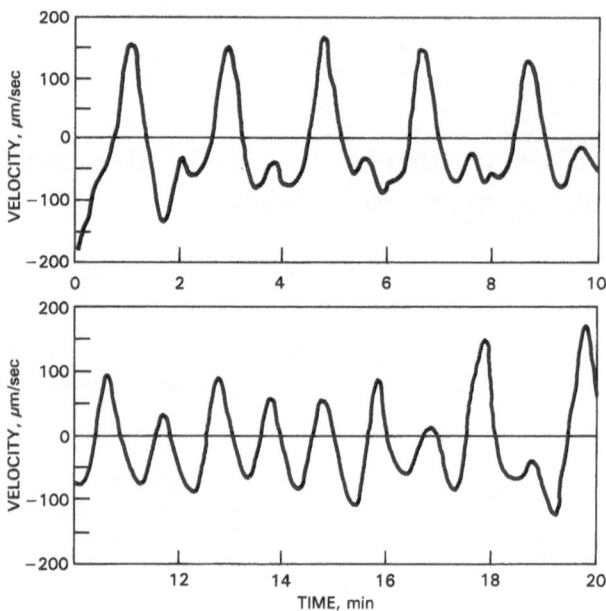

Fig. 1. Real-time recording of the protoplasmic streaming velocity in a 120 µm diameter strand of slime mold *Physarum*

toplasmic strand of *Physarum* plasmodium. The registered shuttle flow of the sol-like endoplasm is the result of periodic distributed contractile activity of actin and myosin structures in the front area of the migrating plasmodium and in the given strand walls. The latter are actually constituted of the same colloid as the endoplasm, but in gel phase. The gel-to-sol transitions can be observed in a long time scale during the migration of the plasmodium. But, we have also registered with the LDM technique very quick and short-time reversible local gel-to-sol transitions in the bulk of endoplasm in the conditions of very low shear stress. These transitions are manifested in Fig. 2 by very quick oscillations of the velocity practically to zero values occurring in the vicinity of the reverse points when shear stress is too low to prevent formation of cross-links leading to local occasional gelation of the streaming endoplasm. The two curves in Fig. 2 were obtained with different time-averaging of the signal yielding different resolution of the velocity oscillations.

2) The directed flows of metabolites and blood in fish embryos were registered at different stages and phases of development of their cardio-vascular system. One of the interesting points of the study was to monitor the development of the pulsations pattern. Two examples of such patterns corre-

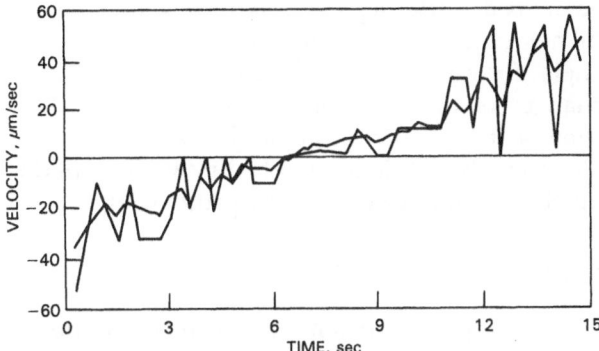

Fig. 2. Velocity recording of the protoplasmic steaming in a strand of *Physarum* at the vicinity of the reverse point. Two curves correspond to different signal processing algorithms

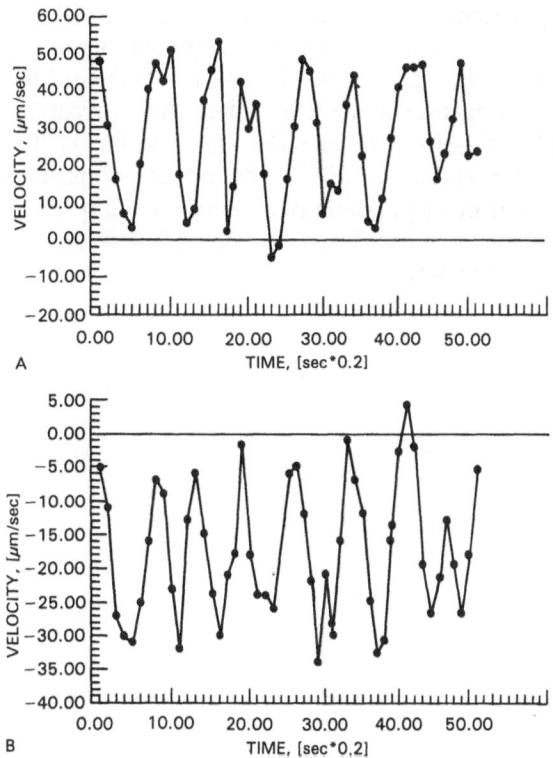

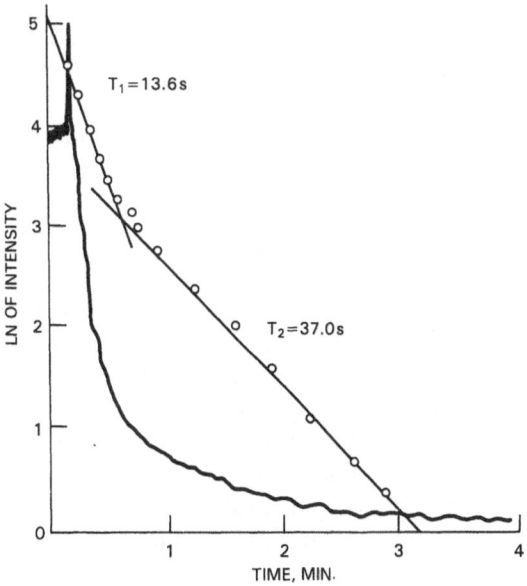

Fig. 3. Time dependencies of the velocities of pulsating: A) arterial and B) venous blood flows in a *Danio rerio* embryo registered under the temperature of 26°C

sponding to the arterial and venous flows and registered in neighboring sites of closely positioned parallel microvessels are presented in Fig. 3. Besides rather high modulation depth of the pulsations, one can notice occasional cases of the reverse of flow. The pulsations periods are not stable and depend on temperature, as well as on amplitude: the maximum values of the velocity decrease with decreasing temperature.

3) Figure 4 shows a plot of the intensity of the backscattered light measured from the normal whole blood undergoing the Coutte flow at high shear rate and after quick stop of the rotating cylinder. After a sharp peak indicating the quick change of the RBCs shape, the intensity monotonically decreases, showing the gradual formation of RBC aggregates of growing dimensions. Plotted in linear scale the corresponding curve has an exponential shape. But plotting in semi-log scale shows that different characteristic times can be distinguished (T_1 and T_2 in Fig. 4). T_1 can be related to the formation of rouleaux, and T_2— to the formation of 3D networks of RBCs.

Figure 5 shows two pathological cases. The pathologies most often lead to the decrease T_1 which, for normal blood, is typically 13.6 ∓ 0.9 s, and, sometimes, to the appearance of the distinguishable third characteristic time T'_1. Our measurements show that blood from 36 patients with psora exhibited a three-exponential process in 14% of cases, blood from 19 patients with pulmonary hypertension in 37%, but blood from 47 patients with inherited cholesteremia only in two

Fig. 4. The normal kinetics of spontaneous aggregation of RBCs of the whole blood in a Couette cell after halting the shear stress

cases. Typically, for all tested patients $T_1' = 20.0 \mp 1.1$ s and $T_1 = 7.4 \mp 0.8$ s. The aggregation process corresponding to T_1' takes from 9 up to 15% of the total aggregation amplitude, while that corresponding to T_1— from 60 to 80%. Our opinion is that the intermediate time T_1' has no self-consistent value but most probably defines the overlap of the processes of the rouleaux and the 3D network aggregates formation.

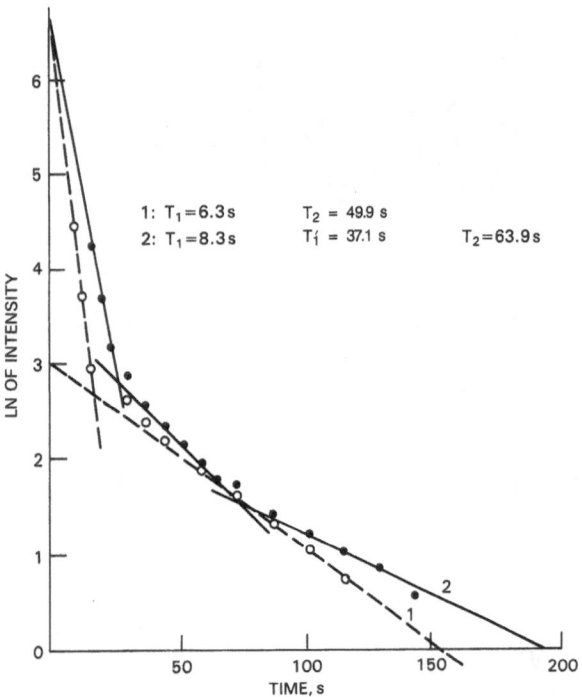

Fig. 5. The kinetics of spontaneous aggregation of RBCs in pathological blood: the cases with two and three characteristic times

Unfortunately, at present we do not have any quantitative model of light scattering from RBCs aggregates of different dimensions. The existing theories describing the aggregation of colloidal particles much smaller than RBCs are not valid in our case. So, the evaluation of the above experimental data still remains an open problem. Nevertheless, the high sensitivity of the measurements perform-

ed in backscattering mode (which would be still higher if the measurements were performed in low-angle forward scattering mode) to pathological changes of the aggregational properties of RBCs, show that even such a complex biologicall colloid as blood can be efficiently studied by light scattering for the diagnostics of medical pathologies.

Conclusions

We have presented some results of application of light-scattering techniques to the study of dynamic and structural properties of protoplasm and blood at *in vivo* and *in vitro* conditions. The complicated structure and composition of these biological colloids as well as the existence of many regulatory processes out of experimental control make the evaluation of the experimental data a very complicated problem. But the possibility of gaining better insight into some basic mechanisms of life and, also, of elaborating new efficient diagnostic procedures makes this work a challenge.

Acknowledgement

The authors thank Prof. Yuri M. Romanovsky and Prof. Otto Glatter for fruitful discussions of the experimental data.

References

1. Bender M (ed) Interfacial phenomena in biological systems (1992) Marcel Dekker, New York, Basel, Hong Kong.
2. Priezzhev AV, Proskurin SG (1991) In: Pryputniewicz RJ (ed) Laser interferometry YI: computer-aided interferometry. Proc SPIE, 1553: 502—514
3. Firsov NN, Priezzhev AV, Stepanian AS (1991) In: Akhmanov SA and Toleutaev BN (eds) Laser applications in life sciences. Proc SPIE, 1403:350—353
4. Beilina SI, Matveeva NB, Priezzhev AV, et al. (1983) In: Proc Int Symp on Self-Organisation, Autowaves and Structures Far from Equilibrium, Puschino, USSR, July 18—23, 1983, pp 218—221

Authors' address:

Dr. Alexander V. Priezzhev
Physics Department
M. V. Lomonosov Moscow State University
Moscow 119899, Russia

Progress in Colloid & Polymer Science   Progr Colloid Polym Sci 93:85—91 (1993)

Inverse problems in neutron and x-ray reflectivity studies

S. H. Chen[1]), X. L. Zhou[1]), and B. L. Carvalho[2])

[1]) Department of Nuclear Engineering and
[2]) Department of Materials Science and Engineering, Massachusetts Institute of Technology,
   Cambridge, USA

Abstract: We present a new method for obtaining the scattering length density (sld) profile perpendicular to a uniform, smooth surface which is consistent with the measured reflectivity over a finite Q range. The method is applicable to a free liquid surface as well as a thin film deposited on a known substrate. As an illustration, we apply the method to obtain: degree of ordering of an intact Langmuir-Blodgett film on a silicon substrate; surface-induced layering of water and oil in bicontinuous microemulsions; and segregation of polymers near air-film and the film-substrate interfaces.

Key words: Reflectivity — Langmuir-Blodgett film — microemulsions — polymer film

Introduction

X-ray and neutron reflectometry has found numerous applications in the study of the structure of complex liquid surfaces and polymers on solid substrates in the past few years [1]. Although the reflectivity measurement itself is straight-forward and efficient, the analysis of the reflectivity data turns out to be a difficult task without a physical model of the surface. This is because there are, so far, no systematic model-independent methods which can directly reconstruct the sld profile of a surface from a set of measured reflectivity data [2]. The cause of the difficulty is that the exact relationship between the Q space reflectance and the real space sld profile is not a simple Fourier transform pair as in the conventional crystallography. Instead, the relationship is highly nonlinear because of the multiple reflection effect, especially near the total reflection edge (Qc). This has the consequence that the conventional multi-parameter optimization schemes such as the simulated annealing [3] and maximum entropy [4, 5] methods do not work efficiently. In this paper, we present two methods: one based on the distorted wave Born approximation (DWBA) applicable to a thin film (< 300 Å) on a substrate of high sld; and one general inversion method based on an algorithm called "groove-tracking" [6] which has been demonstrated to be quite

efficient in converging to the right profile for a given set of reflectivity data [6]. Before discussing the implementation of these inversion schemes, we shall first introduce a rigorous relationship between the reflectivity and the sld profile of the surface based on the solution of Schroedinger equation. This is followed by a derivation of the first consistent approximation, which is the distorted wave Born approximation (DWBA). We shall then give an example of inversion of a reflectivity data from an intact, as-made Langmuir-Blodgett film on a silicon substrate. After that, we show two more examples of the inversion using the "groove-tracking" scheme for cases of thicker films where the DWBA does not apply.

II. Theory of specular reflectivity from a uniform surface

Imagine a uniform surface extending along $x-y$ plane with a sld profile $\rho(z)$ existing along the z-direction. The air-film interface is situated at $z = -d$ and a bulk substrate (or bulk liquid) extends from 0 to $+\infty$. For the geometry of the reflection experiment, the incident plane wave with a wave vector $\bar{k}_a$ ($|\bar{k}_a| = 2\pi/\lambda$) is making an angle $\pi/2 - \Theta$ with respect to the z-axis. It can then be easily shown [7] that the z component of the wave function $U(z)$ in the film satisfies the following one-dimensional Helmholtz equation

$$U''(z) + k^2 U(z) = 0 , \qquad (1)$$

where $k^2 = k_0^2 - 4\pi\rho(z)$ and $k_0 = (2\pi/\lambda) \sin \Theta$ is half of the magnitude of the wave vector transfer Q in the specular reflection. For the x-ray case, $U(z)$ is the z-dependence of the polarized field intensity.

First, for a constant step sld ρ_s of the substrate (or bulk liquid) extending from 0 to ∞, the solution of Eq. (1) in free space and in the substrate together with matching of the boundary conditions at $z = 0$ immediately leads to the familiar Fresnel reflectance:

$$r = \frac{k_0 - k_s}{k_0 + k_s} . \qquad (2)$$

Next, for a constant film of sld ρ of thickness d situated on the top of a substrate of sld ρ_s, the solution of Eq. (1) in the air (region I), the film (region II) and the substrate (region III) together with matching of the boundary conditions at the two interfaces give

$$r = \frac{r_f + r_b e^{2ikd}}{1 + r_f r_b e^{2ikd}} \; e^{-2ik_0 d} , \qquad (3)$$

where $r_f = (k_0 - k)/(k_0 + k)$ and $r_b = (k - k_s)/(k + k_s)$ are Fresnel reflectances for the air-film and film-substrate interfaces, respectively.

Now, we proceed to a more complicated case where the film has an arbitrary sld profile $\rho(z)$. In this case, we can conveniently subdivide the film into N segments each of a width D which is consistent with the spatial resolution of the reflectivity experiment. Numbering the segments from 1 to N starting from the substrate-film to the air-film interfaces and assuming that each segment has a constant sld ρ_i, then a simple extension of Eq. (3) for the one layer case can be made to a layer defined by interfaces $i-1$ and i. The following simple recursion relation is obtained linking the reflectance of segments 1 up to $i-1$ to that of segments 1 up to i. [8]:

$$r_i = \frac{R_i + r_{i-1} e^{2ik_i\Delta}}{1 + R_i r_{i-1} e^{2ik_i\Delta}} , \qquad (4)$$

where $R_i = (k_{i+1} - k_i)/(k_{i+1} + k_i)$ and $r_0 = (k_1 - k_s)/(k_1 + k_s)$ are the Fresnel reflectances of the interface between segment $i + 1$ and segment i and the film-substrate interface, respectively. The usefulness of this formula is that, knowing the individual segmental sld ρ_i, it is possible to rapidly compute the reflectance of the whole film r_N starting from the known substrate reflectance r_0.

On the other hand, Eq. (1) can be rewritten as

$$U''(z) + \bar{k}^2 U(z) = - (k^2 - \bar{k}^2) U(z) , \qquad (5)$$

where $\bar{k}$ is the wave number of a conveniently chosen reference film, i.e., $\bar{k} = \sqrt{k_0^2 - 4\pi\bar{\rho}}$. Solution of Eq. (5) in terms of a Green's function can be put in a simple form [9]

$$r = \bar{R} + C_r \int_{-d}^{0} dz' (k^2 - \bar{k}^{-2}) U(z) [e^{i\bar{k}z'} + R_r e^{-i\bar{k}z'}] , \qquad (6)$$

where

$$C_r = \frac{iC}{2\bar{k}} \; T_l e^{i(\bar{k} - k_0) d} , \bar{R}$$

$$= C (-R_l + R_r e^{2i\bar{k}d}) e^{-2ik_0 d} ,$$

$$C = \frac{1}{1 - R_l R_r e^{2i\bar{k}d}} \; , \quad R_l = \frac{\bar{k} - k_0}{\bar{k} + k_0} ,$$

$$R_r = \frac{\bar{k} - k_s}{\bar{k} + k_s} \; , \quad T_l = \frac{2\bar{k}}{\bar{k} + k_0} .$$

$\bar{R}$ is the reflectance of the reference film. Now, consider the deviation of the actual film from the reference film $\bar{\rho}$ to be $\Delta\rho = \rho - \bar{\rho}$. If we choose the reference film such that $\bar{k}^2 = \dfrac{1}{d} \int_{-d}^{0} dz\, k^2 (z)$, then $\int_{-d}^{0} dz\Delta\rho(z) = 0$. If the wave function $U(z)$ in Eq. (6) is approximated by the wave function inside the reference film, i.e.,

$$U(z) = \frac{2k_0}{k_0 + \bar{k}} \; C \; e^{i(\bar{k} - k_0) d} [e^{i\bar{k}z} + R_r e^{-i\bar{k}z}] , \qquad (7)$$

then, Eq. (6) reduces to

$$r = \bar{R} + \frac{2\pi}{ik_0} \; a^2 \; [\Delta\tilde{\rho}(2\bar{k}) + R_r^2 \, e^{4i\bar{k}d}\Delta\tilde{\rho}(-2\bar{k})] , \qquad (8)$$

which is the DWBA. In this equation, $a = \dfrac{2k_0 C}{k_0 + \bar{k}}$ $\cdot e^{-ik_0 d}$ and $\Delta\tilde{\rho}(2\bar{k}) = \int\limits_{-d}^{0} dz \Delta\rho(z) e^{2i\bar{k}(z+d)}$. This equation can be shown to be accurate up to the critical edge for thin films of less than 300 Å deposited on substrate with a high sld.

On the other hand, if the reference film is chosen to be the air and the interval $(-d, 0)$ is extended to $(-\infty, \infty)$ and the wave function $U(z)$ in Eq. (6) is replaced by the incident wave, then Eq. (6) becomes the well-known Born approximation [10]:

$$r = \frac{i}{2k_0} \int\limits_{-\infty}^{\infty} dz (k^2 - k_0^2) \, e^{2ik_0 z}$$

$$= \frac{4\pi}{Q^2} \int\limits_{-\infty}^{\infty} dz \frac{d\rho}{dz} \, e^{iQz} . \tag{9}$$

The second line in Eq. (9) is obtained from the first line by a partial integration.

Methods of inversion of reflectivity to obtain sld profile

Equation (9) is widely applied for the interpretation of reflectivity data for thin films. From its derivation, it is easily seen that it is only accurate asymptotically in the large Q limit. Furthermore, one shortcoming of Eq. (9) is that it introduces an extra degeneracy into the inversion problem. This can be easily seen by applying it to calculate the reflectivity of a constant film of thickness d a sld ρ situated on top of a substrate of sld ρ_s. A straightforward application of Eq. (9) leads to a result

$$|r|^2 = \frac{\pi^2}{k_0^4} \, [\rho^2 + (\rho_s - \rho)^2$$

$$+ 2\rho(\rho_s - \rho) \cos(2k_0 d)] . \tag{10}$$

It is obvious that Eq. (10) is invariant against an exchange of ρ and $\rho_s - \rho$. Thus, a constant film of sld ρ and that of $\rho_s - \rho$ give an identical reflectivity. This degeneracy is however absent in the exact formula given in Eq. (3). An obvious implication of this simple example is that the uniqueness of inversion relies on the availability of accurate data in the small Q range near the critical edge Qc [6].

The DWBA formula Eq. (8), on the other hand, is more accurate in the small Q region near Qc. In fact, it can be shown that [11] for a liquid surface where

the Qc can be calculated fairly accurately from the sld of the bulk liquid $\bar{\rho}$, the reflectivity $R(Q) = |r|^2$ predicted by Eq. (8) has a small Q expansion

$$R(Q) - \bar{R}(Q) = \left\{ \frac{8\sqrt{\pi}}{\sqrt{\bar{\rho}}} \, M_1 \right\} q$$

$$+ \left\{ \frac{16\pi}{\bar{\rho}} \, M_1^2 - \frac{8}{\bar{\rho}} \, M_1 \right\} q^2 + \\ \cdots$$

$$\tag{11}$$

where $q = \sqrt{Q^2 - 16\pi\bar{\rho}}$ is the reduced Q in the bulk liquid, and $M_i = \int\limits_{-d}^{0} z^i \Delta\rho(z) dz$ is the i-th moment of the sld deviation from the bulk. Thus, an accurate measurement of low Q data allows a model-independent determination of the first moment M_i.

The DWBA can also applied to find out the sld profile of a surface by a conventional least squares fitting scheme [12]. Here, we first imagine the film of thickness d to be divided into N slabs. In this case $\Delta\tilde{\rho}(2\bar{k})$ can be written as the sum

$$\Delta\tilde{\rho}(2\bar{k}) = \sum_{m=1}^{N} \rho_m \left[\frac{e^{i2\bar{k}md/N} - e^{i2\bar{k}(m-1)d/N}}{i2\bar{k}} \right] , \tag{12}$$

where the m-th slab deviates from the average scattering length density, $\bar{\rho}$, by ρ_m. $\Delta\tilde{\rho}(-2\bar{k})$ can be expressed similarly. Equations (12) and (8) allows us to write the reflectivity explicitly in terms of ρ_m, the real space sld deviation from $\bar{\rho}$. The fitting scheme starts with an initial guess of a flat profile of height $\bar{\rho}$ and thickness d on a known substrate. The sequence of N step parameters $\{\rho_m\}$ are then found by using a nonlinear least squares routine that minimizes $\sum[(R_{DWBA} - R_{measure})/\bar{R}]^2$; $\bar{R}$ is the reflectivity of the average film on the known substrate. Setting $N = 48$, the fitting refinement typically takes less than 1 h on a RISC based workstation. N is chosen according to our experimental resolution, which is approximately 5 Å in real space.

For thicker films situated on a substrate, the above method does not work.

Therefore, a new model-independent method was developed [6] to reconstruct the sld profile from the reflectivity data. From the recursion relation Eq.

(4), it is seen that r of the entire film is an analytical function of N variables ρ_i ($i = 1, 2, \ldots N$) though the functional form is complex and highly non-linear for large N. One can write formally

$$r(k_0) = P(k_0, \bar{p}) , \tag{13}$$

where P is an analytical function to be calculated by the recursion relation and $\bar{p} = \{\rho_1, \rho_2, \ldots \rho_N\}$ is an N-dimensional vector representing the scattering length density values at N positions. Taking the squared modulus of Eq. (13) and equating it to the M reflectivity data at k_0^i ($i = 1, 2, \ldots M$), one has a set of M simultaneous equations, which can be written as

$$\varepsilon_i = ||r(k_0^i)|^2 - |P(k_0^i, \bar{p})|^2| = 0 ,$$

$$\text{for } i = 1, 2, \ldots M. \tag{14}$$

Because experimental data contain errors, there exists no solution ρ_i ($i = 1, 2, \ldots N$) which satisfies Eq. (14) simultaneously in the exact sense. Eq. (14) can only be satisfied by a set of ρ_i ($i = 1, 2 \ldots N$) within an error bound commensurate with the experimental errors. This means that one should consider the vector $\bar{p}$ as a solution to Eq. (14) as long as it makes ε_i smaller than the error bars of the data $|r(k_0^i)|^2$ for all the data points.

$$E_i = \left| \frac{|r(k_0^i)|^2 - |P(k_0^i\,\bar{p})|^2}{\sigma_i} \right| \leq 1 ,$$

$$\text{for } i = 1, 2, \ldots M \tag{15}$$

To solve this set of equations, we have devised a method called Groove Tracking Method (GTM) [6] which enables us to approach the solution in a computationally efficient way. It is based on the understanding that there exists a groove in the N-dimensional vector space leading to the solution monotonically. As long as we reach the groove at the beginning and then, as we proceed, bind ourselves to the groove like a roller coaster does itself to the track, we can avoid the necessity of mapping the whole space, which is time-inefficient and unfeasible for the case of neutron reflection (10^N possibilities for N layers even if each layer is allowed to vary for as few as 10 times).

Results of the inversion

To begin with, we use the DWBA to invert x-ray reflectivity data from a Langmuir-Blodgett (LB) film. The LB technique has enjoyed a recent popularity because it seen as a precise way of con-

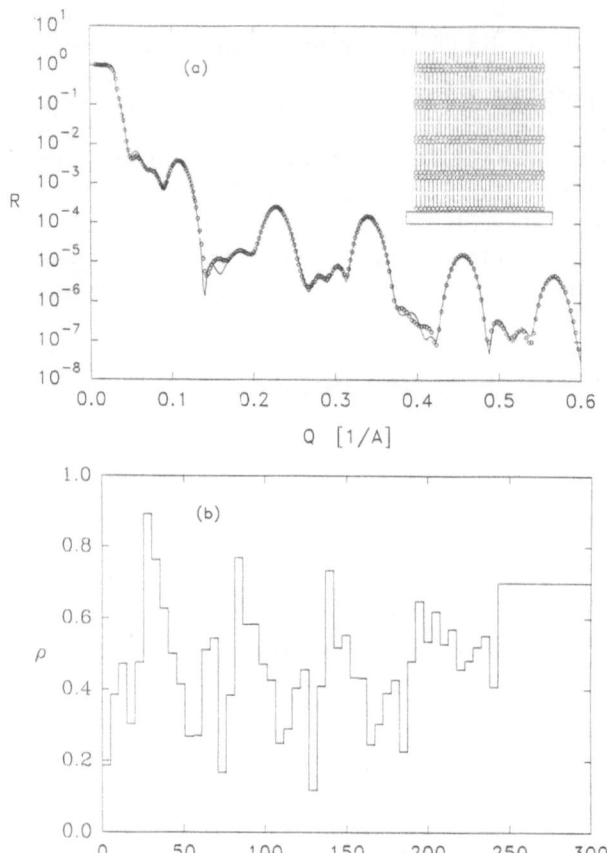

Fig. 1. a) Measured x-ray reflectivity (circles) and the best fit (solid line) for a nine-layer film, using the DWBA scheme as described in the text. The inset is the schematic view of the idealized nine-layer Cd-film as it resides on the Si substrate. The open circles represent the region containing the carboxyl head groups and their cadmium counterions while the sticks stand for hydrocarbon tails. b) The electron density profile (in units electron/Å^3) obtained from a least squares refinement. To convert to sld multiply the ordinate axis by 2.82×10^{-5} Å

trolling the molecular architecture of thin organic films. The LB technique is used for the deposition of organic molecules onto a solid support in a layer-by-layer fashion. However, LB films are rarely perfect — their structure can vary with a variety of fabrication conditions — and any inherent disorder is hidden by the lack of non-invasive techniques for assessing this. The x-ray reflectivity technique may be of help in this regard.

In Fig. 1(a), we show x-ray reflectivity data from a nine-layer Cd-arachidate LB film deposited onto a silicon substrate. Because silicon is moderately reflecting for x-rays, the DWBA can be applied to invert these data. The inverted electron density profile is shown in Fig. 1(b). The striking feature of this profile is the presence of disorder in this, the as-made film. The profile of a perfect film (inset Fig. 1 (a)) is expected to consist of five peaks, each separated by a center-to-center distance of 54 Å and each with a height ranging between 0.7 to 1.0 electron $/Å^3$, depending on the degree of overlap within the counterion regions. The region between the peaks corresponding to the hydrocarbon tails is expected to have a electron density of 0.33 electron $/Å^3$. The actual film, however, is quite disordered. Layers are observed but lose definition near the substrate. The film exhibits intralayer disorder as well. In fact, it appears that some arachidate molecules are reversed in orientation, even within a layer. This conclusion is evidenced by the fact that there are two sets of peaks of equal spacing (56 Å), displaced from each other by 20Å.

Next, we come to the application of the groove-tracking method of inversion using the recursion formula Eq. (4) to find out the surface-induced structure in bicontinuous microemulsions. A microemulsion with equal volume fraction of water and oil is bicontinuous when one can pass from one end of the sample to the other end, either via a water pathway or an oil pathway. A disordered bicontinuous structure can be visualized as an interpenetrating domain of water and oil of some average domain size $d/2$ [13]. The polydispersity of the domain sizes is proportional to a dimensionless quantity d/ξ, where ξ is a coherent length. On the other hand, an ordered bicontinuous structure can, for example, be a lamellar structure with a repeated distance of water domains or oil domains equal to d [14]. In Figs. 2 and 3, we give surface profiles of two three-component microemulsion systems. The reconstruction of each of these profiles took approximately 10 minu on a DEC 5000 workstation.

Figure 2 shows the reflectivity and the corresponding inverted sld profile of a non-ionic microemulsion system consisting of $C_{10}E_4$, D_2O and n-octane at approximately equal volume fractions of water and oil and the surfactant weight fraction 17.40% ($a = 30.60, \gamma = 17.40$). We measured the reflectivity at $T = 18.50°C$. This corresponds to a point near the lower two-phase boundary in the one-phase channel [6]. Figure 2(a) gives the measured reflectivity and the reflectivity calculated from the inverted sld profile in Fig. 2(b). Figure 2(b) is the reconstructed

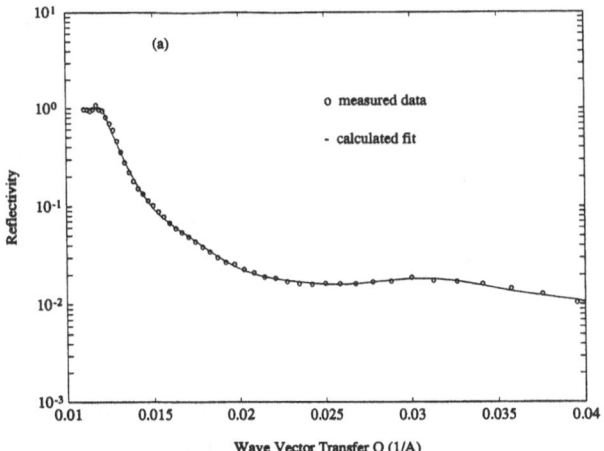

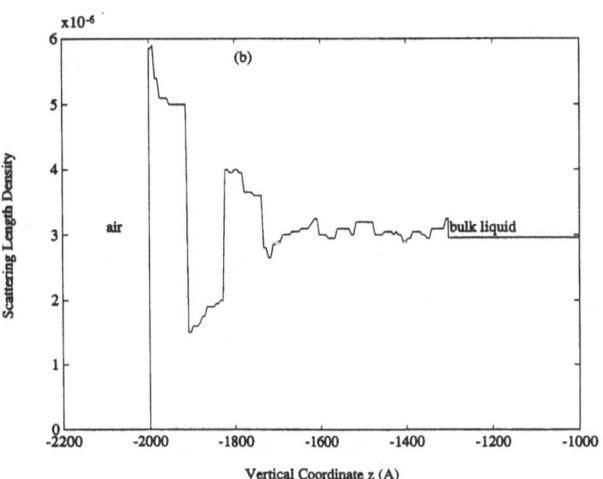

Fig. 2. Inversion of the reflectivity data measured from a one-phase bicontinuous microemulsion consisting of $C_{10}E_4$, D_2O and n-octane at $a = 30.60$, $\gamma = 17.40$ and $T = 18.5°C$. a) The measured reflectivity data (circles) compared with the reflecivity calculated from the inverted profile in (b). b) The inverted sld profile obtained by the model-independent method in [6] from the data in (a)

sld profile of the air-microemulsion interface. A striking feature worth noting is that there is an 85 Å thick layer of D_2O immediately adjacent to the microemulsion-air interface with an interlayer distance of $d = 196$ Å. The surface layering effect seems to penetrate 500 Å into the bulk. Figure 3 shows a similar result for AOT/brine/decane ionic system at $a = 40$, $\gamma = 20$ and $T = 25.2°C$ also in the lower two-phase boundary in the one-phase channel. The surface layering has a repeated distance of $d = 200$ Å penetrating into the bulk also by 500 Å [15].

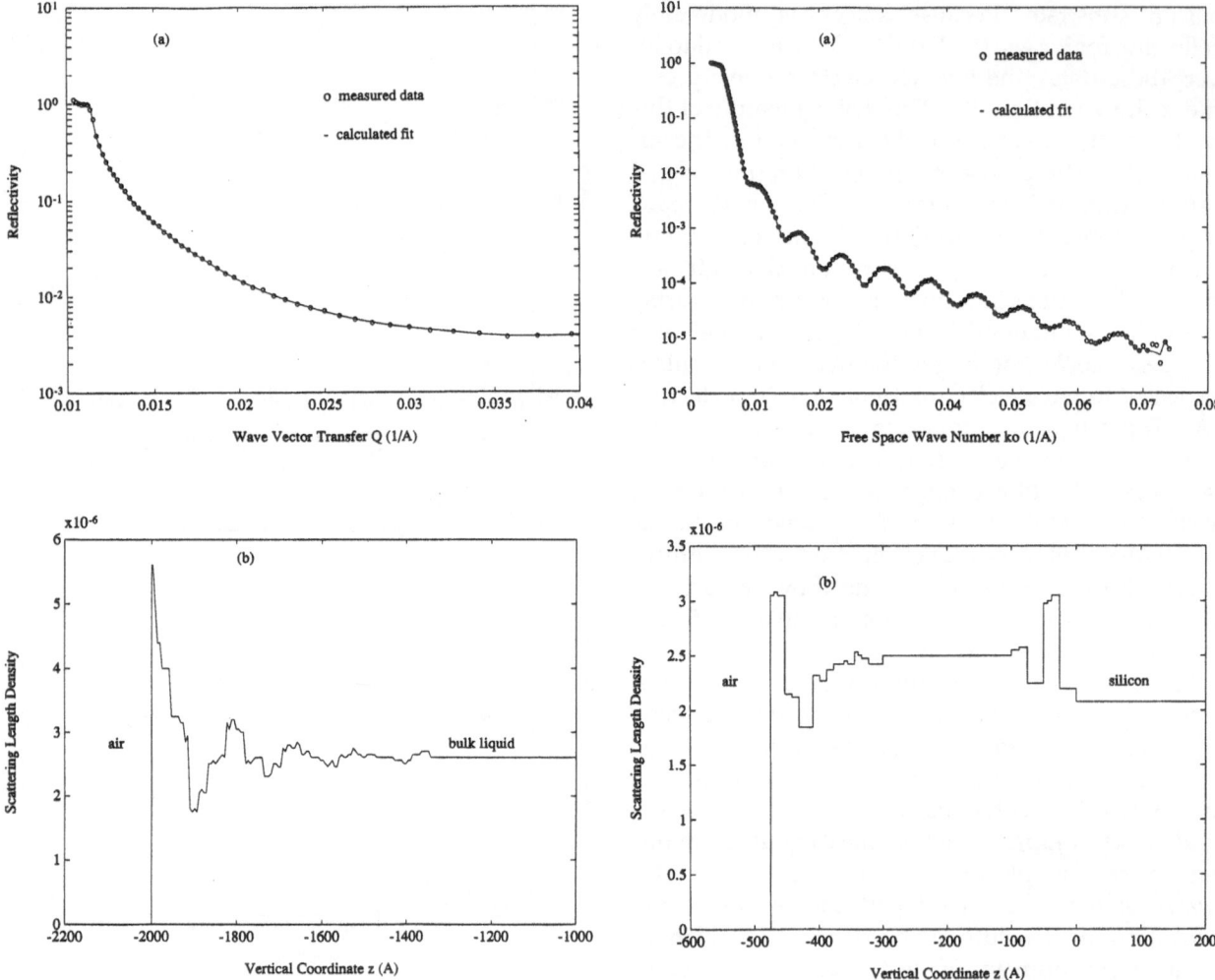

Fig. 3. Inversion of the reflectivity data measured from a microemulsion consisting of AOT, D_2O and decane at a = 40, γ = 20 and T = 25.2 °C. a) The measured reflectivity data (circles) compared with the reflectivity calculated from the inverted profile in (b). b) The inverted sld profile obtained by the model-independent method in [6] from the data in (a)

Fig. 4. Inversion of the reflectivity data measured from a mixture of polymers terminated with fluorine on a silicon substrate. a) The measured reflectivity data (circles) compared with the reflectivity calculated from the inverted profile in (b). b) The inverted sld profile obtained by the model-independent method in [6] from the data in a)

The last example, a more difficult case from the inversion point of view, is a study of polymer adsorption on a silicon substrate. It has been contemplated that neutron reflectivity experiments could determine the effect of the end groups on the surface adhesion properties of polymers. Experiments were carried out on samples consisting of uniform mixtures of polystyrenes. Seventy-five percent of each polystyrene chain is protonated while the other 25% of the chain is deuterated. Then the deuterated end of the chain is terminated with lithium or fluorine tip. The polymers are situated on top of silicon substrate of sld 2.08 ×

10^{-6} Å^{-2}. The sample was annealed for 16 h at 110 °C. Since the sample is a uniform mixture, obviously, the only places where structures can occur are the air the air-polymer interface and the polymer-substrate interface. The middle of the film should be a uniform bulk. For a given chain at the surface, the terminated deuterated end could either be drawn to the surface or remain unaffected by the presence of the surface. The former would result in a high sld density region next to the surfaces followed by a low sld density region corresponding to the protonated segment of the chains. Note that the high density region could be considerably wider than the length of the deuterated segment as the polymers are tangled together. The latter would lead to a uniform region near the surface.

A set of data was measured by Koberstein and coworkers [16] form a polystyrene sample with fluorine termination. The data are plotted in circles in Fig. 4(a). The reconstructed profile is given in solid line in Fig. 4(b). The inversion of the data took about 15 min on a DEC 5000 workstation. In obtaining the profile, the middle of the profile was constrained to be constant to save computation time. It is seen that there is a high sld layer at the front surface and a layer at the rear interface. This indicates that the fluorine tip has an affinity to the interfaces. The correctness of the profile is confirmed by comparing the calculated reflectivity of the profile as given by the solid line in Fig. 4(a) with the data shown by circles in the same graph. Note that the high sld layer in the front is followed by a layer of low density region.

In summary, we have outlined two methods for the inversion of reflectivity data. We have also illustrated these methods through a number of practical examples. Owing to the highly nonlinear relationship between the reflectance and the surface profile, the inversion requires accurate data in the small and intermediate Q range. For a thin film deposited on a highly reflecting substrate, the peturbation approach based on the DWBA is satisfactory for the purpose of inversion. For thick films on an arbitrary substrate the inversion requires the exact recursion relation. Finally, we note that accurate profiles are obtained only when the calculated reflectivity curve and the measured data agree with a $\chi 2$ parameter of order unity.

Acknowledgement

This research is supported by a grant from the Division of Materials Science of the Department of Energy, contract DEFG02-90-ER45429.

References

1. Russell TP (1990) Materials Science Reports 5 57:119
2. Felcher GP, Russell TP (1991) Eds., Proceedings of the Workshop on "Methods of Analysis and Interpretation of Neutron Reflectivity Data", Physica B 173 No. 1 & 2
3. Gelfand SB (1987) Analysis of Simulated Annealing Type Algorithms, Ph.D. Thesis, Electrical Engineering, MIT
4. Fougere P (1990) Maximum Entropy and Bayesian Methods, Klumer, Amsterdam
5. Skilling J (1988) Maximum Entropy and Bayesian Methods, Cambridge
6. Zhou XL, Chen SH (1993) Phys Rev 47E:3174—3190, Zhou XL, Lee LT, Chen SH, Strey R (1992) Phys Rev 46A:6479-6489
7. Zhou XL, Chen SH, Felcher GP (1991) Inverse Problems in Neutron Reflection, in: Bertero M, Pike ER (eds) Inverse Problems in Scattering and Imaging, Hilger Bristol 109
8. Parratt G (1954) Phys Rev 95:359
9. Zhou XL, Chen SH, Felcher GP (1991) J Phys Chem 95:9025—9029; Zhou XL, Chen SH, Felcher GP (1992) Phys Rev A 46:1839—1843
10. Als-Nielsen J (1985) Z Phys B Condensed Matter 61:411—414
11. Carvalho BL, Chen SH (1993) A Low-Wavevector Expansion for Reflectivity, Phys Rev E47:743—745
12. Carvalho BL, Sanyal MK, Sinha SK, Felcher GP, Chen SH "Model Independent Analysis of X-ray Reflectivity Data", in preparation
13. Chen SH, Chang SL, Strey R (1991) J Appl Cryst 24:721-731
14. Chen SH, Chang SL, Strey R, Thiyagarajan P (1991) J Phys Condensed Matter 3:F91—F107
15. We are grateful to Lay-Theng Lee and R. Strey for permission to show this unpublished result.
16. Koberstein JT, unpublished

Authors' address:

Sow Hsn Chen
Dept. of Nuclear Engineering
Massachusetts Institute of Technology
24-211 Mas Ave 77
Cambridge MA 02139-4307, USA

Progress in Colloid & Polymer Science   Progr Colloid Polym Sci 93:92—97 (1993)

Adsorption of alkyltrimethyl ammonium bromide at the air-water interface

J. R. Lu[1]), E. A. Simister[1]), R. K. Thomas[1]), and J. Penfold[2])

[1]) Physical Chemistry Laboratory, Oxford United Kingdom
[2])Rutherford-Appleton Laboratory, Oxon United Kingdom

Abstract: The composition and structure of soluble monolayers of cationic surfactants of alkyltrimethyl ammonium bromide adsorbed at the air-water interface have been studied by the neutron reflection technique. — At the critical micelle concentration (CMC) the area per surfactant molecule was found to be 55 ± 3 Å^2 for decyl trimethyl ammonium bromide (C$_{10}$TAB), 48 ± 3 Å^2 for tetradecyl trimethyl ammonium bromide (C$_{14}$TAB), and 43 ± 3 Å^2 for octadecyl trimethyl ammonium bromide (C$_{18}$TAB). — The determination of the structure of the surfactant layers at their corresponding CMC was based on three differnt isotopic combinations of water and surfactant. The results were analysed using both optical matrix method and the kinematic approach. The thickness of the whole surfactant monolayer was found to be 19 ± 2 Å for C$_{10}$TAB, 22 ± 2 Å for C$_{14}$TAB and C$_{18}$TAB. The thickness of alkyl chain layer out of water was found to be 11 ± 1 Å for C$_{10}$TAB and 12.5 ± 1 Å for the two other surfactants. The amount of alkyl chain immersed in water was found to be 30 — 40%, with shorter chain surfactants being more immersed.

Key words: Neutron reflection — surfactant adsorption — monolayer structure

Introduction

Specular reflection of neutrons is now established as a powerful technique for determining the structure of surfactant layers at both the air-solution and solid-solution interfaces [1, 2]. Recently, we have presented results on the structure of tetradecyl trimethyl ammonium bromide (C$_{14}$TAB) adsorbed at the air-water interface, in which the distribution of alkyl chain, head group and water were measured, and the relative separations of the distributions were also described [2]. The purpose of present work is to extend the study to other cationic surfactants in the same series to determine the effect of the chain length on the structure of the adsorbed layer.

Neutron reflection

Neutron reflectivity data is usually analysed by the optical matrix method, in which the surface is divided into thin layers, each with a different scattering length density [3]. The method gives the exact reflectivity for a given structural model and agreement of calculated and measured profiles is then taken to mean that the structure of the surface is that of the model. The main difficulty in using the optical matrix method is that the structure so obtained may not be unique.

An alternative approach to the analysis of specular reflection profiles is the kinematic approximation outlined by Crowley et al. [4, 5]. In which

$$R(\kappa) = \frac{16\pi^2}{\kappa^4} \mid \hat{\rho}^{(1)}(\kappa)\mid^2 , \qquad (1)$$

where κ is the momentum transfer ($= 4\pi\sin\Theta/\lambda$, Θ is the angle of incidence and λ the wavelength), $\hat{\rho}^{(1)}(\kappa)$ is the one-dimensional Fourier transform of $\rho^{(1)}(z)$ ($= d\rho(z)/dz$, $\rho(z)$ is the scattering length density profile at the height z normal to the surface),

$$\hat{\rho}^{(1)}(\kappa) = \int_{-8}^{\infty} \exp(-i\kappa z)\rho^{(1)}(z)\,dz \ . \tag{2}$$

It should be noted that Eq. (1) is only approximate, but becomes a good approximation if the observed reflectivity R_{obs} is corrected using a formula derived by Crowley [6],

$$R = R_k^0 + \left[\frac{1 + (1 - \kappa_0^2/\kappa^2)^{1/2}}{2} \right] \frac{(R_{obs} - R_f)}{(1 - R_f)} \ , \tag{3}$$

where κ_c is the momentum transfer at total reflection, R is the kinematic reflectivity as in Eq. (1), R_k^0 is the kinematic reflectivity for a sharp interface between the two bulk phases, and R_f is the exact reflectivity for the same sharp interface.

For a surfactant at the air-water interface, the system can be described in terms of the distribution of the surfactant (s) and the water (w). The derivative of scattering length density profile across the interface can then be expressed as [4]

$$\rho^{(1)}(z) = b_s n_s^{(1)}(z) + b_w n_w^{(1)}(z) \ , \tag{4}$$

where b_s, b_w and $n_s^{(1)}$, $n_w^{(1)}$ are the scattering lengths and the derivatives of the number densities of the two species. Substituting Eq. (4) into (2) and (1) gives

$$R(\kappa) = \frac{16\pi^2}{\kappa^4} \left(b_s^2 h_{ss}^{(1)} + 2 b_s b_w h_{sw}^{(1)} + b_w^2 h_{ww}^{(1)} \right) \ , \tag{5}$$

where $h_{ss}^{(1)}$, $h_{ww}^{(1)}$ and $h_{sw}^{(1)}$ are the partial structure factors which are κ dependent, and are given by

$$h_{ii}^{(1)}(\kappa) = |\hat{n}_{ii}^{(1)}(\kappa)|^2 \ , \tag{6}$$

and

$$h_{sw}^{(1)}(\kappa) = Re\,|\hat{n}_s^{(1)}(\kappa)\hat{n}_w^{*(1)}(\kappa)| \ , \tag{7}$$

where i denotes s or w, $\hat{n}_s^{(1)}$ and $\hat{n}_s^{*(1)}$ are the Fourier transforms of $n_s^{(1)}$ and $n_w^{(1)}$ and are given by an equation corresponding to (2) in terms of the number density rather than the scattering length density.

It can be seen from Eq. (5) that to obtain $h_{ss}^{(1)}$, $h_{ww}^{(1)}$ and $h_{sw}^{(1)}$ experimentally three reflectivity profiles with different isotopic contrasts have to be measured. We used chain deuterated surfactant in null reflecting water (n.r.w.), chain deuterated surfactant in D_2O and protonated surfactant in D_2O. The three isotopic combinations of measurements used for each surfactant are such that the deuterated surfactant in n.r.w. gives $h_{ss}^{(1)}$ directly, and $h_{ww}^{(1)}$ and $h_{sw}^{(1)}$ can be determined by solving all three equations derived from Eq. (5).

Experimental

The C_nTABs were made by reacting the 1—bromoalkane with trimethylamine as described by Voeks et al. [7]. The products were recrystallized in a mixture of acetone and ethanol until any minimum in the surface tension — In[concentration] plot disappeared. The value of the CMC was found to be $(67 \pm 3) \times 10^{-3}$ M for C_{10}TAB, $(3.7 \pm 0.3) \times 10^{-3}$ M for C_{14}TAB, and $(0.32 \pm 0.02) \times 10^{-3}$ M for C_{18}TAB, in good agreement with the literature values [8]. Surface tension measurements of the isotopic species of C_{14}TAB indicated no isotope effect on the surface activity [9].

The neutron reflection measurements were made on the reflectometer CRISP at the Rutherford-Appleton Laboratory (Didcot, U.K.) as described previously [10]. The measurements were made at 25°C for C_{10}TAB, and C_{14} TAB,30°C for C_{18}TAB.

Results and discussion

The reflectivity profiles for the adsorption of C_nTAB from the n.r.w. solutions at their corresponding CMC are shown in Fig. 1, those for the chain deuterated surfactants in D_2O in Fig. 2 and those for the protonated surfactants in D_2O in Fig. 3. The important information which can be obtained from measurements in n.r.w. is the surface excess of the surfactant or area per molecule (A). A is obtained by fitting the observed reflectivity with a model of a single uniform layer. The scattering length density of the layer and its thickness (τ) are related to A by

$$A = \frac{\Sigma n_i b_i}{\rho_s \tau} \ . \tag{8}$$

We have shown previously that, although a range of ρ_s and τ may fit a given reflectivity profile with reasonable accuracy, the change in ρ_s almost exactly compensates the change in τ in such a way that A is not affected. A similar situation is found if

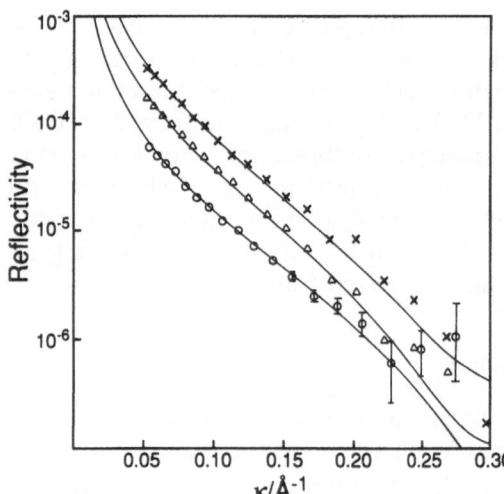

Fig. 1. Reflectivity profiles of chain deuterated $C_{10}TAB(\circ)$, $C_{14}TAB(\triangle)$, and $C_{18}TAB(\times)$ in null reflecting water. The continuous lines are the fits from the two layer model. The corresponding structural parameters are given in Table 3. The error bars are only marked for $C_{10}TAB$

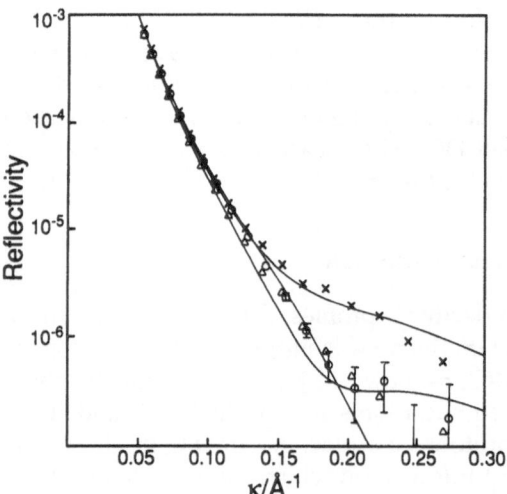

Fig. 2. Reflectivity profiles of chain deuterated $C_{10}TAB(\circ)$, $C_{14}TAB(\triangle)$, and $C_{18}TAB(\times)$ in D_2O. The continuous lines are the fits from the two layer model. The corresponding structural parameters are given in Table 3. The error bars are only marked for $C_{10}TAB$

the surface structure is described by more realistic models with or without surface roughness.

The results of the simple analysis are presented in Table 1 (Scattering lengths for different isotopic groups are given in Table 2). The area per molecule at the CMC was found to be 55 ± 3 Å^2 for $C_{10}TAB$,

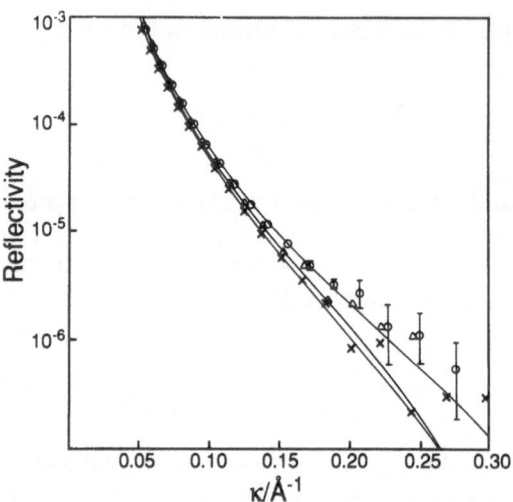

Fig. 3. Reflectivity profiles of protonated $C_{10}TAB(\circ)$, $C_{14}TAB(\triangle)$, and $C_{18}TAB(\times)$ in null reflecting water. The continuous lines are the fits from the two-layer model. The corresponding structural parameters are given in Table 3. The error bars are only marked for $C_{10}TAB$

47 ± 3 Å^2 for $C_{14}TAB$, and 43 ± 3 Å^2 for $C_{18}TAB$. The thickness obtained here is that of the alkyl chain in the layer because of the weak scattering length of the head and counterion.

The measurements in D_2O require a more realistic model because the interfacial scattering length density profiles cannot be approximated as a uniform block. We have previously used a two-layer model, which consists of a fraction of the alkyl chain layer out of the water and a head group region containing the cationic head group, the counterion, water and the remaining fraction of the alkyl chain. The structural parameters used in the fitting are then A, the thickness of the chain (τ_c) and head (τ_h) group regions, and the fraction of the alkyl chain immersed in water f_c. The relations between these variables can be expressed in a set of equations which have been presented previously [2]. The final structural parameters are summarised in Table 3, together with the estimated errors. The corresponding fits are shown as continuous lines in Fig. 1 for the profiles in n.r.w., in Fig. 2 for the chain deuterated surfactants in D_2O and Fig. 3 for the protonated surfactants in D_2O.

In using the kinematic analysis Eq. (5) converts the set of three reflectivity profiles into the partial structure factors $h_{ss}^{(1)}$ (Fig. 4), $h_{ww}^{(1)}$ (Fig. 5) and $h_{sw}^{(1)}$ (Fig. 6). For clarity, error bars are only indicated for $C_{10}TAB$.

Table 1. Parameters from one-layer model fitting for the profiles in n.r.w.

Surfactant	$A/\text{Å}^2$	$\rho \times 10^6/\text{Å}^{-2}$	$\tau/\text{Å}$
dC$_{10}$hTAB	55 ± 3	2.2	17 ± 2
dC$_{14}$hTAB	48	3.2	19
dC$_{18}$hTAB	43	4.5	19

Table 2. Physical constants used in the fitting

Unit	Scattering length $\times 10^5/\text{Å}$	Volume/ Å^3	Fully extended/Å chain length
C$_{10}$H$_{21}$	−12.0	295	14.1
C$_{10}$D$_{21}$	206.6	295	14.1
C$_{14}$H$_{29}$	−15.4	405	19.2
C$_{14}$D$_{29}$	286.5	405	19.2
C$_{18}$H$_{37}$	−18.7	510	24.3
C$_{18}$D$_{37}$	366.5	510	24.3
N(CH$_3$)$_3$Br	2.5	135	—
N(CD$_3$)$_3$Br	96.2	135	—
H$_2$O	−1.7	30	—
D$_2$O	19.1	30	—

The structure factors can, in principle, be Fourier transformed to obtain the distributions of the number density profiles. However, it is obvious that the experimental structure factors have not been determined over a sufficient range of κ to make the Fourier transformation reliable. We here take an alternative route by using model fitting from which some structural features can still be deduced.

For a layer of uniform composition, the partial structure factor $h_{ss}^{(1)}(\kappa)$ is

$$h_{ss}^{(1)}(\kappa) = \frac{4}{(A\tau)^2} \sin^2(\kappa\tau/2) . \tag{9}$$

If the water layer associated with the surfactant head group region is also taken to be a uniform layer, different from bulk water, the water-water partial structure factor is.

$$h_{ww}^{(1)}(\kappa) = 4n_{w0}^2 + n_{w1}(n_{w1} - n_{w0}) \sin^2(\kappa\tau_h/2) , \tag{10}$$

where n_{w1} and n_{w0} are the number densities of water in the layer and the bulk, respectively.

The fitted curves of $h_{ss}^{(1)}$ using Eq. (9) are plotted in Fig. 4 as solid lines and those for $h_{ww}^{(1)}$ using Eq. (10) in Fig. 5. The structural parameters obtained

Table 3. Fitted parameters using two layer model

Surfactant	Substrate	$A/\text{Å}^2$	$\tau_c/\text{Å}$	$\tau_h \text{Å}$	$\rho_c \times 10^6/\text{Å}^{-2}$	$\rho_h \times 10^6/\text{Å}^{-2}$	n	f
dC$_{10}$hTAB	n.r.w.	56 ± 2	11 ± 1	8 ± 1*)	1.97	1.86	—	0.40 ± 0.04
dC$_{10}$hTAB	D$_2$O	55	11	8	2.05	4.98	7 ± 1	0.40
hC$_{10}$hTAB	D$_2$O	55	11*)	8	−0.12	2.89	7	0.40
dC$_{14}$hTAB	n.r.w.	47 ± 2	12 ± 1	10 ± 1*)	3.30	2.13	—	0.35 ± 0.03
dC$_{14}$hTAB	D$_2$O	47	11.5	10	3.45	5.04	7 ± 1	0.35
hC$_{14}$hTAB	D$_2$O	47	12*)	10	−0.18	2.79	7	0.35
dC$_{18}$hTAB	n.r.w.	44 ± 2	12 ± 1	10 ± 1*)	4.92	2.58	—	0.30 ± 0.03
dC$_{18}$hTAB	D$_2$O	44	12	10	4.92	5.22	6 ± 1	0.30
hC$_{18}$hTAB[a]	D$_2$O	44	0	10.5	0	2.54	6	0.30

*) Denotes insensitive variables in the fitting procedure.
[a] The alkyl chain was contrast matched to air.

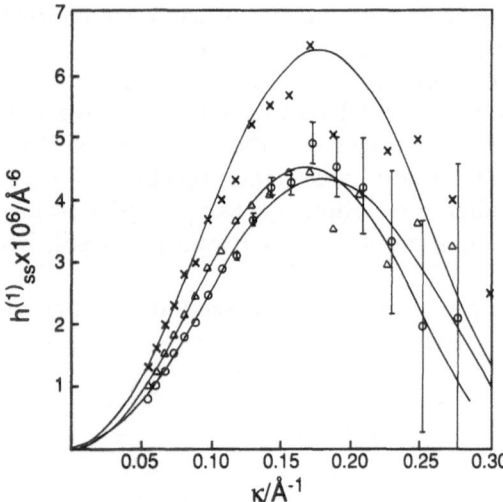

Fig. 4. Surfactant partial structure factors, $h_{ss}^{(1)}$, for (a) $C_{10}TAB(\circ)$, (b) $C_{14}TAB(\triangle)$ and (c) $C_{18}TAB(\times)$. The continuous lines are calculated structure factors for a uniform layer model using a) 17.5, b) 19.0 and c) 18.2 Å. The error bars are only marked for $C_{10}TAB$

are given in Table 4. It can be seen from Eq. (9) that an increase in $A\tau$ decreases the height of the peak position and an increase in τ shifts the peak towards the low κ region. This effect can be seen clearly in Fig. 4.

The thickness of the head group region is obtained by fitting $h_{ww}^{(1)}$. The value for $C_{10}TAB$ was found to be 8 ± 1 Å and that for the other two longer surfactants was found to be 10.5 ± 1 Å, indicating that further increase in the alkyl chain length does not increase the thickness of the layer under water. The number of water molecules per surfactant was calculated from n_{w1} in Eq. (10) and the results are consistent with those calculated from the optical matrix results of Table 3.

The experimentally determined $h_{sw}^{(1)}$ contains information about the relative location of the surfactant chain and water in the layer. For an even distribution of $n_s(z)$ and an odd distribution of $n_w(z)$,

$$h_{sw}^{(1)}(\kappa) = \pm (h_{ss}^{(1)} h_{ww}^{(1)})^{1/2} \sin\kappa\delta_{sw} , \tag{11}$$

where δ_{sw} is the separation of the two distributions. In the present situation the arrangement of surfactant chain in the layer is approximately an even function and the distribution of water in the head group region is approximately an odd function

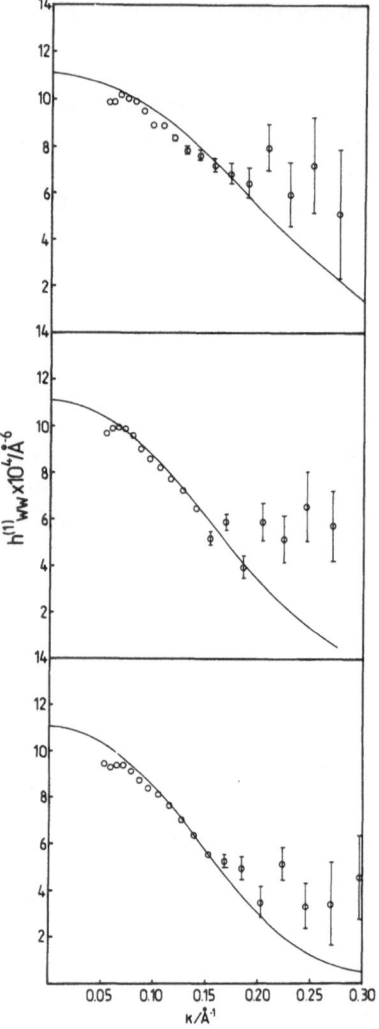

Fig. 5. Solvent partial structure factors, $h_{ww}^{(1)}$, for (a) $C_{10}TAB$, (b) $C_{14}TAb$ and (c) $C_{18}TAB$. The continuous lines are calculated structure factors for a uniform layer model using a) 8.0, b) 10.0 and 10.5 Å

tion and therefore Eq. (11) is appropriate [11]. The calculated $h^{(1)}$ for the different C_nTAB using equation (11) are shown in Fig. 6. The errors in determining δ_{sw} from Eq. (11) are small, better than ± 0.5 Å, but the errors in the measured $h_{sw}^{(1)}$ increase rapidly with κ, just as for $h_{ss}^{(1)}$ and $h_{ww}^{(1)}$. In fitting $h_{sw}^{(1)}$ we have restricted our attention to the low κ range. Values of δ were found to increase for the longer chain C_nTAB, suggesting a larger separation of the chain and the head.

If all the alkyl chain were above the water surface δ_{sw} would be equal to $(\tau + \tau_h)/2$. The fact that δ_{sw} is considerably smaller than $(\tau + \tau_h)/2$ suggests that

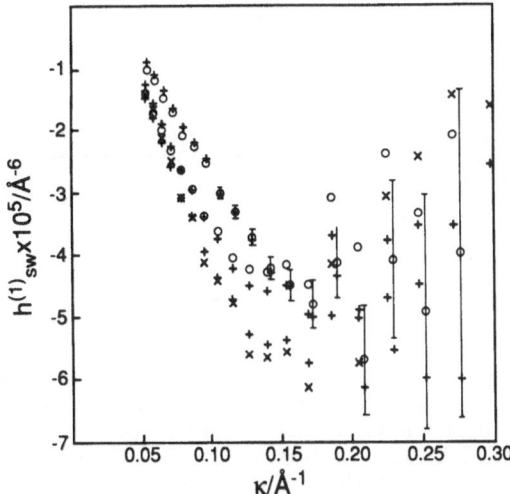

Fig. 6. Partial structure factors $h_{sw}^{(1)}$ for a) $C_{10}TAB(\circ)$, b) $C_{14}TAB(\triangle)$ and C) $C_{18}TAB(\times)$. The fittings using Eq. (11) are indicated as (+) using a) 6.1, b) 8.0 and c) 8.4 Å. The error bars are only marked for $C_{10}TAB$

Table 4. Structural parameters obtained from kinematic analysis

Surfactant	$\tau/Å$	$A/Å$	$\tau_h/Å$
$C_{10}TAB$	17.5 ± 2	55 ± 3	8.0 ± 1
$C_{14}TAB$	19.0	48	10.0
$C_{18}TAB$	18.2	43	10.5

Surfactant	$\delta_{as}/Å$	n	f_c
$C_{10}TAB$	6.1 ± 0.5	7 ± 1	0.38 ± 0.03
$C_{14}TAB$	8.0	7	0.34
$C_{18}TAB$	8.4	6	0.32

ness is comparable with that of the fully extended chain and for $C_{18}TAB$ the layer is no thicker but is more dense.

References

1. McDermott DC, Lu JR, Lee EM, Thomas RK, Rennie AR (1992) Langmuir 8:1204
2. Simister EA, Lee EM, Thomas RK, Penfold J (1992) J Phys Chem 96:1373
3. Born M, Wolf E (1975) Principles of Optics, 5th Ed, Pergamon, Oxford
4. Crowley TL (1984) D Phil Thesis, University of Oxford
5. Crowley TL, Lee EM , Simister EA, Thomas RK (1991) Physika B 173:143
6. Crowley TL, Physica B (in press)
7. Voeks JF, Tartar HV (1955) J Phys Chem 59:1190
8. Murkerjee P, Mysels KJ (1971) Critical Micelles Concentrations of Aqueous Surfactant Systems; National Bureau of Standards: Washington, DC
9. Simister EA, Thomas RK, Penfold J, Aveyard R, Binks BP, Cooper P, Fletcher PDI, Lu JR, Sokolowski A (1992) J Phys Chem 96:1383
10. Lee EM, Thomas RK, Penfold J, Ward RC (1989) J Phys Chem 93:381
11. Lu JR, Simister EA, Lee EM, Thomas RK, Rennie AR, Penfold J (1992) Langmuir 8:1837
12. Tanford CJ (1972) J Phys Chem 76:3020

there must be a significant proportion of alkyl chain immersed in water. For the three C_nTABs studied here, the immersion of the alkyl chain decreases as the chain length increases.

The total thickness of $C_{10}TAB$ at the CMC is 19 ± 2 Å, whilst for the other two longer chain C_nTABs, the value is 22 ± 2 Å. Comparison of the thicknesses of the measured and fully extended alkyl chains suggests that the $C_{10}TAB$ monolayer is staggered. This picture is similar to our earlier observation [10]. For $C_{14}TAB$ the measured thick-

Authors' address:

Dr. J. R. Lu
Physical Chemistry Lab.
South Parks Road
Oxford OX1 3QZ, United Kingdom

Progress in Colloid & Polymer Science   Progr Colloid Polym Sci 93:98—102 (1993)

Layer-by-layer deposited multilayer assemblies of polyelectrolytes and proteins: from ultrathin films to protein arrays

J. D. Hong, K. Lowack, J. Schmitt, and G. Decher

Institut für Physikalische Chemie, Johannes Gutenberg-Universität, Mainz, FRG

Abstract: We have recently introduced a new method of creating ultrathin films of polyelectrolytes based on the electrostatic attraction between opposite charges. Multilayer assemblies are adsorbed in a layer-by-layer fashion from aqueous solutions of the polymers. The total film thickness can easily be adjusted by varying the ionic strength of the solution. Here, we report on the temperature stability and the water content of the multilayer assemblies. Furthermore, we have extended our concept to the incorporation of protein layers into films of synthetic polyelectrolytes. The well established system biotin/streptavidin was used to construct such multilayers, also by biospecific recognition. Adsorption of streptavidin onto previously photostructured precursor films leads to the deposition of the protein on selected areas on the substrate. The films were investigated by small-angle x-ray scattering (SAXS), Fourier Transform Infrared Spectroscopy (FTIR), and by fluorescence microscopy.

Key words: Ultrathin films — polyelectrolytes — adsorption — streptavidin — self-assembly — photo-structuring — biospecific recognition

Introduction

During the last decade much attention has been paid to ultrathin organic films [1—3]. These films are built-up of one or more layers, where the individual layers may consist of different molecules. The large number of molecules with varying functions and structures which can be used for the build-up of the layered assemblies leads to the possibility of a well-directed planning and realization of tailor-made systems with specific properties and justifies the scientific and technological interest in ultrathin films. Examples are applications as in surface-modification, as biosensors or in integrated optics [1—3].

One of the most promising approaches is the deposition of the individual layers by adsorption from solution because, among other advantages, there are no principle limitations as to substrate size or topology.

We have recently introduced a new method of creating ultrathin films which is based on the electrostatic attraction between opposite charges [4—8].

Consecutively alternating adsorption of anionic and cationic polyelectrolytes from aqueous solution leads to the formation of multilayer assemblies. The process of the multilayer build-up is depicted in Fig. 1 where y symbolizes negatively charged ionic groups, x denotes positively charged ionic groups and L is a chemisorbed layer of aminobutyldimethylmethoxysilane. The layer-by-layer deposition is carried out as follows. A solid substrate with a positively charged surface is immersed in a solution containing the anionic polyelectrolyte polystyrenesulfonate (PSS) and a layer is adsorbed by reversing the surface charge (step A). After rinsing with water the substrate is immersed in the solution of the cationic polyelectrolyte polyallylamine hydrochloride (PAH). Again, a polyelectrolyte layer is adsorbed and the original surface charge is restored (step B). By repeating both steps in a cyclic fashion (A, B, A, B, ...) alternating multilayer assemblies are obtained.

Knowledge about the thermal and aging behavior of the multilayer assemblies is important for potential applications as well as for a better understan-

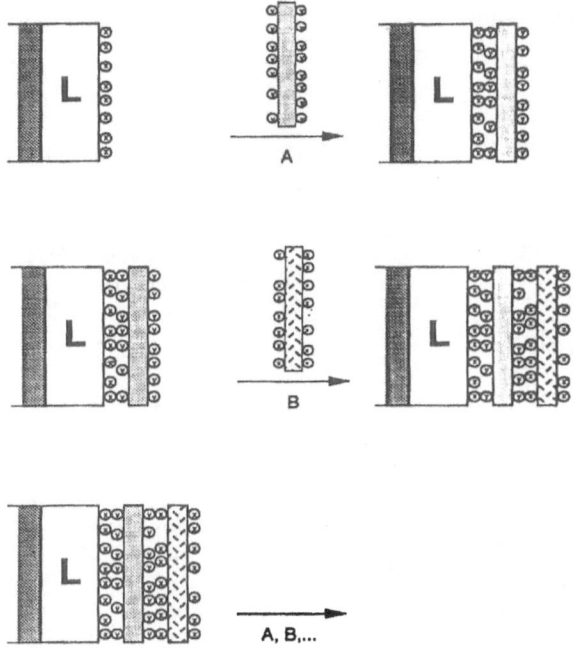

Fig. 1. Building-block approach for the construction of multilayer assemblies. Details are explained in the text.

ding of film stability and dynamics. Since the polyelectrolytes are adsorbed from aqueous solutions, it is interesting to investigate if the multilayer assemblies contain water. In the first part of this report we will discuss the thermal behavior and stability of the self-assembled films.

In the second part, we report on the build-up of sandwiched protein multilayers by biospecific recognition process. Therefore, we have extended our layer-by-layer adsorption technique to functionalized polyelectrolytes. As a first and extremely simple example, poly-l-lysine was biotinylated and subsequently used for the construction of multilayer assemblies by biospecific recognition between biotin groups and streptavidin. The binding constant of the two entities is very high ($k = 10^{15}\ M^{-1}$) which makes the process essentially irreversible [9].

Again the process of multilayer build-up is described in Fig. 1. In the case of protein containing films x symbolizes biotin groups, y symbolizes the binding pockets of the streptavidin and L denotes a 309 Å thick precursor film composed of nine layers of PSS and PAH with a top layer of biotinylated poly-l-lysine (PLB) on a float glass substrate. In a first step a FITC-labeled streptavidin layer (SA) was

deposited on the surface by the biospecific reaction (step A). Then a PLB layer was adsorbed (step B). The last two adsorption steps are then repeated in a cyclic fashion (A, B, A, B, ...) and multilayer assemblies containing sandwiched protein layers are obtained. Finally, we report on the area-selective adsorption of streptavidin after photo-structuring of a precursor film.

Materials and methods

Polystyrenesulfonate (sodium salt, Mr = 100 000) (PSS) was obtained from SERVA, polyallylamine (hydrochloride, Mw = 50 000—65 000) (PAH) was obtained from Aldrich. Both materials were used without further purification. Biotinylated poly-l-lysine was prepared according to standard procedures as already described [10]. Streptavidin was FITC-labeled according to [11].

For all experiments aminobutylsilanized substrates (fused quartz and silicon) were used. The preparation of the substrates is described elsewhere [4].

The adsorption of the polyelectrolyte layers was carried out as follows. A substrate was immersed in 10 ml of an aqueous polyelectrolyte solution for 20 min at ambient temperature. The compositions of the solutions are given in Table 1. After the deposition of each polyelectrolyte layer the substrate was rinsed three times for 1 min in pure water and the surface of the film was blown dry in a stream of nitrogen.

The streptavidin layer was adsorbed from $1*10^{-6}$ molar solution of FITC-labeled streptavidin.

The multilayer assemblies were characterized by small-angle x-ray scattering (Siemens D-500 powder diffractometer using copper $K\alpha$-radiation with a wavelength of 1.541 Å, data aquisition via a DACO-MP interface connected to a personal computer).

The FTIR-spectra were recorded with a Bruker IFS 48 spectrometer in transmission geometry. As a reference a pure silicon-wafer was used.

For the deposition of protein arrays a precursor film composed of seven layers PSS and PAH and a top layer of PLB was used. This film was irradiated through a copper mask from a distance of about 3 mm with UV-light of a pen-ray lamp. After irradiation for 60 min the substrate was immersed in the streptavidin solution for 30 min, washed and dried. The protein arrays were characterized with fluorescence microscopy (Zeiss Axioplan).

Table 1. Composition of solutions used for construction of different films

Experiments	PSS	PAH
Long-term annealing	$1.3 \cdot 10^{-2}$ monomol/l PSS $1.0 \cdot 10^{-2}$ mol/l HCl 0.93 mol/l $MnCl_2$	$1.3 \cdot 10^{-2}$ monomol/l PAH $1.0 \cdot 10^{-2}$ mol/l HCl
Release of water	$9.7 \cdot 10^{-3}$ monomol/l PSS $1.0 \cdot 10^{-2}$ mol/l HCl 0.1 mol/l $MnCl_2$	$9.7 \cdot 10^{-3}$ monomol/l PAH $1.0 \cdot 10^{-2}$ mol/l HCl
Precursor films for protein experiments	$1.5 \cdot 10^{-2}$ monomol/l PSS $1.0 \cdot 10^{-2}$ mol/l HCl 0.5 mol/l $MnCl_2$	$1.0 \cdot 10^{-2}$ monomol/l PAH $1.0 \cdot 10^{-2}$ mol/l HCl 2.0 mol/l NaBr

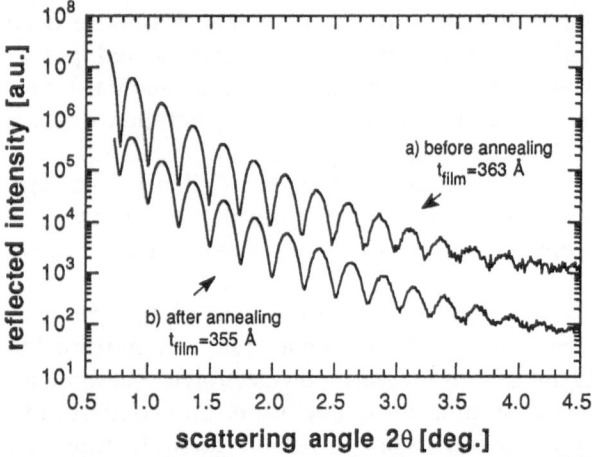

Fig. 2. X-ray reflectivity curves of a film consisting of 20 layer pairs of PSS/PAH a) before and b) after annealing for 3 h at 190°C. The calculated film thicknesses are indicated in the figure

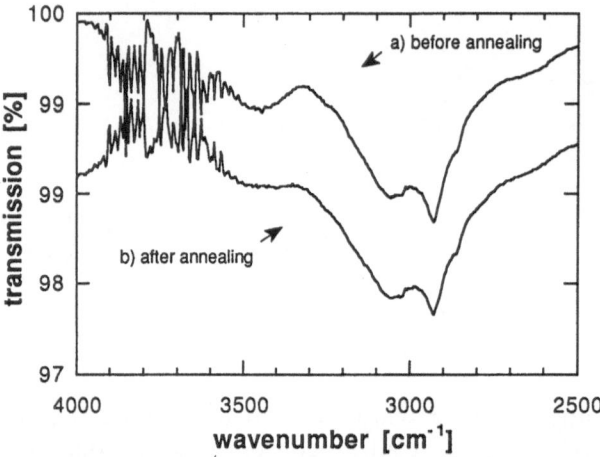

Fig. 3. FTIR-spectra of a film consisting of 20 layer pairs of PSS/PAH a) before and b) after annealing for 3 h at 190°C

were obtained for many different samples. The spectrum after the annealing process shows a marked decrease of the band around 3450 cm⁻¹. From quartz substrate. The film thickness was determined by SAXS and found to be 450 Å. After ageing for 4 month in air, the film thickness was increased by 5% which might be due to a take up of water and / or a rearrangement of the polymers in the film.

For the investigation of the thermal stability of the films, they were annealed for at least one week at temperatures of 104°C, 154°C and 190°C. After cooling down to room temperature, the film thickness was determined immediately. After annealing at 104°C and 154°C the film thickness was unchanged within experimental error. This demonstrates the good thermal stability of the films. In contrast annealing at 190°C leads to a 13% decrease of the film thickness. In order to verify whether this decrease in thickness is connected

Results and discussion

Thermal behavior

A multilayer assembly consisting of 19 individual layer pairs of PSS/PAH was built-up on a fused with a loss of bound water, a multilayer assembly consisting of 20 individual layer pairs of PSS and PAH was constructed as described above. The film thickness was determined by SAXS (Fig. 2 trace a) and a FTIR-spectrum was recorded (Fig. 3, spectrum a). This film was then annealed for only 3 h at a temperature of 190°C. After cooling down to room temperature SAXS- and FTIR-spectra were recorded again (Fig. 2, trace b), (Fig. 3, spectrum b).

The FTIR-spectrum before annealing shows a broad band centered around 3450 cm⁻¹ which is typical for water. The oscillations in the range of 3600 to 3900 cm⁻¹ are due to water-vapor in the atmosphere of the spectrophotometer. Similar results

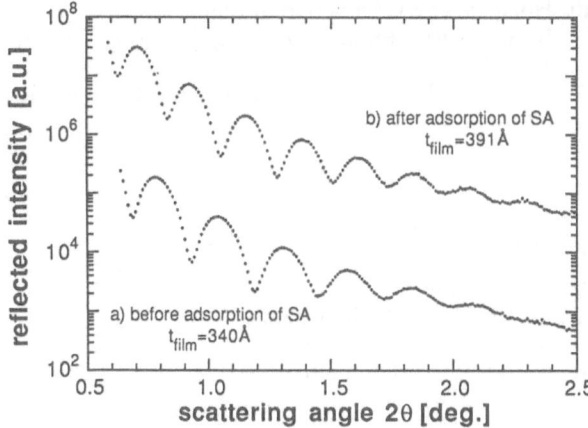

Fig. 4. X-ray reflectivity curves of a precursor film with a PLB surface a) and the subsequently adsorbed streptavidin layer b)

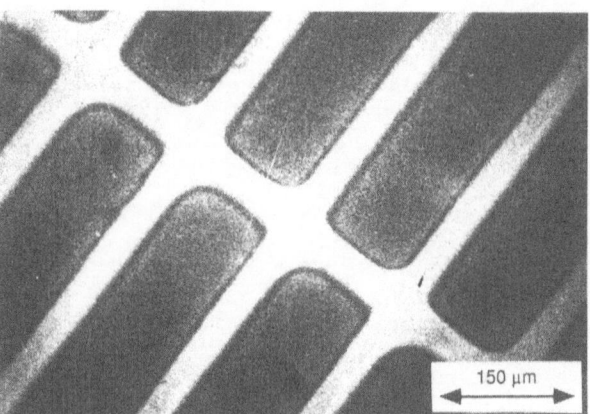

Fig. 5. Fluorescence micrograph of FITC-labeled streptavidin monolayer specifically adsorbed on a laterally photostructured precursor film. Dark areas represent photodesorbed film, bright areas correspond to the intact biotinylated surface, a prerequisite for streptavidin binding

the corresponding SAXS-spectra a decrease in film thickness of approximately 2% after annealing is observed. This shows that the decrease in film thickness is connected with a loss of bound water.

Protein multilayers and protein arrays

As already mentioned in the introduction, a polyelectrolyte precursor film was used for the construction of protein containing films. Film growth was monitored by x-ray reflectivity and an average thickness of 51.9 Å was determined for each streptavidin layer which is in good agreement with the

dimensions of the protein. Figure 4 shows an example for x-ray reflectivity curves before (trace a) and after adsorption of a streptavidin layer (trace b) on the underlying PLB layer.

UV-irradiation of layer-by-layer deposited polyelectrolyte films leads to a destruction of the film surface and a decrease of film thickness. It can be used to photostructure polyelectrolyte films in a similar way as reported in [12]. Figure 5 shows a fluorescence micrograph of a precursor film with a top layer of PLB which was UV-irradiated through a copper mask. Subsequently, FITC-labeled streptavidin was adsorbed from solution. The areas which were protected from UV-light show the strong fluorescence of the selectively adsorbed protein, because only there the biotinylated surface remained intact.

Summary and conclusions

Multilayer assemblies of PSS deposited from $MnCl_2$ solution and PAH contain bound water, which is lost at temperatures above 150°C. They possess a good thermal stability and the loss of water does not destroy the films. We have demonstrated that multilayer build-up is also possible by biospecific recognition. The recorded SAXS diffractograms show well developed interference fringes from which the thickness of each individual protein layer within the multilayer film was determined to be 51.9 ± 5.5 Å. Photostructuring of the films opens the possibility to adsorb proteins at selected areas of the substrate.

Acknowledgement

We thank the "Bundesministerium für Forschung und Technologie" for financial support, Wacker GmbH, and Boehringer Mannheim for the generous gift of the silicon substrates and streptavidin. We are grateful to H. Riegler and A. Leuthe for help with the x-ray measurements, to P. Quint for the introduction in photostructuring, to F. Klinkhammer for help with and to R. Stadler for the use of the FTIR spectrometer, and to H. Möhwald for stimulating discussions.

References

1. Advanced Materials (1991) Special Issue: Organic Thin Films 3
2. Ulman A (1991) "An Introduction to Ultrathin Organic Films" Academic Press Inc.

3. Swalen JD, Allara DL, Andrade JD, Chandross EA, Garoff S, Israelachvili J, McCarthy TJ, Murray R, Pease RF, Rabolt JF, Wynne KJ, Yu H (1987) Langmuir 3:932
4. Decher G, Hong JD (1991) Macromol Chem Macromol Symp 46:321
5. Decher G, Hong JD (1991) Ber Bunsenges Phys Chem 95:1430
6. Decher G, Hong JD, Schmitt J (1992) Thin Solid Films 210/211:831
7. Europ Patent No.: 0472990 A2
8. Decher G, Schmitt J (1992) Progr Colloid Polym Sci 89:160
9. Green NM (1975) Protein Chem 29:85
10. Hong JD (1991) Dissertation Mainz
11. Nargessi RD, Smith DS (1986) Methods Enzymol 122:67
12. Möhwald H, Höhne U, Quint P (1991) Polymer Journal 23:583

Authors' address:

Dr. Gero Decher
Institut für Physikalische Chemie
Johannes Gutenberg-Universität
Welder-Weg 11, 55128 Mainz, FRG

Progress in Colloid & Polymer Science   Progr Colloid Polym Sci 93:103—104 (1993)

Effects of surface fluctuations studied in a two-dimensional emulsion

E. van Faassen

Dept. of Molecular Biophysics Utrecht State University, The Netherlands

Abstract: In a soluble model we account for the effects of geometrical shape fluctutations on the equilibrium properties of a polydisperse two-dimensional emulsion. The emulsion consists of closed loops moving in a plane. Interactions within a droplet include bending rigidity and spontaneous curvature. Interactions between droplets are omitted. The free energies of the droplets take rigorous account of all possible droplet shapes. After a grand canonical summation the droplet size distribution is found for arbitrary temperature. Calculations show that the fluctuation dominated regime extends to temperatures far lower than expected from a mean field calculation.

Key words: Emulsions — entropy — polydispersity — surface fluctuations

Recent years have seen intense theoretical and experimental interest in the equilibrium properties of amphiphilic systems, emulsions, and other realizations of large interfaces between two separate phases [1]. Special attention is devoted to the role of geometrical fluctuations of the interface for two principal reasons: First, many amphiphilic systems exhibit low or vanishing surface tension and interface bending energies of the order of thermal excitations at room temperature, allowing deformations of the interface to occur. Second, in view of the low dimensionality of the interface and the embedding space, the effects caused by surface fluctations should be far from negligible. In particular, the equilibrium thermodynamic properties of the interface should be strongly affected.

Theoretically, this problem has been mostly studied in planar configurations of unbounded sheets. It was concluded [2] that the microscopic parameters of the interface Hamiltonian (like the bending rigidity) are strongly dependent on the length scale under consideration, this renormalization being caused by the surface fluctuations taking place on a smaller scales. Finite systems have been considered [3, 4] in time-consuming dynamics calculations on single isolated vesicles. An interesting analytical study [5] considered the implications of small distortions of spherical vesicles in a polydisperse emulsion. However, progress on the problem of incorporating large shape distortions has been notably absent. An obvious reason lies in the difficulty of accounting for the constraint of spherical topology, as this constraint introduces effective long range interactions between all molecules in the vesicle.

We therefore formulated a highly idealized model for a two-dimensional emulsion that rigorously accounts for arbitrarily large shape fluctuations and polydispersity. We treat the suspension as a collection of closed chains of arbitrary shape and size, moving in a plane. Each chain consists of a finite number of sticks of equal length, connected at the endpoints. For the Hamiltonian of such a chain we take the sum of the bending energies arising from the bending angles $\Delta_i \in [-\pi, \pi]$ at the ith vertex:

$$H_N (\Delta_1, ..., \Delta_N) = \kappa \sum_{i=1}^{N} 1 - \cos(\Delta_i - a) , \qquad (1)$$

where κ and a are the bending rigidity and spontaneous curvature respectively. From this N-chain Hamiltonian we define the free energy F_N in terms of a sum over all configurations consistent with the requirement that the chain be closed. However, we will not exclude the possibility that the droplet intersects itself (ghost interface):

$$e^{-\beta F_N} = \int_0^1 da \, e^{ia2\pi} \, Z_N (a) \qquad (2a)$$

$$Z_N(a) = \int_{-\pi}^{\pi} \prod_{i=1}^{N} \frac{d\Delta_i}{2\pi} \, e^{-\beta H_N} \, \delta\left(\sum_{i=1}^{N} \cos\sigma_i\right)$$

$$\cdot \, \delta\left(\sum_{i=1}^{N} \sin\sigma_i\right) e^{-ia(\sigma_N-\sigma_1)} \tag{2b}$$

$$\sigma_i \equiv \sum_{n=1}^{i} \Delta_n . \tag{2c}$$

The integration over the topological angle a serves to select those configurations that have unit winding number $\sigma_N - \sigma_1 = 2\pi$. This restriction serves as a crude approximation to the more desirable self avoidance property [6] lacking in our formulation. Using a formal analogy with the Berlin-Kac model [7] for ferromagnetic spin systems, the N-2 fold integral may be manipulated into a single integral over the trace of a transfer matrix $T(a, \rho)$:

$$Z_N(a) = \frac{1}{2\pi} \int_0^{\infty} \rho \, d\rho \, \text{tr} \, T^N(a, \rho) . \tag{3}$$

Therefore, the successive $Z_N(a)$ may be computed iteratively. The resulting free energies may subsequently be used in a grand canonical sum over all numbers of the various droplet sizes, giving the droplet size distribution:

$$P_N(\beta, \mu) = D^{-1} e^{-\beta(F_N-\mu N)} . \tag{4}$$

The chemical potential μ is adjusted to set the average droplet size $\overline{N}$.

In the predictions from this model effects from the shape fluctuations manifest themselves most prominently in the cross-over region $\beta\kappa \cong 1$. Whereas it is not surprising that these entropic contributions dominate at high temperatures, it is remarkable that this dominance extends down to temperatures where $\beta\kappa = 1$. This may be seen from Fig. 1, showing the temperature dependence of the spread ΔN in droplet sizes. The exact calculation shows a marked absence of any temperature effect as soon as $\beta\kappa < 1$, indicative of a total insensitivity of the size distribution to the details of the Hamiltonian.

The soluble model outlined here will be applied to a study of the correlations along the interfaces, and serve as a test case for a number of approximations found in the literature.

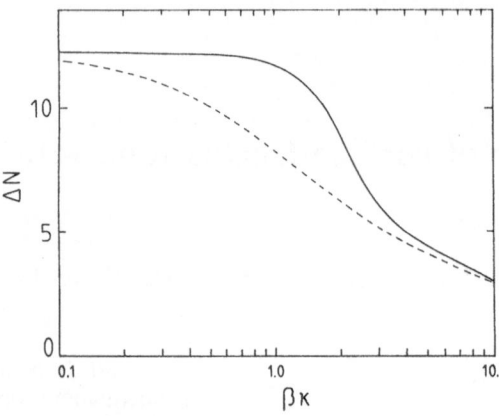

Fig. 1. Spread ΔN in droplet size distribution as a function of temperature. The solid curve is the exact result. The dashed curve gives the mean field approximation, where the configurational integral is replaced by the contribution from regular polygons only. The spontaneous curvature $a = .5$ and average droplet size $\overline{N} = 15$. The lack of temperature dependence of the exact curve shows that the fluctuation dominated regime extends from high temperatures right into the cross-over regime $\beta\kappa \cong 1$

References

1. cf., e.g., Nelson D, Piran T, Weinberg S (eds) (1989) Statistical Mechanics of Membranes and Surfaces. World Scientific, Singapore
2. Peliti L, Leibler S (1985) Phys Rev Lett 54:1690
3. Baumgartner A, Ho J (1990) Phys Rev A 41:5747
4. Leibler S, Singh R, Fisher M (1987) Phys Rev Lett 59:1989
5. Safran S (1982) J Chem Phys 78:2073
6. Kardar M, Nelson D (1988) Phys Rev A 38:966
7. Berlin T, Kac M (1952) Phys Rev 86:821

Author's address:

E. van Faassen
Dept. of Molecular Biophysics
Buys Ballot laboratory
Utrecht State University
3508 TA Utrecht, Netherlands

Progress in Colloid & Polymer Science

Progr Colloid Polym Sci 93:105—107 (1993)

Polydispersity of microemulsions droplets. Role of surfactant film bending elasticity

F. Sicoli[1]), D. Langevin[1]), L. T. Lee[2]), and M. Monkenbusch[3])

[1]) Laboratoire de Physique Statistique de l'Ens, Paris, France
[2]) Laboratoire Léon Brillouin, CE Saclay, Gif/Yvette, France
[3]) Institut für Festkörperforschung, KFA, Jülich, FRG

Abstract: The surfactant film bending elasticity can be described by a spontaneous curvature C_0 and two elastic constants K and $\bar{K}$, associated with the mean curvature and the Gaussian curvature, respectively. These parameters are very important in the determination of the structure of the dispersions stabilized by the surfactant (droplets or sponge-like structures). This is discussed in relation to former experiments. Recent neutron scattering determinations of the microemulsion droplet polydispersity are presented. It is shown that, in combination with results from ellipsometric experiments, these data can allow the estimation of $\bar{K}$.

Key words: Microemulsion — bending energy — Gaussian curvature elasticity — neutron scattering — droplet polydispersity

Microemulsions [1] are dispersions of oil and water stabilized by surfactant molecules. They are frequently made of droplets (oil in water (o/w) microemulsions, water in oil (w/o) microemulsions) surrounded by a surfactant monolayer and dispersed in a continuous phase (water or the oil respectively). Sponge-like structures where oil and water microdomains are multiply connected also exist, but they are less frequent than droplet structures; they are frequently discussed in the literature, because they correspond to a maximum in the solubilization power. When the composition of the medium is known, the droplet radius can be predicted quite accurately by using the following relation

$$R = \frac{3\phi}{c_s \sum} \, , \tag{1}$$

where ϕ is the dispersed volume fraction, c_s the number of surfactant molecules per unit volume, and $\sum$ the area per surfactant molecule. For sponge-like structures, this equation can be generalized to $\xi = 6\phi_0\phi_w/c_s\sum$, where ξ is the dispersion size, ϕ_0 and ϕ_w the oil and water volume fractions. These relations express the fact that virtually all the surfactant molecules sit at the oil-water interface and that each of them occupies a well defined area, independent of the composition. This is because, in order for the microemulsion to be thermodynamically stable, the surfactant monolayer must reduce the oil-water interfacial tension to about zero: its surface pressure must balance the tension of the bare interface, thus fixing the value of $\sum$. In the following, we will assume, as in recent microemulsion models, that the interfacial tension is exactly zero [2].

The type of microstructure is closely related to the sign of the spontaneous curvature of the surfactant layer C_0. In many droplet microemulsions, the magnitude of C_0 also determines the maximum droplet size (maximum solubilization power) [3]. This can be simply established by intoducing the surfactant film bending energy [4]:

$$F = \frac{1}{2} \, K \, (C_1 + C_2 - 2C_0)^2 + \bar{K} \, C_1 \, C_2$$

(per unit area), $\tag{1}$

where C_1 and C_2 are the two principal curvatures of the surfactant layer and K and $\bar{K}$ the mean and

Gaussian bending elastic constants [3]. By convention $C_0 > 0$ for aqueous dispersions and $C_0 < 0$ for reverse systems. The second term is present, because the system can change its topology: positive $\bar{K}$ favors saddle-splay structures as in bicontinuous cubic or sponge phases, while negative $\bar{K}$ favors lamellar or spherical structures. F is a surface energy usually negligible compared to the interfacial tension contribution. But in microemulsion systems, the interfacial tension being small or even zero, the bending energy becomes a very important term. For instance, the maximum droplet size can be shown to be close to:

$$R_m = \frac{2K}{2K + \bar{K}} \, R_0 \, , \qquad (2)$$

where $R_0 = C_0^{-1}$ [5]. When by increasing the dispersed phase volume fraction ϕ, R exceeds R_m, the system phase separates into two phases: a microemulsion with droplet size R_m and an excess phase (emulsification failure) [3]. In sponge-like structures, C_0 is close to zero, and the maximum dispersion size ξ_m is related to the persistence length ξ_k of the surfactant layer, which is an exponential function of K/kT (the finite persistence length results from the competition between the bending energy and the thermal energy kT; $k =$ Boltzmann constant, T absolute temperature) [1], [2]. In this case, when the system is above its solubilization limit, it separates into three phases: the microemulsion and two excess phases.

Experiments show that K is typically of order kT in microemulsion systems and that ξ_m is small ($\sim 100 - 500$ Å) and close to ξ_k, as predicted [6]. This also means that in droplet microemulsions, the droplets are substantially distorted due to thermal motion. This has been theoretically analyzed by describing the droplet deformation with an expansion of spherical harmonics [7]. It can then be shown that the main contribution comes from droplet size fluctuations, which are equivalent to what is usually called droplet polydispersity. When using a Schultz distribution [8] to describe this polydispersity, one finds at the emulsification failure limit ($R = R_m$):

$$\frac{\langle (R - R_m)^2 \rangle}{R_m{}^2} = \frac{1}{z + 1} = \frac{kT}{8\pi \, (2K + \bar{K})} \, . \qquad (3)$$

$\bar{K}$ plays a role here, because when changing the droplet radius at constant $\sum$, one changes the total

number of droplets and, thus, the topology. The polydispersity would, of course, also depend on the droplet surface tension γ_d. Recent spin-echo neutron-scattering experiments suggest that $\gamma_d = 0$, i.e., the surfactant film is truly incompressible (constant $\sum$) [9].

We have measured the droplet polydispersity by static small-angle neutron-scattering experiments in Saclay (PAXE spectrometer). We have studied ternary oil-water-nonionic surfactant mixtures, on which previous determinations of the modulus K have been done with ellipsometry [10]. The surfactants are alkyl polyethylene glycol either surfactants with alkyl chains of n carbon atoms and polar parts of m ethoxy groups: $C_{12}E_5$ and $C_{10}E_4$. They have been mixed with deuterated water and either octane or hexane. Their composition was such that, at a given temperature, a dilute microemulsion phase ($\phi \sim 1\%$) was in equilibrium with an excess phase; o/w systems were obtained at low temperature, w/o systems at high temperature. Details on the phase diagrams can be found in [10, 11]. Figure 1 shows two typical spectra, together with the fit with a

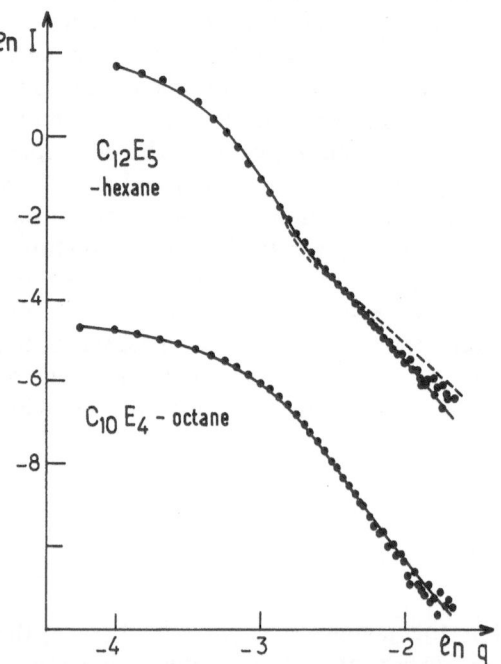

Fig. 1. Neutron scattered intensity I versus wave vector q (in Å^{-1}) for two different w/o microemulsions; $T = 34\,°C$. The lines are fits: upper full line $R_m = 63$ Å, $z = 20$, dotted line $R_m = 65$ Å, $z = 25$; lower line $R_m = 35$ Å, $z = 10$

Schultz distribution of spheres [8]. The fit includes the convolution with a Gaussian resolution function.

Although the quality of the spectra is very good, the accuracy on the polydispersity is limited. Within the accuracy, z is independent of the temperature (i.e., on the type and radius of the droplets) and on the nature of oil. The data analysis leads to $z = 25 \pm 5$ for $C_{12}E_5$ and $z = 12 \pm 8$ for $C_{10}E_4$. We have previously found that for $C_{10}E_4$-octane microemulsions, K was 0.52 kT, independent of temperature, and that for $C_{12}E_5$-hexane microemulsions K was larger : $K \sim kT$. When using eq. 3, this leads to $\bar{K}$ values respectively of $-kT$ and $-0.56kT$. A more complete data analysis will be presented elsewhere [12]. We plan to improve the accuracy of these determinations by performing experiments with different contrasts (deuterated oils) and to study other oils and surfactants of other chain lengths. Neutron spin-echo experiments are also projected.

In conclusion, we have shown that there is a close relationship between the surfactant film bending properties and the bulk structures in microemulsion systems. By analyzing the polydispersity of droplet systems and using independent measures of the bending modulus K, we have shown that it is possible to obtain at least a rough estimate of the Gaussian bending elastic modulus $\bar{K}$. Further investigations to improve the accuracy on these determinations are currently under way.

Acknowledgement

We are grateful to J. Teixera for his help during the neutron-scattering experiments and for many suggestions concerning the data analysis. We also thank S. A. Safran for making available to us his calculations prior to publication.

References

1. de Gennes PG, Taupin C (1982) J Phys Chem 86:2294—2304
2. Safran SA, Roux D, Cates ME, Andelman D (1986) Phys Rev Lett 57:491—493
   Cates ME, Andelman D, Safran SA, Roux D (1988) Langmuir 4:802—806
3. Safran SA, Turkevich LE (1983) Phys Rev Lett 50:1930—33
4. Helfrich W (1973) Z Naturforsch 28:693
5. Safran SA in "Modern Amphiphilic Systems" Ed Ben-Shaul A, Gelbart W, Roux D Springer (1993)
6. Binks BP, Meunier J, Abillon O, Langevin D (1989) Langmuir 5:415—421
7. Milner ST, Safran SA (1987) Phys Rev A 36:4371
8. Kotlarchiyk M, Chen SH (1983) J Chem Phys 79:2461
9. Huang JS, Milner ST, Farago B, Richter D (1987) Phys Rev Lett 59:2600—2603
   Farago B, Richter D, Huang JS, Safran SA, Milner ST (1990) Phys Rev Lett 65:3348—51
10. Lee LT, Langevin D, Meunier J, Wong K, Cabane B (1990) Prog Colloid Polym Sci 81:209—214
11. Kahlweit M, Strey R, Firman P, Haase D, Jen J, Schomäcker R (1988) Langmuir 4:499—511
    Kahlweit M, Strey R, Firman P (1986) J Phys Chem 90:671—677
12. Sicoli F, Lee LT, Langevin D, J Chem Phys, to appear

Authors' address:

D. Langevin
Laboratoire de Physique
Statistique de l'ENS
24 rue Lhomond
75231 Paris cedex 05, France

Progress in Colloid & Polymer Science   Progr Colloid Polym Sci 93:108—111 (1993)

A novel surface induced phase transition in normal alkane fluids

J. C. Earnshaw and C. J. Hughes

The Department of Pure and Applied Physics
The Queen's University of Belfast, Northern Ireland

Abstract: Light scattering from thermally excited capillary waves has been used to study normal liquid alkanes (15 to 18 carbon atoms in length) as a function of temperature. In all cases the frequency and damping of the surface waves exhibit well-defined discontinuities at a temperature near, but distinct from the melting point. Above that temperature the data behave as expected for a clean liquid surface; below it they are consistent with predictions for a fluid having a structured surface layer through which the surface excess viscoelastic modulus is negative in magnitude. The phenomena suggest the occurrence of a surface-induced phase transition, in which the conformational degrees of freedom of molecules close to the surface are inhibited.

Key words: Surface phase transition — alkane — capillary waves — light scattering

Introduction

The normal alkanes are among the most fundamental series of molecules, being important in both technological and scientific fields. These molecules, such as hexadecane $CH_3(CH_2)_{14}CH_3$, are simple flexible chain molecules, some 4 Å in thickness and roughly 20 Å in length. They exhibit a rich variety of phase transitions [1], which continues to yield novelties [2]. These transitions strongly affect the behavior of molecules incorporating alkane moieties, and thus have a wider significance. Most of the observed complexity of phases is confined to states below the melting points [1], where the materials are in the solid state. The present report concerns an investigation using light scattered by thermally excited capillary waves on free liquid surfaces wich reveals a new phase transition in the fluid state for normal alkanes of moderate chain lengths (between 15 and 18 carbon atoms in length). It appears to involve a surface induced reduction in molecular degress of freedom.

Methods

Before use the alkane liquids were purified by passage through an alumina column packed with silver nitrate [3]. The melting points of our samples agreed with literature values to within the precision of the thermometry ($\pm 0.05\,°C$), confirming the lack of contamination. Spectroscopic analysis by capillary gas chromatography suggested that any contaminants (present in trace quantities, $\ll 0.1\%$) were other normal alkanes of chain lengths comparable to the main constituents.

Heterodyne photon correlation of light scattered by thermally excited capillary waves of selected wavenumber (q) provided unbiased estimates of the wave frequency (ω_0) and damping (Γ) [4, 5]. The well-known dispersion equation for capillary waves on the free surface of a liquid [6]:

$$(\omega + 2\nu q^2)^2 + gq + \frac{\gamma q^3}{\rho} = 4\nu^2 q^2 \left(1 + \frac{\omega}{\nu q^2}\right)^{1/2},$$

$$(1)$$

relates the complex frequency ω ($= \omega_0 + i\Gamma$) to q. Here, γ is the surface tension, and ν ($= \eta/\rho$) and ρ are respectively the kinematic viscosity and density of the liquid.

Results

A typical set of data for one of the fluids studied (hexadecane) is shown in Fig. 1. For temperatures

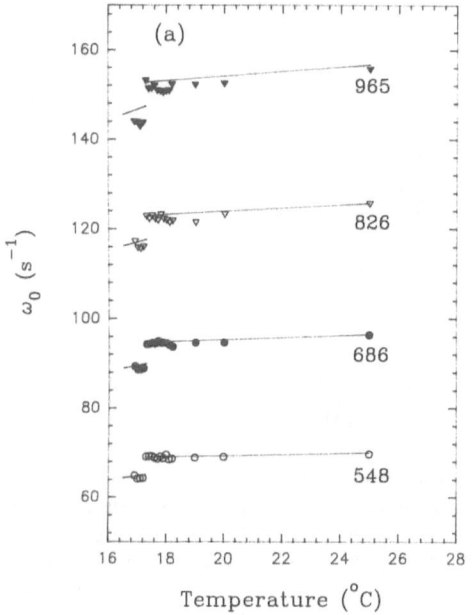

a

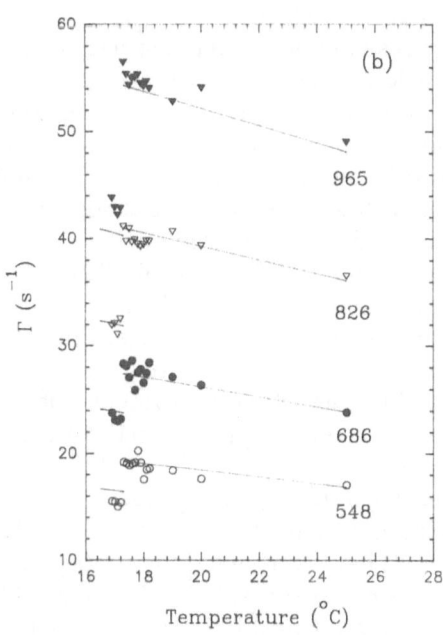

b

Fig. 1. The temperature dependence of the frequency (a) and damping (b) of capillary waves on *n*-hexadecane. Data are shown for several different wavenumbers, indicated in cm^{-1}. The lines are discussed in the text

above a critical value (17.3 °C for hexadecane) ω_0 and Γ are in excellent accord with the wave frequency and damping predicted from the dispersion

equation (Eq. (1)), the liquid properties being taken to have their accepted values. However, at 17.3 °C both properties undergo significant discontinuities, and at lower T lie well below the expected variation. This behaviour was found for all q studied. The changes in the light scattering data were entirely reversible, showing no signs of hysteresis, and occurred over a temperature range too narrow to be resolved by our thermometry.

Similar phenomena were observed at different temperatures for normal alkanes from pentadecane to octadecane, the only members of the homologous series whose melting points were easily accessible with our apparatus. While the changes in ω_0 and Γ were observed for hexadecane only on super-cooling (melting point = 18.2 °C), for the other materials they appeared above the melting point.

Estimates of the values of the apparent γ and η affecting the capillary waves can be derived by solving the dispersion equation using ω_0 and Γ at a given q as input data (assuming ρ to have its literature value). The results of this procedure are shown in Fig. 2 for hexadecane. Values found at different q agreed with each other within the errors. As expected, above 17.3 °C these data accord well with accepted values. The discrepancies below that temperature are quite large, that for γ increasing as T is reduced. Both tendencies were also observed for the other materials. Classical measurements of the surface tension [7] confirmed the positive gradient $d\gamma/dT$ at low temperatures, but did not show the step discontinuity evident in the light scattering values. The classically measured viscosity showed no change in behaviour, except at the melting point, for any material examined. Both frequency and damping of the capillary waves on pure alkane fluids thus depart from expectation based on the thermodynamic tension and viscosity at low temperatures.

These phenomena appear to be intrinsic to the alkane fluids, only being observed for purified materials. Various aspects of the data were incompatible with the presence of a contaminant film at the liquid surface [8]. The observed systematic dependences of various features upon alkane chain length [7] are difficult to understand if the effect arises from contamination.

Discussion

To understand these unprecedented results it is convenient to start with the classical tension meas-

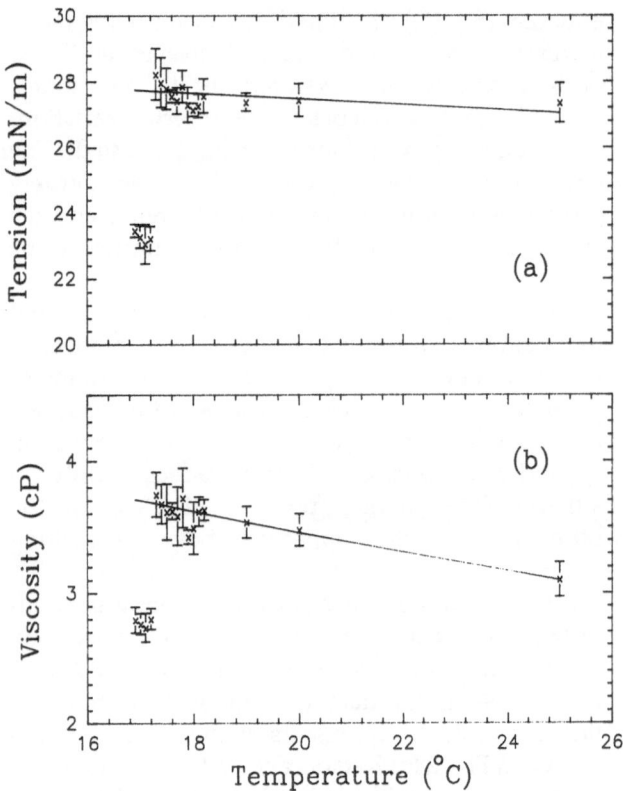

Fig. 2. Values of apparent tension and viscosity deduced from the observed propagation of capillary waves on the surface of hexadecane. The lines show the literature variations, extrapolated smoothly into the supercooled state from above the melting point

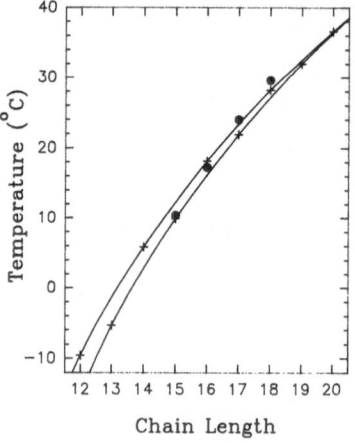

Fig. 3. The temperatures at which the new transition occurred (T_k, ●) for alkanes of different chain lengths. The lines indicate separately the variations of the melting points (+) of alkanes of even and odd chain lengths [1]

urements [7], which show a change in temperature gradient from negative to positive as the liquid is cooled through the temperature of the discontinuity in the light scattering data. The present systems each comprise a single component so that the tension is just the surface excess free energy [9]. Changes in its temperature dependence thus indicate a change of phase affecting the surface layers. The data thus indicate a new phase transition in the fluid phase of these alkanes, occurring at a temperature which we will call T_k (Fig. 3). For such one component systems the surface excess entropy density $S_s = -d\gamma/dT$ [9]. The present transition seems to be first-order in nature: S_s, while constant above and below T_k, shows a discontinuous change at that temperature [7].

The positive surface excess entropy density above T_k indicates the somewhat greater freedom of molecules near the surface compared to those in the bulk [9]. However, the negative S_s found below T_k implies a marked reduction in the possible molecular degrees of freedom in the surface layers. Below T_k the field exerted by the surface apparently suffices to restrict the molecular freedom, probably by inhibiting chain flexing etc. The transition thus appears to be surface-induced in nature, and we associate it with a cooperative reduction in the internal molecular degrees of freedom.

This new transition occurred for all alkanes examined. The observed values of T_k vary systematically with the chain length (n) of the alkane, falling significantly faster than the melting point as n decreases (Fig. 3). They show no evidence of the well-known odd-even effect apparent in the melting points [10].

Thus the alkane fluids studied possess a structured surface layer, differing from the bulk fluid. Tejero and Baus [11] have considered the propagation of surface waves on a viscoelastic fluid having such a structured surface. Considering only capillary waves, and restricting the viscoelastic behaviour to the surface layer alone (i.e. the bulk is simply a viscous liquid), their results reduce to making the simple substitution (Eq. (1)):

$$\gamma \rightarrow \gamma_{eq} + \gamma_s(\omega) + i\omega\gamma'(\omega) , \qquad (2)$$

in the dispersion equation (Eq. (1)). Here, γ_{eq} is the thermodynamic, classically measured tension and

$(\gamma_s + i\omega\gamma')$ represents the surface excess of the viscoelastic modulus governing shear normal to the surface plane. The elastic part of this surface excess transverse shear modulus is $\gamma_s(\omega)$, and the viscous part $\gamma'(\omega)$; both may be frequency dependent. To first order the well-known Kelvin and Stokes results become

$$\omega_0 = \sqrt{\frac{[\gamma_{eq} + \gamma_s(\omega)]\, q^3}{\rho}} \qquad (3)$$

and

$$\Gamma = 2\nu q^2 + \frac{\gamma'(\omega)q^3}{2\rho} \, . \qquad (4)$$

More exact predictions of the capillary wave propagation on such a surface can be found by numerical solution of the modified dispersion equation.

The effects observed in the light scattering experiments accord with the predictions of this theory. Equations (3) and (4) show that, to first order, the changes in ω_0^2 and Γ at the transition should scale as q^3, provided that the surface excess viscoelastic properties are essentially constant over the frequency range studied. The observed discontinuities do display such scaling [8]. Figure 1 shows that both ω_0^2 and Γ fall as T decreases through T_k, implying that $\gamma_s < 0$ and $\gamma' < 0$. At all q, the absolute values of the changes at T_k were very similar for all the alkane fluids studied. It bears emphasis that, as with the entropy density, the elastic and viscous properties of the fluid are actually everywhere positive; it is only the surface excesses of these quantities which are negative.

The magnitudes of γ_s and γ' can be estimated by comparing the observed changes in ω_0^2 and Γ with those predicted using the modified dispersion equation for assumed values of these quantities. In this way, we find $\gamma_s \approx -3.2$ mN/m, and $\gamma' \approx -2.6 \times 10^{-5}$ mN s/m, averaged over all q and all the alkanes. There is no clear evidence of any quantifiable frequency dependence. The lines in Fig. 1 show the temperature variations predicted by the modified dispersion equation using the known properties (γ_{eq}, η and ρ) of hexadecane (extrapolated below the melting point where necessary) and these values for the surface excess viscoelastic properties below T_k. The slight differences between these predictions and the data may arise for some small frequency dependence of the latter properties. Further analysis seems unwarranted at present.

The fact that the surface excesses of the viscous and elastic parts of the transverse shear viscoelastic modulus are negative, as for the entropy density, is as expected. In the one density van der Waal's theory [9] the variation of any quantity through the interface is imagined to follow that of the quantity whose density is used as the defining property (here entropy). Thus a negative surface excess for one quantity necessarily implies similar surface excesses for others.

Conclusions

A novel phase transition has been observed for normal alkane fluids between 15 and 18 carbon atoms in length. This transition involves the liquid surface, which apparently induces a reduction in the conformational freedom of molecules in a thin surface layer. The surface-induced phase is characterized by negative values of the surface excesses of the entropy density and of the viscoelastic modulus governing shear transverse to the surface phase.

Acknowledgement

One of us (C. J. H.) thanks the AFRC and Unilever Research for support in the form of a CSA studentship.

References

1. Reviewed in Small DM (1988) The Physical Chemistry of Lipids. Plenum, New York
2. Sirota EB, King HE, Hughes GJ, Wan WK (1992) Phys Rev Lett 68:492—5
3. Murray EC, Keller RN (1969) J Org Chem 34:2234—5
4. Earnshaw JC, McGivern RC (1987) J Phys D: Appl Phys 20:82—92
5. Earnshaw JC, McGivern RC (1988) J Colloid Interf Sci 123:36—42
6. Lamb H (1945) Hydrodynamics. Dover, New York, pp 625—628
7. Earnshaw JC, Hughes CJ (1992) Phys Rev A 46:R4494—6
8. Hughes CJ (1992) Ph.D Thesis, Queen's University of Belfast
9. Rowlinson JS, Widom B (1982) Molecular Theory of Capillarity. Clarendon Press, Oxford
10. Farkas A (1950) Physical Chemistry of the Hydrocarbons. Academic, New York, pp 318—325
11. Tejero CF, Baus M (1985) Mol Phys 54:1307—1324

Authors' address:

Prof. J. C. Earnshaw
Department of Pure and Applied Physics
The Queen's University of Belfast
Belfast BT7 1NN, Northern Ireland

Progress in Colloid & Polymer Science   Progr Colloid Polym Sci 93:112—117 (1993)

Kinetics of the gel-to-liquid phase transition of binary lipid bilayers using volume perturbation calorimetry*

R. L. Biltonen and Q. Ye

Departments of Biochemistry & Pharmacology, University of Virginia Health Sciences Center, Charlottesville, Virgnia, USA

Abstract: The relaxation kinetics of the gel-liquid crystalline transition of multilamellar vesicles (MLV) made of binary mixtures of dimyristoylphosphatidylcholine (DMPC), dipalmitoylphosphatidylcholine (DPPC) and distearoyl-phosphatidylcholine (DSPC) have been studied with volume perturbation calorimetry. The temperature and pressure relaxations following a volume perturbation were used to monitor the transition time-course. Data collected in the time domain were converted into and analyzed in the frequency domain using Fourier series representations of the perturbation and response functions. In binary phosphatidylcholine mixtures, the relaxation process consists of more than one exponential decay. The overall relaxation rate of a binary lipid system is increased relative to a single component lipid system. The mean relaxation rate of DMPC-DSPC MLV containing 6 mole% DSPC is nearly three orders of magnitude greater than that of DMPC MLV. But addition of 6 mole% DMPC into DSPC had an effect on the mean relaxation rate at least one order of magnitude smaller. Our data for DMPC-DSPC, DMPC-DPPC and DPPC-DSPC systems show that mixtures with a small amount of lipid with higher melting temperature, T_m, in a dominant amount of lipid with lower T_m have faster mean relaxation rates than those with a small amount of lipid with lower T_m in a dominant amount of lipid with higher T_m. Two relaxation time maxima were observed in the phase transition region of all three 1:1 phosphatidylcholine mixtures studied, even though the corresponding heat capacity function may have only one maximum. These two relaxation time maxima correspond to temperatures where the two maxima of the equilibrium heat capacity function occur for 1:1 DMPC-DSPC mixture, or the onset and completion edges of the phase transition for 1:1 DMPC-DPPC and DPPC-DSPC mixture. These results suggest that the long relaxation times (~ sec.) for MLV of pure lipid is not the result of bilayer-bilayer interactions; that binary lipid systems may exhibit dynamic lateral phase separation; and that diffusion may play a role in the dynamics of the relaxation process.

Key words: Lipid volume perturbation — kinetics — calorimetry — phase transition, temperature — phase transition, pressure

Introduction

Binary lipid systems are the simplest multi-component systems with which to model the biological membrane. The detailed bilayer structure and thermodynamics of binary mixture membranes provide information about the molecular interactions between unlike lipid pairs. This information is crucial for the eventual expansion of the interpretation of the model membrane to a more realistic many-component membrane [1]. Unfortunately, the phase transition properties of binary lipid mixtures are greatly complicated, both statically and dynamically, by the presence of the second lipid component [2, 3, 4].

In previous studies [5], we have found that the primary relaxation kinetics of a single component

*) Supported by Grants from NIH (GM37658), NSF (DMB 9005374), and ONR (N0014-88-K-03260)

phospholipid system can be well described using a single exponential decay. Essentially identical relaxation time profiles with a common maximum value of 2—4 s occurring near the gel-liquid crystalline transition temperature have been found for a homologous series of saturated phosphatidylcholines containing 14 to 18 carbon units in the diacyl chains. On the other hand, kinetic experiments with binary lipid mixtures are much more complex than single component systems and a greater challenge to methodology, experimental design and interpretation. Difficulties with most existing techniques in studying binary lipid systems may be the main reason why little work on the phase transition kinetics of multicomponent bilayers has been reported. It is the unique feature of the volume-perturbation calorimeter to monitor the kinetic energetics of the transition that makes the kinetic experiments of binary lipid systems possible [6, 7]. In this communication, we will summarize the results of kinetic experiments with multilamellar vesicles (MLV) made of binary mixtures of dimyristoyl-phosphatidylcholine (DMPC), dipalmitoyl-phosphatidylcholine (DSPC) and distearoyl-phosphatidylcholine (DSPC) using volume-perturbation calorimetry. Detailed results of the entire series of this study will be reported elsewhere [4, 8].

Methods and materials

Kinetic studies were conducted with a home-built volume-perturbation calorimeter as described previously [6, 9]. In brief, the instrument uses a small volume change generated by a stack of piezoelectric crystals to perturb the equilibrium poise of a physical-chemical process. The temperature and pressure of the sample are measured in the time domain and then converted into the frequency domain in order to obtain the power spectrum of the heat capacity function, from which the relaxation properties of the sample can be characterized. Phosphatidylcholines (DMPC, DPPC and DSPC) were purchased from Avanti Polar Lipids (Birmingham, Alabama). The dispersions of binary lipid mixtures were prepared and characterized as described elsewhere [5, 7].

Results

Figure 1 shows the frequency-dependent relaxation amplitude, $A(\omega)$, for pure DMPC and a

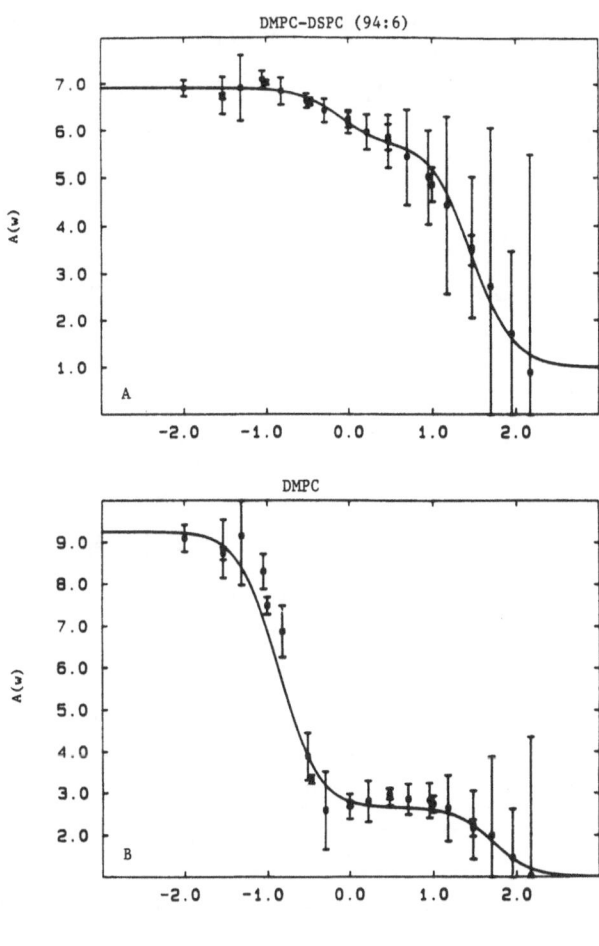

Fig. 1. The relaxation amplitude, $A(\omega)$, as a function of $\log_{10}$ of the perturbation frequency for A) 94:6 DMPC-DSPC, and B) DMPC at their respective T_m. The ordinate is in units of the response due to water at the respective temperature. The solid line in each graph represents the best fit of the data to a sum of multiple (two for DMPC and three for DMPC-DSPC) exponential decay processes of dimensionality one, the fastest relaxation process containing the baseline Joule-Thompson effect (Ye et al., 1991). The standard errors in the data are indicated by the vertical bars. Note that the frequency spectrum of 94:6 DMPC-DSPC is significantly shifted toward the high-frequency side relative to that of DMPC, indicating a shorter relaxation time in the DMPC-DSPC mixture than in DMPC

DMPC-DSPC mixture containing 6 mole% DSPC, at their respective transition temperature, T_m. The ordinate is normalized by the response due to water to correct for the instrumental effects which are not relevant to the relaxation dynamics of the lipid [6].

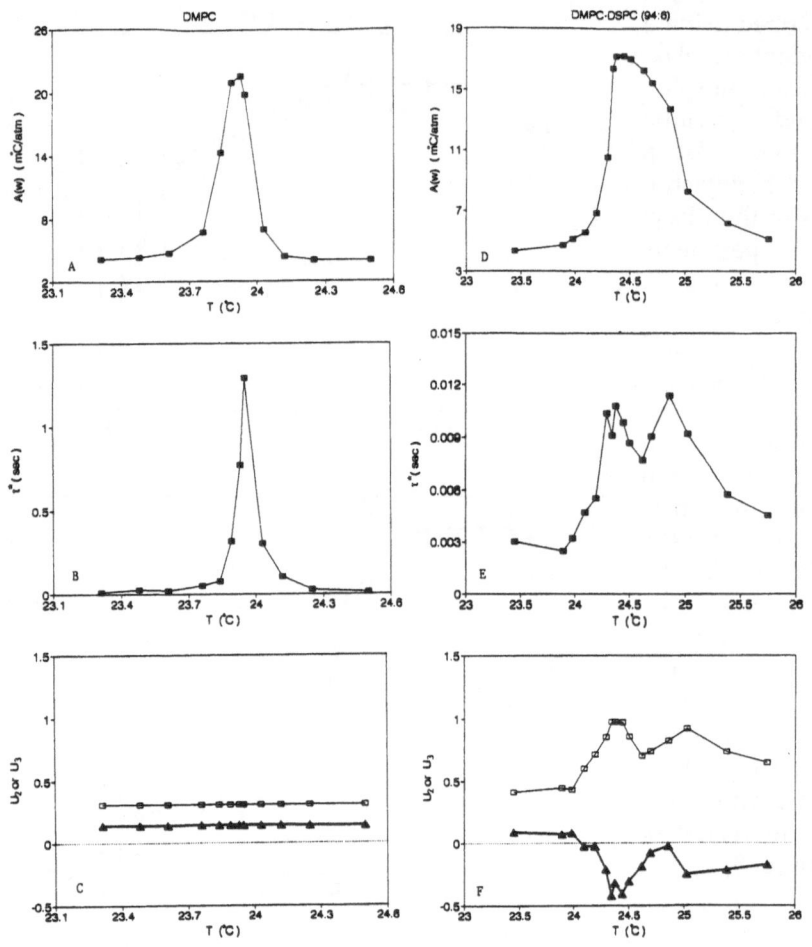

Fig. 2. (Upper) The relaxation amplitude at a perturbation frequency of 0.01 Hz, (middle) the mean relaxation time, and (lower) the second (□) and third (▲) moments of the derivative relaxation spectrum, as a function of the scaled temperature for pure DMPC (graphs A, B, C) and 94:6 DMPC-DSPC (graphs D, E, F). The scaled temperature is defined as $T = T_0\text{-}(P\text{-}1) \cdot (dT_m/dP)$, where T_0 is the actual mean temperature of the experiment, P the mean pressure, and dT_m/dP the pressure dependence of the lipid phase transition temperature. This scaling procedure reduces the equilibrium transition profile to a common temperature scale for all pressures (van Osdol et al., 1991)

It can be seen that the frequency spectrum of 94:06 DMPC-DSPC is shifted to the high frequency side relative to that of pure DMPC, indicating a faster average relaxation rate in the lipid mixture than in the pure lipid. Since the relaxation process of binary lipid mixtures cannot be well fit with a single exponential decay of any dimensionality [7], all data presented for the binary lipid systems have been analyzed as a sum of several independent, one-dimensional exponential decays (solid curves in Fig. 1).

In the phase transition region, the major relaxation process of pure DMPC can be characterized by a single relaxation time, with a pronounced maximum of ~ 2 s occurring near T_m [5]. The two major relaxation components characterizing the phase transition of the 94:6 DMPC-DSPC mixture at T_m (Fig. 1) have relaxation times of ~ 0.004 and 0.021 s, with respective relaxation amplitudes of ~ 0.6

and 0.3 relative to the total relaxation amplitude. These results show that incorporation of 6 mole% DSPC into DMPC causes an increase of the average relaxation rate of the phase transition by two to three orders of magnitude. On the other hand, addition of 6 mole% DMPC into DSPC has an effect on the relaxation rate at least one order of magnitude smaller [7]. Note that the dynamic response time of the thermistor in the sample cell of the volume-perturbation calorimeter is about 0.003—0.006 s so that the true relaxation time of 94:6 DMPC-DSPC could be overestimated.

It is difficult to define a single characteristic time to describe the overall trend of the entire relaxation process of a multi-step exponential decay. Therefore, we have developed a moment analysis procedure [4], in analogy to that of Sturgill and Biltonen [10], to generate characteristic parameters for complex relaxation processes. In this method, the first, second, and third moments of the derivative relaxation spectrum with regard to the $\log_{10}$ of the perturbation frequency are calculated and used to describe the multi-exponential decay process. These moments are defined as follows [4]:

first moment:

$$u_1 = \int y \cdot A'(\omega) \cdot dy / \int A'(\omega) \cdot dy$$

i^{th} moment $(i > 1)$:

$$u_i = \int [y - u_1]^i \cdot A'(\omega) \cdot dy / \int A'(\omega) \cdot dy \, ,$$

where ω = perturbation frequency, $y = \log_{10}(\omega)$, and $A'(\omega) = dA(\omega)/d\log_{10}(\omega)$.

The mean relaxation time is defined as:

$$\tau^* = 1/(2 \cdot \pi \cdot 10^{u_1})$$

With the above definitions, the first moment describes a mean relaxation frequency, which can be used to define a mean relaxation time; the second moment characterizes the distribution of the relaxation times or the fractional dimensionality of the relaxation process [11]; and the third moment characterizes the asymmetry of the distribution of the relaxation times. An example of the application of moment analysis to the study of binary lipid mixtures is given in Fig. 2. The mean relaxation time and the second and third moments of the derivative relaxation spectra of pure DMPC are those corresponding to the dominant slow relaxation step, analyzed in terms of a single exponential decay process of dimensionality one [11]. These results serve as a reference for comparison of the data obtained for the binary lipid mixtures. Thus, for a single relaxation process $u_2 = 0.31$ and $u_3 = 0.15$. For any complex process requiring more than one exponential decay, $u_2 > 0.31$. If $u_2 < 0.31$, the relaxation process can be described in terms of a single exponential decay of dimensionality > 1. For example, if the dimensionality $n = 2$, then $u_2 = 0.10$. With regard to u_3, the relaxation process is dominated by the faster steps if $u_3 < 0.15$ and by the slower steps if $u_3 > 0.15$ [4]. Note that further information about the relaxation characteristics of pure DMPC can be extracted from the moment analysis if it is done based on a better fit of the DMPC data using an exponential decay of dimensionality > 1 [11]. However, the reference for comparison of complex data established above still stands.

For both pure DMPC and 94:6 DMPC-DSPC, the relaxation amplitude profile at 0.01 Hz perturbation frequency (upper panels of Fig. 2) is very similar to the shape of the corresponding heat capacity curve of the transition [7]. The relaxation time (middle panels of Fig. 2) reaches a pronounced maximum value at a temperature near T_m and declines quickly as the temperature deviates from T_m. The increase of the second moment value of 94:6 DMPC-DSPC relative to those of pure DMPC (lower panels of Fig. 2) implies that more exponential decay terms are is required to describe the relaxation process. The negative sign of the third moment in the phase transition region of 94:6 DMPC-DSPC is an indication that the phase transition process is dominated by the faster steps [4]. These results indicate that addition of 6 mole% DSPC into DMPC results in a decrease of the maximal relaxation time relative to that of pure DMPC by more than two orders of magnitude and that the relaxation process is more complex.

In Fig. 3, the relaxation amplitudes (upper panels) of 1:1 DMPC-DSPC and DMPC-DPPC mixtures at 0.01 Hz perturbation frequency are shown as a function of the scaled temperature. At a perturbation frequency of 0.01 Hz, the shape of the relaxation amplitude as a function of temperature is very similar to the corresponding equilibrium heat capacity curve as obtained by differential scanning calorimetry [5, 7]. The frequency dependent relaxation spectra of these lipid mixtures were least-square fit to a sum of three exponential decays. The middle panels of Fig. 3 show the longest relaxation time components as a function of the scaled temperature for 1:1 DMPC-DSPC and DMPC-DPPC. A similar profile is obtained for the mean relaxation time (lower panels of Fig. 3). The kinetic characteristics of 1:1 DPPC-DSPC (data not shown; see [7]) are very similar to those of 1:1 DMPC-DPPC. The most interesting finding here is the occurrence of two relaxation time maxima in the phase transition region of all three 1:1 phosphatidylcholine mixtures studied, even though the corresponding heat capacity function may have only one maximum. This is true whether the relaxation process is characterized by the slowest step or by the mean relaxation time. Furthermore, the presence of bi-maximal relaxation times is also observed if the second slowest relaxation step is examined [7]. These two relaxation time maxima correspond to temperatures where the two maxima of the equilibrium heat capacity function occur for 1:1 DMPC-DSPC, or the onset and completion edges of the phase transition for 1:1 DMPC-DPPC and DPPC-DSPC.

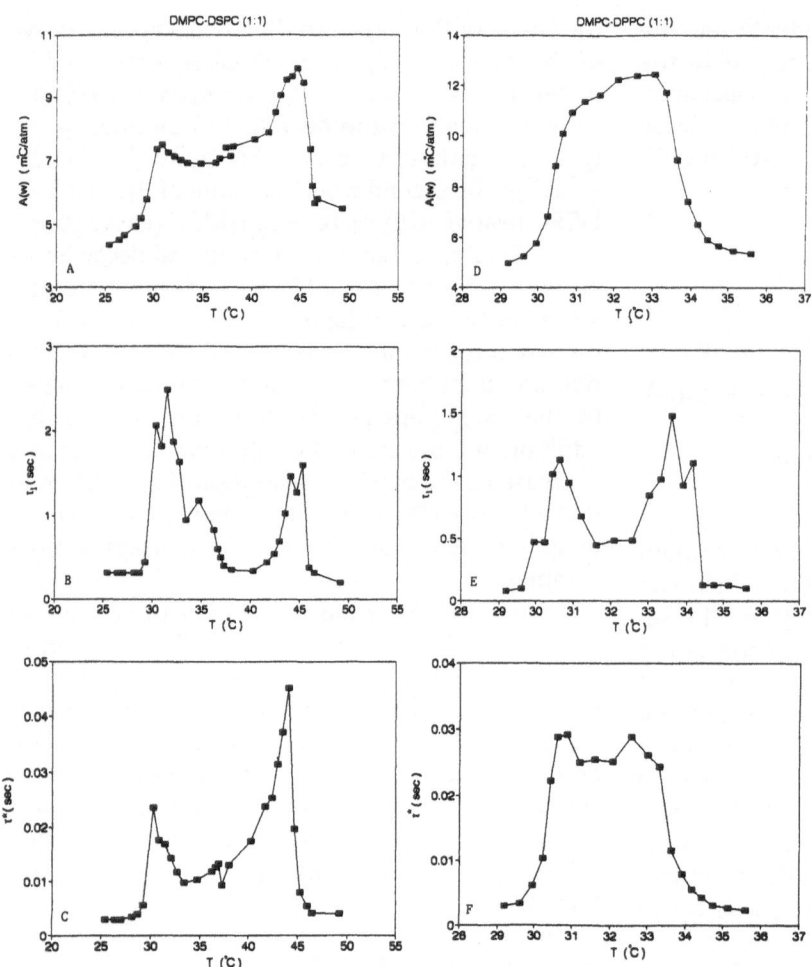

Fig. 3. (Upper) The relaxation amplitude at a perturbation frequency of 0.01 Hz, (middle) the longest relaxation time component obtained from the least-square fitting of the data by a sum of three exponential decays, and (lower) the mean relaxation time, as a function of the scaled temperature for 1:1 DMPC-DSPC (graphs A, B, C) and 1:1 DMPC-DPPC (graphs D, E, F) mixtures. Note that the 1:1 DMPC-DSPC data represent results obtained from two kinetic experiments which account for the small discontinuity in the amplitude vs. temperature profile (graph A) in the mid-range of the transition. Error bars have been omitted for clarity

Discussion

We have studied the phase transition kinetics of binary lipid mixtures with volume-perturbation calorimetry. Much more complicated relaxation characteristics were observed in a binary lipid system than a single component system. The relaxation kinetics in a binary lipid system can be described using a sum of two or more exponential decays rather than a single exponential decay as was found for a one-component system. A dramatic change in the relaxation kinetics of the gel-liquid crystalline transition of the bilayer membranes has been found upon addition of a small amount of second lipid component. The relaxation rate of 94:6 DMPC-DSPC was found to be almost three orders of magnitude greater than that of pure DMPC. All binary phospholipid mixtures (DMPC-DSPC, DMPC-DPPC and DPPC-DSPC) studied have faster relaxation rates of the gel-liquid crystalline transition than their compositional lipids (DMPC, DPPC and DSPC).

There are several major points we wish to make. First, the long relaxation time on the order of seconds for multilamellar vesicles of pure phosphatidylcholines [5] is unlikely to be result of bilayer-bilayer interactions since almost three orders of magnitude change in the relaxation time has been observed with addition of only 6 mole% DSPC into DMPC. This conclusion supports our earlier suggestion that the rate-limiting step is not solely determined by bilayer-bilayer interactions based on the kinetic results of dibucaine-DPPC mixture system [5]. In that case, addition of 1—5 mole% of dibucaine into DPPC causes a decrease of the maximal relaxation time by a factor of ~ 2 and an increase of the second moment of the derivative relaxation spectrum by a factor of > 3 [4, 12].

Second, the appearance of two relaxation time maxima in the phase transition region of all 1:1 binary lipid mixtures studied suggests the existence of dynamic lateral phase separation. Recall that in the one-component phosphatidylcholine systems, the maximal relaxation time, τ_{max}, has been found to approximately correspond to the temperature of the maximal heat capacity, $C_{p,\ max}(T)$, of the transition [5]. This relationship between τ_{max} and $C_{p,\ max}(T)$ has also been observed in some binary lipid systems with one of the two lipids as a minor component [7]. If the two maxima in the transition heat capacity of 1:1 DMPC-DSPC reflect the existence of two separated DMPC-rich and DSPC-rich phases due to nonideal mixing of DMPC and DSPC molecules, the kinetic behavior of the DMPC-rich and DSPC-rich phases would likely be similar to that of a binary lipid mixture with one of the two lipids as a minor component. In this case, it would not be surprising that the 1:1 DMPC-DSPC mixture shows two relaxation time maxima corresponding to the two peaks in the equilibrium heat capacity function. For 1:1 DMPC-DPPC and DPPC-DSPC mixtures, similar arguments could also be applied except that the two peaks in the transition heat capacity are presumably not well separated as in 1:1 DMPC-DSPC so that only a plateau in the transition region is observed.

Third, if the above argument of dynamic lateral phase separation is not applicable to the nearly ideal-mixing systems of DMPC-DPPC and DPPC-DSPC [2, 13, 14], an alternative interpretation for the appearance of bi-maximal relaxation times in 1:1 DMPC-DPPC and DPPC-DSPC mixtures would be that molecular diffusion plays a significant role in determining the phase transition kinetics of 1:1 binary lipid mixtures. This interpretation is also consistent with the kinetic data of 1:1 DMPC-DSPC. The question of whether dynamic phase separation, molecular diffusion, or both play major roles in the relaxation behavior of binary lipid systems will only be answered by further study.

References

1. Gennis RB (1989) Biomembranes. Molecular structure and function. Springer-Verlag, London 533 pp
2. Mabrey S, Sturtevant JM (1976) Proc Natl Acad Sci USA 73:3862—3866
3. Lee AG (1978) Biochim Biophys Acta 507:433—444
4. Ye Q, Biltonen RL (1992a) (in preparation)
5. van Osdol WW, Johnson ML, Ye Q, Biltonen RL (1991) Biophys J 59:775—785
6. van Osdol WW, Biltonen RL, Johnson ML (1989) J Biochem Biophys Methods 20:1—46
7. Ye Q (1992) Phase transition kinetics of multicomponent lipid membranes. Ph D dissertation. University of Virginia, Charlottesville VA
8. Ye Q, Biltonen RL (1992b) (in preparation)
9. Johnson ML, Winter TC, Biltonen RL (1983) Anal Biochem 128:1—6
10. Sturgill T, Biltonen RL (1976) Biopolymers 15:337—354
11. Ye Q, Van Osdol WW, Biltonen RL (1991) Biophys J 60:1002—1007
12. van Osdol WW, Ye Q, Johnson ML, Biltonen RL (1992) Biophys J 63:1011—1017
13. Knoll W, Ibel K, Sackmann E (1981) Biochemistry 20:6379—6383
14. Shimshick EJ, McConnell HM (1973) Biochemistry 12:2351—2360

Authors' address:

Professor Rodney L Biltonen
Dept. of Biochemistry & Pharmacology
University of Virginia
Health Science Center
Box 448
Charlottesville, VA 22908, USA

Progress in Colloid & Polymer Science

Progr Colloid Polym Sci 93:118—122 (1993)

Application of titration calorimetry to study binding of ions, detergents, and polypeptides to lipid bilayers

A. Blume, J. Tuchtenhagen, and S. Paula

Department of Chemistry, University of Kaiserslautern, FRG

Abstract: The binding of ions, the incorporation of detergents and polypeptides into lipid bilayers, and the CMC and heat of micellization of detergents were studied by titration calorimetry. The heat of dissociation of dimyristoylphosphatidic acid ($DMPA^- + OH^- \rightleftharpoons DMPA^{2-} + H_2O$) was investigated as a function of temperature covering the phase transition of singly and doubly charged DMPA. The intrinsic pK_0 for the dissociation was determined from the tiration curves applying the Gouy-Chapman theory. pK_0 decreases with temperature from ca. 6.2 at 11°C to 5.4 at 54°C. The temperature dependence of the dissociation enthalpy ΔH_{Diss} was combined with DSC data on the transition enthalpies ΔH_{Trans} for DMPA in its two ionization states to construct a complete enthalpy vs. temperature diagram. Titration calorimetry was also used to determine the CMC and the heat of micellization of the detergents SDS, octylglucoside, Na-cholate, and Na-deoxycholate. From the temperature dependence of the CMC and the heat of micellization the thermodynamic functions ΔG and ΔS were determined as a function of temperature. The interaction of Na-deoxycholate with lipid bilayers was studied at a temperature where the heat of mizellization was zero. Complex titration peaks were observed, indicating incorporation reactions on different time scales. The heats of reaction depended on the nature of the phospholipid. Titration calorimetry was also used to study the binding of melittin to lipid vesicles. The sign of the heat of incorporation depended on whether the lipid was in the gel or the liquid-crystalline state and on the chemical nature of the phospholipid.

Key words: Titration calorimetry — lipid bilayers — ion binding — detergents — peptide binding

Introduction

Differential scanning calorimetry has been widely used to study phase transitions in lipid bilayers [1, 2]. Reaction and titration calorimetry has only recently become applicable due to the availability of sensitive instruments [3]. We have used titration calorimetry in the titration and the batch mode employing the Microcal OMEGA titration calorimeter to test this method for its suitability to study a variety of reactions using lipid vesicles and detergents.

Heat of Dissociation of DMPA

Phospholipids such as dimyristoylphosphatidic acid (DMPA) show a pH-dependent ionization and

transition behavior, which has been studied in quite some detail [4, 5]. At lipid concentrations of ca. 1 mM the apparent pK for the dissociation $DMPA^- + OH^- \rightleftharpoons DMPA^{2-} + H_2O$ is ca. 10—10.5. This apparent pK can be shifted by changing the ionic strength and thus the extent of counter ion condensation at the lipid bilayer surface [5]. The dissociation of the second proton can be induced by the addition of NaOH. As the permeation of H^+/OH^- through lipid bilayers is within the dead time of the titration calorimeter, all head groups can be titrated. The observed heats of reaction ΔH_R will be the sum of several heat effects, namely, the dissociation enthalpy ΔH_{Diss}, the enthalpy of neutralization of water ΔH_{Neutr}, the heats of dilution of NaOH, and

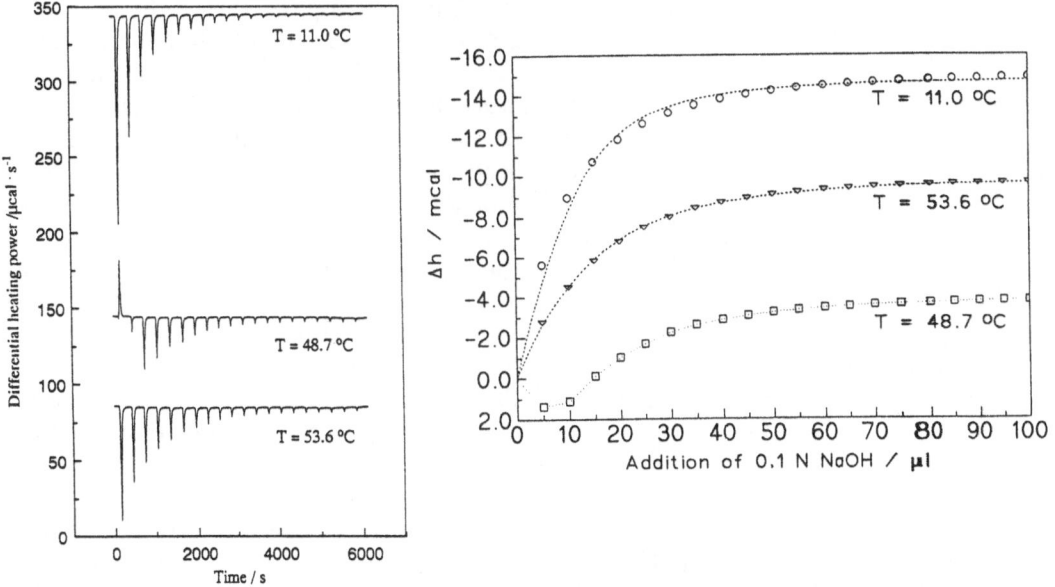

Fig. 1. (Left) Calorimetric heat signals observed by titrating a 1 mM DMPA$^-$ vesicle suspension with 5 μL additions of 0.1 M NaOH. (Right) Total integrated heat of reaction calculated from these curves after subtraction of the heat of dilution of NaOH. The dashed curves at 11° and 53.6 °C were calculated using the Gouy-Chapman theory with pK_0-values of 6.2 and 5.4, respectively [6]

additional enthalpic contributions caused by changes in head-group hydration, head group interactions, and rearrangements of the fatty acyl chains. The heat of dilution of NaOH and the heat of neutralization of water can be measured separately or taken from tabulated values. After subtraction of these from ΔH_R, the residual value for ΔH_{Diss} still contains the non-separable contributions. Figure 1 shows the calorimetric heat signals observed when titrating a 1 mM vesicle dispersion of DMPA$^-$ with 0.1 M NaOH [6]. The integrated heat of reaction is also shown in Fig. 1. The curves for 11 °C and 53.6 °C correspond to the dissociation of DMPA$^-$ in the gel state and liquid-crystalline state, respectively. At 48.7 °C the dissociation induces a transition from the gel into the liquid-crystalline state. This contribution leads to the effect that the first heat signals are endothermic. The titration curves can be simulated using the Gouy-Chapman theory [6, 7]. At low temperature the intrinsic pK_0 for the DMPA$^-$ dissociation is 6.2, whereas at higher temperature it was determined to 5.4. At intermediate temperatures no simulations could be performed due to the pH induced gel to liquid-crystalline phase transition. The total dissociation enthalpy was also determined by a batch experiment titrating a 0.01 M NaOH solution with a 1 mM

DMPA vesicle dispersion. The results of the temperature dependence of ΔH_{Diss} are shown in Fig. 2. The sudden jump in the dissociation enthalpy between 23 and 52 °C is caused by the OH$^-$ induced phase transition. Thus, in this temperature range ΔH_{Diss} includes the transition enthalpy ΔH_{Trans}. The dissociation enthalpies are positive above 52 °C and below 23 °C. In the intermediate temperature range the dissociation enthalpy increases with temperature, indicating a positive ΔC_p. This shows that during the transition into the doubly charged liquid-crystalline state more hydrophobic surface of lipid molecules is exposed to water [6]. The titration data could be combined with the differential scanning calorimetry data to construct a complete enthalpy vs. temperature diagram of DMPA in its two ionization states (see Fig. 2) [6, 8].

Heat of demicellization of detergents

The CMC and the heat of demicellization of detergents can be easily studied by titration calorimetry, providing the CMC is not too low, i.e., below $10^{-4} - 10^{-5}$ M. The experiments are carried out by diluting a micellar solution of the detergent in buffer or water to concentrations below the CMC. The observed enthalpic effects are then

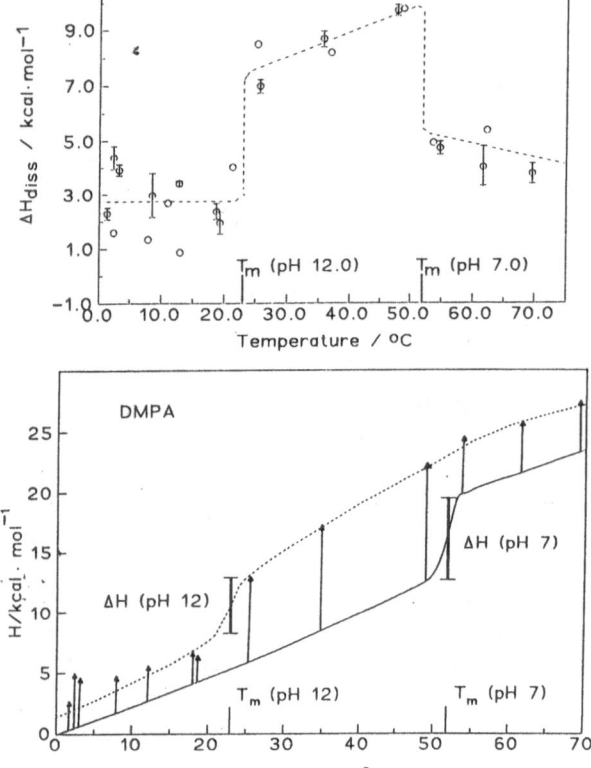

Fig. 2. (Top) Enthalpy of dissociation of DMPA$^-$ as a function of temperature. The sudden changes in the dissociation enthalpy at 23 and 52°C are caused by the pH-induced transition from a gel into a liquid-crystalline state in the intermediate temperature range. (Bottom) Enthalpy vs. temperature diagram for DMPA$^-$ and DMPA^{2-} constructed from DSC and titration calorimetry. The arrows are values determined by titration calorimetry, whereas the bars represent the transition enthalpies determined by DSC. The slopes are determined by the apparent molar heat capacities [6]

caused by dilution of micelles and demicellization. An example for this dilution experiment performed at two different temperatures is shown in Fig. 3 for the neutral detergent octylglucoside. The heat of demicellization is clearly temperature dependent. This is due to the hydrophic effect [9]. Exposure of hydrophobic groups of the chains is connected with a large positive ΔC_p, leading to an increase in ΔH with temperature. ΔH passes through zero at a temperature where the CMC has its minimum. The advantage of the calorimetric experiment is that the temperature dependence of the CMC and the heat of demicellization can be obtained from the

same experiment. Thus, the thermodynamic data ΔG and ΔS can be determined with more reliability, as ΔH is measured directly and not from the temperature dependence of the CMC. The first experiments have been performed with the detergents SDS, octylglucoside, Na-cholate, and Na-deoxycholate at different temperatures. The latter two detergents are extensively used in solubilizing and reconstituting membrane proteins. They are different from SDS and octylglucoside in that they have no alkyl chains. Thus, their micellar aggregation numbers are quite low and the CMC is not very well defined. ΔC_p of the heat of demicellization for these two detergents is much lower than for SDS and octylglucoside, as less hydrophobic surface is exposed when transferring the molecule into water. ΔH becomes zero at temperatures between 20° and 30°C for the negatively charged detergents SDS, cholate and deoxycholate. For octylglucoside this occurs at higher temperature. Figure 4 shows the thermodynamic functions for the four detergents as a function of temperature.

Interaction of deoxycholate with bilayers

Preliminary experiments have been carried out on the interaction of deoxycholate with lipid bilayers in the liquid-crystalline state using DMPC and a 1:1 mixture of POPC/POPG. Addition of deoxycholate to DMPC at 28°C leads to an enthalpic signal indicating the incorporation of the detergent into the lipid bilayer. The titration signals are biphasic with opposite signs arising from two contributions. The demicellization has a slightly positive ΔH at this temperature, whereas the incorporation has a negative heat of reaction. Addition of deoxycholate to POPC/POPG bilayers leads to positive titration signals which are twice as high as the heat signals observed on diluting deoxycholate. This indicates that there is apparently no or only little incorporation of the detergent into the lipid bilayer. Dynamic light scattering results on the vesicle size support this conclusion. The underlying reason is probably of electrostatic origin, as POPG is negatively charged.

Incorporation of polypeptides into bilayers

Polypeptides, such as melittin, incorporate readily into lipid bilayers when they are in the liquid-

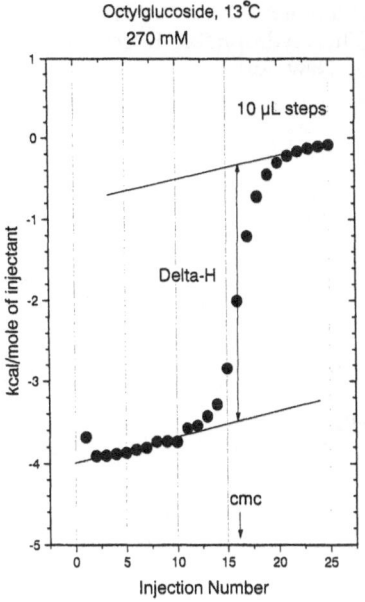

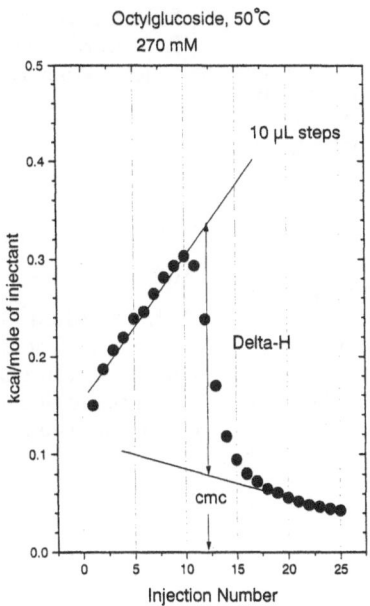

Fig. 3. Enthalpy of dilution of octylglucoside (270 mM) into 1.3 mL of water at two different temperatures. The sudden jump in the reaction enthalpy is caused by the effect that at high injection numbers only the heat of dilution of micelles is measured. The height of the jump determined at the CMC corresponds to the heat of demicellization.

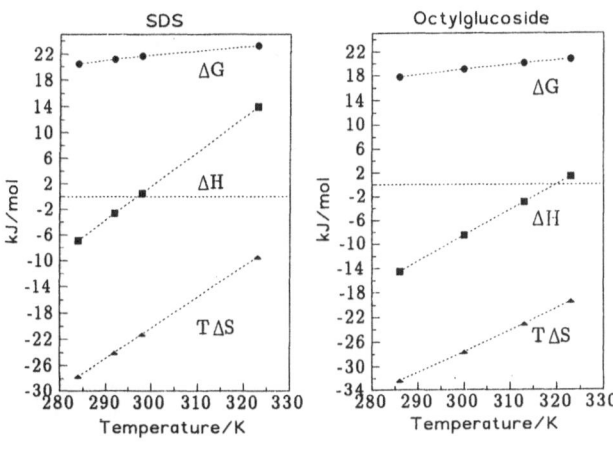

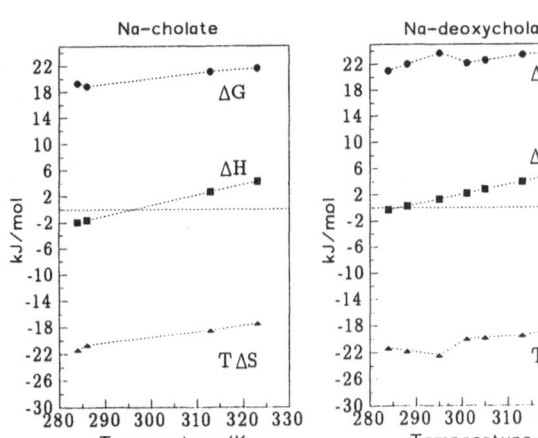

crystalline state. Melittin incorporation can be easily studied by titration calorimetry [1, 2]. Above the phase transition of the lipid, the incorporation enthalpy is negative. This indicates that the polypeptide induces some ordering of the fatty acyl chains next to the helical peptide. Incorporation into zwitterionic lipids, such as phosphatidylcholines, does not lead to a large shift in transition temperature, whereas with negatively charged lipids, such as phosphatidylglycerol and phosphatidic acid, the transition temperature is shifted to higher values, indicating electrostatic interactions between the positively charged lysine residues of melittin and the negatively charged head groups. The enthalpic effects decrease with increasing temperature, indicating that the ordering of the lipid chains due to the polypeptide decreases with temperature. Below the transition temperature of the lipid melittin incorporation is slow and the reaction enthalpy is positive due to a perturbation of the packing of the acyl chains [1, 2].

Fig. 4. Thermodynamic parameters ΔH, ΔG, and ΔS for demicellization as a function of temperature as obtained from titration calorimetry for the detergents SDS ($\Delta C_p = 534 \pm 25$ J $\cdot$ mol^{-1} $\cdot$ K^{-1}), octylglucoside ($\Delta C_p = 440 \pm 20$ J $\cdot$ mol^{-1} $\cdot$ K^{-1}), Na-cholate ($\Delta C_p = 140 \pm 50$ J $\cdot$ mol^{-1} $\cdot$ K^{-1}), and Na-deoxycholate ($\Delta C_p = 150 \pm 40$ J $\cdot$ mol^{-1} $\cdot$ K^{-1})

References

1. Blume A (1988) In: Hidalgo C (ed) Physical Properties of Biological Membranes and Their Functional Implications. Plenum Press, New York, pp 71—121
2. Blume A (1991) Thermochim Acta 193:299—347
3. Wiseman A, Williston S, Brandts JF, Lin LN (1989) Anal Biochem 179:131—137
4. Träuble H, Eibl H (1974) Proc Natl Acad Sci USA 71:214—219
5. Blume A, Eibl H (1979) Biochim Biophys Acta 558:13—21
6. Blume A, Tuchtenhagen J (1992) Biochemistry 31:4636—4642
7. Träuble H, Teubner M, Woolley P, Eibl H (1976) Biophys Chem 4:319—342
8. Blume A (1983) Biochemistry 22:5436—5442
9. Tanford C (1980) The Hydrophobic Effect, 2nd ed. John Wiley & Sons, New York

Authors' address:

Prof. Dr. Alfred Blume
Fachbereich Chemie
Universität Kaiserslautern
Erwin-Schrödinger-Straße
67663 Kaiserslautern, FRG

Progress in Colloid & Polymer Science   Progr Colloid Polym Sci 93:123—129 (1993)

Critical behavior of smectic liquid crystals

D. Gazeau, Th. Zemb, and M. Dubois

Service de Chimie Moléculaire DRECAM/DSM, CE Saclay, Gif sur Yvette C, France

Abstract: The occurrence of a critical point in smectic liquid crystal is experimentally evidenced by i) broadening of the Bragg peak of the coexisting lamellar phases, and ii) strong scattering at low angle in the vicinity of the critical point, where the concentration and periodicity difference between the two phases in equilibrium vanishes. First approximate values of the critical exponents are given.

Key words: Phase equilibrium — lamellar phase — smectic phase — interaction between bilayers — critical behavior

Introduction

Critical phenomena have been extensively studied and numerous critical points have been observed in fluids of particles such as surfactant micellar solutions and microemulsions. Attractive interactions between the particles are responsible for the critical demixion in two separate phases: a gas of independent particles and a condensed phase. The possibility to observe a similar feature in smectic liquid crystals has been predicted theorically [1], and the observation of a critical point between two thermotropic smectic liquid crystalline phases S_{Ad} and S_{a2} was reported by Shashidar [2]. The microstructure of lyotropic lamellar phases is characteristic of smectic liquid crystals: parallel bilayers of molecules separated by solvent. Solid crystalline order propagates only in one direction. Low spontaneous curvature of water-oil interface [3] as well as long-range repulsive interaction forces such as steric or electrostatic forces induce a large stability of this lamellar structure versus concentration and temperature variations [4]. Miscibility gaps between two lamellar phases have been reported in several types of systems as different as mixtures of surfactants [5, 6], one surfactant with a mixture of monovalent and divalent counterions [7], or pure surfactant in water [8]. The origin of such miscibility gaps may be a critical point in the phase diagram. We describe in this paper the behavior of lamellar liquid crystals made with didodecyldimethylam-

monium bromide (DDAB): This "missing link" between the extensively studied single-chain surfactants and biologically relevant phospholipids or glycolipids shows a very large region of lamellar phase existence [8]. In the presence of water, the chain melting temperature is in the range of 17° to 20°C, so all the experiments reported here are in the fluid state of the hydrocarbon chains. Osmotic pressure experiments have shown that a true equilibrium between the two lamellar phases exists [4]. We show in this paper that very strongly enhanced repulsive forces near the boiling temperature of the mixture allow the true coexistence of La and La' phases with vanishing concentration difference.

Experimentals

Phase diagrams were determined by preparing the samples of DDAB (from Kodak) and keeping them in a thermostat for visual inspection between crossed polarisers. Neutron scattering experiments were performed on the PAXE experiment (Orphée reactor, Saclay) using a wavelength of 6 Å with $\Delta\lambda/\lambda$ of 10% selector and a sample to detector distance of 5 m.

The small-angle x-ray scattering experiments have been performed on a laboratory built small-angle x-ray camera in pinhole symmetry. The optical part is made of a bent asymmetric cut germanium crystal following a bent glass mirror. The sample-to-detector distance is 2.1 m. No deconvolution of the observed signal is needed. The detector is an "image plate" scanned with a Phosphorimager device (Molecular Dynamics, USA). Image processing procedures used to extract I ($q = 4\pi \sin\Theta/\lambda$) expressed in cm^{-1} are described elsewhere [9]. The characteristics of this "Huxley-Holmes" camera are: total flux through the sample $1.2.10^7$ CuK$_a$ photons per second; background on 43*43 cm detector: 600 counts/s due to x-ray background and self exposure of the plates due to the cosmic background. From the edge of the beam-stop to the edge of the image the accessible q values range from $q = 0.01$ to $q = 0.65$ Å^{-1}.

Some precautions have to be taken during the scattering experiments to avoid hysteresis effects as well as too long equilibration times at high temperature in very viscous samples. In order to avoid hydrolization of samples, the total time to make a phase diagram is kept to less than a week.

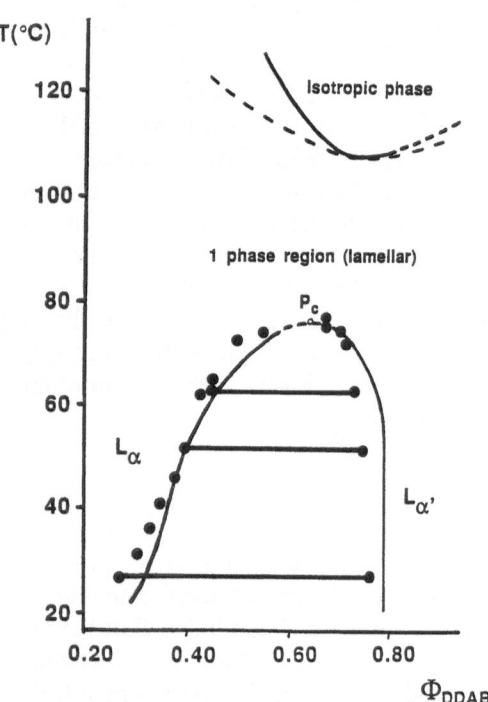

Fig. 1. Partial (ϕ-T) phase diagram of degased solutions of DDAB in H$_2$O; La is the collapsed lamellar phase with a periodicity of 30 Å including the thin water layer (6 Å). La is the swollen lamellar phase which can be diluted with water to 3% DDAB content. P_c is the critical point. The two-phase region between the lamellar and the isotropic phase present at higher temperature is too narrow to be experimentally observed

The scattering experiments are always made by lowering the temperature within a few minutes from the homogeneous isotropic high-temperature phase. Experiments could not be done by increasing the temperature because once a macroscopic phase separation has occurred, temperature jumps alone are not efficient to induce mixing in the sample cell.

Results

Figure 1 shows the partial phase diagram of the binary system water/DDAB in the range of concentrations where a thermodynamic equilibrium between two lamellar liquid crystalline phases is observed. In the diluted region, between 3% and 30%, a swollen lamellar phase exists [10]: this phase-called hereafter the La phase-shows up to four sharp Bragg peaks and a linear swelling behavior in the range of 100 to 700 Å period for a bilayer thickness of 24 Å. The area per headgroup is independent of the temperature and has been measured independently of the peak position (using Porod's limit) by neutron scattering. It has been found to be about 68 +/− 3 Å^2 per molecule. In the concentrated region (75% to 85% weight), the structure of the collapsed phase (La') is also a 24 Å-thick bilayer separated by 6 Å water plus the counter-ion layer. This phase is clear, extremely viscous and birefringent. Samples of intermediate content are extremely turbid, due to the coexistence of the two structures, as shown below. The continuous line boundary between the two lamellar phases La and La' has been determined by visual inspection of sealed, degased and thermostated samples between crossed polarisers. The single-phase region is easily detected by disappearance of the turbidity of the samples, and static birefringence is observed between crossed polarizers. The points in the phase diagram represent the composition of the two lamellar phases coexisting at the ends of a tie line. These compositions are deduced from the periodicities measured by the small-angle scattering experiments, assuming the area per head group is constant (68 Å^2): previous studies have shown experimentally the area per DDAB molecule does not vary by more than 10% in the whole temperature range. Therefore, the possible but small variations in the range 60°C—70°C have been neglected.

At room temperature, the collapsed La' phase coexists with the swollen La phase with a large miscibility gap. At higher temperature, the miscibility gap vanishes, and there is a critical point located at 62.2 *wt* % DDAB and $T_c \approx 348$ K. Note that fluctuations of concentration near this point involve the same unidirectional symmetry in the coexisting fluids.

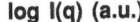

log I(q) (a.u.)

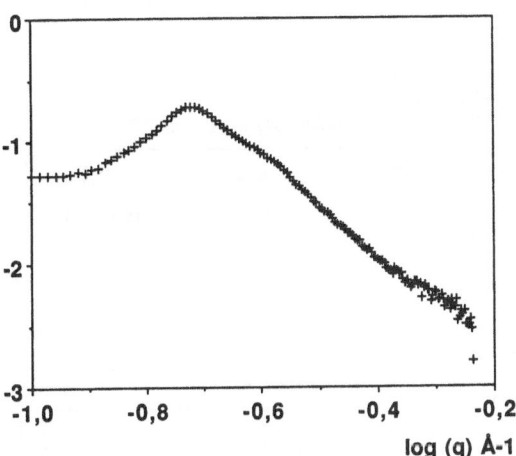

log (q) Å-1

Fig. 2. The scattered intensity versus q in the high temperature isotropic phase

At very high temperature (> 373 K), an isotropic phase which exhibits all the characteristics of a disordered lamellae phase (11) is visible. This phase is not birefringent and shows a very low viscosity compared to the viscosities of the lamellar phases observed at lower temperature. We added methanol to a mixture of water and DDAB at the critical concentration in order to make it turn isotropic at 90 °C. The scattering curve measured for this isotropic phase is plotted in Fig. 2. The scattering spectrum exhibits liquid-like pattern with a very broad peak. This phase is too concentrated in DDAB to allow us to observe the characteristic q^{-2} behavior expected for the scattering by a bilayer in the intermediate q-range. This behavior is certainly hidden by the correlation peak, so that we cannot discriminate between the scattering by a concentrated solution of reverse micelles and the scattering by an L_3 phase.

Near the critical point, the concentration fluctuations involve the coexistence of well-defined regions containing a given concentration of surfactant and, hence, sharp Bragg peaks with a very well defined periodicity. Such "crystallites" of given periodicity coexist with adjacent regions of different periodici-

ty. Figure 3 shows the small-angle x-ray scattering powder patterns obtained upon slow cooling ($\approx$ 1K/hour) from the monophasic domain. At 348 K, one obtains one single broad scattering peak in the monophasic lamellar domain. At 343 K, the first order peaks relative to the "collapsed" and the brackets "swollen" phase are broad and now separated. At 335 K and below, the fluctuations between the collapsed and the swollen liquid crystal do not broaden the peaks in the scattering pattern. At room temperature, one observes the first orders of the two coexisting lamellar phases, with periodicities of around 30 Å and 100 Å, respectively. The second order of the collapsed structure is enhanced on the figure by a factor of 10 for clarity.

I(q) u.a.

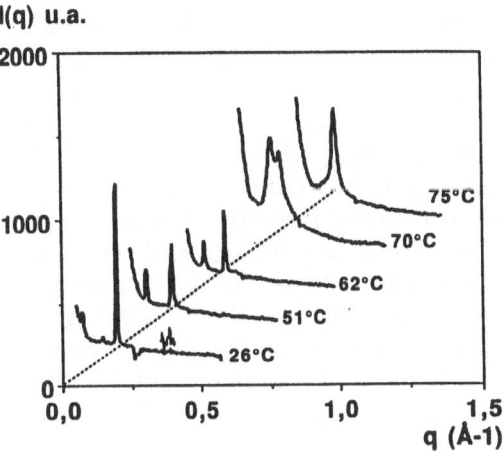

q (Å-1)

Fig. 3. Small-angle x-ray scattering spectra obtained for a DDAB/water solution (62.2 wt% DDAB). At 26 °C, two orders of each of the coexisting lamellar structures are observed. At room temperature, the two corresponding periods are 100 Å and 30 Å. At 70 °C, the two scattering peaks are broadened due to *La* versus *La'* concentration fluctuations. At 75 °C, just above the critical point, the transparent monophasic birefringent solution gives only one sharp Bragg reflection

Using now the binary system DDAB/D$_2$O, we have followed by SANS the small-angle scattering approaching the critical point from the one-phase domain, lowering the temperature. The temperature and the composition of the critical point of DDAB are somewhat different in D$_2$O and in H$_2$O: due to some uncertainty with the temperature control in the SANS experiment, our best guess is a 5—10 degree temperature shift between the phase diagrams. The critical scattering shown in Fig. 4a is clearly increased when the critical point is approached. In Fig. 4b, the observed scattering is fitted with the Ornstein-Zernicke formula expected to be valid in the critical region: $I(q) = I(0)/(1 + q^2\xi^2)$. The correlation lengths derived at each temperature are also shown in this figure. The correlation length of this mixture is of the order of a few periods of the smectic liquid crystal. Two observations can be immediatly made:

i) A characteristic divergent shape for the correlation length variation is observed in the critical region. In more diluted samples (30% weight), an increase in the small-angle scattering can be measured even at 40 K from the critical temperature.

ii) In the two-phase region, the value of the intensity $I(0)$ at zero scattering angle allows the evaluation of the crystallite size, because the two concentrations corresponding to the extremities of tie lines are known. At about 2° below the critical temperature, $I(0)$ is about 10 cm^{-1}. The scattering length density contrast between crystals containing respectively 50% and 75% DDAB is $\Delta b = (b_{50\%} - b_{75\%}) \approx 2.10^{10}$ cm^{-1}. This allows to estimate the size of the coexisting La' and La domains: $I(0) \approx (\Delta b)^2 \cdot \langle V \rangle^2 \phi \cdot (1 - \phi)$ [11]. The knowledge of the phase boundary gives an estimate of the volume fraction of the collapsed phase 5°C below the critical point: $\phi \approx 25\%$.

The typical volume of an individual collapsed crystallite is therefore about 10^{15} Å^3. The one dimensional extension of the crystallites is in the micron range, i.e., about 1000 times the correlation length: our results at the present time do not allow us to determine if the crystallites are flat and parallel to layers or not. The size of the condensed or diluted domains is much larger than the correlation length: the energy associated to the defects occurring at the interface between two crystallites with different periodicities are probably not important in this critical demixion.

The scattered intensity at zero angle $I(0)$ is related the isothermal osmotic compressibility of the sample $I(0) \approx (dc/d\beta\Pi)_T$ where c is the concentration of lamella imposing the period $D^* \approx 1/c$. The divergence of both the scattered intensity at zero angle $I(0)$ and the correlation length are plotted in Fig. 5 in reduced coordinates. The continuous line

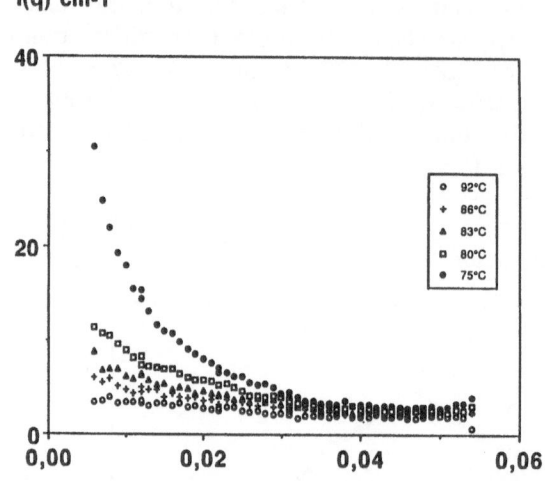

a

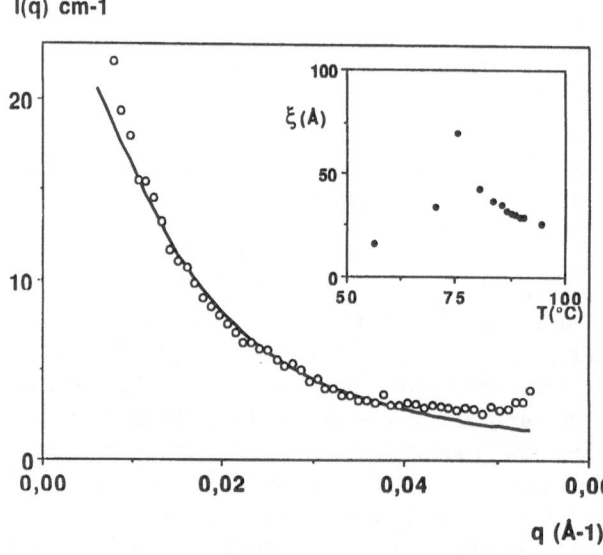

b

Fig. 4. a) Small-angle neutron scattering obtained with a DDAB/D$_2$O solution 62.2 wt% (PAXE, Orphée reactor in Saclay). The q range scanned by SANS is at lower q than the first order peak of the swollen phase which is located at $q \approx 0.2$ Å^{-1}. b) the scattering measured at $T = 75$°C fitted by the Ornstein-Zernicke expression. Inside: the values of the correlation length ξ versus temperature, diverging when approaching the critical point

is the best fit obtained when setting T_c at 344.35 K. The corresponding critical exponents for the correlation length of the fluctuations of concentration and the osmotic compressibility are respectively: $\nu = 0.57 +/- 0.10$ and $\gamma = 1.13 +/- 0.10$.

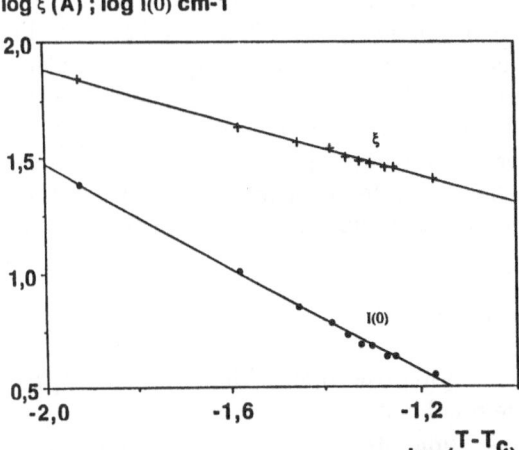

log ξ (Å) ; log I(0) cm-1

Fig. 5. Determination of the exponents v related to the correlation length ξ of the critical fluctuations, and γ related to the osmotic compressibility. From these data the best estimates for these exponents are: $v = 0.57 +/- 0.10$ and $\gamma = 1.13 +/- 0.10$. The critical temperature is set at 344.35 K (71.2 °C)

Discussion

A pioneering work by Sackmann [5] showed that critical fluctuations may exist in ternary systems with two surfactants (DMPC and DSPC) forming a mixed lamellar liquid crystal and exhibiting a miscibility gap. In the bilayer plane, the two surfactants can either mix ideally, giving then a bilayer of average thickness, or laterally phase separate, inducing coexistence in the same sample of two different compositions. Therefore, for mixed surfactants, in case of coupling between lateral phase separation and crystallite growth, there may be a miscibility gap in the lateral direction, inducing the coexistence of two periodicities in the same sample. This miscibility gap may vanish for a certain concentration and temperature. Sackmann et al. have shown that some excess scattering occurs when the miscibility gap is reduced, but the diluted lamellar liquid crystal shows a phase transition toward an isotropic phase (which is probably a disordered lamellar phase) before the critical temperature could be reached. Fortunately, in the case of DDAB the lamellar-isotropic phase transition occurs 20 °C higher than the critical point, so we can obtain SANS data for a very small miscibility gap and ap-

proach the critical point. Bryant and Wolfe [6] have similarly investigated a mixture of phospholipids (DOPE-DOPC), showing that two lamellar phases of periodicities 52 and 60 Å can coexist. However, it was not possible with this system to reduce the periodicity difference between the two coexisting lamellar liquid crystals to zero. Finally, Shashidar and coworkers (2) have reported what seems to be the first observation of a critical point between two smectic phases in a binary mixture of 11OPCBOB and 9OBCD.

The interplay of repulsive electrostatic, steric as well as Van der Waals forces explain the coexistence of *La* and *La'* structures. This was experimentally confirmed using osmotic pressure measurements [4]. At room temperature, the undulation forces are negligible in this system [4]. However, approaching the boiling point of the solvent, undulation forces or some other repulsive force become important and reduce the concentration gap between the two lamellar phases. Evans, Parsegian and coworkers have shown on a very similar surfactant system (DHDA) that undulation forces are a small but relevant part of the repulsive forces at intermediate temperatures [13].

The approximate values for the critical exponents γ and v measured on this system of two coexisting lamellar phases, are close to those expected for the 3D Ising model ($\gamma = 1.24$ and $v = 0.63$), as observed in liquid binary mixtures, micellar solutions and microemulsions (14, 15) or to those predicted by Prost and coworkers [1]. We cannot say where the phase transition takes place — inside the bilayers, that is to say, in a two-dimensional space as previously reported by Sackmann and coworkers [5], or in a three-dimensional space if the transition involves two macroscopically different lamellar phases. One also should take into account the fact that these critical exponents are deduced from powder averaged spectra. The critical exponents may be different if the scattered intensity is measured upon approaching the critical point on an oriented sample [1].

We propose the following driving force for the observed critical demixion. At room temperature, the combination of electrostatic, steric and Van der Waals forces explains quantitatively the miscibility gap between the *La'* and *La* phases. At equilibrium the osmotic pressure is equal in both lamellar phases. And this equilibrium pressure is such as the free energies for both phases are equal, i.e., that the hatched surfaces (on a linear representation) are

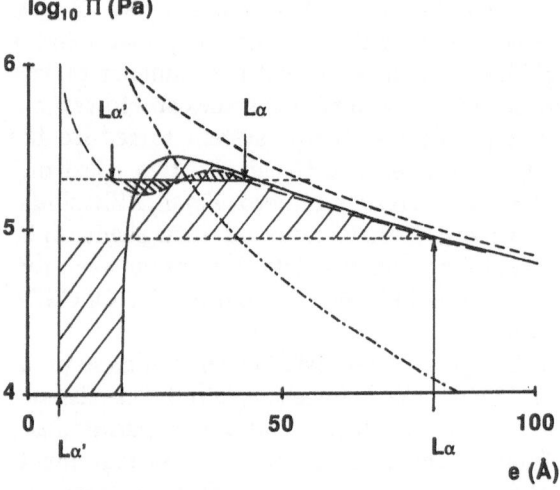

Fig. 6. Sketch of the effect of temperature on the pressure — distance diagram. At low temperature the Vander Waals and the electrostatic pressures quantitatively explain the existence of the equilibrium between the two lamellar phases (La and La'). (—·—) (-1) × $F(d)$ Van der Waals. (—·—) $F(d)$ electrostatic. The continuous line represents the total pressure at low temperature. Arrows indicate the points representing the two lamellar phases in equilibrium. The interlamellar distances (t) increased of the lamellar thickness (24 Å) are the Bragg distances measured in SAXS expreiments. (——) represents the total pressure at high temperature. Due to the short range repulsive pressure arising upon increasing temperature the collapsed La' phase swells slightly. The system finds a new equilibrium state to equalize the two hatched surfaces on a linear plot. As a result, the interlamellar distance in the swollen phase decreases

equal. Increasing temperature does not modify the electrostatic or Van der Waals effects. But now a third and repulsive force between bilayers come into play: it could be either protrusion or undulation forces. As a result of this repulsive interaction the interlamellar distance in the La' phase is increased. In order to find a new equilibrium state the interlamellar distance in dilute phase decreases to equalize again the two hatched surfaces, as shown in Fig. 6. At the critical temperature the two interlamellar spacings are equal and the two surfaces equal to zero. Above the critical temperature only one lamellar phase exists and the osmotic pressure monotonically decreases with the periodicity. Experimental evidencing of these effects will require osmotic pressure measurement close to the critical point.

Finally, the coexistence of micronic size crystallites of two lamellar phases with similar, but different periodicities, is a puzzling question: defects or even tearing off or holes in the lamellar phases are needed for the intertwining of one structure into the other: thus, some aspects of the critical demixion may be very different with lamellar phases than with particle fluids due to topological constraints.

Conclusion

We have shown the first evidences for an upper consolute point between two smectic liquid crystals. The macroscopic phase diagram has been confirmed by small-angle scattering experiments: x-ray experiments show the two phases in equilibrium as well as broadening due to fluctuations in the system when approaching the critical point. The critical scattering shows a typical Ornstein-Zernicke behavior. The corresponding critical exponents γ and ν have been evaluated to be similar to, but may be lower than, the universal values obtained for binary fluids made of interacting particles. Each of the two lamellar liquid crystal in equilibrium below the critical temperature exhibits sharp Bragg peaks reflecting one direction of solid type order.

Acknowledgement

Initial suggestions made by Didier Roux are acknowledged, as well as helpful discussions with Luc Belloni and Stephan Marcelja. Small-angle neutron scattering was made at the instrument (PAXE) with the advices and help of José Teixeira.

References

1. Park Y, Lubensky TC, Barois P, Prost J (1988) Phys Rev A 37:2197
2. Shashidar R, Ratna BR, Krishna Prasad S, Somasekhara S, Heppke G (1987) Phys Rev Lett 59:11, 1209
3. Israelachvili JN, Mitchell DJ, Ninham BW (1976) J Chem Soc Faraday Trans II 72 p 1525
4. Zemb TH, Dubois M, Belloni L, Marcelja S (1992) Prog in Colloid and Polymer Science 89
5. Knoll W, Schmidt G, Sackmann E, Ibel K (1983) J Chem Phys 79:3439
6. Bryant G, Pope JM, Wolfe J (1992) Eur Biophys J 21:223
7. Khan A, Jönsson B, Wennerström H (1985) J Phys Chem 89:5180
8. Fontell K, Ceglie A, Lindman B, Ninham BW (1986) Acta Chem Scand A 40:247—256

9. Né F, Gazeau D, Lambard J, Lesieur P, Zemb Th, Gabriel A J Appl Cryst, Accepted or in press
10. Dubois M, Zemb Th (1991) Langmuir 7:1352—1360
11. Barnes IS, Dérian PJ, Hyde ST, Ninham BW, Zemb Th (1990) Journal de Physique Paris 51 pp 2605—2628
12. Glatter O, Kratky O (1982) in "Small angle x-ray scattering", Academic press (1982)
13. Tsao YH, Evans DF, Rand RP, Parsegian VA, Langmuir (in press)
14. Bellocq AM, Gazeau D (1990) J Phys Chem 94 pp 8933—8938
15. Beysens D, Bourgou A, Calmettes P (1982) Physical Review A 26:3589—3609

Authors' address:

Dr. D. Gazeau
Service de Chimie Moleculaire
DRECAM/DSM
Bât. 125
CE Saclay
91191 Gif sur Yvette, France

Progress in Colloid & Polymer Science   Progr Colloid Polym Sci 93:130—134 (1993)

Dynamics of water in nonionic amphiphile systems: raman scattering

D. Majolino[1]), F. Mallamace[1]), N. Micali[2]), M. Corti[3]), and V. Degiorgio[3])

[1]) Dipartimento di Fisica dell'Universita' di Messina
[2]) Instituto di Tecniche Spettroscopiche del C N R Messina, Italy
[3]) Dipartimento di Elettronica, Universita' di Pavia, Italy

Abstract: The structure of water in aqueous solutions of polyoxyethylene nonionic amphiphiles $C_{10}E_5$ is studied by Raman scattering along an isothermal path crossing the isotropic one-phase region from 0 to 1 amphiphile volume fraction ϕ. The isotropic OH stretching vibration spectrum evidences a behavior that depends on the amphiphile concentration. The interpretation of such scattering data leads to the following structural picture for the water in the systems: for ϕ lower than 0.75, water is partially bound to the oxyethylene groups of the amphiphile; above $\phi = 0.75$ all water present in the system is bound. The structure of such bound water presents a local, low dense, four coordinated environment.

Key words: Water — amphiphile — raman scattering — vibrational dynamics

Introduction

Aqueous solutions of non-ionic polyoxyethylene amphiphile, in order to clarify the structural and dynamical properties of the supramolecular aggregates) i. e., micelles and microemulsions) formed above the critical micelle concentration (CMC), have been the subject of many studies [e. g., 1, 2]. Their chemical formula is $C_mH_{2m + 1}(OCH_2CH_2)_n$ OH or C_mE_n for short. We consider the $C_{10}E_5$ solution above the mesophase regions and below the cloud-point curve, where it is possible to follow a continuous isothermal path which crosses the isotropic one-phase region from 0 to 100% volume fraction ϕ of amphiphile. In order to explain its structural properties along this path (it is well-known that above the CMC $C_{10}E_5$ forms globular micelles, whereas it is not known up to what amphiphile concentration the solution can still be described as a water continuous dipersion of amphiphile aggregates, and how the system evolves towards the pure liquid amphiphile phase) the system has recently been the subject of careful studies x-ray and neutron scattering [3]. In particular, it has been shown that along this isothermal path ($T = 35°C$) the solution is structured for all

concentrations ranging from the micellar region to the pure liquid amphiphile. As the volume fraction ϕ increases, the micellar structure becomes less and less sharp, but some orientational correlations between neighboring amphiphile molecules are preserved even at high concentrations. From the small angle neutron data (SANS) [3] a structure peak is clearly observable up to $\phi = 0.95$, but the pure $C_{10}E_5$ neutron spectra were absolutely flat. This is due to the fact that the neutron scattered intensity is mainly determined by the large contrast between the deuterated water and the hydrogenated amphiphile. More precisely, the interpretation of scattering data (SANS and SAXS) leads to the following conclusions: large orientational correlations exist among neighboring amphiphile molecules, and at high surfactant concentrations ($\phi > 0.7$) the system behaves essentially as a block-copolymer melt. The existence of a structure peak ($\phi = 0.95$ for neutrons, and $\phi = 1$ (pure amphihile) for X-ray) is a direct result of the block structure of the surfactant monomer with attractive head-head and tail-tail interactions and repulsive head-tail interactions. This phenomenon, proposed by P. G. de Gennes [4], is known as the correlation-hole effect.

It is important to stress that this latter phenomenon can be observed in our system because the hydrophilic groups are hydrated. The use of deuterated water enhance via the hydration the contrast between hydrophilic and hydrophobic groups of the amphiphile molecule, allowing for the observation of the correlation hole effect.

Therefore, from the picture proposed by SANS and SAXS in these non-ionic amphiphile solutions, it turns out that water molecules play a significant role. A part of water is hydrogen-bounded (HB) to the polyoxyethylene head groups of the amphiphile. More precisely, the analysis of SANS data, in terms of Porod's scattering invariants $K =$

$$\int_0^\infty k^2 \, I(k) \, dk \text{ and } d = \pi \, K^{-1} \int_0^\infty k \, I(k) \, dk \text{ [5] (where } k \text{ is}$$

scattered wavevector and $I(k)$ the scattered intensity) reveals that: the oxyethylene group are hydrated, the average number of bound water molecules for group is $n_w \simeq 1.5$ for $\phi < 0.7$, and $n_w \simeq 2$ for $\phi > 0.7$; furthermore, for $\phi > 0.7$ all water molecules are bound via HB to the oxyethylene chains. The fact that, at high concentrations, there is no free water in the system is consistent with the low values of the electric conductivity measured in the solution.

On this basis, we considered that Raman scattering can give confirmation, from the study of the vibrational stretching of water, of this structural model proposed from the non-ionic amphiphile solutions by the analysis of SANS and SAXS data. Considering that Raman scattering constitutes a powerful tool in order to study vibrational dynamics, we tried to investigate the structural properties of water in non-ionic amphyphile solutions through the analysis of the spectral region of O — H stretching vibrations. On the other hand, it is well known that O — H stretching is very sensitive to the molecular organization of water; in particular, for pure water it can provide, as a function of the thermodynamic variables (T, P, etc.), detailed information on the structure corresponding to a particular state (solid, liquid, vapour, supercooled) [6]. Here, we present results obtained at 35°C with $C_{10}E_5$ solutions. Our data confirm the model proposed for SANS and SAXS experiments, giving new detailed information about the dynamics of water molecules hydrogen-bound with the polyoxyethylene head groups. In particular, we show that the O — H stretching vibrations of such bound water are analogous to glassy water [7].

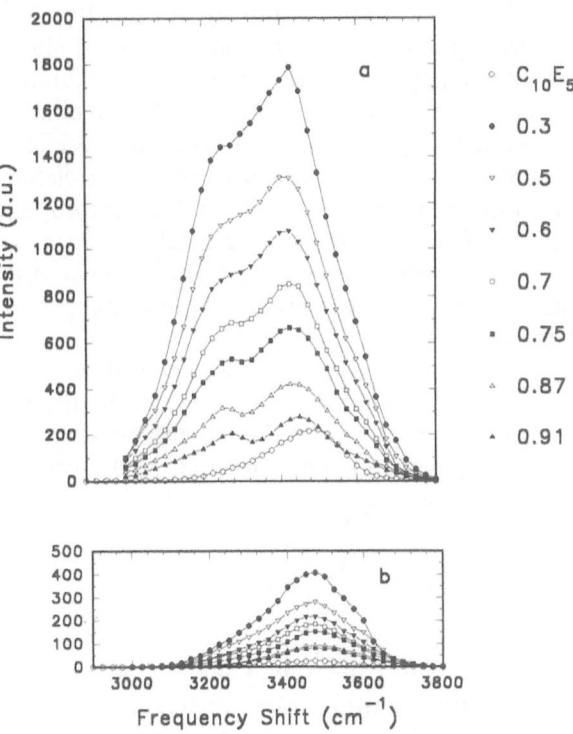

Fig. 1. Raman spectra of the OH stretching of water at different amphiphile volume fractions. a) represents the parallel and b) the anisotropic contribution

Experimental

Non-ionic amphiphile was provided by Nikko Chemical, Japan, and was used without further purification. Triple distilled deionized gas-free water was used. Great care was taken, in the sample preparation, to avoid dust contamination. The phase diagram of $C_{10}E_5$ in H_2O [1] shows a cloud-point curve with a minimum at about 45°C and liquid crystalline region in the range of amphiphile volume fraction between 0.5 and 0.85 with temperature between 0^0 and 20°C; therefore, we worked along a isothermal path ($T = 35$°C), which no hit mesophase regions. The studied concentrations are: $\phi = 0.3, 0.5, 0.6, 0.7, 0.75, 0.87, 0.91$, and the pure amphiphile.

Raman scattering measurements were performed in the usual scattering geometry of 90°C with a high-resolution triple monochromator (Spex Ramalog V), together with an Ar$^+$ laser operating at a 5145 Å wavelength with a mean power of 300 mW. The sample was thermostated in a Harney-Miller cell. The cell temperature was controlled by a thermostat to within 0.02°C. The measured spectra was taken in the range 2900 — 3800 cm^{-1}, with a resolution of 4 cm^{-1}, both in the parallel (VV) and orthogonal (VH) polarizations. Figure 1 shows the obtained spectra of the OH stretching vibration in both the polarization geometries. We can observe that we have a good experimental detection of this vibrational contribution because the CH's stretching contribution to the spectra of the amphiphile molecule falls in a different spectral region of the OH band. Furthermore, pure $C_{10}E_5$ shows a stretching contribution due to the OH terminal groups of the amphiphile.

Considering the theory of Raman scattering, we analyzed the obtained spectra, taking into account the isotropic part of the Raman intensity $I_{iso}(w)$ that, as is well known, depends only on the molecular vibrations [8] and is written as:

$$I_{iso}(w) = I_{vv}(w) - \frac{4}{3}\ I_{VH}(w) \ .$$

In order to study the true dynamic of water in the solution, the isotropic OH stretching spectrum measured for the sample with a volume fraction ϕ, $I_{is}^{\phi}(w)$, was considered as the sum of two contributions: one due to the water $I_{is}^{\phi,\ wat}(w)$, and the second to the pure amphiphile $I_{is}^{amph}(w)$, both weighed for the corresponding concentrations. Therefore, we calculated, for the different samples studied in the present work, the OH stretching spectrum of water in the amphiphile solution using the following relation:

$$I_{is}^{\phi}(w) = I_{is}^{\phi,\ wat}(w)(1 - \phi) + I_{is}^{amph}(w)\phi \ .$$

Figure 2 shows the OH stretching spectra obtained with such a procedure for all studied solutions.

Discussion

Before discussing the results shown in Fig. 2, it is important to briefly consider the interpretation of the OH stretching behavior as dictated by the different theoretical models proposed for water structure. Historically, two classes of models are considered [9]:

a) continuous models, where water is pictured as a continuous network of tetrahedrally bound molecules, in which the bonds are more or less distorted, giving rise to a continuous distribution of structures.

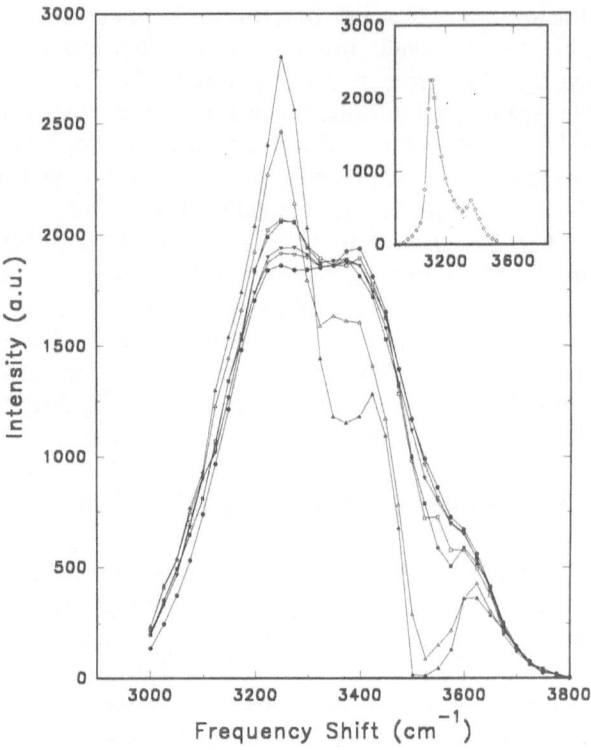

Fig. 2. Isotropic OH stretching contributions of water at all the different amphiphile concentrations. In the inset is reported the isotropic OH stretching contribution for glassy water evaluated from the data of Li PC and Delvin JP [7].

b) Discrete models, according to which a discrete number of species exist that differs from one another according to the specific structural arrangement.

However, both have in common the fact that a local four-coordinated environment with low density is preferred for the water structural arrangement.

Recently, a new model has been proposed which can be considered intermediate between the above-mentioned model [10]. In such a model, each water molecule is assigned to one of five species, according to the number (from zero to four) of HB. Then, using percolation concepts (site percolation calculations) it is shown that tetrabound molecules tend to clusterize, giving rise to finite regions or patches, whose structural properties are different from those of the remainder. In these terms, water molecules can be divided into two classes: "open" water in which a regular tetrahedral structure exists, and

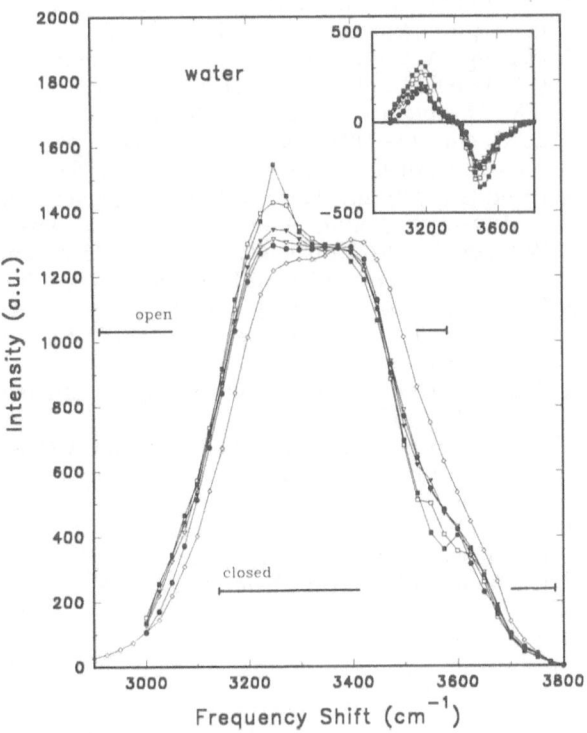

Fig. 3. OH stretching spectra for the amphiphile volume fractions $\phi \leq 0.75$. The spectrum of pure water at the same temperature (35 °C) is shown for comparison. In the inset are reported the differences between the spectra of the different solutions and the spectrum of pure water

"closed" water that behaves like a continuum, being the mixing of all the remainder molecules. This model, stressing that "open" water is related with low density structures, explains quite well the anomalous properties of water. In particular, it explains also the measured spectra of OH stretching in supercooled [11] and amorphous water [7]. Contributions of "open" and "closed" water fall in two different regions of frequency of the OH stretching Raman spectra [11]; while the "open" contribution has a mean peak centered at about 3150 cm^{-1}, the corresponding peak for "closed" structures is centered at about 3500 cm^{-1} [11]. Figure 3 show $I_{is}^{\phi,\ wat}(w)$ for $\phi = 0.3$, 0.5, 0.6, 0.7 and 0.75; in the same figure is also reported, for comparison, the OH stretching of pure water at the same temperatue of the amphiphile solutions ($T = 35$ °C), and the frequency intervals corresponding respectively to "open" and "closed" water. As it can be observed, the spectrum of pure water differs from the spectrum of water in the mixture also for the less concentrated solution ($\phi = 0.3$). In particular, the com-

parison of pure water spectrum with the spectrum for $\phi = 0.3$ evidences that, in the amphiphile solution, a larger amount of water molecules are bound in structures of low density. Increasing the amphiphile content results in an increase in the open water contribution to the spectrum. This behavior can be observed in the inset of Fig. 3 where the differences between the OH stretching spectra of water in the amphiphile solutions and the corresponding spectra of pure water are reported. This result can be rationalized with SANS and SAXS [3] findings which show that for all values of ϕ below a saturation value ϕ_S ($\phi_S \sim 0.75$), we can have a certain quantity of water bound to the polyexyetylene head groups, and the remainder of water is completely free. This ϕ range is the concentration interval for which well-defined micellar structures are present in the system. More specifically, in the concentration region where stable micelles are present, SANS and SAXS [3] data can be well described considering a three-component model: the hydrocarbon region (hydrophobic chain of the amphiphile), the hydrophilic region (polyoxyethylene head groups together with bound water, with an average number of water molecules bounded per oxyethylene group $n_w = 1.5$) and a region of free water. The value of n_w can be roughly calculated by evaluating the "open" water contribution to the area of the OH stretching spectrum, once the latter has been normalized so as the cover a unit area. The obtained value ranging (within experimental uncertainty) between 1.4 and 1.7 agrees with the SANS value.

About the molecular organization of the solution for high surfactant concentration, $\phi > \phi_S$, we can only give qualitative confirmation of the main results of the model proposed for small-angle scattering data [3]: all the water present in the system is bound to oxyethylene groups. From the spectra shown in Fig. 2, the significant difference between the spectra with ϕ above and below the saturation value, and the spectrum corresponding to the pure water is evident. This is a further suggestion that for $\phi > \phi_S$ water molecules are arranged in the amphiphile solution in a different way in comparison with the concentrations where the surfactant molecules are aggregated in micellar structures. In particular, the OH stretching spectra for solutions with $\phi = 0.87$ and 0.91 are entirely located in the region of "open" water; the percentage of "closed" water, in comparison with solutions of lower amphiphile volume fractions is irrelevant. The dominant spectral contribution is located at the frequency of about 3200 cm^{-1}. Such a result indicates that a very large amount of water molecules is bound with the amphiphile, and the corresponding structure reflects a local environment with low density in comparison to the bulk water. As a proof of this result, on the structure of water bound to the oxyethylene groups of the amphiphile, we show in the inset of Fig. 2 the isotropic OH stretching spectrum of amorphous solid water in film with a thickness of $\sim$ 1 µm, prepared by vapor deposition, at $T = 100$ °K [7]. As can be observed, the spectra of solutions at high volume fraction are similar to those of glassy water; the relevant difference is in the frequency value of the mean peak, but this is a temperature effect (in glassy water this frequency is temperature dependent: increasing with increasing T [7]).

In conclusion, the data reported in this work agree with the structural picture proposed for nonionic amphiphile-water solutions. Water molecules are partially bound to the oxyethylene head groups of the surfactant for amphiphile volume fractions ϕ lower than 0.75; above such a value all the water present in the system is bound to the oxyethylene groups. In addition, the water structure around the surfactant has a local structure corresponding to an environment with low density.

References

1. Degiorgio V (1985) In: Degiorgio V, Corti M (eds) Physics of Amphiphiles, Micelles, Vesicles and Microemulsions. North-Holland, Amsterdam p 303; and refs. cited therein
2. Magid LJ (1987) In: Schick MJ (ed) Nonionic surfactants: physical chemistry. Dekker, New York
3. Degiorgio V, Corti M, Piazza R, Cantu' L, Rennie AR (1991) Colloid and Polym Sci 269:501; Barnes IS, Corti M, Degiorgio V, Zemb T (1992) (in press)
4. de Gennes PG (1969) Scaling concepts in polymer physics. Cornell University Press, Ithaca p 65
5. Porod G (1982) In: Glatter O, Kratky O (eds) Small-angle X-ray scattering. Academic Press, New York
6. Walrafen G (1972) In Franks F (ed) Water a comprehensive treatise. Plenum Press, New York 1:161
7. Li PC, Derlvin JP (1973) J Chem Phys 59:547; Sivakumar TC, Rice AS, Sceats MG (1978) J Chem Phys 69:3468
8. Berne BJ, Pecora R (1976) Dynamic light scattering. Wiley, New York
9. See for example: Eisemberg DE, Kauzmann W (1969) The structure and properties of water. Oxford University Press, Oxford
10. Stanley HE, Teixeira J (1980) J Chem Phys 73:3034
11. D'Arrigo G, Maisano G, Mallamace F, Migliardo M, Wanderlingh F (1981) J Chem Phys 75:4264

Authors' address:

Prof. F. Mallamace
Dipartimento di Fisica
Universita' di Messina
98166 Vill. S. Agata C.P. 55, Messina

Progress in Colloid & Polymer Science   Progr Colloid Polym Sci 93:135—136 (1993)

Non-equilibrium surface phenomena in fine-porous charged diaphragms in mixed electrolyte solutions

A. E. Yaroshchuk

Institute of Bio-Colloid Chemistry, Ukrainian Academy of Sciences, Kiev, Ukraine

Abstract: Apparent osmotic pressure and diffusion at zero electric current have been predicted for mixed electrolyte solutions within the scope of irreversible thermodynamics.

Key words: Apparent osmotic pressure — diffusion — fine-porous charged diaphragms — electrolyte mixtures — anomalous transport phenomena

As examples of non-equilibrium surface phenomena, I consider apparent osmotic pressure and solute diffusion at zero electric current in fine-porous charged diaphragms in dilute mixed electrolyte solutions. The purpose is to predict anomalies, i.e., noticeable deviations from the behavior characteristic of semipermeable membranes.

The equations of irreversible thermodynamics can be written in this way:

$$J_w = l_{ww}\Delta\mu_w + \sum_i^n l_{iw}\,\Delta\mu_i \, ,$$

$$J_i = l_{iw}\Delta\mu_w + \sum_j^n l_{ij}\,\Delta\mu_j \, , \tag{1}$$

where J_w, J_i are the molar fluxes of water and ions (in mole/cm$^2 \cdot$ s), n is the number of ions, μ_w, μ_i are the chemical (electrochemical) potentials of water (ions) in RT units. The condition of zero water flux yields:

$$\Delta\mu_w = -\frac{1}{l_{ww}} \cdot \sum_i^n l_{iw}\Delta\mu_i \, , \tag{2}$$

$$J_i = \sum_j^n \bar{l}_{ij}\Delta\mu_j \, ,$$

$$\bar{l}_{ij} = l_{ij} - l_{iw}l_{jw}/l_{ww} \, . \tag{3}$$

At zero electric current conditions

$$\sum_i^n Z_i \sum_j^n \bar{l}_{ij}\,(\Delta\tilde{\mu}_j + Z_j\Delta\phi) = 0 \, , \tag{4}$$

where μ_j are the ionic chemical potentials, ϕ is the electric potential. Equation (4) yields:

$$\Delta\phi = -\sum_i^n Z_i \sum_j^n \bar{l}_{ij}\Delta\tilde{\mu}_j \Big/ \sum_i^n Z_i \sum_j^n Z_j\bar{l}_{ij} \, , \tag{5}$$

substituting Eq. (5) into Eq. (2) yields:

$$J_i = \frac{1}{l_e} \sum_j^n \bar{l}_{ij} \sum_k^n B_k\,(Z_k\Delta\tilde{\mu}_j - Z_j\Delta\tilde{\mu}_k) \, , \tag{6}$$

$$\Delta\mu_w = -\frac{1}{l_{ww}l_e} \cdot \sum_i^n \sum_j^{l-1} (l_{iw}B_j - l_{jw}B_i)$$

$$(Z_j\Delta\tilde{\mu}_i - Z_i\Delta\tilde{\mu}_j) \, , \tag{7}$$

$$B_i = \sum_j^n Z_j\bar{l}_{ij} \, , \quad l_e = \sum_i^n Z_iB_i \, . \tag{8}$$

l_e is the membrane electric conductivity. A normal situation characteristic of binary electrolyte solutions is small J_i, and $\Delta\mu_w$ determined by coions whose concentration in the pores is low.

It can be shown that at sufficiently high diaphragm electrochemical activity anomalously large contributions to Eqs. (8) and (9) can be made

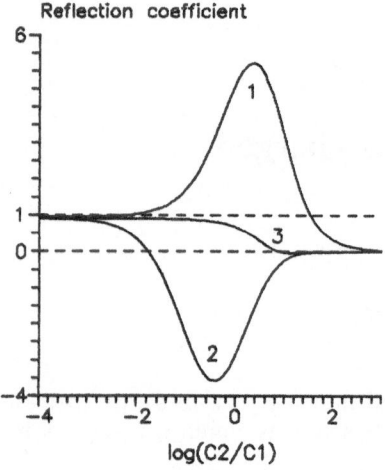

Fig. 1. Reflection coefficient vs relative concentration of an electrolyte added in equal concentrations to both compartments; KCl/HCl type mixture ($D_{H^+} = 7D_{K^+}$); $Z_x c_x / c_1 = -20$; 1) KCl is added; 2, 3) HCl is added (3 — $Z_x c_x / c_1 = 20$)

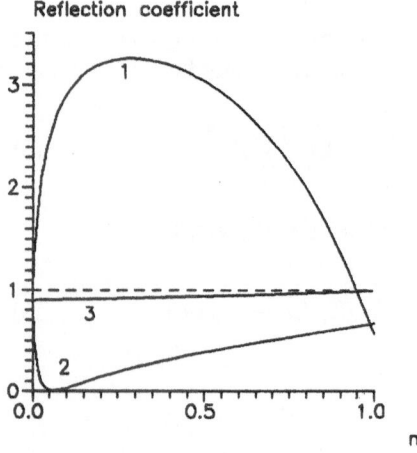

Fig. 2. Reflection coefficient vs mixture's composition; m is the mole fraction of the electrolyte with double-charged counterion; $Z_x c_x / c_1 = -20$; 1 — HCl/MgCl$_2$ type mixture ($D_{H^+} = 7D_{Cl^-}$; $D_{Mg^{2+}} = 0.5 D_{Cl^-}$); 2,3 — LiCl/CaCl$_2$ type mixture ($D_{Li^+} = D_{Ca^{2+}} = D_{Cl^-}$; 3 — $Z_x c_x / c_1$ 20)

only by coefficients $\bar{l}_{ij}$ whose two subscripts correspond to counterions. This enables to make the following conclusions.

1) Anomalous phenomena can occur only if the solution contains more than one counterion; the number of coions does not matter.
2) Pairs of counterions contribute to the difference of water chemical potentials, $\Delta\mu_w$, provided that $l_{iw}B_j \neq l_{jw}B_i$, which is true, in particular, at different mobilities and/or charges.
3) The magnitude of anomalous phenomena may increase without bound with growing electrochemical activity, while that of the normal ones tends to a saturation characteristic of semipermeable membranes.
4) The coion fluxes are always small, therefore, those of different counterions must have alternating signs.
5) In the case of symmetrical mixtures (all counterions have the same charge) anomalous phenomena occur only if there is a difference in solution composition.
6) In the case of nonsymmetrical mixtures multiple-charged counterions with higher mobility than single-charged ones may give rise to a negative apparent osmotic pressure, whereas those with lower mobility give rise to a "supernormal" one (reflection coefficient larger than unity).

Figures 1 and 2 exemplify the conclusions within the scope of the TMS model (homogeneous approximation).

Author's address:

Dr. A. E. Yaroshchuk
Institute of Bio-Colloid Chemistry
Ukrainian Academy of Sciences
42, Vernadskij Ave.,
252142 Kiev, Ukraine

Progress in Colloid & Polymer Science

Progr Colloid Polym Sci 93:137—145 (1993)

Perforated vesicles in ternary surfactant systems of alkyl-dimethylaminoxides, cosurfactants and water

U. Munkert [1]), H. Hoffmann[1]), C. Thunig[1]), H. W. Meyer[2]), and W. Richter[2])

[1]) Lehrstuhl für Physikalische Chemie I, Universität Bayreuth, FRG
[2]) Abteilung für Elektronenmikroskopie der Friedrich-Schiller-Universität Jena, FRG

Abstract: The phase diagrams of the ternary surfactant systems dodecyl-dimethylaminoxide (C_{12}DMAO)/hexanol/water and tetradecyldimethyl-aminoxide (C_{14}DMAO)/heptanol/water are reported for small surfactant concentrations. With increasing concentration we find a L_1-, a L_1^*-, a L_{al}-, a L_3^*-, a L_{ah}- and a L_3-phase. The L_1-, L_1^*-, L_3^*- and the L_3-phase are optically isotropic, while the L_a-phases are birefringent. — Besides some macroscopic properties like rheological and conductivity data, we also report on freeze fracture electron micrographs. These were made from the samples in the actual phase regions and show the different microstructures. In the L_a-phase at low cosurfactant concentration, we found uni- and multilamellar vesicles with a large polydispersity. The bilayers of the vesicles show defects that can be seen best after etching the fracture plane for a short time. These vesicles we have called perforated.

Key words: Alkyldimethylaminoxides — lamellar phases — vesicles — phase diagrams — FFEM

Introduction

Dilute lamellar phases have been the subject of theoretical and experimental investigations. They have been found in nonionic [1], ionic [2] or zwitterionic systems [3, 4]. Excess salt and cosurfactant are sometimes necessary for the formation of these phases [5—8]. A lot of interest has been devoted during recent years to uncharged singlechain surfactants. The shape of the micelles and their curvature can be influenced by the addition on n-alcohols as cosurfactants [9]. The curvature of the micellar interface controls not only the formation and the shape of the micelles, but has also some effect on the properties of mesophases [10].

With increasing concentration of n-alcohol to a zwitterionic alkyldimethylaminoxide [11, 12], one usually observes several phase transitions. We have investigated two ternary systems in the dilute region. The sequence of the different macroscopic phases is L_1, L_1^*, L_{al}, L_3^*, L_{ah} and L_3.

The purpose of our present work was to give further evidence for the existence of different macroscopic phases by conductivity and rheological measurements and, most of all, by electron microscopy which gives us direct information about the microstructures of the systems.

Results and discussion

1. Phase diagrams and phase behavior

The phase diagrams were established by observing the mixtures in calibrated test tubes under temperature-controlled conditions for several weeks. All phases were characterized by visual inspection with and without crossed polarizers. The surfactants were a gift of HOECHST AG, Gendorf. They were used after purification by recrystallization twize from acetone.

The surfactants were characterized by their melting points and CMC values (C_{12}DMAO: mp 119—120°C. cmc 1.7 mmol/l; C_{14}DMAO: mp 130—131°C, cmc 0.14 mmol/l).

The n-alcohols C_6OH and C_7OH were Fluka p. a. chemicals and used without further purification.

1.1 The system tetradecyldimethylaminoxide, heptanol and water

The ternary phase diagram tetradecyldimethyl-aminoxide, heptanol and water is reported for low surfactant concentrations in Fig. 1. It consists of several macroscopic single phases. The isotropic phases are L_1, L_1^*, L_3^*, and L_3. Furthermore, there are two birefringent lamellar phases, L_{al} for lower alcohol content and L_{ah} for higher alcohol content at a constant surfactant concentration. Between them is a third birefringent region called L_{al-h}.

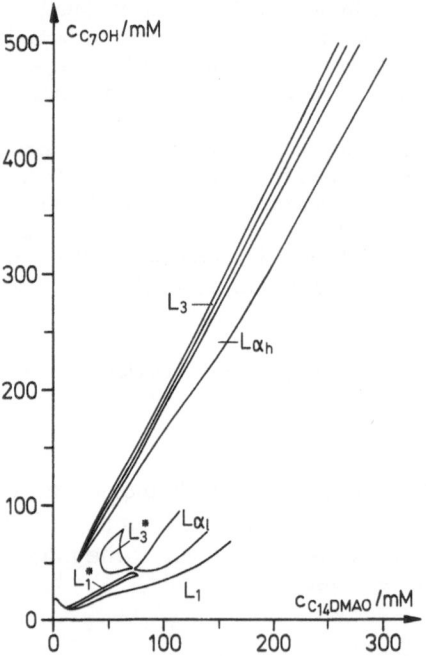

Fig. 1. Phase diagram of the system tetradecyldimethyl-aminoxide/heptanol/water for small surfactant concentrations at 25 °C. The isotropic single-phase regions are L_1, L_1^*, and L_3^*. The anisotropic single phase regions are the lamellar phases L_{al} and L_{ah}

The following macroscopic properties were observed:

L_1-phase: This is an isotropic micellar phase. It is somewhat viscous and consists of long rodlike micelles, which give rise to electric and flow birefringence.

L_1^*-phase: A slightly turbid, optically isotropic phase below the L_{al}-phase which shows strong flow birefringence. It is separated from the L_1-phase by a two-phase region which separates macroscopically upon standing. In the investigated system,

the L_1^*-phase extends from the very dilute system on a straight line to the L_{al}-phase.

L_3^*-phase: Isotropic, viscoelastic region between the two lamellar phases that exists only below a surfactant concentration of less than 70 mM. It is conceivable that it is not in a thermodynamically stable state, because after a long time the boundary towards the L_{al-h} region shifts to smaller surfactant concentration.

L_{al}-phase: This is a viscoelastic birefringent phase with long structural relaxation times. It shows strong shear waves between crossed polarizers when samples with the phase are abruptly rotated a few degrees around the axis.

L_{al-h}-phase: This looks like a mixture of L_{al} and L_{ah}, that macroscopically does not separate into two single phases.

L_{ah}-phase: This is a clear typical lamellar phase with low viscosity which shows bright iridescent colors for low surfactant concentrations [16, 17].

L_3-phase: This is an isotropic clear phase [13—15] with low viscosity. It shows flow birefringence under shear.

All the macroscopic single phases are surrounded by multiphase regions. The whole multiphase region is of triangular shape because the stability of the existing phases is determined by a specific cosurfactant/surfactant ratio, which is independent on the total surfactant concentration. For small concentrations the behavior is more complicated and, because of the small difference of the densities of the different phases, we were not able to determine the phase boundaries exactly.

1.2 The system dodecyldimethylaminoxide, hexanol and water

In comparison with the system $C_{14}DMAO/C_7OH/H_2O$ the shorter chainlength system $C_{12}DMAO/C_6OH/H_2O$ shows nearly the same phase behavior (Fig. 2). The main difference is that the L_{al}-phase can also be diluted to iridescent colors at low surfactant concentrations, but the colors of the L_{ah}-phase are much brighter than the colors of the L_{al}-phase. When the dilute L_{al}-phase in the visible colored range is strongly shaken the colors disappear, the

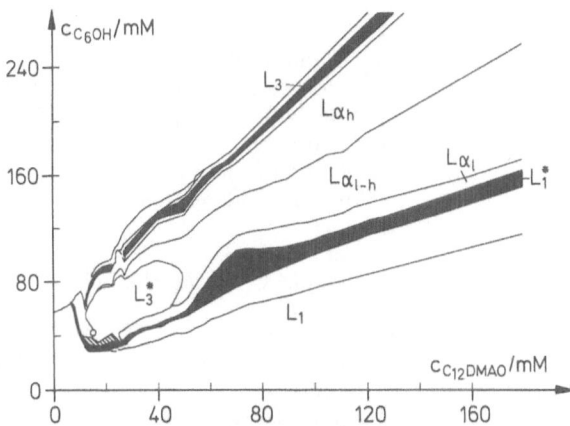

Fig. 2. Phase diagram of the system dodecyldimethyl-aminoxide/hexanol/water for small surfactant concentrations at 25 °C. The isotropic single-phase regions are L_1, L_1^* and L_3^*. The anisotropic single-phase regions are the lamellar phases L_{al} and L_{ah}

L_{al}-phase is destroyed and it takes a long time to recover. The bilayers of the two lamellar phases probably have a different thickness [18].

In Fig. 3 two samples of the different L_a-phases are viewed between crossed polarizers. The left one is a sample in the L_{ah}-range and the right one is a sample in the L_{al}-range. The polarizers are arranged in such a way that the axis of one of them is parallel to the axis of the cylindrical test tube.

In this position the L_{ah}-phase shows a bright texture with narrow spots of brillant colors while the L_{al}-phase shows a very weak birefringence. The intensity of the anisotropy is constant over the whole range of the sample. There are no domains emerging with more or less birefringence, as is the case in the L_{ah}-phase.

The birefringence in the L_{al}-phase is strongly influenced by small shearing forces. When the test tube with a L_{al}-phase is slightly tilted a large birefringence appears, as seen in Fig. 4. A large birefringence is also observed when the test tube remains in the upright position and the crossed polarizers are turned 45°. This experiment demonstrates convincingly that the microstructure of the L_{al}-phase is totally orientated in cylindrical glasstubes and arranged in concentric bilayers which are built up in the tube on a macroscopic scale.

2. Rheological measurements

The rheological measurements were made with a Rheometrics RF 7800 fluid rheometer at 25 °C, the measurement unit was a cone-plate system. The angular frequency was varied from 0.01 to 100 s⁻¹.

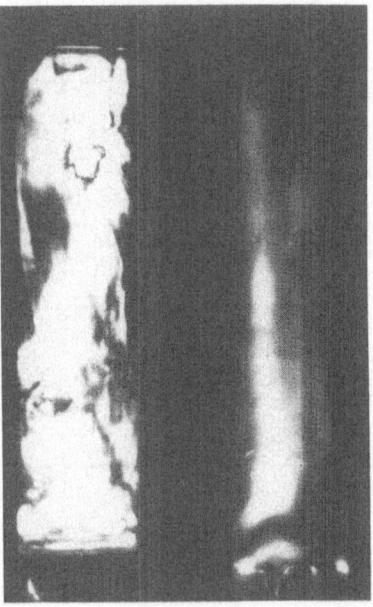

Fig. 3. Images of samples in the L_a-range shown through crossed polarizers without shearing. The left image shows a sample in the L_{ah}-phase with a composition of 100 mM C_{12}DMAO/200 mM C_6OH; the right one shows a sample in the L_{al}-phase with the composition of 100 mM C_{12}DMAO/120 mM C_6OH

Fig. 4. Images of samples in the L_a-range shown through crossed polarizers after tilting. The left image shows a sample in the L_{ah}-phase with a composition of 100 mM C_{12}DMAO/200 mM C_6OH; the right one shows a sample in the L_{al}-phase with the composition of 100 mM C_{12}DMAO/120 mM C_6OH

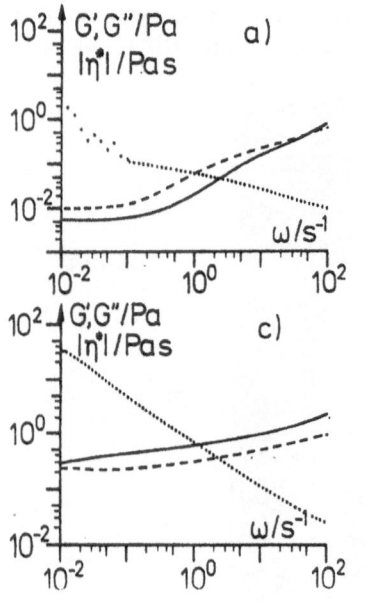

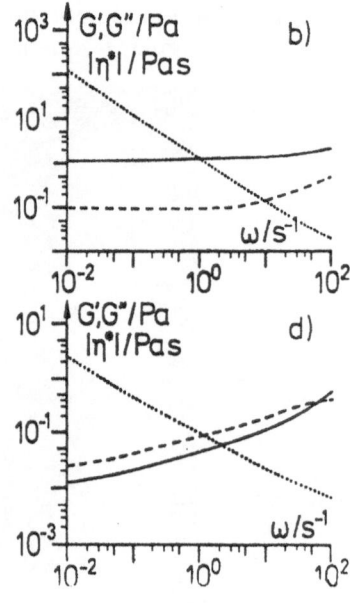

Fig. 5. Plot of the storage modulus G' (—), loss modulus G'' (---), and the complex viscosity $|\eta^*|$ ($\cdots$) as a function of the angular frequency at a constant surfactant concentration of 100 mM for different alcohol concentrations.

a) composition:
100 mM $C_{12}DMAO$/100 mM C_6OH
L_1^*-phase
b) composition:
100 mM $C_{12}DMAO$/110 mM C_6OH
L_{al}-phase
c) composition:
100 mM $C_{12}DMAO$/160 mM C_6OH
L_{al-h}-phase
d) composition:
100 mM $C_{12}DMAO$/200 mM C_6OH
L_{ah}-phase

The rheological behavior of a L_1^*-, L_{al}-, L_{al-h}- and L_{ah}-phase was measured for 100 mM $C_{12}DMAO$ with various amounts of n-hexanol (Fig. 5).

Figures 5a and d show that in the L_1^*- and L_{ah}-phase the storage moduli G' are smaller than the loss moduli G'' over the whole measured frequency range, while for both the L_{al}- and the L_{al-h}-phase (Fig. 5b and c) the storage moduli are larger than the loss moduli. The rheological results are in agreement with the observation of shear waves in the L_{al}- and L_{al-h}-phase. It is remarkable that both the storage and the loss moduli are nearly constant over the whole oscillating frequency range over several decades. The L_{al}- and the L_{alh}-phases have finite long structural relaxation times, but do not have a yield stress value. Bubbles that are mechanically dispersed in the samples rise in a few days.

In contrast to these two L_a-regions where the elastic properties dominate are the isotropic L_1^*- and the lamellar L_{ah}-phase. These solutions show only viscous properties.

Therefore, we cannot propagate shear waves in the L_1^*- and L_{ah}-phase, because their viscous moduli are higher than their elastic moduli. As we know from electron micrographs, the L_1^*-phase consists of vesicles which are not connected and lead to a low viscosity which we get also in micellar solutions [19]. When the vesicles are not isolated, but rather are connected, we would expect to

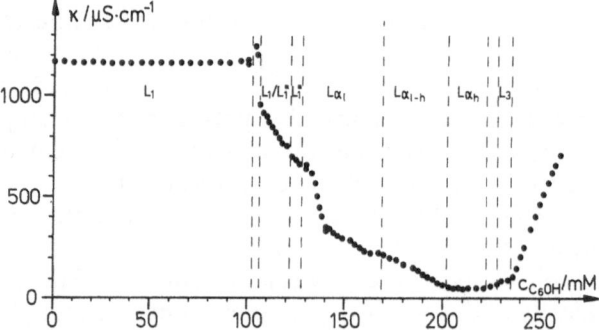

Fig. 6. Plot of the conductivity against the hexanol concentration at a constant surfactant concentration of 100 mM $C_{12}DMAO$ in the presence of 10 mM NaCl. The dashed lines show the phase boundaries detected by visual inspection after stopping the stirrer for a short time. Temperature 25 °C

observe a behavior like in entanglement networks with viscoelastic properties [20, 21]. The L_{ah}-phase has a low viscosity as does the L_3-phase; these results are not shown in the rheological plot [4, 22].

3. Conductivity measurements

Figure 6 shows a plot of the conductivity against the hexanol concentration at a temperature of 25 °C.

The total concentration was 100 mM of C_{12}DMAO. 10 mM NaCl was added to the solutions to get a constant background of ionic strength what is useful to detect conductivity differences in the low conducting zwitterionic surfactant system. The phase boundaries change only a little by addition of 10 mM NaCl. However, the phase boundaries change drastically under shear, most of all the L_1/L_1^* and the L_{ah}/L_3 transition. The experiment was carried out under shear in oder to obtain average values for the two phase regions.

In the micellar L_1-phase the conductivity does not change with the hexanol concentration. This means that both the number of charge carriers and their mobility is not affected by growth of the micelles. The growth of the micelles is reflected in the macroscopic viscosity, but the ions remain in the continuous aqueous phase and monitor only the water viscosity [19]. During shear the L_1/L_1^* region is shifted to higher hexanol content. The L_1^* region is detected as a slight turbid, shearinduced birefringent phase between crossed polarizers. There, it was surprising that the first two-phase region at the L_1-phase boundary shows two isotropic liquid phases after stopping the stirrer, and none of them belongs to the L_1^* region. By adding more hexanol the typical L_1/L_1^* phase was reached. Therefore, at a hexanol concentration of 106 mmol/l, we observe a drop in the conductivity which shows that part of the ions become trapped and do not contribute any more to the conductivity. Electron micrographs prove that the ions are trapped by vesicles.

In the further range of the lamellar phases the conductivity changes little but continuously with increasing hexanol concentration. In the L_{ah}-phase the conductivity is the lowest and is comparable to the conductivity of the L_3-phase under shear. Without shearing it is much higher than in the L_{ah}-phase. Then, the value is around 650 μs. This shows that the L_3-phase does not seem to be stable under shear and is transformed to the L_{ah}-phase under shear.

4. Electron microscopy

The freeze-fracture electron microscopy (FFEM) technique was used to investigate the typical microstructure in the different phase regions. A small amount of the sample was placed on a 0.1-mm-thick copper disc, covered with a second copper disc. The probe frozen by plunging the sandwich into liquid propane, which was cooled by liquid nitrogen. Fracturing and replication were carried out in a freeze-fracture apparatus BIOTECH 2005 (Leybold — Heraeus, Germany) at a temperature of −100°C. Pt/C was evaporated under an angle of 45°. The replicas were examined in a CEM 902 electron microscope (Zeiss, Germany).

The TEM images were prepared from samples that contained about 20% glycerol. The glycerol was added to suppress the formation of ice crystals during the freezing process. Figures 7 — 10 show TEM images of the system C_{14}DMAO/C_7OH at a total surfactant concentration of 100 mM with increasing cosurfactant concentration.

In Fig. 7 TEM images of samples in the L_{al}-range are shown. In this phase uni- and multilamellar

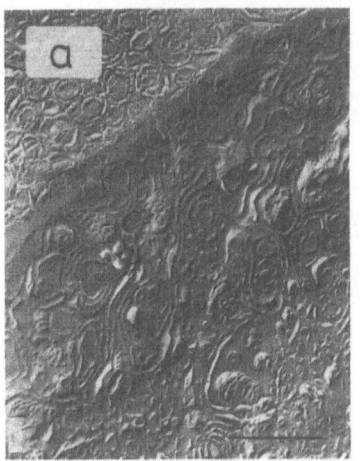

Fig. 7a, b

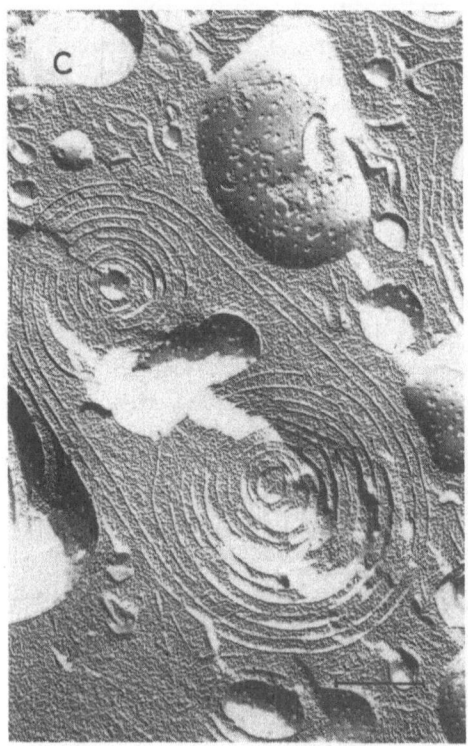

Fig. 7. TEM images of samples in the L_{al}-phase.
a) 100 mM C$_{14}$DMAO/70 mM C$_7$OH, bar = 1 µm
b) 100 mM C$_{14}$DMAO/80 mM C$_7$OH, bar = 1 µm
c) 100 mM C$_{14}$DMAO/80 mM C$_7$OH, bar = 0.5 µm

vesicles [23—27] are present, with a wide distribution of diameters that extends from about 100 nm to several µm. Mostly only in the larger ones the multilamellar structure can be seen, but it has to be taken into account that also the smaller ones may consist of several bilayers. The curvature of the bilayers and also defects in the bilayers can influence the fracture behavior. This could be a reason that the real multilamellar structure cannot be represented by this preparation method.

The micrographs, Figs. 7b and d, show that the bilayers of the vesicles are full of defects, so that some of them look more like skeletons of vesicles. We call these vesicles perforated vesicles. They can be observed best when the fractured samples are etched for a short time under controlled conditions. During the etching process some of the water is removed and the vesicular structures stick out of the fracture plane and cast sharp shadows during the evaporation with platinum [12, 24].

Images of samples of the same concentration without glycerol also show the defects, but they

seem to be a little larger. Therefore, it seems that the size of the defects depends on the conditions of the freezing process and does not reflect the size "in situ". The typical fracture for bilayer structures will follow the midplane of the bilayers. We recognize that, due to the increase of the size of the defects, more fractures occur perpendicular to the bilayers.

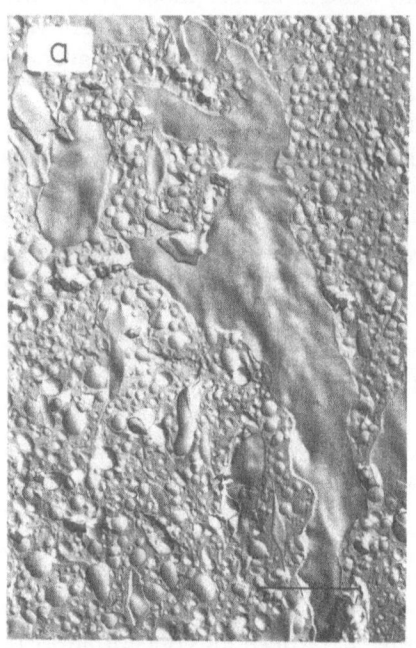

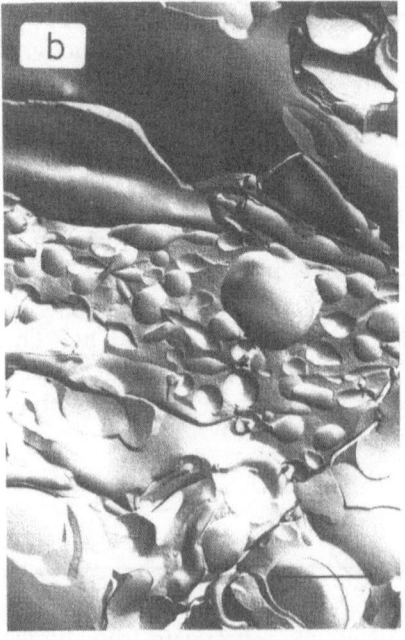

Fig. 8. TEM images of samples in the L_{al-h}-phase.
a) 100 mM C$_{14}$DMAO/100 mM C$_7$OH, bar = 3 µm.
b) 100 mM C$_{14}$DMAO/160 mM C$_7$OH, bar = 1 µm

We believe that the holes are no artefacts because many preparations show the same result and this result is also in agreement with the conductivity data that decrease at the phase boundary, but are still to high for a structure of vesicles with closed bilayers.

With increasing alcohol concentration we reach the L_{al-h}-phase (Fig. 8) where the vesicular structures are still present. The vesicles consist of closed bilayers so that the bilayers have a very strong curvature. In comparison to the micrographs in the L_{al}-phase the micrographs in the L_{al-h}-phase show a remarkable difference. The micrograph shows practically no multilamellar vesicles. For nearly all vesicles (with very few exceptions) the fracture does not occur through vesicles but follows the midplane of the vesicles. Beside these vesicular structures the micrographs show large planar bilayers with a small curvature.

In the single-phase region of L_{ah} no more vesicles are present, so that the structure consists only of flat, opened bilayers. The TEM image (Fig. 9) shows the two significant lamellar structures.

These are, on the one hand, the steps that are the result of the fracture that occurs perpendicular to the bilayers through the water-rich region and, on

Fig. 10. TEM image of a sample in the L_3-phase. The bar represents 0.5 µm. Composition: 100 mM $C_{14}DMAO$/200 mM C_7OH

the other hand, the large planar surfaces that can be seen when the fracture follows the midplane of the bilayers.

The L_3- or so-called sponge phase [28] consist also of bilayers, but in this case they are arranged in such a way that they form a bicontinuous network (Fig. 10).

The above described microstructures of the L_3-, L_{ah}- and L_{al-h}-phases were also found in the system $C_{12}DMAO/C_6OH$, as shown in Fig. 11. The L_{al}-phase is still under investigation because of the sensitivity of preparation.

Conclusions

For the tetradecyl- and dodeyl-dimethylaminoxide surfactants and the cosurfactants heptanol and hexanol the dilute section of the phase diagrams was established by observing the macroscopic behavior of the phases. The sequence of the phases for a 100 mM surfactant solution of $C_{12}DMAO$ or $C_{14}DMAO$ with increasing amount of alcohol is: L_1, L_1^*, L_{al}, L_{al-h}, L_{ah}, and L_3. Some phases are separated by macroscopic two-phase regions, others are not. The rheological measurements show that the two lamellar phases are of different rheological behavior.

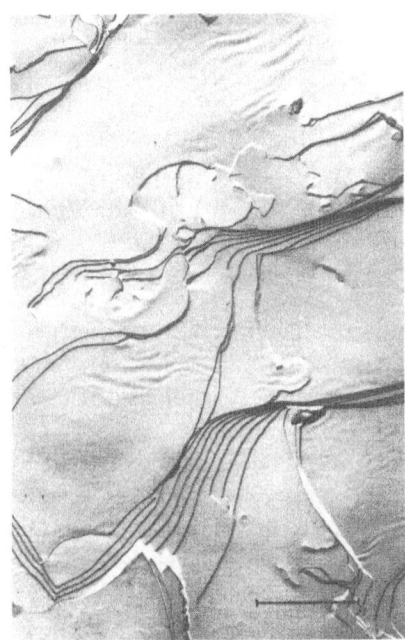

Fig. 9. TEM image of a sample in the L_{ah}-phase. The bar represents 3 µm. Composition: 100 mM $C_{14}DMAO$/190 mM C_7OH

The elastic shear moduli are higher in the L_{al}- and the two-phase region $L_{al\text{-}h}$ than the viscous shear moduli. This is the evidence for the existence of shear waves in these two phases, whereas in the L_{ah}- and the L_1^*-phases the viscous shear moduli dominate. Because of their rheological properties it

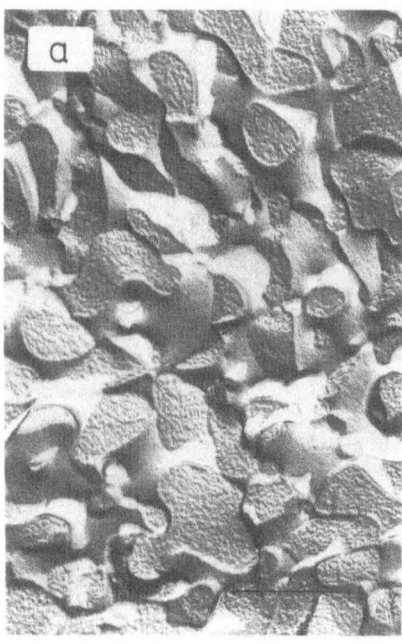

Fig. 11. TEM images of the system C_{12}DMAO/C_6OH in the different phase regions.
a) composition:
   100 mM C_{12}DMAO/220 mM C_6OH L_3-phase, bar = 0.3 µm
b) composition:
   100 mM C_{12}DMAO/200 mM C_6OH L_{ah}-phase, bar = 2 µm
c) composition:
   100 mM C_{12}DMAO/150 mM C_6OH $L_{al\text{-}h}$-phase, bar = 1 µm

is not surprising that in the $L_{al\text{-}h}$-region the two L_a-phases do not separate macroscopically.

The transition in the micellar structure is reflected in the conductivity/hexanol concentration plot. There is a jump in the conductivity where the two-phase region L_1/L_1^* starts.

The difference in the macroscopic properties is evident on a microscopic level. The electron micrographs give us proof for the existence of the different phases. Both the L_{al}- and the L_{ah}-phase consist of bilayer structures. The difference lies in the curvature of the bilayers. In the lowest lamellar structure L_{al}, we find uni- and multilamellar vesicles so that the conductivity must decline. This is indeed the case, but the conductivity does not decrease to values one would expect for closed multilamellar vesicles. The reason, therefore, might be the holes found in the bilayers. The lamellar structure with a higher amount of alcohol consists of flat bilayers which are typical for lamellar phases.

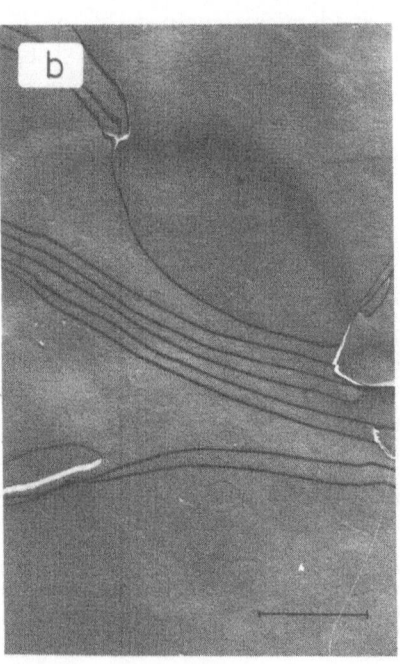

FFEM shows, furthermore, that the sponge phase consists of a bicontinuous network that is in agreement with the increase of the conductivity at the phase boundary L_{ah}/L_3.

By the FFEM, we were able to show the different microstructures that belong to the various macroscopic phases of the phase diagrams. It is obvious that it is important to know whether there is a difference or not in various lamellar and isotropic phases.

Acknowledgement

Financial support of this work by grants of the Deutsche Forschungsgesellschaft, DFG (SFB 213) is appreciated.

References

1. Strey R, Winkler J, Magid L (1991) J Phys Chem 95:7502
2. Fontell K, Khan A, Lindström B, Maciejewska D, Puang-Ngern S (1991) Colloid Polym Sci 269:727—742
3. Marignan J, Gauthier-Fournier F, Appell J, Akoum F, Lang J (1988) J Phys Chem 92:440
4. Miller CA, Gradzielski M, Hoffmann H, Krämer U, Thunig C (1990) Colloid Polym Sci 268:1066—1072
5. Miller CA, Gosh O (1986) Langmuir 2:321
6. Ghosh O, Miller CA (1987) J Phys Chem 91:4528
7. Gomati R, Appell J, Bassereau P, Marignan J, Porte G (1987) J Phys Chem 91:24
8. Bassereau P, Marignan J, Porte G (1987) J Phys Chem 48:673—678
9. Hoffmann H (1990) Progr Colloid Polym Sci 83:16—23
10. Wang ZG, Safran SA (1990) Europhys Lett 11:425-430
11. Platz G, Thunig C, Hoffmann H (1990) Progr Colloid Polym Sci 83:167
12. Hoffmann H, Thunig C, Munkert U, Meyer HW, Richter W (1992) Langmuir 8:2629
13. Balinov B, Olsson U, Söderman O (1991) J Phys Chem 95:5931—5936
14. Cates ME, Roux D (1990) J Phys Condens Malter 2:SA339
15. Cates ME, Roux D, Andelmann D, Milner ST, Safran SA (1988) Europhys Lett 5:733—739
16. Satoh N, Tsujii K (1987) J Phys Chem 91:6629
17. Thunig C, Hoffmann H, Platz G (1989) Progr Colloid Polym Sci 79:297
18. Platz G, Thunig C, Hoffmann H (1992) Ber Bunsenges Phys Chem 96:667—677
19. Valiente M, Thunig C, Lenz U, Hoffmann H in preparation
20. Candau J, Hirsch E, Zana R, Delsanti M (1989) Langmuir 5:1225
21. Brye TJ, Cates ME (1992) J Chem Phys 96:1367
22. Miller CA, Gradzielski M, Hoffmann H, Krämer U, Thunig C (1991) Prog Colloid Polym Sci 84:242—249
23. Hoffmann H, Thunig C, Valiente M, Munkert U, in preparation
24. Lenz U, Munkert U, Hoffmann H in preparation
25. Kaler E, Murthy AK, Rodriguez BE, Zasadzinski JAN (1989) Sci 245:1371
26. Kaler E, Herrington KL, Murthy AK, Zasadzinski JAN (1992) J Phys Chem 96:6698—6707
27. Laughlin RG, Munyon RL, Burns JL, Coffindaffer TW, Talmon Y (1992) J Phys Chem 96:374—383
28. Strey R, Jahn W, Porte G, Bassereau P (1990) Langmuir 6:1635—1639

Authors' address:

Prof. Dr. H. Hoffmann
Lehrstuhl für Physikalische Chemie I
Universität Bayreuth
Universitätsstr. 30, 95447 Bayreuth, FRG

Progress in Colloid & Polymer Science   Progr Colloid Polym Sci 93:146—149 (1993)

Counterion effects in the self-assembly of didodecyldimethylammonium halides and sulphate surfactants

A. Khan and C. Kang

Division of Physical Chemistry 1, Chemical Centre, University of Lund, Sweden

Abstract: Binary phase diagrams of the aqueous didodecyldimethylammonium surfactant with four different counterions — Cl^- (DDAC), OH^- (DDAOH), CH_3COO^- (DDAAc) and SO_4^{2-} (DDAS) are determined at 298K. A lamellar liquid crystalline phase is identified for all systems. DDAOH and DDAAc also form normal micellar solution phases in addition to lamellar liquid crystals. The swelling of the lamellar phase at maximum water is much reduced with SO_4^{2-} counterion compared to that with other monovalent counterions. The ternary phase diagram determined for the system DDAS-water-dodecane at 298K shows that the system forms a water-rich solution phase with oil-in-water type structure, a normal hexagonal liquid crystalline phase at intermediate water contents, and an oil-rich cubic phase with a bicontinuous-type structure, in addition to the lamellar phase formed in the binary system.

Key words: Didodecyldimethylammonium surfactant; mono- and divalent counterions; phase diagrams; liquid crystals; microstructures

Introduction

It is a well known fact that the long-range electrostatic effects play a dominant role in the self-assembly processes of ionic surfactants [1]. Increasing the valency of the counterion from one to two drastically changes the electrostatic interactions in the ionic surfactant system. Moreover, the counterion hydration may also play an important role in surfactant aggregation process. In a series of studies [2], we have observed the effect of counterions and the counterion valency on the phase behavior of several anionic and cationic surfactant systems. However, no such systematic studies have been reported for the double tailed quaternary ammonium type surfactants. The purpose of this study is to provide an insight into the self-assembly processes of a double-tailed quaternary ammonium surfactant with different counterions. To this end, we have studied the phase equilibria of four binary systems water-didodecyldimethylammonium chloride (DDAC), hydroxide (DDAOH), acetate (DDAAc) and sulphate (DDAS)

and one ternary system DDAS-water-dodecane by water deuteron NMR and optical microscopy methods at 298K and the microstructures of single solution phases are studied by the multicomponent pulsed gradient spin echo (PGSE) FT-NMR technique.

Materials and methods

Preparation of surfactants

The surfactants used in this work are prepared from didodecyldimethylammonium bromide (DDAB) by an ion-exchange method as described [3]. Strongly alkaline solution of didodecyldimethylammonium hydroxide is obtained by exchanging the Br^- ion of the purified DDAB (Fluka Ltd.) with OH^- ion through the anionic ionexchange resin Dowex SBR (Dow Chemical company). DDAS, DDAC, DDAAc are, respectively, obtained by neutralizing the hydroxide solution with dilute H_2SO_4 solution at pH = 6, dilute HCl solution at pH = 6 and dilute CH_3COOH solution at pH = 9 and the solid surfactants are obtained by lyophilizing the resultant surfactant solutions.

Sample preparation

The samples are prepared by weighing appropriate amounts of substances into glass tubes which are flame sealed. The samples are mixed by centrifigution and equilibriated as described previously [4].

^2H NMR and Optical microscopy

The water ^2HNMR in combination with optical polarizing microscopy have been used to determine the phase diagrams of the surfactant systems. Anisotropic liquid crystalline materials exhibit their characteristic microscopic textures [5] which are used to identify lamellar and hexagonal liquid crystalline phases, whereas the isotropic phases such as micellar solutions or cubic liquid crystalline phases do not exhibit any microscopic texture, instead they produce a black background when viewing through the microscope.

The ^2HNMR line-shapes [6] are used to identify single anisotropic liquid crystalline phase as well as multi-phase heterogeneous regions consisting of isotropic and anisotropic phases. Analysis of ^2HNMR line shapes and their intensity vs sample compositions yield the stability regions of single phases and heterogeneous regions; hence, the complete phase diagrams of binary and ternary surfactant systems can be determined. The ^2HNMR spectra are recorded at 15.71 MHz on the Brukar MSL 100 spectrometer.

Self-diffusion measurements

The self-diffusion coefficients of water, surfactant and oil in isotropic solutions are measured at 303 K by the pulsed gradient spin echo FT-NMR technique [7], monitoring the ^1HNMR spectra on a modified Jeol FX-60 spectrometer using an external ^2H lock.

Results and discussion

Binary surfactant systems

The binary phase diagrams for the aqueous systems-DDAC, DDAOH, DDAAc and DDAS deter-

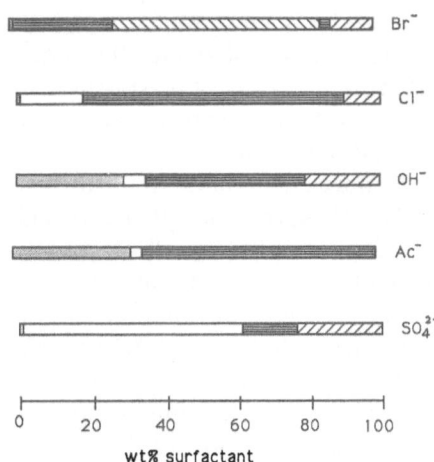

Fig. 1. Binary phase diagrams for the aqueous systems of didodecylammonium — bromide (Br⁻), (at 20°C), chloride (Cl⁻), hydroxide (OH⁻), acetate (Ac⁻) and sulphate (SO_4^{2-}) (at 25°C). Phase notations: ▨, isotropic solution phase; ▤, lamellar liquid cryatalline phases; ▭, ◺, and ▨, appropriate two-phase regions

mined experimentally are shown in Fig. 1. The binary phase diagram for the DDAB system reported previously [8] is redrawn and also shown in Fig. 1. At room temperature, the systems DDAC and DDAS are practically insoluble in water, instead they swell in water, giving rist to lamellar dispersions in water. At high surfactant concentrations, a single lamellar phase is produced in both the systems. On the other hand, DDAOH and DDAAc systems are soluble in water forming micellar solution phases followed by homogeneous lamellar liquid crystalline phases. The DDAB system is shown to form two lamellar phases — one, at water-rich part, and the other, at water-poor region and the two lamellar phases coexist over an extensive twophase region.

One significant difference among the systems is that the swelling capability of the single lamellar phase within its stability range with water (n = number of molecules of water per surfactant ion) is drastically reduced with divalent SO_4^{2-} counterion (n = 14) compared to systems with monovalent Cl⁻ (n = 102), OH⁻ (n = 37) and CH_3COO^- (n = 42) ions. However, the minimum number of water molecules per surfactant ion required to form the homogeneous lamellar phase does not differ (n = 3—5) significantly for all five systems.

The formation of lamellar phase by the double-tailed quaternary ammonium surfactants in water may be understood from surfactant geometrical constraints [9]. The Poisson-Boltzmann (PB) equation which predicts only repulsive electrostatic interaction is used satisfactorily to explain experimentally obtained phase behavior for the surfactant systems with monovalent counterions in one coherent model [1]. However, the PB theory fails completely to rationalize the swelling of the lamellar phase with divalent counterions. The Monte Carlo simulations [10] show that one should expect an attractive electrostatic interaction between lamellar with divalent and not with monovalent counterions. This is precisely the implication obtained from the phase behavior of the DDAS system. With SO_4^{2-} counterion, a concentrated lamellar phase is in equilibrium with a dilute solution. This can only occur if the lamellar phase is held with some attractive force.

Ternary surfactant systems

The single phases identified for the isothermal ternary system DDAS-water-dodecane are shown in Fig. 2. For comparision, the single phases reported [8] for the corresponding surfactant system with Br^- counterion are also reproduced. As can be seen, the isotropic and anisotropic single phases formed in the two systems are the same, but the location of identical phases, their existence of stability ranges and aggregate structures of some single phases are drastically different between the two systems.

The isotropic solution phase, L_1, forms in the water-rich part with SO_4^{2-} counterion, whereas with Br^- ion, the isotropic solution phase, L_2, exists in the oil-rich corner and the region becomes wider as it extends to the water-surfactant axis. The bicontinuous cubic phase for the sulphate system exists in the water-poor region and the phase can incoporate about 49% of oil, but the cubic phase, which is also of the bicontinuous type, for the bromide system appears at high water and low oil contents. The reversed hexagonal phase forms with high amounts of water with Br^- counterion, but with SO_4^{2-} ion, normal hexagonal phase can exist only in the intermediate water contents. The lamellar phase appeared in the water-poor part of the bromide system can solubilize about 8% of dodecane against about 3% in the lamellar phase with SO_4^{2-} as a counterion. Like binary systems,

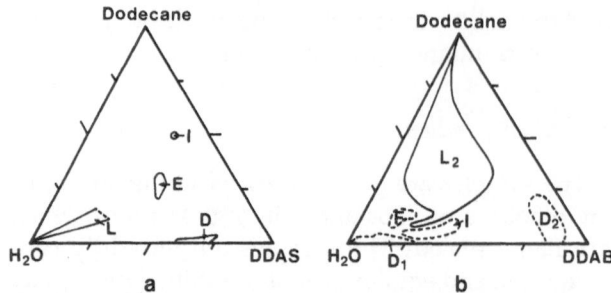

Fig. 2. Isothermal ternary phase diagrams for the systems (a) DDAS-dodecane-water at 25 °C and (b) DDAB-dodecane-water at 20 °C. Phase notations: L_1 and L_2, isotropic solution phases; D, D_1 and D_2, lamellar liquid crystalline phase, E and F, hexagonal liquid crystalline phases; I, isotropic liquid crystalline phases

the phase behavior of the ternary systems can be rationalized qualitatively from the consideration of electrostatic effects and surfactant packing parameter.

Solution microstructures

Self-diffusion coefficients of water, D^w, surfactant, D^s and oil, D^o obtained by the PGSE FT-NMR method for the isotropic solution phases of the binary DDAOH-H_2O, DDAAc-H_2O and the ternary DDAS-H_2O-dodecane are given in Fig. 3. A complete analysis of self-diffusion data will be given elsewere. Here, a qualitative description is given briefly.

For the hydroxide and acetate systems, the water self-diffusion coefficient, D^w, is fast and the D^w-value is not much different from that of the neat water for the dilute samples. The observed reduction of D^w-values with increased surfactant concentration can be accounted for by obstruction and hydration effects [11]. On the other hand, the surfactant self-diffusion coefficient, D^s, is much smaller than that of water. For example, at about 0.01 molal DDAOH solution, $D^s \sim 6 \cdot 10^{-11}$ m^2 s^{-1} against $D^w \sim 2 \cdot 10^{-9}$ and at about 0.2 molal solution, $D^s \sim 8 \cdot 10^{-12}$ against $D^w \sim 1.5 \cdot 10^{-9}$. For the acetate system, the values are $D^s \sim 8.8 \cdot 10^{-11}$, $D^w \sim 2.2 \cdot 10^{-9}$ at 0.03 molal solution and $D^s \sim 1.5 \cdot 10^{-11}$ against $D^w \sim 1.3 \cdot 10^{-9}$ at about 0.2 molal solutoin. Thus, both the systems form micelles with normal aggregate structure. Neglecting electrostatic effects and intermicellar interactions for very dilute

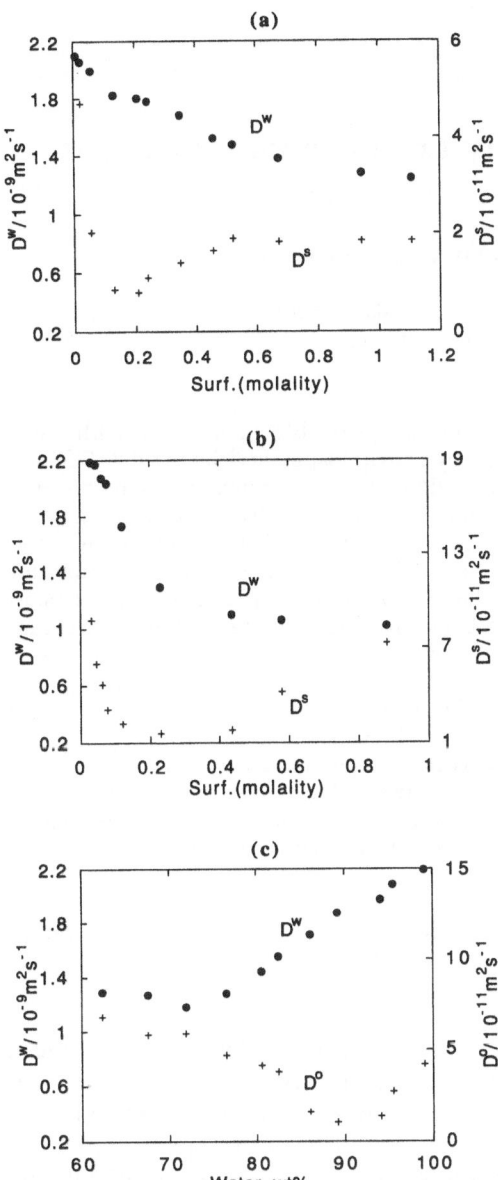

Fig. 3. Components' self-diffusion coefficients at 30°C for the systems a) DDAOH-water, b) DDAAc-water and c) DDAS-water-dodecane at a constant molar ratio of X(oil)/X(surf.) = 0.56

micellar growth in both the systems with increased surfactant concentration. However, the increased D^s-values observed at higher surfactant concentrations (> 0.2 molal) are difficult to explain at present. More data are needed to understand the micellization process at higher surfactant concentrations.

Like acetate and hydroxide systems, D^w-values obtained (Fig. 3) in the L_1-phase for the ternary DDAS-water-dodecane system is high. The relative diffusion coefficient, D^w/D_0^w, where D_0^w is the diffusion coefficient of the neat water, varies from ~ 0.9 at 98% of ^2H_2O to 0.5 at 60% of ^2H_2O whereas D^o-values (here, $D^o \approx D^s$) are very low and vary 10 — 68 $\cdot$ 10^{-12} m^2 s^{-1} between 60 and 99% of ^2H_2O. Clearly, the system forms oil-in-water type structure. However, the droplets are of asymmetric shapes.

References

1. Jönsson B, Wennerström H (1987) J Phys Chem 91:338
2. Wennerström H, Khan A, Lindman B (1991) Adv in Colloid and Interface Sci 34:433 and refs. therein
3. Chen SJ, Evans DF, Ninham BW (1984) J Phys Chem 88:1622
4. Kang C, Khan A (1993) J Colloid Inter Sci 156:218
5. Rosevear FB (1968) J Soc Cosmet Chem 19:581
6. Khan A, Fontell K, Lindblom G, Lindman B (1982) J Phys Chem 86:4266
7. Stilbs P (1987) NMR Spectroscopy 19:1
8. Fontell K, Ceglie A, Lindman B, Ninham BW (1986) Acta Chem Scand A 40:247
9. Israelachvili JN, Mitchell DJ, Ninham BW (1976) J Chem Soc Faraday Trans 2 72:1525
10. Guldbrand L, Jönssonm B, Wennerström H, Linse P (1984) J Chem Phys 80:2221
11. Bell GM (1965) Trans Faraday Soc 60:1752

systems, the radius of a micelle, $R \sim 40$ Å and 25 Å are, respectively, obtained for the hydroxide and acetate systems at 0.01 molal solution by Stokes-Einstein equation. Furthermore, the results indicate that the initial formation of micelles with CH_3COO^- counterion is approximately spherical and that with OH^- ion is non-spherical. There is a

Authors' address:

Ali Khan
Division of Physical Chemistry 1
Chemical Centre, Box 124, University of Lund
22100 Lund, Sweden

Progress in Colloid & Polymer Science

Progr Colloid Polym Sci 93:150—152 (1993)

Thermodynamics and volume compressibility of phosphatidylcholine liposomes containing bacteriorhodopsin

T. Hianik[1]), B. Piknova[1]), V. A. Buckin[2]), V. N. Shestimirov[2]), and V. L. Shnyrov[2])

[1]) Department of Biophysics and Chemical Physics, Comenius University, Bratislava, Slovakia
[2]) Institute of Biological Physics, Russian Academy of Sciences, Pushchino, Russia

Abstract: The effects of bacteriorhodopsin (BR) interaction with large unilamellar vesicles (LUV) of dimyristoylphosphatidylcholine (DMPC) and dipalmitoylphosphatidylcholine (DPPC) were examined at various BR/lipid ratios, using differential scanning calorimetry (DSC) and ultrasound velocimetry (USV). On DSC, a shift of T_c towards higher temperatures and considerable dynamics of structural transition of membranes were found for DMPC proteoliposomes. The changes of thermodynamical parameters suggest long-distance interactions between regions of altered bilayer structure which arise around each BR molecule. The changes of ultrasound absorbance and velocity were in coincidence with DSC results. The concentration increment of ultrasound velocity that characterizes volume compressibility of vesicles, increased with the increasing BR concentration in DPPC LUV in gel state $(T < T_c)$. Saturation of concentration increment was observed at BR/LUV ratio 1/0.5 mol/mol. No significant changes of compressibility were observed at liquid-crystal state $(T > T_c)$. This means that, in the gel state of lipid bilayer, one BR molecule is enough to change the physical properties of one LUV.

Key words: Bacteriorhodopsin — phosphatidylcholine liposomes — phase transitions — calorimetry — ultrasound velocimetry

Introduction

The study of protein-lipid interactions represents an important step in understanding processes of energy transfer, reception, etc. In this respect interaction of integral proteins with lipids are of special interest. Integral proteins span the membrane transversely, establishing contacts between their hydrophobic moieties and the hydrocarbon chains of the lipids. Owing to the different geometry of the hydrophobic moiety of proteins and that of lipids, as well as due to the action of electrostatic and elastic forces, region of altered structure may arise around protein molecules. The formation of similar regions may represent one of the reasons for the occurrence of long-distance interactions in a membrane (see [1, 2] for review). Several questions arise in this respect: 1) What is the nature and dynamics of protein-lipid interactions? 2) Do structurally and mechanically nonequilibrium states of the membrane occur during the process of protein incorporation in the lipid bilayer? To obtain quantitative information concerning the influence of integral protein on their lipid environment, the present investigation was focused on the effects of bacteriorhodopsin (BR) on thermodynamics and mechanical properties of large unilamellar vesicles (LUV) from dimyristoylphosphatidylcholine (DMPC) and dipalmitoylphosphatidylcholine (DPPC). The thermodynamical parameters were estimated using differential scanning calorimetry (DSC). The effect of BR on mechanical properties of liposomes was studied using differential ultrasound velocimetry.

Materials and Methods

Experiments were performed on LUV (approx. 100 nm in diameter), prepared with the method of detergent dialysis as described elsewhere [3]. DMPC and DPPC (Calbiochem) were purity grade and used without further purification. Fragments of purple membranes (PM) containing BR isolated from Halobacterium halobium (strain 353P) were used and added into the LUV suspension. All experiments were done with dark-adapted BR, i. e., the sample was kept in dark throughout the measurements which started 30 min after the chambers had been filled with liposome suspension; this is an interval sufficient for BR to become dark-adapted [4]. DSC measurements were made with DASM 4 differential scanning microcalorimeter [5]. Mechanical properties (volume compressibility) were studied by measurements of ultrasound velocity using RADA 4 or RADA S differential ultrasound scanning velocimeters [3, 6].

Results

DSC measurements of LUV containing BR allow to study the dynamics of structural changes of lipid bilayer. Figure 1 shows changes of specific heat capacity ΔC_p per mole of lipids from pure DMPC (a) and for those in a mixture with BR-containing PM for two characteristics BR/DMPC ratio: 1/5.8 (b) and 1/0.4 mol/mol (c). Here, we do not include the lipids around the BR in PM. Similar changes of ΔC_p as for pure DMPC were characterized also for DMPC proteoliposomes up to ratio BR/DMPC = 1/93. However, the shift of transition temperature T_c toward high values occur. A considerable time development of thermodynamical parameters were observed at BR/DMPC molar ratio from 1/77 to 1/0.4. In addition, beginning from BR/DMPC = 1/32 after a certain time of cyclic scan we start to observe a second peak at higher temperatures (Fig, 1b). At higher BR concentration we have observed only peaks at higher temperatures (Fig. 1c). In contrast with DMPC, T_c for DPPC proteoliposomes practically does not shift with BR concentration and any time changes of thermograms were not observed in this case.

Figure 2 shows the temperature dependence of ultrasound velocity change $\Delta u/u_0$ (curves 1,1') and absorption change per wavelength unit $\Delta(\alpha\lambda)$ (curves 2,2') for LUV from DMPC containing BR in ratio BR/DMPC = 1/0.7. The absorption curve of the first scan (curve 2) has a marked maximum at $T = 35\,°C$. This maximum is, in comparison with pure DMPC, shifted toward higher temperatures on

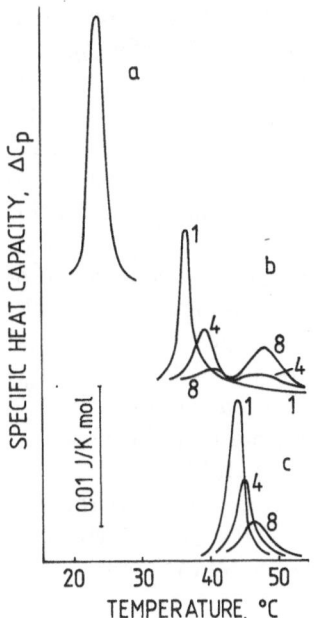

Fig. 1. Changes of specific heat capacity ΔC_p for LUV from pure DMPC and BR/DMPC recombinants. BR/DMPC ratios: 0 (a); 1/5.8 (b); 1/0.4 mol/mol (c). Time development of thermograms 1,4 and 8 h after the beginning of experiments is marked by corresponding numbers at the curves. Scans have had baselines subtracted from the curves

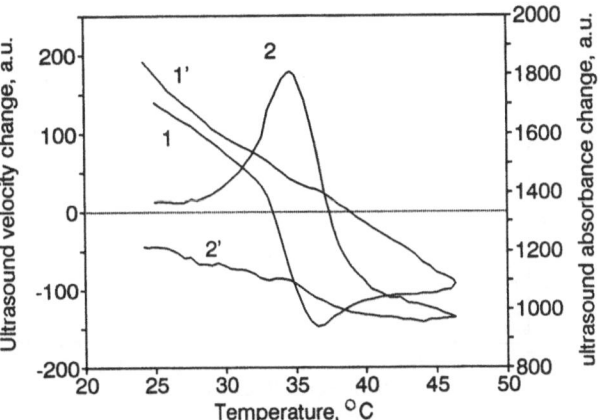

Fig. 2. Temperature dependence of relative ultrasound velocity changes (curves 1,1') and relative ultrasound absorbance changes (curves 2,2') of DMPC LUV containing BR (BR/DMPC = 1/0.7 mol/mol). Scanning from low to high (1,2) and from high to low (1', 2') temperatures

about 5°C. Strong ultrasound absorption in the phase transition region is associated with a strong growth of thermal fluctuations [7]. Dependence of $\Delta u/u_0$ (T) in the first scan (curve 1) dramatically decreases within the gel ($T < T_c$) to liquid-crystal

$(T > T_c)$ phase transition interval, which is associated with a change of the lipid ordering and, thus, with different volume compressibility of LUV. As is seen from Fig. 2, the changes of ultraso und parameters have considerable hysteresis. We have also observed similar behavior but less expressed hysteresis for smaller BR concentrations in DMPC and DPPC LUV.

To check the possible changes of mechanical properties of membranes at the incorporation of BR, we have measured the changes of ultrasound velocity concentration increment $\delta A = \delta (\Delta u/u_0 c)$ (c is the concentration of lipid) of the BR/LUV system in different phase state of lipid bilayer of DPPC for $T = 25°C$ ($T < T_c$) and $T = 50°C$ ($T > T_c$). Addition of BR leads to growth of parameter δA only at the gel state of DPPC (Fig.3). We have not observed any changes of δA at the liquid crystal state of lipid bilayer.

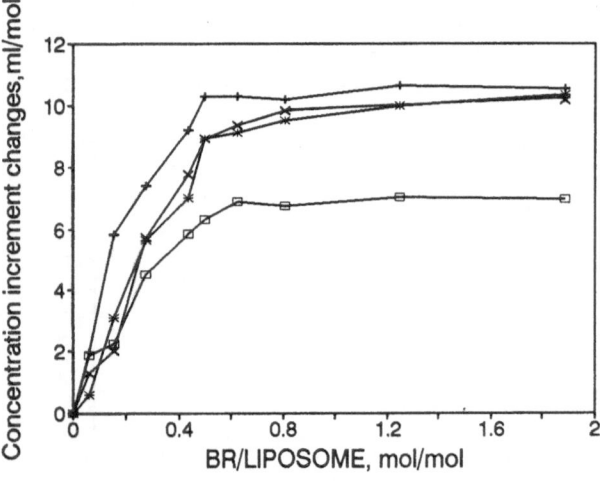

Fig. 3. Dependence of concentration increment of ultrasound velocity changes δA on molar ratio of BR and DPPC LUV for several independently prepared samples of LUV of identical composition at gel state of DPPC ($T = 25°C$)

Discussion

Our results show evidence about considerable changes of lipid bilayer physical properties under the influence of BR. The shift of T_c of DMPC LUV

toward high temperatures with increasing of BR concentration support the mattress model of protein-lipid interaction [8]. According to this model, we can imagine BR as a rigid body that, thanks to different lengths of the hydrophobic part of BR and membrane, has a disordered membrane structure caused by formation of wide regions of varying thickness. These changes are less expressed for DPPC because the average thickness of its hydrophobic part is similar to BR hydrophobic moiety.

Considerable changes of TD parameters of DMPC proteoliposomes can be due to rigidization of the membrane as well as to origin a new phase of the lipid bilayer. At higher BR concentration this phase plays a dominant character. BR probably structuralizes the lipid chains and extends them. It makes the DMPC hydrophobic length similar to that of DPPC. This assumption is also supported by considerable increasing of T_c of BR/DMPC system that becomes similar to T_c for BR/DPPC proteoliposomes ($T_c \sim 41°C$).

According to our analysis, the main reason for the changes of δA in dependence on BR concentration in gel membrane state is influence of BR on the large membrane region. From critical ratio BR/LUV ~ 0.5 mol/mol, we can assume that one BR molecule is able to change the physical properties of one LUV. In contrast to gel state, probably a considerably disordered membrane in liquid-crystal phase does not allow BR to change the structural state of lipid bilayer.

References

1. Riegler J, Möhwald H (1986) Biophys J 49:1111—1118
2. Lee AG (1991) Prog Lipid Res 30:323—348
3. Piknova B, Hianik T, Shestimirov VN, Shnyrov VL (1991) Gen Physiol Biophys 10:395—409
4. Stockenius W, Lozier RH, Bogomolni RA (1979) Biochim Biophys Acta 505:215—278
5. Privalov PL (1980) Pure Appl Chem 52:479—497
6. Sarvazyan AP (1982) Ultrasonics 20:151—154
7. Mitaku S, Date T (1982) Biochim Biophys Acta 688:411—421
8. Mouritsen OG, Bloom M (1984) Biophys J 46:141—153

Authors' address:

Prof. Dr. Tibor Hianik
Department of Biophys. Chem. Phys.
Fac. Math. Physics
Comenius University
Mlynska dol. F1
84215 Bratislava, Slovakia

Thermodynamic stability of lipid phases in low-molecular solutions

J. G. Brankov[1]) and B. G. Tenchov[2])

[1]) Institute of Mechanics and Biomechanics, Bulgarian Academy of Sciences, Sofia, Bulgaria
[2]) Central Laboratory of Biophysics, Bulgarian Academy of Sciences, Sofia, Bulgaria

Key words: Lipid phases — low-molecular solutes — phase transitions — Clapeyron-Clausius equation

The ability of biomembrane lipids to undergo phase transformations into non-bilayer states, or into a variety of crystalline and gel phases, is a well-known phenomenon. It is thought to be involved in various cellular events such as fusion, cell permeability, biomembrane damage, etc. Since, in many cases, these events are strongly influenced by the composition of the aqueous phase, it is of interest to develop a thermodynamic description of the stability of the lipid phases in aqueous solutions of low-molecular solutes. To this end, we consider a model thermodynamic system Σ consisting of an aggregate of N_l lipid molecules together with N_w molecules of bound water and N_c solute molecules. The system Σ is kept at mechanical and thermal equilibrium at pressure p and temperature T, and in material contact with a reservoir R of water and solute molecules, the chemical potentials of which are μ_w and μ_c, respectively. In order to simplify the arguments, we neglect the lipid solvability and assume that N_l remains constant. The state of Σ is completely described by one extensive quantity, the number of lipid molecules N_l, and four intensive variables: pressure p, temperature T, and chemical potentials μ_w, μ_c. Its thermodynamic potential is

$$G^{(1)}(p, T, N_l, \mu_w, \mu_c) = N_l \mu_l (p, T, \mu_w, \mu_c) .$$

All quantities of interest can be obtained by taking the appropriate derivatives.

Consider a first-order phase transition point, where two phases of the lipid aggregate Σ coexist. The condition for the phase equilibrium reads

$$\mu_l' (p, T, \mu_w, \mu_c) = \mu_l'' (p, T, \mu_w, \mu_c) , \tag{2}$$

where the superscripts distinguish between the high- and low-temperature phase. Equation (2) defines the transition temperature T_t as an implicit function of p, μ_w and μ_c. Model expressions for μ_w and μ_c can be obtained by considering the reservoir as an ideal solution with solute concentration c. An extended version of the Clapeyron-Clausius equation follows then,

$$\Delta V^T dp - \Delta S^T dT$$
$$+ k_B T [\Delta N_w/(1-c) - \Delta N_c/c] dc = 0 , \tag{3}$$

where ΔV^T and ΔS^T are the changes in the volume and entropy of the total system $\Sigma + R$, while $\Delta N = N' - N''$ refer to Σ. At constant pressure the derivative of the transition temperature $T = T_t$ with respect to the concentration c can be determined from Eq. (3). If we assume that c is small, denote by $q_t = T_t \Delta S^T/N_l$ the latent heat of the transition per lipid molecule, and introduce for each phase a of Σ the corresponding fraction of bound water per lipid molecule, $x^a = N_w^a/N_l$, and solute concentration in the bound water, $c^a = N_c^a/N_w^a$, we obtain

$$dT_t/dc = (k_B T_t^2/q_t)[(1 - c'/c) x' - (1 - c''/c) x''] . \tag{4}$$

A useful simplification of Eq. (4) follows if we neglect the change of the solute concentration in the bound water at the phase transition and set $c' = c'' = c^l$:

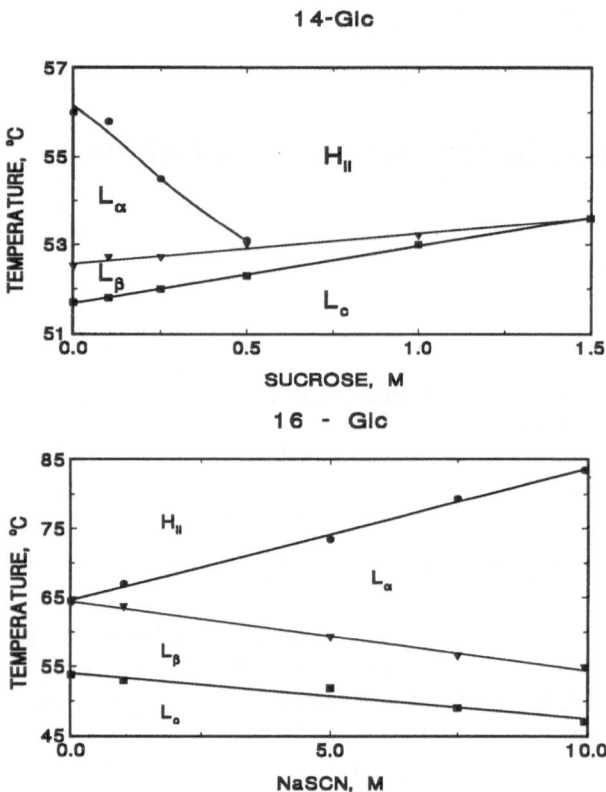

Fig. 1. Phase transition temperatures of glycoglycerolipids in sucrose (top) and NaSCN (bottom) solutions. Sucrose is a kosmotropic and NaSCN a chaotropic solute. The two lipids are 1,2-di-*O*-tetradecyl-3-*O*-β-D-glucosyl-sn-glycerol (14-Glc) and 1,2-di-*O*-hexadecyl-3-*O*-β-D-glucosyl-sn-glycerol (16-Glc).

$$dT_t/dc = (k_B T_t^2/qt)(1 - c^l/c)(x' - x'') . \qquad (5)$$

Some general features of the transition temperature behavior can be readily deduced from the above relations:

1) An important result evident from Eq. (4) is that the absolute value of dT_t/dc is inversely proportional to the phase transition latent heat q_t.
2) When bound water solutions behave as ideal ones, the ratios c'/c and c''/c depend on the pressure and temperature only. Hence (see Eq. (4)) at small c the transition temperature T_t is a linear function of c.
3) One may define kosmotropic solutes by the inequalities c', $c'' < c$, and *chaotropic solutes by the inverse inequalities* c', $c'' > c$. Then, from the simplified version in Eq. (5), it follows that:
   a) For kosmotropes the slope of T_t is positive when the low-temperature phase is the less

hydrated one ($x'' < x'$), and is negative in the opposite case ($x'' > x'$).
   b) For chaotropes the slope of T_t is negative when the low-temperature phase is the less hydrated one ($x'' < x'$), and positive in the opposite case ($x'' > x'$).

These conclusions have been checked by means of a comparison with experimental data on a number of different lipids and solutes obtained by differential scanning calorimetry. These data will be published and discussed in detail elsewhere (see also [1—5]), while here we show in Fig. 1 two representative examples of the effect of a kosmotropic and a chaotropic solute on the transition temperatures of the lamellar and inverted hexagonal lipid phases. It is clear from this figure that: 1) transitions of smaller latent heat ($L_a - H_{II}$ in Fig. 1) are more sensitive to addition of solutes, 2) T_t shifts linearly with the solute concentration, and 3) kosmotropic and chaotropic solutes have opposite effects on the phase transition temperatures, in agreement with the conclusions based on Eqs. 4 and 5. It should be noted, however, that in some lipid-water systems (results not shown here) deviations from linearity have been found, which indicate that the ideal solution assumption we are using here is not valid in these systems at higher solute concentrations.

Acknowledgement

We gratefully acknowledge the contribution of Dr. R. Koynova who performed the differential scanning calorimetry measurements.

References

1. Oku N, MacDonald R (1983) J Biol Chem 258: 8733—8738
2. Yeagle P, Sen A (1986) Biochemistry 25:7518—7522
3. Bryszewska M, Epand R (1988) Biochim Biophys Acta 943:485—492
4. Koynova R, Technov B, Quinn P (1989) Biochim Biophys Acta 980:377—380
5. Tsvetkova N, Koynova R, Tsonev L, Quinn P, Tenchov B (1991) Chem Phys Lipids 60:51—59

Authors' address:

Prof. Boris LG. Technov
Central Laboratory of Biophysics
Bulgarian Academy of Sciences
Acad. Bonchev Str. — Bl. 21
1113 Sofia, Bulgaria

Anomalous diffusion in PDMS melts studied by a NMR field-gradient method

E. Rommel[1]), R. Kimmich[1]), M. Spülbeck[1]), and N. F. Fatkullin[2])

[1]) Sektion Kernresonanzspektroskopie, Universität Ulm, Federal Republic of Germany
[2]) Kazan State University, Physics Department Kazan, Tatarstan, Russia

Abstract: Diffusion in PDMS melts was measured by an NMR-field-gradient technique using the fringe field of a supercon magnet. Anomalous diffusion was detected at molecular weights above $M_c \approx 24000$. The dependence of the mean-square displacement of the molecular weight and on the diffusion time is described by the Doi-Edwards theory. The length scale of detectable displacements with the applied technique is above 10^{-8} m.

Key words: Anomalous diffusion — PDMS — NMR

Introduction

Current theories of polymer dynamics [1—3] predict a diffusion behavior of segments which tends to be anomalous in length scales within the polymer coil. The time dependence of the mean-square displacement (in the Euclidean space) is then expected to follow a power law $\langle r^2(t) \rangle = at^\kappa$ where $\kappa < 1$. Elementary quantities determining the length scales in which anomalous diffusion can be expected are the length of a Kuhn segment, b, the step length of the so-called primitive path, a, and the radius of gyration R_g. The Doi/Edwards tube model [1] in particular suggests the following limits for the average mean-square segment displacement

$$b^2 < \langle r^2 \rangle < a^2 : \quad \langle r^2 \rangle \approx \beta_1 N^0 t^{1/2} \quad \text{(limit I)} \tag{1}$$

$$a^2 < \langle r^2 \rangle < \sqrt{6}R_g a: \quad \langle r^2 \rangle \approx \beta_2 N^0 t^{1/4} \quad \text{(limit II)} \tag{2}$$

$$\sqrt{6}R_g a < \langle r^2 \rangle < 6R_g^2: \quad \langle r^2 \rangle \approx \beta_3 N^{-1/2} t^{1/2} \quad \text{(limit III)} \tag{3}$$

$$6R_g^2 < \langle r^2 \rangle: \quad \langle r^2 \rangle \approx \beta_4 N^{-2}t \quad \text{(limit IV)}, \tag{4}$$

where β_1, β_2, β_3, β_4 are constants and N is the number of Kuhn segments per chain.

The purpose of this paper is to investigate the mean-square displacement behavior in PDMS melts in length scales involving the above limits as far as possible. It is an attempt to verify the combined chain length and time dependencies characterizing the diverse cases. A suitable technique for this objective is the supercon fringe field (SFF) NMR technique permitting investigations of particularly small diffusive displacements [4, 5]. First indications of anomalies with respect to the chain-length and time dependencies were already reported in previous papers [4—8].

Results

The experiments of this study were carried out with the magnet of a tomograph Bruker BMT 47/40 (4.7 T) by the aid of a stimulated echo sequence

$$\left(\frac{\pi}{2}\right) - \delta - \left(\frac{\pi}{2}\right) - \tau_2 - \left(\frac{\pi}{2}\right) - \delta - \text{echo} .$$

The magnetic flux density and the gradient at the measuring position in the fringe field was $B = 2.14$ T, $G = 9.04$ T/m.

The attenuation of the echo amplitudes was measured in dependence on the diffusion time $\Delta = \delta + \tau_2$ for two fixed δ intervals, $\delta = 4$ ms and $\delta = 10$ ms. The influence of the spin-lattice relaxation in the τ_2 interval was corrected using T_1 values determined in separate experiments. A series of different molecular weights, 53 900, 110 000 and 335 000, all above the critical value of $M_c \approx 24000$, was investigated.

To avoid multiexponential decays, samples with a polydispersity of less than 1.1 were used.

Taking the data of all experiments together (Fig. 1), the set of decays could definitely not be described within the experimental error if only the normal diffusion was assumed to be valid in the whole range. Rather, the crossover between the limits II, III and IV had to be taken into account, whereas there was no experimental evidence for limit I.

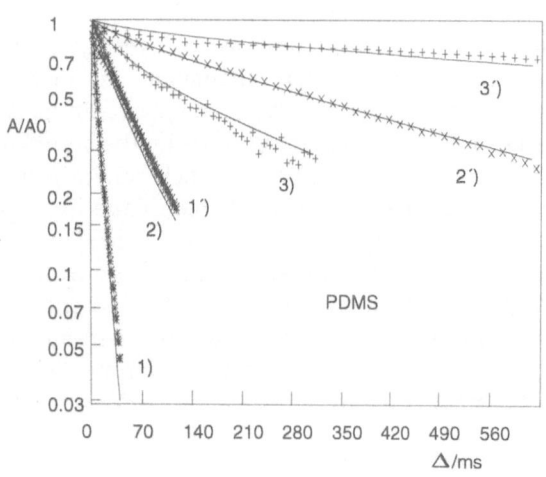

Fig. 1. Stimulated echo attenuation versus Δ. The data were recorded with PDMS melts at 20°C using the SFF NMR technique. The decays are corrected for spin-lattice relaxation. The weight average molecular weights are M_w = 53900 (data sets 1 and 1'), 110 000 (data sets 2 and 2'), and 335 000 (data sets 3 and 3'). The phase encoding intervals δ were 10 ms (data sets 1, 2, 3) and 4 ms (data sets 1', 2', 3',). The solid lines were fitted to the data.

For limits II, III, and IV the Doi-Edwards theory yields for the attenuation of the stimulated echo amplitudes by diffusion:

$$A = A_0 \exp \left\{ \frac{k^4 a^2 D^*(t) t}{36} \right\}$$

$$\cdot \, \text{erfc} \left\{ \frac{(k^2 a (D^*(t) t)^{1/2}}{6} \right\} \exp \{- D k^2 t\} , \quad (5)$$

where

$$D^* (t) = \frac{D_0}{N} + b \left(\frac{D_0}{3 \pi t} \right)^{1/2} \quad (6)$$

is the curvilinear diffusion coefficient,

$$D = \frac{D_0 a^2}{3 N^2 b^2} , \quad (7)$$

is the self-diffusion coefficient of the macromolecule,

$$k = \gamma \delta G , \quad (8)$$

and D_0 is the segmental diffusion coefficient.

Discussion and conclusion

The anomalous diffusion behavior expected on the basis of the Doi/Edwards model for segment displacements within the tube and for reptation along the primitive path has been verified for PDMS melts. Especially the molecular weight dependence of D and D^* and the time dependence of the diffusion predicted by the model is in very good agreement with the data. However, the values for b and R_g are significantly larger than predicted. This may be due to the fact that the condition $M >> M_c$ which is assumed in the tube model is not strictly fullfilled. The length scale of the segment displacements studied in this work is roughly 100 Å and more. Processes taking place in dimensions of the tube diameter therefore are not entirely accessible by this method in the case of PDMS melts. There is, however, an indirect way to study translational diffusion in length scales of about 10 Å: Nuclear magnetic relaxation spectroscopy using the field-cycling technique in particular is sensitive to molecular reorientations mediated by translations. The special and universal power laws observed for the frequency and molecular weight dependences of the NMR relaxation times [9] can be deduced from the anomalous time dependencies of the mean-square displacements predicted by the Doi/Edwards theory for small length scales.

References

· 1. Doi M and Edwards SF (1986) The Theory of Polymer Dynamics, Clarendon Press, Oxford
2. de Gennes PG (1979) Scaling Concepts in Polymer Physics, Cornell University, Ithaca
3. Lodge TP, Rotstein NA, Prager SP (1990) Advan Chem Phys 79:1
4. Kimmich R, Unrath W, Schnur G, Rommel E (1991) J Magn Reson 91:136
5. Unrath W, Klammler F, Kotitschke K, Kimmich R, Rommel E (1990) In: M Mehring, JU von Schütz, and C Wolf, Extended Abstracts of Congress Ampere on Magnetic Resonance and Related Phenomena, Springer, Berlin, p. 312
6. Zupančič I, Lahajnar G, Blinc R, Reneker DH, Vanderhart DL (1985) J Polym Sci: Polym Phys 23:387
7. Coy A, Callaghan PT (1991) 1st International Conference on NMR Microscopy, Heidelberg
8. Fleischer G, Fujara F, private communication
9. Weber HW, Kimmich R, Köpf M, Ramik T, Oeser R (1992) Progr Colloid Polym Sci (90:104)

Authors' address:

Prof. Dr. Rainer Kimmich
Sektion Kernresonanzspektroskopie
Universität Ulm
89069 Ulm, FRG

Progress in Colloid & Polymer Science   Progr Colloid Polym Sci 93:158 (1993)

Dynamic light scattering study of concentrated dispersions of charged colloidal particles

T. Bellini[1]), R. Piazza [1]), V. Degiorgio[1]), and T. Palberg[2])

[1]) Dipartimento di Elettronica — Sezione di Fisica Applicata, Università di Pavia, Italy
[2]) Dep. of Physics, University of Konstanz, FRG

Key words: Dynamic light scattering — charged colloidal particles

We have investigated the phase diagram and the melting transition of a colloidal suspension of charged optically anisotropic spherical particles. The ionic strength of the samples has been carefully measured and controlled. We have observed that it is possible to desorb surface charges from the paticles and we have sampled the phase diagram for different ionic strengths, particle charges and particles concentrations. Depolarized dynamic light scattering has been used to measure the mean square amplitude of the Brownian oscillation of the particles about their crystalline sites as a function of the distance from the melting transition boundary, and preliminary results show that this method can be a test of the Lindemann melting criterion.

Authors' address:

T. Bellini
Dipartimento di Elettronica
University of Pavia
Via Abbiategrasso 209
I-27100 Pavia, Italy

Progress in Colloid & Polymer Science   Progr Colloid Polym Sci 93:159—166 (1993)

Determination of size distributions of submicron particles by dynamic light scattering experiments taking into account normalization errors

H. Ruf, W. Haase, W. Q. Wang, E. Grell, P. Gärtner*), H. Michel and J.P. Dufour[1])

Max-Planck-Institut für Biophysik, Frankfurt, FRG
[1]) Unité de Brasserie et des Industries alimentaires, Université Catholique de Louvain, Louvain-la-Neuve, Belgium

Abstract: The determination of continuous size distributions from dynamic light scattering measurements requires data of extremely high accuracy. The ill-conditioned nature of the inversion of experimental data is illustrated. It is shown that large differences in size distributions are only poorly expressed by the corresponding autocorrelation functions and thus difficult to extract, even from relatively low noise data. A high statistical accuracy of data is a prerequisite, but often may not suffice. It is demonstrated with data of high statistical accuracy from calibrated polystyrene latex particles that baseline errors of less than 0.1% can distort the resulting size distribution and mimic non-existent side peaks. Using a new method, where the characteristics of normalization errors are incorporated into the inversion algorithm, and which fits for the relative baseline error as additional parameter, these distortions can be avoided. This technique is applied to determine size distribution of phospholipid vesicles prepared by gel filtration and sonication using a newly synthesized phospholipid. The results indicate that both methods yield small vesicles of essentially the same size, the average radius of which is about 10 nm. In a second application, two stages of a crystallization experiment are investigated with dynamic light scattering and electron microscopy and the results compared. There is a very good agreement in the particle sizes determined from the two methods, which confirms the ability of the extended inversion algorithm to provide stable and reliable solutions, and hence also to discern fine details of such particulate suspensions.

Key words: Dynamic light scattering — normalization errors — latex beads — lipid vesicles — H$^+$-ATPase

Introduction

The determination of size distributions of particles in liquid suspensions from dynamic light scattering (DLS) measurements involves the inversion of a system of integral equations

$$| g^{(1)}(\tau) | = \int_0^\infty S(\Gamma)e^{-\Gamma\tau}d\Gamma, \qquad (1)$$

where $| g^{(1)}(\tau) |$ is the magnitude of the normalized first order or field autocorrelation function of the delay time τ, $\Gamma = q^2 D$ the characteristic decay constant for particles of a given diffusion constant D, and $S(\Gamma)$ the distribution function. $q = (4\pi n/\lambda)$ sin $(\Theta/2)$ denotes the scattering vetor, with n being the refractive index of the medium, λ the wavelength of the incident light and Θ the scattering angle. Estimating $S(\Gamma)$ from noisy experimental data is an ill-conditioned problem [1], because marked differences in size distributions often appear only as rather small differences in the corresponding autocorrelation functions. Conversely, there exists a variety of partly rather different size distributions,

---

*) present adress: Philipps Universität Marburg, Fachbereich Biologie, Karl von Frisch Straße, D-3350 Marburg, FRG

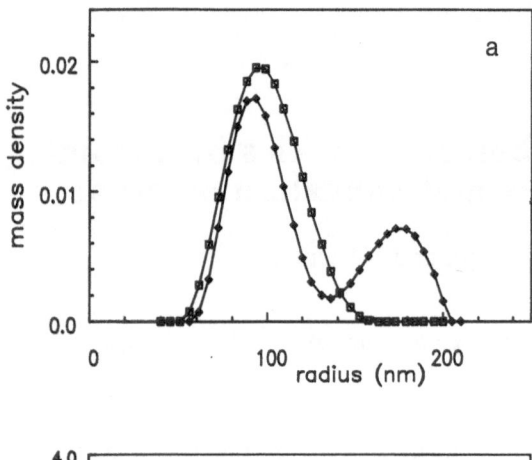

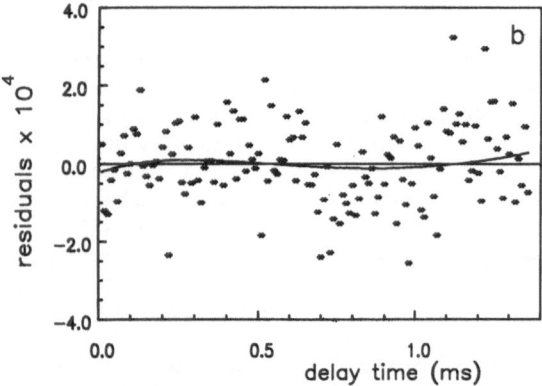

Fig. 1. Illustration of the ill-conditioned nature of the inversion of Eq. (1): The considerable differences of the size distributions in a are only weakly expressed in the corresponding autocorrelation functions, the differences of which are shown in b (solid line). Comparision with the noise (*) in DLS data from a measurement of 2h duration ($N = 7.2 \times 10^8$, $\delta\tau$, = 10μs, $\bar{n} = 0.5$; cf [4], Fig. 4c) illustrates the difficulties in making a decision in favour of one of the two solutions

which describe the experimental data nearly equally well. An illustration of the damping effect of the integral equation (1) is given in Fig. 1. While the two size distributions are rather different, the differences in the corresponding field autocorrelation functions (Fig. 1b) are only of the order of 10^{-5}. When comparing these really small differences with the noise in data of a DLS measurement of fairly long duration, the difficulty in making a decision in favour of one of the two distributions from such data becomes obvious. Figure 1 also illustrates why the result of the inversion are so susceptible to even small experimental errors arising from sources other than the statistical ones. One that introduces

systematic errors is normalization of the experimental photocount or second-order autocorrelation function by its baseline. The severe effects of these errors have been demonstrated on simulated [2, 3] and experimental data [4]. It has been shown that the distorting effects are due to the characteristics of normalization errors in field or first-order autocorrelation functions [3]. Unlike with second-order autocorrelation functions, the errors increase exponentially with the delay time as a consequence of the square root extraction involved in Siegert's equation [5]. In [3] it was shown that normalization errors can be described by an analytical expression in terms of the relative baseline error $\Delta B/\hat{B}$. Its linear approximation in $\Delta B/\hat{B}$, called normalization error function (NEF), has been incorporated into Provencher's inversion algorithm CONTIN [6—8], and the effectiveness of the extended inversion algorithm was demonstrated on simulated [3] and experimental data [4]. Here, we apply it to data of high statistical accuracy obtained from two sizes of calibrated polystyrene latex bead and a mixture of them, and compare the results obtained with and without NEF. Furthermore, two applications are presented: the investigation of the size distributions of lipid vesicles prepared from a newly synthesized lipid by two different methods, and the study of two stages in an attempt to crystallize an integral membrane protein, where the results are compared with those obtained from electron microscopy.

Materials and methods

Polystyrene latex beads LB1 and LB3 with nominal diameters of 90 nm (standard deviation 24 nm) and 303 nm (standard deviation 4 nm) were purchased from SIGMA. Aliquots from the stock solutions were diluted with water purified on Milli-Q columns (Millipore, Bedford, mass.) to final concentrations of 0.1 mg/ml (LB1) and 0.02 mg/ml (LB3). The samples were filtered through 0.4 μm or 0.8 μm Nuclepore filters. The mixture was adjusted such that the intensity scattered by the two populations was roughly the same (LB1: 44%, LB3: 56%).

Lipid vesicles were prepared by gel filtration [9] and sonication [10] from 1-octadecyl-2dodecyl-lecithin, which was synthesized in our laboratory. The transition temperature of this ether lipid was determined to be around 14°C. The lipid (20 mM) was allowed to swell overnight in 5 mM phosphate buffer, pH 7.2 at room temperature. Sonication was performed with a titanium micro-tip (MSE, Sonifier) for 30 min at 20°C. The samples used for gel filtration on a Sephadex G50 column contained 30 mM sodium cholate.

H^+-ATPase from the plasma membrane of Schizosaccharomyces pombe was purified as described in [11]. The buffer used for all steps was 20 mM MES/KOH, 1 mM EDTA, 10% (w/v) glycerol, 0.05% dodecyl maltoside. Size exclusion chromatography was performed on Superose 6 in the presence of 100 mM NaCl.

Dynamic light scattering: Experimental set-up and methods for data analysis were described in [4]. All measurements were carried out in batch mode. Each measurement (3—5 min duration) was normalized by the corresponding baseline value calculated from the total number of counts. The autocorrelation functions consisted of 136 linearly spaced data points. Some of the inversions were carried out on an IBM compatible PC using 40 to 41 grid points for integration. The distribution shown here were expressed in terms of the hydrodynamic radius calculated from the Stokes-Einstein equation, and were normalized so that the integral of all values of the radius in one. In the case of latex particles and lipid vesicles the model of full or hollow spheres and Rayleigh-Gans-Debye scattering form factors were included in data inversion and the mass density distribution was determined. In the case of the samples from the crystallization experiments the distribution of the scattered intensity is determined since the polarizabilities of the different particle species were unknown.

Negative staining: A drop of the protein suspension was placed on a Formvar-coated grid. After 3 min most of the suspension was removed by suction with filter paper and a drop of 1% uranyl formate was added. Two minutes later all fluid was sucked up by filter paper and after drying the grid was analyzed under an electron microscope (Philips CM12).

Results

The size distributions obtained from DLS measurements on different samples of calibrated polystyrene latex beads and a mixture of them are shown in Figs. 2 to 4. Measurement durations were 5.5 h for LB1 (total number of samples $N = 3.3 \times 10^9$, sample time interval $\delta\tau = 6$ μs, mean number of counts per sample time interval $\bar{n} = 0.3$), 26.6 h

for LB2 ($N = 5.32 \times 10^9$, $\delta\tau = 18$ μs, $\bar{n} = 0.9$) and 17.8 h for the mixture ($N = 4.28 \times 10^9$, $\delta\tau = 15$ μs, $\bar{n} = 0.7$), which yielded data of reasonably high statistical accuracy. In the case of the monodisperse

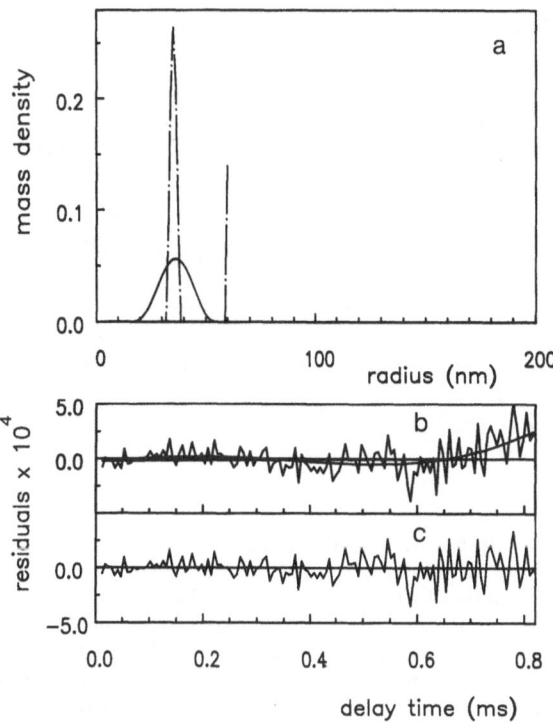

Fig. 2. Polystyrene latex beads LB1 (nominal radius 45 nm, standard deviation 12 nm). a) Mass density distributions obtained from inversions carried out without (dotted line) and with NEF (solid line) plotted versus the hydrodynamic radius (integration range: 15—60 nm). b, c) Residuals of the corresponding best-fit autocorrelation functions, the differences of which are depicted in b (solid line)

samples the inversions without NEF give size distributions with unexpected side peaks (Figs. 2a, 3a). These solutions are quite unstable. The positions of the side peaks coincide with the right integration limit and the residuals show correlated deviations (Figs. 2b, 3b). Both features are typical of nomalization errors. When the inversions are carried out with NEF the correlated deviations disappear (Figs. 2c, 3c), and the expected monomodal size distributions are obtained. The differences of the corresponding best-fit autocorrelation functions (Figs. 2b and 3b) largely trace the correlated deviations of the noise, which also shows that these are

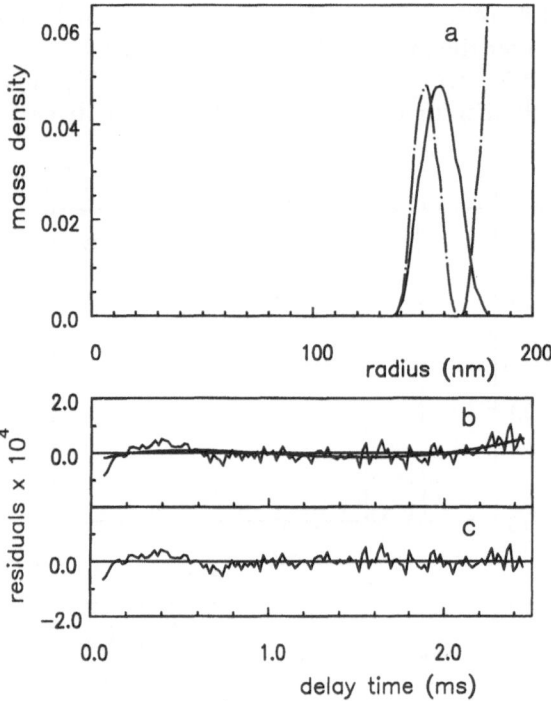

Fig. 3. Polystyrene latex beads LB3 (nominal radius 151.5 nm, standard deviation 2 nm). a) Mass density distributions obtained from inversions carried out without (dotted line) and with NEF (solid line) plotted versus the hydrodynamic radius (integration range: 135—180 nm). b, c) Residuals of the corresponding best-fit autocorrelation functions, the differences of which are depicted in b (solid line)

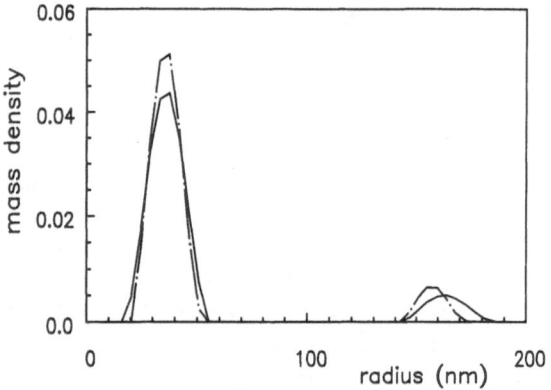

Fig. 4. Mixture of polystyrene latex beads LB1-LB3. Mass density distributions obtained from inversions carried out without (dotted line) and with NEF (solid line) plotted versus the hydrodynamic radius (integration range: 15—180 nm)

mainly due to normalization errors. In the case of the data from the mixture, both inversions with and without NEF yield a bimodal distribution (Fig. 4). Although the baseline error is of a similar magnitude as before, the effects are not as drastic here, yet the analysis with NEF yields a better fit. The corresponding sums of residual least squares are reduced about three fold from 2.3×10^{-7} to 6.6×10^{-8}. The mean radii and the widths of the distributions resulting from the analyses with NEF are summarized in Table 1.

The size distributions of 1-octadecyl-2-dodecyl-lecithin vesicles are depicted in Fig. 5. From these it follows that the two different preparation methods, gel filtration on a Sephadex G50 column, which starts from a suspension of small lipid-detergent mixed micelles, and sonication that starts from large lipid aggregates, yield essentially vesicles of the same size. A mean mass-weighted radius of 9.8 nm and standard deviation of 3.6 is obtained for vesicles from gel filtration ($N = 1.04 \times 10^{10}$, $\delta\tau = 2$ μs, $\bar{n} = 0.09$, $\Delta B/\hat{B} = -6.7 \times 10^{-3}$), and a radius of 10.9 nm and standard deviation of 4.9 nm for vesicles from sonication ($N = 5 \times 10^{8}$, $\delta\tau = 5$ μs, $\bar{n} = 0.05$, $\Delta B/\hat{B} = 5.2 \times 10^{-3}$). The analyses with the cumulant method [12] yield intensity-weighted averages of 13 and 20 nm respectively. The sonicated dispersion still contained a fraction of larger particles, as seen from the side peak. This represents here real particles (the exact size of which was not determined), which follows from the fact that a larger sample time was needed to measure the autocorrelation function appropriately, and that a larger overall average size of 20 nm was obtained from the cumulant method.

Two different stages in an attempt to crystallize H+-ATPase from the plasma membrane of Schizosaccharomyces pombe were investigated with DLS and electron microscopy. This enzyme, which plays an essential role in intracellular pH regulation and in generation of the electrochemical proton gradient necessary for ion and nutrient transport, consists of a single subunit of approximately 100 kDa as revealed by SDS-Page [13]. It was purified with the aim of crystallizing it for x-ray crystallographic studies. Purification involves the solubilization of membranes with lysolecithin and subsequent separation on a sucrose gradient. After this step the ATPase is present as a complex of several monomers which presumably interact at their hydrophopic regions. This was assyed by size exclusion chromatography, DLS (Fig. 6a) and electron microscopy (Fig. 6b). The

Table 1. Mass-weighted (m) and number (n) average radii and standard deviations of calibrated polystyrene latex beads determined from DLS measurements

	r_m [nm]	s_m [nm]	$\Delta B/\hat{B}$, 10^{-4}	r_n [nm]	s_n [nm]
LB1	36	6.5	−9.2	33	6.4
LB3	158	7.6	−3.3	157	7.6
Mixture	36	7.2	8.1		
LB1—LB3	163	8.2			

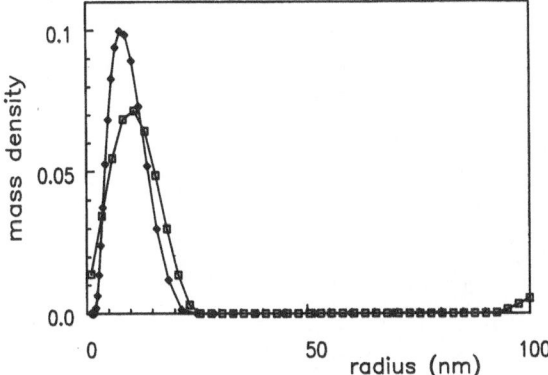

Fig. 5. Mass density distributions of phospholipid vesicles prepared by gel filtration (♦) and sonication (■) plotted versus the hydrodynamic radius

elution profile of the Superose 6 column showed a maximum with an apparent molecular weight of 1500 kDa and a second maximum at the exclusion limit of the column (>40.000 kDa). Most ATPase activity was associated with the 1500 kDa maximum. The electron micrograph of this fraction (Fig. 6b) reveals two types of particles, rod-like particles with lengths of 10—20 nm, which are supposed to represent the protein complexes, and nearly spherical particles with radii in the range 20—40 nm. Similarly, the corresponding DLS measurements ($N = 1.5 \times 10^9$, $\delta\tau = 4$ μs, $\bar{n} = 0.1$, $\Delta B/\hat{B} = -5 \times 10^{-2}$) yield a relatively broad size distribution (Fig. 6a) with a maximum at about 30 nm and a shoulder in the size range of the rods. The spherical structures are presumably impurities. They were found in the protein-free solution, too. To separate these from the rods, a second exclusion chromatography run was carried out and the corresponding fraction was characterized by DLS and electron microscopy (Figs. 6c and d). The electron micrograph shown a

more homogeneous suspension of rod-like structures, which in the meantime appeared to have grown. One also observes some faint, large structures and, correspondingly, the size distribution resulting from the DLS measurement ($N = 3.1 \times 10^9$, $\delta\tau = 4$ μs, $\bar{n} = 0.1$, $\Delta B/\hat{B} = -2.3 \times 10^{-3}$) exhibits a second peak at about 100 nm. Here, the distributions are plotted versus a size parameter instead of the diffusion constant. For long, thin rods an equation similar to the Stokes-Einstein relation exists [14]. In this equation the hydrodynamic radius is replaced by an effective size, that being the half-length of the rod multiplied by a factor that is constant for a given ratio of length to width. The intensity-weighted average size of the first peak in Fig. 6c is 21 nm, and its corresponding number average about 7 nm. For a length-to-width ratio of 4:1 the multiplication factor is 1.44, and the length of a corresponding rod becomes about 20 nm. This calculated value for the average rod length compares very well with the sizes observed in the electron micrograph.

Discussion

A very low noise level in the experimental data is a prerequisite for being able to reliably determine size distributions from DLS measurements. Generally, this necessitates experiments of rather long duration [15, 16] for reducing noise arising from various sources, such as correlated "diffusion" noise [17], random noise of photon detection [18] and correlated deviations due to the sampling scheme of the correlator [18, 19]. Unfortunately, unlike the case of monodisperse scatterers [18, 20], at present there is no quantitative relationship available from which the necessary accuracy for resolving specified details of continuous size

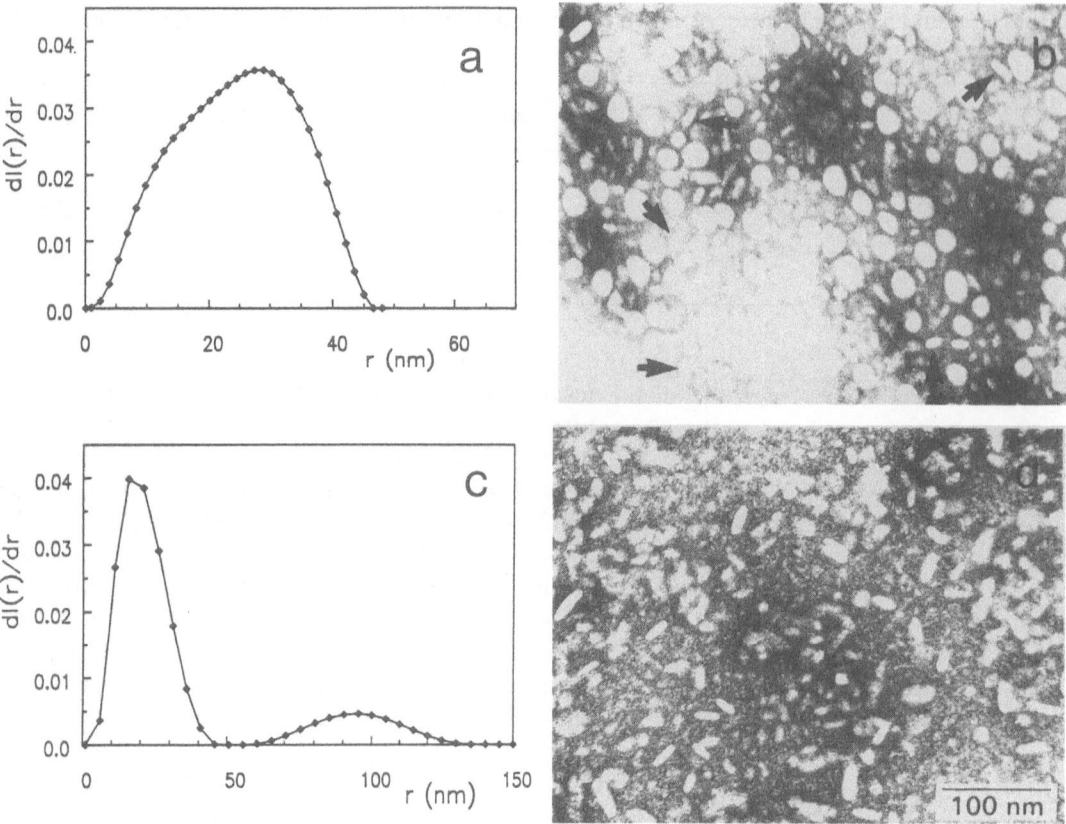

Fig. 6. Two stages of a crystallization experiment with the integral membrane protein H+-ATPase from the plasma membrane of the yeast Schizosaccharomyces pombe. Intensity-weighted size distributions of the particles after first (a) and second run (c) of size exclusion chromatography. The corresponding electron micrographs obtained from these suspensions are shown in b and d

distributions can be derived. The procedure outlined in Fig. 1 to calculate for given size distributions the corresponding differences in the field-autocorrelation functions, however, shows one way of determining the accuracy of data necessary to resolve the differences in these size distributions.

In [4] it was shown that normalization errors due to drifts in the intensity of the light source can be widely reduced by measuring the normalized photocount autocorrelation function in the batch mode. Here, the experiments with polystyrene latex beads show that even in batch data of rather high statistical accuracy small baseline errors of nearly 0.1% occur, and in two cases have drastic effects on the resulting size distribution. Inversions carried out with NEF completely remove the distortions, which indicates that a significant part of the errors in the data are in fact normalization errors. The use of Provencher's "dust" term, which searches for a constant background in all data from the field autocorrelation function, also gives some improvements (this is to be expected since normalization errors can be thought of as being composed of a constant and an exponentially rising part). However, the "dust" term option yields results with larger ambiguities. The values obtained for the dust term are often higher than would be concluded from the contents of the delayed channels. In some cases the dust value is negative, which is physically unrealistic. If the inversion algorithm is forced in these cases to restrict its solutions to positive background values, the background associated with the solution chosen by the program is zero, and the same distribution is obtained as if no additional functions were specified.

In the case of LB3 particles there is quite a good agreement between nominal size and the average size determined from the DLS measurements, whereas in the case of LB1 particles a significant difference is found. Interestingly, this has been observed serveral times before in the cases of latex beads with diameters around 90 nm (eg., [21—23]). To clarify this discrepancy, we began determining sizes of LB1 latex beads from electron microscopy. First results suggest that both the average size and the width of the distribution of these particles are actually smaller than specified. In the case of LB3 particles a width broader than 2 nm was obtained. To determine size distributions of such small widths would require DLS measurements of much longer duration than used here. A comparison of the autocorrelation function of our solution with the corresponding autocorrelation function of an approximately 25% narrower size distribution, on the other hand, suggests that it should be possible to distinguish between these two size distributions on the level of noise of our data. This raises the question as to wheter the latex particles in this sample are in fact of such high homogeneity.

The investigations on phospholipid vesicles and complexes of H^+-ATPase show that DLS together with the extended inversion algorithm is a very useful method that allows one to study rather fine details of such particle systems. But the analyses also revealed rather large baseline errors in these data, which raises the question of their physical origin. The drift in the light source is unlikely because the measurements were carried out in the batch mode. The statistical errors in the experimental values of the baseline, which in the case of the polystyrene particles have more than likely contributed to the baseline errors, are only small here. At present there is no final answer. However, we have some hints that stray light from dust particles that have not been removed from the dispersion could be responsible for these large normalization errors. From experiments with particle dispersions which were not so clean, or where the small volume available did not allow extensive filtering, we always obtained quite large baseline errors from the analysis. Some of our current theoretical considerations appear to support this conclusion. If this were proven to be true, this would mean that the extended inversion algorithm can succesfully treat dynamic light scattering data also from samples where extensive cleaning is not possible, as is often the case, for example, with biological material.

The data presented here confirm the conjecture that an erroneous baseline value is one of the main causes of severe distortions in size distributions derived from the inversion of dynamic light scattering data. The importance of the accuracy of the experimental baseline for obtaining valuable results has been recognized many times, and serveral procedures have been devised to minimize errors in that baseline [24—26]. The method used here is an alternative approach that allows one to determine estimates of the baseline error itself and, subsequently, to partly correct the data. In most cases two or three itterations are sufficient to remove normalization errors almost completely and, as shown here, to obtain valuable, results.

References

1. Phillips DL (1962) J Ass Comput Mach 9:84—97
2. Weiner BB, Tscharnuter WW (1987) In: Provder T (ed) Particle Size Distribution. ACS Symposium Series, vol 332. American Chemical Society, Washington DC, pp 48—61
3. Ruf H (1989) Biophys J 56:67—78
4. Ruf H, Grell E, Stelzer EHK (1992) Eur Biophys J 21:21—28
5. Siegert AJF (1943) MIT Rad Lab Rep No 465
6. Provencher SW (1982) Comput Phys Commun 27:213—227
7. Provencher SW (1982) Comput Phys Commun 27:229—242
8. Provencher SW (1982) CONTIN, Users Manual. EMBL technical report DA 05. Heidelberg: European Molecular Biology Lab
9. Brunner J, Skrabal P, Hauser H (1976) Biochim Biophys Acta 455:322—331
10. Huang CH (1969) Biochemistry 8:344—352
11. Dufour JP, Goffeau A (1978) J Biol Chem 253:7026—7032
12. Koppel DE (1972) J Chem Phys 57:4814—4820
13. Goffeau A, Coddington A, Schlesser A (1989) In: Nasim A, Young P, Johnson BF (eds) Molecular Biology of the Fission Yeast. Academic Press, San Diego, pp 397—429
14. Flamberg A, Pecora R (1984) J Phys Chem 88:3026—3033
15. Pike ER, Watson D, McNeil Watson F (1983) In: Dahneke BE (ed) Measurement of Suspended Particles by Quasi-Elastic Light Scattering. Wiley, New York, pp 107—128
16. Morrison ID, Grabowski EF, Herb CA (1985) Langmuir 1:496—501
17. Saleh BEA, Cardoso MF (1973) J Phys A Math Nucl Gen 6:1897—1909
18. Jakeman E, Pike ER, Swain S (1971) J Phys A: Gen Phys 4:517—534
19. Jakeman E (1974) In: Cummins HZ, Pike ER (eds) Photon Correlation and Light Beating Spectroscopy. NATO Adv Study Inst Series B, vol 3. Plenum Press, New York London, pp 75—149

20. Schätzel K (1990) Quant Opt 2:287—305
21. Trotter CM, Pinder DN (1981) J Chem Phys 75:118—127
22. Lee SP, Tscharnuter W, Chu B (1972) J Polymer Sci 10:2453—2459
23. Stelzer EHK (1982) Diplomarbeit. JW v Goethe Universität, Frankfurt
24. Glatter O, Sieberer J, Schnablegger H (1991) Part Part Sys Charact 8:274—281
25. Schnablegger H, Glatter O (1991) Appl Opt 30:4889—4896
26. Schätzel K, Drewel M, Stimac S (1988) J Mod Opt 35:711—718

Authors' address:

Dr. Horst Ruf
Max-Planck-Institut für Biophysik
Kennedy-Allee 70
60596 Frankfurt, FRG

Progress in Colloid & Polymer Science

Progr Colloid Polym Sci 93:167—174 (1993)

Charged o/w microemulsion droplets

M. Gradzielski and H. Hoffmann

Lehrstuhl für Physikalische Chemie I der Universität Bayreuth, FRG

Abstract: Structure and properties of charged oil-in-water (o/w) microemulsions have been investigated. A particular system made up from a zwitterionic surfactant and hydrocarbon which becomes charged upon the addition of either cationic or anionic surfactant has been studied. A particular feature of this microemulsion system is that the charge density on the droplets can be fixed at a desired value in an easily controllable fashion. These systems have been characterized by means of light scattering, SANS, interfacial tension measurements, electric conductivity and viscosity measurements. The experiments showed that the aggregates remain constant in size over a large concentration region, i.e., from 0.1 to 30 wt%. Moreover, their size is not much changed by the addition of the ionic surfactant. The interactions in this microemulsion system could be described by a hard sphere interaction with an additional DLVO — potential term that accounts for the electrostatic repulsion.

Key words: O/W microemulsions — electrostatic interactions — light scattering — surfactant mixtures

Introduction

Microemulsions [1] are colloidal systems made up of structures typically in the size range of 20—500 Å. They have been subjected to numerous investigations and they are also of great interest because of their practical applications [2—7]. They are normally composed of a surfactant, a cosurfactant, hydrocarbon, and water, where the cosurfactant (usually an intermediate chain alcohol) is required in order to lower the interfacial tension of the interface: hydrocarbon/surfactant solution because a low interfacial tension is a prerequisite for the formation of microemulsions. In principal, one can distinguish three different structural types of microemulsions: hydrophobic aggregates in water (oil-in-water, o/w), hydrophilic aggregates in oil (water-in-oil, w/o) and, finally, bicontinuous structures. So far, mostly non-ionic surfactants have been employed for the formation of O/W microemulsions, but they can also be produced with ionic surfactants, although those normally possess a much higher interfacial tension than their uncharged counterparts [8]. Therefore, in ionic

systems the influence of cosurfactants is of much higher importance.

In our investigations the influence of charges on the structural properties of o/w microemulsions has been studied. Zwitterionic alkyldimethylamine oxides were employed as surfactant, especially the tetradecyl- and the oleyldimethylamine oxide (TDMAO and ODMAO). These surfactants display a behavior which could be described as being somewhere in between that of a typical nonionic and that of a typical cationic surfactant, for instance, with respect to their CMC values, which are 0.121 and $5.6 \cdot 10^{-3}$ mmol/l, respectively [9, 10]. Just above the CMC the binary surfactant systems at first contain spherical micellar aggregates, but start to form rodlike micelles at higher concentrations; the length of the rods increases with increasing concentration [11]. This effect is much more pronounced for the ODMAO where the rods already overlap in a 10 mM solution and which, therefore, displays viscoelastic properties already at this very low concentration [12]. Upon solubilization of hydrocarbon into these rodlike micelles they become transformed into globular microemulsion droplets. This transition can be monitored by means of light scattering, viscosity, or electric birefringence experiments [13]. Further addition of hydrocarbon causes the formed droplets to grow gradually until the solubilization capacity of the systems in reached.

Experimental results

First, we investigated microemulsion systems with a constant hydrocarbon to surfactant ratio, which was kept close to the solubilization capacity and well above the one of the rod-to-sphere transition. The chosen molar ratio of decane to TDMAO was 1:3.2 and for decane to ODMAO 1.:1.11. These microemulsions were investigated by means of static light scattering up to concentrations of ~ 300 g/l and the obtained scattering intensity is shown as a function of the total concentration of surfactant plus hydrocarbon in Fig. 1. For both systems a maximum is observed around 10 wt% which is indicative of hard sphere interactions. From light scattering and electric birefringence experiments it is evident that the aggregates in these solutions have to be of spherical shape. Their size is much smaller than the wave length λ of light, i.e. one can assume to be in the low q-limit. Therefore, the form factor $P(q)$ becomes equal to unity and only $S(O)$ is required in order to describe the scattering data adequately (Eq. (1)). Because a nonionic surfactant was employed the main interaction between the aggregates should result from their mutual steric hindrance. Therefore, we tried to fit the data with the well-known Carnahan-Starling expression for the structure factor of a hard sphere system [14] (Eq. (2)), whereby we assumed the aggregates to remain constant in size. With that only two parameters were needed to describe the whole curve: the radius R of the particles (derived from the molecular weight) and their effective hard sphere interaction radius R_s (derived from the effective volume fraction ϕ). The obtained fitted curves are plotted as solid lines and are in very good agreement with the experimental data. The values for R and R_s are 31.8, 34.3 Å for the TDMAO system and 50.2, 53.4 Å for the ODMAO system. In both cases the interaction radius R_s is about 2.5—3 Å larger than R, which is a very reasonable value for the thickness of the hydration shell of the particles. This value is also consistent with viscosity measurements and NMR self-diffusion studies [15].

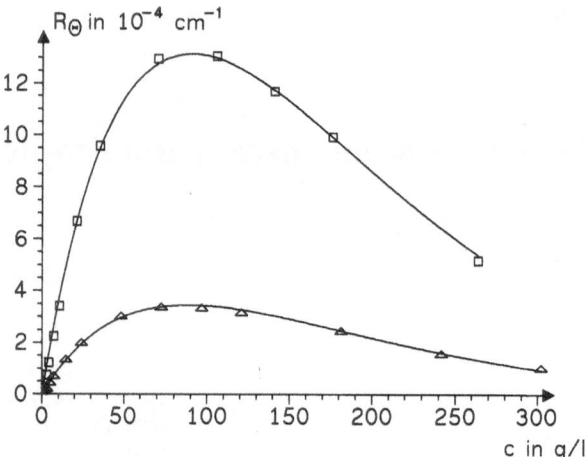

Fig. 1. Rayleigh factor R_Θ for microemulsions with TDMAO (Δ; x (TDMAO)/x (decane) = 3.2 : 1) and ODMAO ($\square$; x (ODMAO)/x (decane) = 1.11 : 1) as a function of the total concentration of surfactant and hydrocarbon (at 25°C).

Therefore, it is clear that these o/w microemulsions are made up from spherical aggregates which are of constant size irrespective of their overall concentration and their interactions can be explained via a simple hard sphere model.

$$R_\Theta = \frac{4 \cdot \pi^2 \cdot n_0^2}{N_A \cdot \lambda^4} \left(\frac{dn}{dc}\right)^2$$
$$\cdot c_g \cdot M_w \cdot P(q) \cdot S(q) \qquad (1)$$

n_0:   refractive index of the solvent
N_A: Avogadro constant
dn/dc: refractive index increment
c_g: weight concentration of the solute
q: magnitude of the scattering vector

$$\left(q = \frac{4 \cdot \pi \cdot n}{\lambda} \sin(\Theta/2)\right)$$

$P(q)$:   form factor (accounts for intraparticle interferences)
$S(q)$:   structure factor (accounts for interparticle interferences)

$$S(0) = \frac{(1 - \Phi)^4}{(1 + 2\Phi)^2 - 4 \cdot \Phi^3 + \Phi^4} \qquad (2)$$

The same conclusion could also be drawn from SANS experiments performed on samples of constant TDMAO to decane ratio, but now with D_2O as solvent [16]. They confirmed that the individual aggregates do not change in size upon changes in the total concentration and the scattering curves could well be described by a simple hard sphere model. In those scattering curves for samples of higher concentration a second maximum of the scattering intensity was observed. The occurrence of such a second maximum is uncommon for surfactant systems and can be viewed as an indication of a fairly small polydispersity of the aggregates, because a higher polydispersity would wipe out such a maximum.

In order to elucidate the degree of polydispersity in these systems in more detail a SANS experiment was performed on a diluted sample. Here, one can expect the effects of the structure factor to be negligible, i.e., only the particle form factor $P(q)$ should be responsible for the angular dependence of the scattered intensity. The corresponding scattering curve for a 1 wt% sample of ODMAO and decane (molar ratio 1.11:1) in D_2O is plotted in Fig. 2 and one observes a minimum of the intensity at 0.1 Å which is followed by a maximum again. The relative extent of this maximum should be related to the degree of polydispersity for the respective particles. These experimental data could be described by means of a model of spherical particles with a Schulz distribution [17] (Eq. (3)) of the radii, where $P(q)$ of a homogeneous sphere of radius R is given by Eq. (4). This model enables one to calculate the scattering intensity, i.e., the differential cross — section $d\Sigma/d\Omega$, according to Eq. (5) and a fit to the data is in good agreement with the experimental values.

If one takes into consideration the experimental wave length distribution of $\Delta\lambda/\lambda$ (= 9% FWHM), one obtains a radius of 49.9 Å for the aggregates (good agreement with the light scattering data) and a value for the standard deviation of the distribution function of less than 10% of the mean radius ($\sigma \leq 0.1 \cdot R$). This is still an upper limit for the polydispersity of the aggregates because all other experimental effects (like collimation effects, the finite size of the scattering volume and detector) will also tend to smear out the maximum in the scattering curve [18, 19]. Furthermore, here it was assumed that the aggregates posses a sharp boundary to the solvent, whereas in reality this interface will be dynamic and therefore will have some roughness. But it is known that the influence of

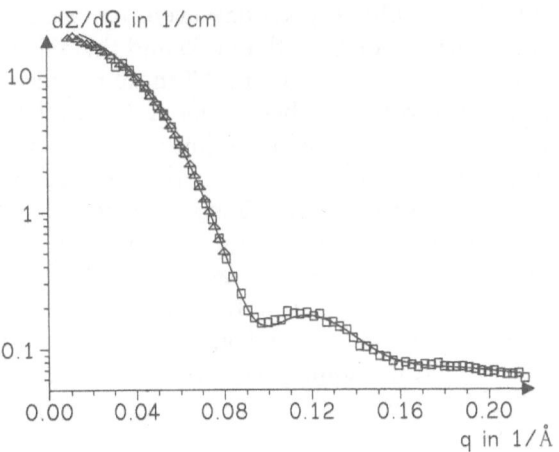

Fig. 2. SANS intensity for a 1 wt% solution of ODMAO/decane (molar ratio 1.11 : 1) in D_2O with the fit curve as solid line (at 25 °C).

such a diffuse interfacial region on the scattering pattern would be the same as that of the polydispersity [20].

$$f(r) = \left(\frac{t+1}{R}\right)^{t+1} \frac{r^t}{\Gamma(t+1)} \, \exp\left(-\frac{t+1}{R} \, r\right) \tag{3}$$

with $t = \left(\frac{R}{\sigma}\right)^2 - 1$ σ : standard deviation

$$P(q,r) = \left(3 \frac{\sin(q \cdot r) - q \cdot r \cdot \cos(q \cdot r)}{(q \cdot r)^3}\right)^2 \tag{4}$$

$$\frac{d\Sigma}{d\Omega} = {}^1N \cdot (\rho_a - \rho_s)^2 \cdot \langle V^2 \cdot P(q)\rangle \cdot S(q) \tag{5}$$

^1N: particle number density
ρ_A: scattering length density of the aggregate
ρ_s: scattering length density of the solvent

According to these investigations the microemulsion droplets in the amine oxide/hydrocarbon system exhibit a fairly low degree of polydispersity. In addition, their interactions can be described by a simple hard sphere potential even up to high volume fractions of more than 0.35 and their size is independent of concentration. All these properties render them to be suitable model systems for the study of the effects of changing interactions in such microemulsions. For instance, the interactions could become varied by the substitution of the zwitterionic amine oxide by an ionic surfactant. By perturbing the system in such a way, one would expect to switch from the formerly uncharged microemulsion droplets to charged aggregates interacting via a screened Coulomb potential.

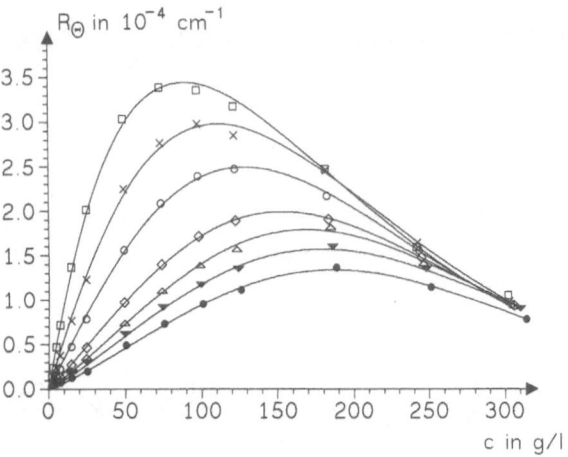

Fig. 3. Static light scattering data at 25°C for microemulsions formed from various mixtures of TDMAO and TTABr with decane. The molar ratio of surfactant to decane was always 3.2 : 1. Molar content of TTABr (with respect to the total surfactant): □: 0%; ×: 1.25%; ○: 2.5%; ◇: 5%; △: 7.5%; ▼: 10%; ●: 15% (fitted curves as solid lines)

For the investigation of this effect the cationic trimethyltetradecylammonium — bromide TTABr, and the anionic sodium — tetradecylsulfate TDS were employed as ionic surfactants. In order to study the structural properties of these charged microemulsion systems static light scattering experiments were performed on microemulsions based on TDMAO as surfactant [21]. As in the case of the uncharged systems, concentration series were investigated where the molar ratio of surfactant to decane was kept constant at 3.2:1.

First, the effect of the cationic surfactant TTABr on the microemulsion systems shall be discussed. In Fig. 3 the scattering intensity as a function of the total concentration is given for systems that contained 0, 1.25, 2.5, 5, 7.5, 10, and 15 mol% TTABr (in relation to the total surfactant). In comparison to the uncharged system one observes basically similar scattering curves, but the higher the content of ionic surfactant the lower the scattering intensity and its maximum is shifted to higher concentrations. This behavior is due to the electrostatic repulsion between the droplets, which increases the ordering in the microemulsion, leads to a diminished structure factor and therefore to a decreased scattering intensity. Finally for very high concentrations the intensity becomes almost independent of the content of ionic surfactant in the mixtures. This only means that for these high volume fractions of about 0.35 the main contribution to the repulsion between the aggregates simply arises from the excluded volume, i.e., the hard sphere interaction, because here the system is already densely packed. Therefore, the structure factor becomes practically independent of the particle charge and the similar scattering intensity at this point is an indication that the size of the aggregates should be the same for all the mixtures.

As in the case of the uncharged microemulsion systems, we attempted to account for the experimental scattering data by means of a theoretical model that includes the electrostatic interaction of the aggregates. A variety of approximations are known to calculate the structure factor of such a system and we chose a particularly simple model, the random phase approximation (RPA) [22, 23]. For the case of charged colloids with an effective diameter D_{eff}, one may simply take a hard sphere fluid as reference system and account for the additional interaction by means of a screened Coulomb potential (Eq. (6)) [24]. Then in the RPA the additional term can simply be obtained by a Fourier transform of the potential [20]. Now the structure factor can be calculated and because the wave length of the light is much larger than the size of the aggregates, again only $S(0)$ is required, as given by Eq. (7). This model was used to fit the experimental data. The fit parameters in this procedure were the radius of the aggregates R, the effective interaction radius R_s ($= D_{eff}/2$), and now in addition the charge of the aggregates (via the effective surface potential $\Psi_{0\,eff}$). The Debye screening length κ was assumed to be fixed by the concentration of the ionic surfactant.

$$U(r) = \pi \cdot \varepsilon \cdot D_{eff}^2 \cdot \psi_{0\,eff}^2 \frac{\exp\left(-(r - D_{eff})/\kappa\right)}{r} \quad (6)$$

$$S(0)^{-1} = \frac{(1 + 2\Phi) - 4 \cdot \Phi^3 + \Phi^4}{(1 - \Phi)^4}$$

$$+ \ 24 \cdot \Phi \ \frac{\pi \cdot \varepsilon \cdot X \cdot \psi_{0\,eff}^2}{k \cdot T}$$

$$\cdot \ \frac{(1 + D_{eff}/\kappa)}{D_{eff}/\kappa} \qquad (7)$$

$$\psi_{0\,eff} = z \cdot e_0/(\pi \cdot \varepsilon \cdot D_{eff} \cdot (2 + D_{eff}/\kappa))$$

$$\kappa = \sqrt{\frac{\varepsilon \cdot k \cdot T}{N_A \cdot e_0^2 \cdot I}}$$

z: number of charges on the colloidal aggregate
I: ionic strength
ε: dielectric constant of the medium

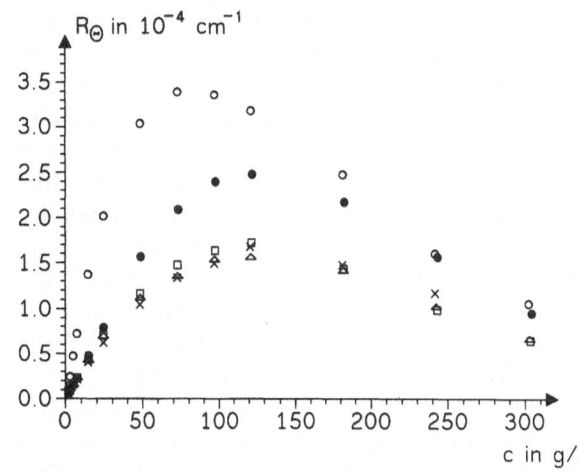

Fig. 4. Rayleigh factor R_Θ at 25°C for microemulsions (x (surfactant)/x (decane) = 3.2:1) that contain 2.5 mol% (with respect to the total surfactant) of various anionic surfactants (□: TDS; ▲: SDS; ×: Na-myristate). For comparison: ●: 2.5 mol% TTABr; ○: pure TDMAO

Table 1. Fit parameters R, R_s, and charge z (RPA) for various mixtures of TDMAO and TTABr. In addition, the charge z_0 as calculated from the composition is given for an assumed radius of 31.7 Å

$\dfrac{x\ (TTABr)}{x\ (Ten.)}$	z_0/e_0	$\dfrac{z\ (RPA)}{e_0}$	R in Å (RPA)	R_s in Å (RPA)
0	0	0	31.7	34.0
0.0125	2.87	2.44	32.6	35.3
0.050	11.50	11.40	32.3	34.9
0.075	17.3	17.7	33.3	36.2
0.10	23.0	21.5	31.6	33.8
0.15	34.5	30.2	31.0	33.5

The theoretical fit curves are plotted as solid lines in Fig. 3 and are in good agreement with the experimental data, which shows that the charged aggregates also remain of constant size irrespective of the concentration. The corresponding fit parameters are given in Table 1. The radii R and R_s remain fairly constant and the obtained effective charges z(RPA) are close to the ones predicted from the composition of the aggregates assuming a constant radius of 31.7 Å. Above 10 mol% ionic content the obtained charges start to deviate towards smaller values than those predicted according to

the composition. This effect should be associated with the onset of the counterion condensation as was further corroborated by measurements of the electric conductivity [26].

In a next step, static light scattering experiments were performed on concentration series where a certain amount of TDMAO was now substituted by the anionic surfactant TDS. In principal, the obtained scattering curves are similar to those of the cationic case, but if one looks more carefully at them one observes that for the same degree of ionic substitution the scattering intensity is significantly lower than in the case of the cationic TTABr. Moreover, it was found that even for the highest concentrations the scattering intensity still decreases with increasing TDS content, in contrast to the observation in the cationic case [21]. These differerences may best be seen in a direct comparison with the data for a TTABr containing microemulsion, where always 2.5 mol% of the TDMAO was substituted by an ionic surfactant (Fig. 4). In addition, one observes that the effect of lower scattering intensities is not particular to the surfactant TDS, but is similarly observed for other anionic surfactants like SDS and Sodium — myristate.

Of course, the different scattering behavior results in the fact that the description of the experimental data of the TDS system is by far not as satisfactorily possible as in the case of the cationic substitution. The fits are significantly worse and tend to lead to too small radii and, also, too large charges of the aggregates are obtained. How may one account for this different behavior? Usually, very valuable information regarding the properties of micellar solutions and especially of their microemulsions may be gained from the knowledge of the interfacial tension of the respective surfactant system against the hydrocarbon to be solubilized. Upon the admixture of the cationic surfactant one observes a continuous increase of the interfacial tension. However, in the case of mixtures with the anionic SDS a much different behavior is found and here a broad minimum of the interfacial tension occurs at roughly equimolar ratios [21]. This minimum is an indication of strong synergistic effcts that occur on mixing the zwitterionic amine oxide with the anionic surfactant. The amine oxide itself has a weak basicity that will be enhanced by the presence of the electric field of the anionic head groups [9, 27]. From the theory of microemulsions it is well established that the value of the interfacial tension should be directly related to the ability of the respective surfactant mixture to solubilize the corresponding hydrocarbon. The lower the value of the interfacial tension the more hydrocarbon can be solubilized and the larger will be the microemulsion droplets that are formed [28].

light scattering intensity would make on inclined to assume the presence of smaller aggregates in these solutions. In order to clarify this point, we made some light scattering measurements of samples with constant total concentration (36 mM surfactant/11.25 mM decane) in 100 mM KCl as a function of the content of ionic surfactant. The presence of the electrolyte should result in a shielding of the electrostatic repulsion and at this high KCl concentration one will basically return to the hard sphere case. In Fig. 5 the scattering intensity for the various mixtures is plotted and one finds only a small decrease of the scattering intensity upon the addition of the TTABr, whereas a fairly significant increase occurs with increasing TDS content. From this experiment it is clear that the presence of the anionic TDS leads to an increase in the droplet size, whereas the addition of the TTABr has only little influence on the particle size.

But how does that agree with the deviation to lower scattering intensities for the electrolyte-free concentration series as displayed in Fig. 4? Some increased repulsion has to be present here and one possibility for that would be a larger hard sphere interaction radius which could be due to a larger hydration shell. Such an effect appears to be quite likely because the sulfate head group of the TDS should be more strongly hydrated than, in comparison, the amine oxide group or the trimethylammonium group. This ought to be that way because only in the case of the sulfate group (or for that matter, the carboxyl group) is strong hydrogen bonding of water molecules is possible.

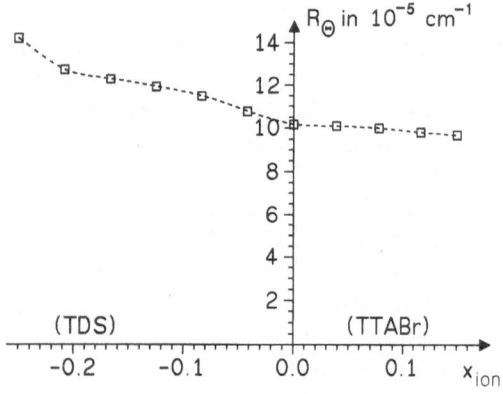

Fig. 5. Static light scattering intensity for microemulsions (36 mM surfactant/11.25 mM decane) as a function of the mixing ratio of TDMAO with TDS (negative sign) and TTABr (positive sign) in 100 mM KCl (at 25°C)

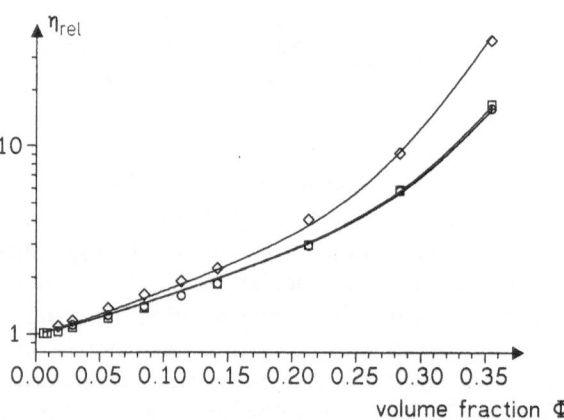

Fig. 6. Relative viscosity η_r at 25°C as a function of the volume fraction Φ for microemulsions of constant surfactant/decane molar ratio of 3.2 : 1. □: pure TDMAO; o: 7.5 mol% TTABr; ◇: 7.5 mol% TDS. (fit curves as solid lines)

From this, one would conclude that in the case of the TDS mixtures the droplets might be larger than in the pure TDMAO system, whereas the lower

In order to check this assumption viscosity measurements were performed on the same systems used for the light scattering experiments. The relative viscosities for the pure nonionic system and for systems containing 7.5 mol% of TTABr or TDS are shown in Fig. 6 as a function of the volume fraction of the surfactant plus hydrocarbon. However, whereas the viscosities of the uncharged and the cationic system are almost exacly equal, much larger values are observed in the anionic case. This behavior is evidence for a much higher effective volume fraction in the anionic system, i.e., a significantly larger hydration shell of the particles. The viscosity data also compare well with an expression (Eq. (8)) found by Thomas [29] for the viscosity of hard sphere suspensions. The theoretical curves are plotted as solid lines and from the fits a value for the ratio of the effective volume fraction to that of the "dry" aggregate, or, respectively, a ratio between the hydrodynamic radius and the particle radius was obtained (Table 2). In the case investigated here the hydrodynamic radius for the TDS system is about 5% larger than for the other two systems.

$$\eta_r = 1 + 2.5 \cdot \Phi + 10.05 \cdot \Phi^2 + 0.00273$$

$$\cdot \exp{(16.6 \cdot \Phi)} . \tag{8}$$

Table 2. Fit parameter Φ_{eff}/Φ_E for the Thomas model for differently substituted microemulsion systems with constant molar surfactant/decane ratio of 3.2:1. In addition, the ratio of hydrodynamic radius to particle radius is given

System	Pure TDMAO	x(TTABr)	x(TDS)
		x(Ten.) = 0.075	x(Ten.) = 0.075
Φ_{eff}/Φ_E	1.418	1.408	1.600
R_h/R_0	1.123	1.121	1.170

However, this larger interaction radius would still not be sufficient to account fully for the larger repulsive forces that must be present in the TDS containing microemulsions. If it is considered in the description of the scattering curves there still remains some additional repulsion which in the RPA model results in a too large estimate of the charges on the aggregates. We can only speculate how this additional repulsion in the case of the anionic surfactant comes about. One possibility would be that the more hydrophilic head group of the anionic surfactant protrudes further into the surrounding aqueous solution, which would locate the charges further outside the microemulsion droplets. Another possibility could be that the counter cations are less effective than the counter anions in shielding the charged head groups. Both effects could explain the deviating behavior in the case of anionic substitution. The assumption that this additional repulsion is not due to a specific interaction between TDS and the zwitterionic surfactant was corroborated by similiar investigations on o/w microemulsions based on a different nonionic surfactant, the Brij 96, a decaethylene-oxide-oleylether [26]. Therefore, the lower scattering intensity in the case of the anionic surfactant should first be specific for it and secondly be a result of an increased effective electrostatic repulsion.

Conclusions

From SANS and light scattering experiments it can be concluded that the pure alkylamineoxide/hydrocarbon system is composed of uncharged microemulsion droplets. Their size remains constant up to very large concentrations of $\sim$ 30 wt% and their interactions can be described by a simple hard sphere model. Furthermore, only little polydispersity was detected for the droplets ($<$ 10%). All these properties render these o/w microemulsions to be suitable model systems to study the effect of changing interactions between the droplets.

These interactions may be changed by the substitution of the amine oxide by an ionic surfactant. By doing so charged microemulsion droplets with a charge density according to the surfactant composition are formed. Again, it could be shown that these droplets are of constant size over the whole existence region of the L_1-phase. It is interesting to note that the scattering properties differ for the substitution by the cationic (TTABr) or the anionic surfactant (TDS). In the cationic case the interactions are well described by a random phase approximation for the structure factor, whereas deviations occur for the anionic substitution. The different behavior in the anionic case is due to a growth of the particles, a larger interaction diameter of the aggregates which is due to an increased hydration of the anionic head group and, finally, to a higher effective electrostatic repulsion between the aggregates. However, it could be shown that one is able to fix the charge (and thereby the surface potential) of the self-aggregating microemulsion droplets at a desired value simply by choosing the proper surfactant mixture, without changing the other properties of the system substantially. Here, the charges of the colloidal aggregates are easily controlled by the composition of the system, whereas other similar colloidal systems, e.g., latex or silica particles have to be synthesized specifically for this purpose.

References

1. Hoar TP, Schulman JH (1943) Nature 152:102
2. Prince LM (ed) (1977) Microemulsions, theory and practice, Academic Press Inc., New York, San Francisco, London
3. Robb IE (ed) (1982) Microemulsions, Plenum Press, New York
4. Gillberg G (1984) In: Lissant KJ (ed) Emulsions and Emulsion Technology, Surfactant Science Series 6, Marcel Dekker, New York, pp 1—43
5. Shah DO (ed) (1981) Surface Phenomena in Enhanced Oil Recovery, Plenum Press, New York
6. Tadros TF (1984) In: Mittal KL, Lindman B (eds) Surfactants in Solution vol. 3, Plenum Press, New York
7. Miller CA, Mukherjee S, Benton WJ, Natoli J, Qutubuddin S, Fort Jr T (1982) AlChE Symp Ser 78:28
8. Oetter G, Hoffmann H (1989) J Dispers Sci Technol 9:459
9. Pößnecker G (1991) Dissertation, Universität Bayreuth
10. Imae T, Ikeda S (1984) J Colloid Interface Sci 98:363
11. Hoffmann H, Oetter G, Schwandner B (1987) Prog Colloid Polym Sci 73:95
12. Hoffmann H, Rauscher A, Gradzielski M, Schulz SF (1992) Langmuir 8:2140
13. Oetter G, Hoffmann H (1989) Coll Surf 38:225
14. Carnahan NF, Starling KE (1969) J Chem Phys 51:635
15. Walther KL, Gradzielski M, Hoffmann H, Wokaun A, Fleischer G (1992) J Colloid Interface Sci 153:272
16. Gradzielski M, Hoffmann H, Oetter G (1990) Colloid Polym Sci 268:167
17. Schulz GV (1939) Z Phys Chem B43:25
18. Skov Pedersen J, Posselt D, Mortensen K (1990) J Appl Cryst 23:321
19. Wignall GD, Christen DK, Ramakrishnan V (1988) J Appl Cryst 21:438
20. Cabane B, Duplessix R, Zemb T (1984) In: Mittal KL, Lindman B (eds) Surfactants in Solution Vol. 1, Plenum Press, New York, pp 373—404
21. Gradzielski M, Hoffmann H (1992) Adv Coll Interface Sci 42:149
22. Hansen JP, Mc Donald IR (1985) Theory of Simple Liquids, Academic Press, New York
23. Andersen HC, Chandler D (1970) J Chem Phys 53:547
24. Verwey EJW, Overbeek JTG (1948) Theory of Stability of Lyophobic Colloids, Elsevier, Amsterdam
25. Baba-Ahmed L, Benmouna M, Grimson MJ (1987) Phys Chem Liq 16:235
26. Gradzielski M (1992) Dissertation, Universität Bayreuth
27. Rosen MJ, Zhu BY (1984) J Colloid Interface Sci 99:427
28. Murphy DS, Rosen MJ (1988) J Phys Chem 92:2870
29. Thomas DG (1965) J Colloid Sci 20:267

Authors' address:

Prof. Dr. H. Hoffmann
Physikalische Chemie I
Universität Bayreuth
Postfach 101251
95412 Bayreuth, FRG

Progress in Colloid & Polymer Science   Progr Colloid Polym Sci 93:175—177 (1993)

Field-induced structure in magnetorheological suspensions

Y. Grasselli, G. Bossis and E. Lemaire

Laboratoire de Physique de la Matière Condensée — URA 190 Université de Nice — Sophia Antipolis, France

Abstract: We have measured the yield stress of a magnetic suspension as a function of the external field. To explain the existence of this yield stress, we report a comparison between two models, one based on isolated chains of particles and the other taking into account that the structure is formed by aggregates of particles.

Key words: Yield stress — magnetorheological fluid — field — induced aggregation

Magnetorheological as well as electrorheological suspensions exhibit an important change of structure when submitted respectively to a magnetic or an electric field. The external applied field induces a dipolar moment on each particle. The mutual interaction, which tends to align two polarized particles in the direction of the field, is responsible for the formation of a fiber structure in the suspension, with the axis of the fibers aligned in the direction of the external field. This change of structure produces an important change in the rheology of these suspensions; in particular, with the appearance of a yield stress which is needed to break the fibers connecting the two disks of the rheometer. We have studied a magnetic suspension composed of polystyrene particles with inclusions of magnetite (62% in weight), suspended in water. The particles are polydisperse with an average radius, $a = 0.4$ μm. We have experimentally determined the yield stress, by the use of a controlled stress rheometer, in a plane-plane configuration, with the suspension placed between two disks and the magnetic field parallel to the axis of the disks (Fig. 1). Figure 2 represents a typical rheogram obtained with this suspension at a volume fraction, $\phi = 10\%$, and for a magnetic field of 900 Oe.

A simple model used to predict the value of the yield stress, τ_s, was proposed by Klingenberg et al. in 1989 [1, 2]. The structure created by the external field is modelized by isolated chains of particles which link the two disks. In this situation, the yield stress can be obtained by looking at the restoring force between two paricles, which were initially

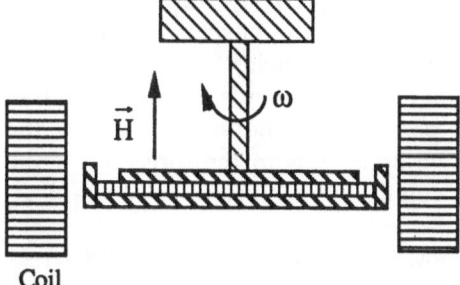

Fig. 1. Experimental apparatus using the controlled stress rheometer for the determination of the yield stress in the plane-plane configuration

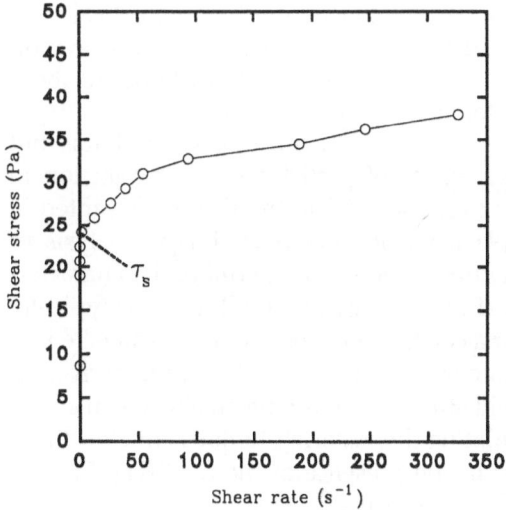

Fig. 2. Rheogram obtained with the magnetic suspension at a volume fraction $\phi = 10\%$ and for a magnetic field of 900 Oe. τ_s represents the position of the yield stress

Progress in Colloid & Polymer Science, Vol. 93 (1993)

aligned in the direction of the field and supposing that one is pulled in the direction of the shear. We found that the yield stress can be written as [3]:

$$\tau_s = \frac{3\phi}{2\pi a^2} F_r^m , \tag{1}$$

where ϕ is the volume fraction and F_r^m the maximum value of the restoring force. Its expression is [1]:

$$F_r^m = 3\mu a^2 H^2 \beta^2 f_m \tag{2}$$

with $\beta = (\mu_i - \mu)/(\mu_i + 2\mu)$, $\tag{3}$

where μ_i and μ are, respectively, the permeabilities of the particles and of the suspension. Both are functions of the magnetic field; f_m is the maximum of the function f which depends on the separation between the particles, on the angle made by the axis of the two spheres with the direction of the field and on coefficients tabulated by Klingenberg [1].

Figure 3 shows the behavior of the yield stress versus the magnetic field. The dashed curve represent the experimental results. As it could be expected from Eq. (2), the yield stress is not proportional to the square of the applied field because the permeability of the particles depends on the magnetic field and there is a saturation of μ_i as the intensity of the field increases. The theoretical yield stress obtained with the help of Eq. (1) is also plotted in Fig. 3 (solid curve). The predictions are about 60% greater than the experimental results.

In general, we do not have isolated chains but rather aggregates of particles whose projection in the plane perpendicular to the field is presented in Fig. 4. We have performed an image analysis of these pictures in order to determine the characteristic size of the aggregates [4], [5] which is typically 60 μm for a cell thickness of 1 mm. We can take into account the size of the aggregates in the derivation of the yield stress, by assuming an ellipsoidal shape and calculating the restoring torque acting on these ellipsoids in the presence of the field (Fig. 5). The torque, K, is given by:

$$K = -\frac{\partial W}{\partial \theta} , \tag{4}$$

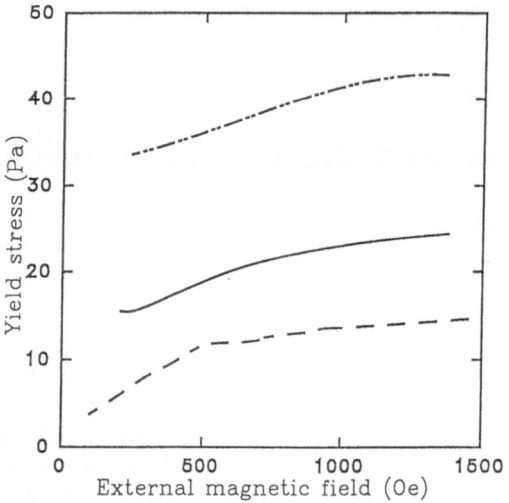

Fig. 3. Behavior of the yield stress versus the magnetic field. ----- Experimental results. ——— Theoretical results obtained from the magnetic restoring force between two particles. —··— Theoretical result obtained calculating the restoring torque on an ellipsoidal aggregate in a magnetic field.

where W is the total magnetic energy:

$$W = -\frac{1}{2} N_a \frac{M_a \cdot H_0}{a} , \tag{5}$$

N_a and M_a being, respectively, the number and the moment of the aggregates present in the suspension. Since the external field is modified by the presence of the aggregates, M_a must be deduced from a self consistent equation:

$$M_a = a \cdot \left(H_0 - 4\pi \frac{N_a}{V} M_a \cdot nn \right) , \tag{6}$$

a is the polarizability tensor of an ellipsoid and n is the normal to the surface.

Using Eq. (4), we get the following expression for the restoring torque:

$$K = -\frac{1}{8\pi} \frac{\phi}{\phi_a} \frac{1}{(1 + \lambda)} \frac{\partial \lambda}{\partial \theta}$$

with

$$\lambda = (\mu_a - 1)$$

$$\cdot \left\{ \frac{\cos^2\theta}{1 + n_\parallel(\mu_a - 1)} + \frac{\sin^2\theta}{1 + n_\perp(\mu_a - 1)} \right\} , \tag{7}$$

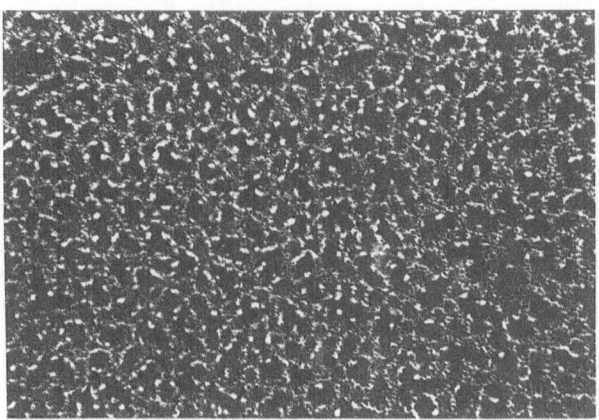

Fig. 4. Optical microscopy observation of the structure in a plane perpendicular to the magnetic field

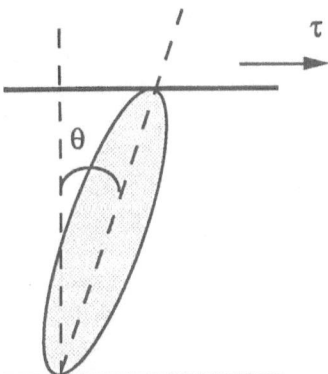

Fig. 5. Schematic representation of an inclined ellipsoid in the shear

where ϕ_a is the volume fraction inside the aggregates, μ_a is the permeability of an aggregate, and $n_\parallel$ and $n_\perp$ are the demagnetizing factors, respectively, in the direction of the large axis and in the direction of the small axis of the ellipsoids [6].

Now, the yield stress is then $\tau_s = N_a K_m / V$, where K_m is the maximum value of the torque. The results are plotted in Fig. 3 (dash-dotted curve). This model surestimates the experimental values by about four times. This model is based on the assumption that we have homogeneous ellipsoids and it does not take into account the discrete separation of the particles inside the aggregates which are extended by the strain as in the model of isolated chains. This separation will lower the restoring torque. A future investigation should introduce both the ellipsoidal model and its discrete internal structure.

References

1. Klingenberg DJ (1990) Ph D Thesis, University of Illinois
2. Klingenberg DJ, Zukoski CF (1990) Langmuir 6:15
3. Bossis G, Lemaire E (1991) J of Rheology 35:1345
4. Lemaire E, Grasselli Y, Bossis G (1992) J Phys II 2:359
5. Field induced structure in magneto and electrorheological fluids Grasseli Y, Bossis G, Lemaire E unpublished
6. Electrodynamics of continuous media Landau & Lifshitz, T 8, Pergamon Press

Authors' address:

Y. Grasselli
Laboratoire de Physique
de la Matière Condensée
URA 190
Universite de Nice — Sophia Antipolis
Parc Valrose
06108 Nice Cedex 02, France

Progress in Colloid & Polymer Science

Progr Colloid Polym Sci 93:181 (1993)

The structure of endoglucanase I (Trichoderma resei) in solution

P. M. Abuja), M. Hayn[2]), H. Chen[2]),
and H. Esterbauer[2])

[1]) Institut für Biophysik und Röntgenstrukturforschung der Österreichischen Akademie der Wissenschaften, Graz, Austria
[2]) Institut für Biochemie, Universität Graz, Austria

Abstract: The structure of endoglucanase I (EG I) from Trichoderma reesei has been investigated in solution, using small-angle x-ray scattering. The enzyme showed the typical shape ("tadpole") and domain arrangement (catalytic core-spacer-cellulose binding domain) which is obviously common to all celluloses degrading solid substrate. The molecular size parameters resemble cellobiohydrolase I closely: radius of gyration is 4.74 nm, overall length 18.0 nm, diameter 5.3 nm. The main difference lies in the spacer region, which contains much more scattering mass, probably due to a higher extent of glycosylation which extends towards the core domain.

Key words: Cellulases — endoglucanase — small-angle x-ray scattering — solution structure

Small-angle x-ray scattering (SAXS) has been used previously to obtain information on the structure in solution of the extracellular exoglucanases (CBH I, CBH II) of *T, resei* [1, 2] showing a peculiar tadpole-like shape. In these studies, together with information obtained by partial proteolysis, a structure-function relationship emerged which is assumed to hold true for all cellulases capable of adsorption to and degradation of crystalline cellulose: a large so-called core domain, which contains the active site, is connected to the cellulose-binding domain (CBD) by a long spacer sequence. The core domain is a large ellipsoid and can be separated from the other parts of the enzyme by partial proteolysis [3, 4], yielding a protein that is still capable to degrade soluble cellulose derivatives, but has only limited ability to adsorb onto crystalline cellulose. The spacer is a highly glycosylated extended amino acid strand — 10/14 nm long (CBH I/CBH II), whereas the CBD is small and compact.

Endoglucanase I, from the same microorganism, however, has up to now resisted all efforts to separate the core domain from the other parts of the enzyme, although from the amino acid sequence a similar organization as with the exocellulases CBH I and CBH II might be expected. The degree of glycosylation of endoglucanase I is about twice as high as for CBH I and its spacer sequence contains approximately twice the number of putative glycosylation sites compared to the analogous domain CBH I, especially near the side where the core is located.

By model calculations, using MULTIBODY [5], it can be shown that the core is partially embedded in what is thought to be sugar residues from glycosylation of the spacer. This shielding of the core-spacer junction site (which is also the site where papain digestion occurs) is probably the explanation for the failures of separating a core from EG I.

The size parameters, such as the radius of gyration (4.74 nm), the overall length (18.0 nm) and diameter (5.3 nm), as well as the estimated sizes of the core and the CBD (as found by modeling) match quite well those of CBH I, whereas the spacer contains far more scattering mass — a finding that may be attributed to a higher degree of glycosylation since its length and total number of amino acid residues are comparable to that of CBH I.

References

1. Abuja PM, Schmuck M, Pilz I, Tomme P, Claeyssens M, Erstbauer H (1988) Eur Biophys J 15:339—342
2. Abuja PM, Pilz I, Tomme P, Claeyssens M (1988) Biochem Biophys Res Comm 156(1):180—185
3. van Tilburgh H, Tomme P, Claeyssens M, Bhikhabai R, Pettersson LG (1986) FEBS Lett 204:223—227
4. Tomme P, van Tilbeurgh H, Pettersson G, van Damme J, Vanderkerckhove J, Knowles J, Teeri T, Claeyssens M Eur J Biochem 170:579—581
5. Glatter O (1980) Acta Phys Austr 52:234—256

Authors' address:

P. M. Abuja
Institut für Biophysik und Röntgenstrukturforschung der Österreichischen Akademie der Wissenschaften
Steyrergasse 17
8010 Graz, Austria

Progress in Colloid & Polymer Science   Progr Colloid Polym Sci 93:182 (1993)

Characterization of the acid-base properties of alumina surfaces

M. Airiau and N. Lebel

Rhône-Poulenc Recherches, Aubervilliers, France

Abstract: Three aluminas have been characterized by electrophoretic mobilities measurements in polar nonaqueous solvents, differential thermal analysis, and infrared spectroscopy, in order to study the acid-base properties of their surfaces. The solvents were chosen according to the Gutmann donor/acceptor concept to cover a wide range of electron acceptor and electron donor numbers. The electron donor properties of the three alumina surfaces are different and correlate with water desorption and OH abundance as seen by DTA and IR spectroscopy. Conversely, the electron acceptor properties are identical for the three surfaces. These facts are interpreted as results of the presence of electropositive (Al) or electronegative (O) atoms in different quantities at the surface of the oxides.

Key words: alumina — surface — electrophoretic mobility — acid-base properties — non-aqueous medium

Three aluminas have undergone elektrokinetic measurements in non polar media, thermogravimetric analysis, and infra-red spectroscopy in order to study their surface acid-base properties. One of these alumina is an alpha alumina. Elektrokinetic measurements were made in seven solvents which were chosen according to their electron-donor or electron acceptor properties [1]. This is a way to define an electron-donor or electron-acceptor number for the alumina surface, hence, to study its surface acid-base character [2—4].

The experimental results shown no difference between the three solids in the variation of the mobility vs. acceptor number of the liquid, whereas there are important differences in the mobility vs. donor number curves. On the other hand, infra-red spectroscopy and thermogravimetric analysis show various hydroxylation states of the three aluminas: the hydroxyl groups on the alpha alumina are barely seen, but they are easily detected on the two other solids. There is a good correlation between these results and the position of the curves on the graph of mobility vs. donor number. This points to a dominant effect of hydroxyl groups on the basicity of the alumina surfaces, which is related to the electronegativity of the oxygen atom. On the contrary the graph of mobility vs. acceptor number shows there is no difference between the acid properties of the surface, which are dominated by the abundance of electropositive (A1) atoms.

Table 1. Comparison between electrophoresis. Differential thermal analysis and infra-red spectroscopy results

	Alumina A	Alumina B	Alpha alumina
Electrophoresis	Positive and highest mobilities. e.g., at DN = 17: $2.5\ 10^{-8}$ m²/Vs	Positive mobilities. Intermediate values. e.g., at DN = 17: $1.1\ 10^{-8}$ m²/Vs	Presence of an isoelectric point. Lowest mobility values. e.g., at DN = 17: $-0.3\ 10^{-8}$ m²/Vs
Differential thermal analysis	-8.4% weight loss due to OH condensation between 20 and 325°C	-5.4% weight loss due to OH condensation between 20 and 280°C	$+0.3\%$ weight increase between 20 and 1000°C
IR spectroscopy	Important OH peak. At 3400 cm^{-1} and for 20 mg: absorbance = 2.1	Important OH peak, with lower absorbance. At 3400 cm^{-1} and for 20 mg: absorbance = 2.0	No OH peak. At 3400 cm^{-1} and for 20 mg: absorbance = 1.2

References

1. Gutmann V (1988) The Donor-Acceptor Approach to Molecular Interactions. Plenum, New York
2. Siffert B, Kuczinski J, Papirer E (1990) J Colloid Interface Sci 135(1):107
3. Labib, Mohamed E (1984) J Colloid Interface Sci 97(2):356
4. Labib, Mohamed E "The origin of the surface charge on particles suspended in organic liquids", no ref.

Authors' address:

M. Airiau
Rhône-Poulenc Recherches
52, rue de la Haie Coq
93308 Aubervilliers, France

Vesiculation of single-chain surfactant mixtures

M. Ambühl[1]), F. Bangerter[2]), P. L. Luisi[1])
P. Skrabal[2]), and H. J. Watzke[1])

[1]) Institut für Polymere, ETH-Zentrum, Zürich, Switzerland
[2]) Laboratorium für Technische Chemie, ETH-Zentrum, Zürich, Switzerland

Abstract: Mixtures of single-chain surfactants usually form micellar aggregates. However, catanionic mixtures exhibit spontaneous reorganization to vesicles under proper conditions. The strong electrostatic interactions of the headgroups cause the formation of a closed-shell bilayer structure. We have synthesized long-chain acylated amino acid salts to study the influences of charge, polarity, size, rigidity, and chirality of the anionic headgroups on the self-assembling processes when being mixed with N-dodecyl pyridinium chloride (DPC). — Vesicles are formed upon gently mixing of the parent micellar solutions. Investigating the whole range of catanionic compositions by light scattering, we could distinguish areas of mixed micellar and unilamellar vesicular aggregates, which were evidenced by various microscopic techniques. — By ^{1}H-NMR the electrostatic interactions between the cationic DPC and anionic SLS (Sodium N-lauroylsarcosinate) headgroups were found to cause small but significant changes in the chemical shift of the methylene protons adjacent to the carboxyl group. Spin-spin relaxation (T_2, T_2^*) measurements reveal a strong immobilization of the anionic headgroups due to membrane formation. We also found significant changes in the configuration distributions of SLS accompanying the membrane formation in the vesicles. — These findings evidence the importance of

configurational and conformational states of the amphiphiles for the micellar-vesicular transformations.

Key words: Self-vesiculation — bilayer organization — single-chain surfactants — proton NMR spectroscopy

Mixtures of single-chain surfactants usually form micellar aggregates. However, catanionic mixtures exhibit spontaneous reorganization to vesicles under proper conditions [1—3]. The strong electrostatic interactions of the headgroups cause the formation of a closed-shell bilayer structure producing unilamellar vesicles. We studied the influences of charge, polarity, size, rigidity, and chirality of the anionic headgroups on the self-assembling processes. We report on the aggregational behavior of mixtures of N-dodecylpyridinium chloride (DPC) with the following synthesized amino acid surfactants: Sodium N-lauroylsarcosinate (SLS), sodium N-lauroylglycinate (SLG), sodium N-lauroyl-alaninate (SLA), sodium N-caprinoyl-glycylsarcosinate (SCGS). The composition is indicated by Ψ, the molar ratio of the negatively charged component.

Using proton NMR, chemical shift assessment indicates that the amino acid headgroups are not strongly perturbed by their interaction with the aromatic pyridinium head of DPC. However, the N-methylene protons of the E-configurated SLS molecules are significantly affected by the ion-pair formation, leading to a line-crossing of the resonance lines of E- and Z-configurations.

The line-width of the resonance lines can be used to estimate the relaxation time T_2^*. Comparing T_2^* values of micellar with vesicular solutions, a strong immobilization of the anionic headgroups is indicated. The hydrophobic alkyl chain protons are mostly affected by the reorganization of the amphiphilic aggregates. Generally, the T_2^* decrease is stronger, the closer the composition approaches the 1:1 ratio. Spin-spin relaxation measurements of SLS/DPC confirm the results of the T_2^* estimations. Spin-lattice relaxation times T_1 are practically independent of the composition Ψ.

N-methylated amino acid surfactants exhibit two configurations with respect to the amide bond. It was found that the ratio of configuration changes with aggregation [3—4]. Mixing SLS with DPC increases the E-configuration population. Interestingly, in the coexistence range of micelles and vesicles, a significant increase can be found depending on the extent of vesicles formed.

Employing freeze-fracture electron microscopy,

video-enhanced light microscopy and quasi-elastic light scattering, we characterized the translucent vesicular samples obtained by mixing the parent micellar solutions. Using freeze-fracture electron microscopy, we found unilamellar vesicles in solutions in the composition range $0.4 \leq \Psi \leq 0.6$. From the electron micrographs, the vesicles appear to be polydispersed, having radii of 7—30 nm (Fig. 1). Half-neutralization of SLS-solutions produce vesicles of up to more than 500 nm radius as characterized by video-enhanced light microscopy. Hydrodynamic radii and size distributions were determined by quasi-elastic light scattering. The aggregational behavior appeares different for three composition regions: low and high Ψ: mixed micelles; intermediate Ψ: vesicles plus mixed micelles (except for $\Psi = 0.5$, a lamellar phase). On the conrary SCGS only forms mixed micelles with DPC over the whole composition range.

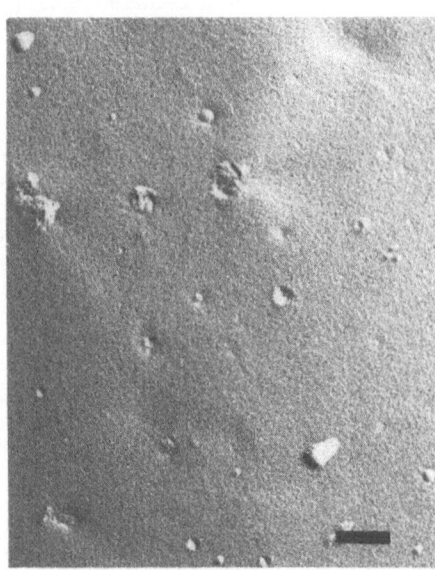

Fig. 1. Freeze-fracture electron micrograph of vesicular SLA/DPC with composition $\Psi = 0.6$, total surfactant concentration 0.05 moles/dm. Bar indicates 100 nm

References

1. a) Kaler EW, Kamalakara Murthy A, Rodriguez BE, Zasadzinski JAN (1989) Science (Washington) 245:1371. b) Kaler EW, Kamalakara Murthy A, Rodriguez BE, Zasadzinski JAN (1992) J Phys Chem 96:6698
2. Szönyi S, Camabon A, Watzke HJ, Schurtenberger P, Wehrli E (1992) In: Structure and conformation of amphiphilic membranes. Springer proceedings in physics (vol 66), Springer-Verlag Berlin, p 198
3. Ambühl M, Bangerter F, Luisi PL, Skrabal P, Watzke HJ (1993) Langmuir 9(1):36
4. Takahashi H, Nakayama Y, Hori H, Kihara K, Okabayashi H, Okuyama M (1976) J Coll Interf Sci 54(1):102

Authors' address:

Dr. Heribert J. Watzke
ETH-Zürich
Institut für Polymere
Universitätsstr. 6
8092 Zürich, Switzerland

Interaction of tertiary amine anesthetics with phosphatidylcholine bilayers

P. Balgavý, D. Uhríková, J. Gallová, K. Lohner*, and G. Degovics*

Katedra fyzikálnej chémie, Farmaceutická fakulta, Univerzita J. A. Komenského, Bratislava, SR
* Institut für Biophysik und Röntgenstrukturforschung, Österreichische Akademie der Wissenschaften, Graz, Österreich

Key words: Local anesthetics — phosphatidylcholine bilayer — ESR spectroscopy — NMR spectroscopy — x-ray diffraction

Local anesthetics N-[2-(alkyloxy)-phenyl]-2-(1-piperidinyl)ethyl esters of carbamic acid (C_nA, $n = 1 \div 10 =$ number of alkyloxy chain carbons) partition between phosphatidylcholine (PC) and aqueous phases. The partition coefficient between small unilamellar egg yolk PC (EYPC) liposomes and aqueous phase exponentially increases with n: log $K_p^+ = 0.52 + 0.37 \cdot n$. The C_nA molecules disorder the bilayer core: the number of gauche conformers in EYPC acyl chains increases with decreasing n at a constant bilayer C_nA concentration. In equimolar C_nA: EYPC mixtures and EYPC: $H_2O = 1:1$ (w:w), the short chain C_nA homologs ($n \leq 4$) induce a hexagonal phase formation as observed by ^{31}P-and ^{2}H-NMR spectroscopy and x-ray diffraction (XRD). In equimolar mixtures with the long chain ($n \geq 5$) homologs only the lamellar phase has been observed. At equal sample C_nA concentrations, the depression of the gel-liquid crystal phase transition (ΔT_m) of dipalmitoyl-PC (DPPC) liposomes displays a quasi-parabolic dependence on n with a maximum at $n = 9$. The values of ΔT_m at equal bilayer C_nA concentrations indicate a diminishing effect of C_nA with the increasing of n.

Due to the mismatch between the lengths of C_nA and PC hydrocarbon chains, the intercalation of C_nA in the bilayer creates free volume in the bilayer below the C_nA alkyloxy methyl group. At equal C_nA bilayer concentrations, the total free volume decreases with increasing n. At equal C_nA sample concentrations, the number of C_nA molecules in the bilayer increases exponentially with n and, thus, the total free volume shows a quasi-parabolic dependence on n. The free volume formation could cause the observed bilayer disordering and disruption. The dependence of the total free volume on n shows a trend similar to the anesthetic and antimicrobial activities of C_nA.

Experimental

C_nAs prepared as described in [1] were from Prof. Dr. J. Čižmárik. DPPC was from Fluka, doxylstearic acid spin probes (DSA) were from Syva, and EYPC was isolated according to [2]. Unilamellar EYPC liposomes were prepared as in [3], samples for ESR, NMR, XRD and DSC were prepared by mixing of C_nA + lipid in $CHCL_3/CH_3OH$, evaporation of solvents, hydration, and homogenization. Partition coefficients were determined by ultraviolet difference spectroscopy at $\bar{v} = 34640\ \psi\mu^{-1}$, pH 5.0, 25°C according to [4]. ESR spectra measurements and evaluation of the number of gauche conformers are described in [5], NMR spectra were obtained and evaluated as in [6] and the differential scanning calorimetry as in [7]. Small angle x-ray diffraction was carried out in a modified Kratky compact camera (A. Paar, Graz, Austria) with Ni-filtered CuK_a radiation ($\lambda = 0.154$ nm). The camera was equipped with a Peltier-controlled variable-temperature cuvette and a linear one-dimensional position-sensitive OED 50-M (M.Braun, FRG) installed at a sample to detector distance of 21 cm.

References

1. Čižmárik J, Borovanský A (1975) Chem Zvesti 29:119—123
2. Singleton WS, Gray MS, Bown LN, White JL (1965) J Amer Oil Chem Soc 42:53—56
3. Barenholz Y, Gibbes D, Litman BJ, Goll J, Thompson TE, Carlson FD (1966) Biochemistry USA 16:2806—2810
4. Welti R, Mullikin LJ, Yoshimura T, Helmkamp Jr GM (1984) Biochemistry 23:6086—6091
5. Gallová J, Devínsky F, Balgavý P (1990) Chem Phys Lipids 53:231—241
6. Balgavý P, Gawrisch K, Frischleder H (1984) Biochim Biophys Acta 772:58—64
7. Gallová J, Bágel'ová J, Čižmárik J, Balgavý P (1992) Pharmazie 47:873—874

Authors' address:

Dr. P. Balgavý, Department of Physical Chemistry, Faculty of Pharmacy, J. A. Comenius University, Odbojárov 10, 832 32 Bratislava, Slovakia

Elucidation of osmotic water transport through emulsion liquid membranes

H.-J. Bart and H. Jüngling

Christian Doppler Laboratorium für Modellierung Reaktiver Systeme in der Verfahrenstechnik Institut für Thermische Verfahrenstechnik und Umwelttechnik Technische Universität Graz, Austria

Key words: Liquid membranes — water transport — surfactants and water solubilization

Emulsion liquid membrane permeation is a process for the recovery of solutes from very dilute solutions [1]. It is usually accompanied by an osmotic water flux [2] that depends on the presence of surface active agents such as Span 80 and ion-exchanger (D2EHPA). The aspect of water solubilization by surface active agents such as Span 80 and liquid ion-exchangers such as D2EHPA in n-dodec-

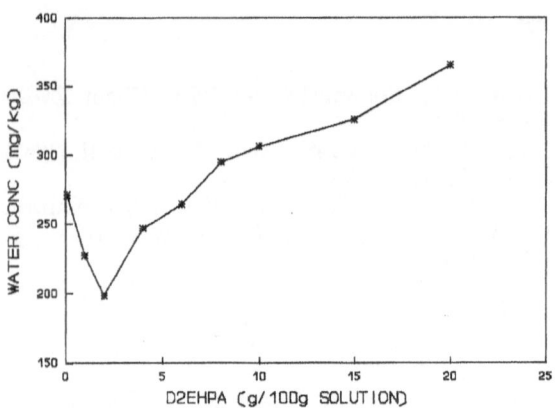

Fig. 1. Solubilization by D2EHPA

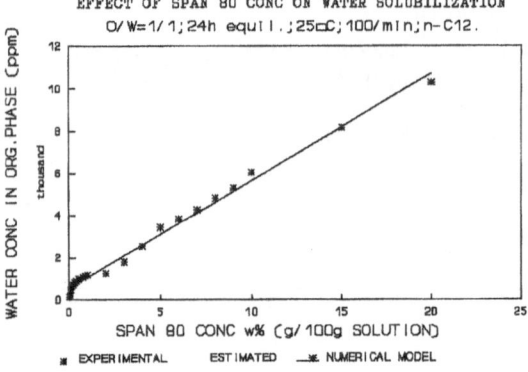

Fig. 2. Model for the estimationof c(H$_2$O) solubilized by Span 80

ane has been investigated. The results have shown a linear correlation between solubilization of water by Span 80. This correlation can be applied to predict the amount of solubilized water in analogous systems with high accuracy. However, when D2EHPA was used instead of the surfactant a minimum [3] in water solubilization was found and the amount of the solubilizate was decisively higher in the former than in the later case. Analytical and instrumental analysis have been exploited to study the physicochemical properties. The investigations on the combined effect imposed by the surfactant and the carrier are in progress and will be the subject of a forthcoming paper.

References

1. Marr R, Kopp A (1980) Chem Ing Techn 52:399—410
2. Matsumoto S, Inoue T, Kohda M (1980) J Colloid Interface Sci 77(2):555
3. Bart HJ, Ramaseder C, Haselgrübler T, Marr R (1992) Hydrometallurgy 28:253—257

Authors' address:

H.-J. Bart
Institut für Thermische Verfahrenstechnik
und Umwelttechnik
Technische Universität Graz
Inffeldgasse 25
8010 Graz, Austria

Optical investigations of disperse systems

A. G. Bezrukova

Technical University, St. Petersburg, Russia

Disperse systems could be considered as sensors of environmental condition changes. Due to their instability they need expressive methods of control. Optical methods providing the possibility of automatization for both measurements and results calculations are the most convetient for this purpose. We have the experience of investigating characteristics for the systems with different nature of particles such as aggregates of proteins, lipoproteins, viruses, blood substitutes, lipid emulsions, liposomes, cells with various forms and sizes, dispersions of erythrocyte diagnosticums, latex and metallic particles, clay dispersions, and natural waters. Among these characteristics are mean size of particles, their refractive index, form, internal structure and distributions in size and in mass. For the investigation of disperse systems, we use the following optical methods: refractometry, absorbency, fluorescence and light scattering (elastic and quasi elastic, unpolarized, and polarized singular and multiple). For each of the investigated systems the complex set of optical parameters that allows registration of small changes in the system state could be suggested. Depending on the research purposes and the peculiarities of the system the optimal algorithm for express analysis of the system under investigation — even with polymodal size distribution and with different nature of particles — could be elaborated. Taking into account the optical theory and information on the system features helps us to rapidly monitor the presence of the component or the quotas of each of the components under analysis. For example, after investigating model mixed systems consisting of kaolin particles and cells of *E. coli*, we could suggest a set of parameters by which the volume share of mineral and organic particles could be evaluated.

Author's address:

Dr. Alexandra G. Bezrukova
Department of Biophysics
Physico-Mechanical Faculty,
St. Petersburg State
Technical University,
195251 St. Petersburg, Russia

separations. The existence of fluid-fluid phase separations has now been confirmed experimentally [4].

The presence of sugars can reduce the incidence of membrane damage and phase separations during dehydration. We show that the ability may be largely explained by non-specific osmotic effects and the volume of the solute, without recourse to specific interactions invoked in other studies [5]. The presence of sugars increases the intermembrane separation at any water content, and this in turn reduces the membrane lateral stress. Thus, the various deleterious phase changes are inhibited, resulting in a decrease in membrane damage. Preliminary experiments appear to confirm this interpretation [6].

Membrane damage at low hydrations: Lipid phase behaviour and the effect of solutes

G. Bryant* and J. Wolfe

School of Physics, University of New South Wales, Kensington, Australia
* Currently at the Helmholtz-Institut für Biomedizinische Technik, Aachen FRG

Key words: Phase separations — membrane dehydration — solute effects — cryobiology — anhydrobiology

Loss of semipermeability in dehydrated cell membranes has been associated with phase separation of membrane components and the formation of non-lamellar phases (such as the inverse hexagonal (H_{II}) phase) [1, 2]. The topology of the H_{II} phase cannot provide semipermeability, leading to cell death. Most membranes have only small quantities of H_{II} forming lipids, and so in order for such a phase to form, these lipids must be concentrated. Possible mechanisms for this include dehydration-induced gel-fluid phase separations [1] and fluid-fluid phase separations [3]. Phase behaviour of both of these types can be induced dehydration; the repulsive hydration force that exists between lipid bilayers at low hydration gives rise to a compressive lateral stress in the plane of the bilayer. This lateral compressive stress elevates the temperature of fluid-gel transitions and also mediates fluid-fluid

References

1. Crowe LM, Crowe JH (1982) Arch Biochem Biophys 217:582
2. Gordon-Kamm WJ, Steponkus PL (1984) Proc Nat Acad Sci 81:6373—6377
3. Bryant G, Wolfe J (1989) Eur Biophys J 16:369—374
4. Bryant G, Pope JM, Wolfe J (1992) Eur Biophys J 21:223—232
5. Crowe JH, Crowe LM (1986) In: Leopold AC (ed), Cornell Univ Press NY
6. Wolfe J, Bryant G (1992) In: Karalis TK (ed), NATO ASI-H, 64:205—244

Authors' address:

Gary Bryant
Helmholtz-Institut für Biomedizinische
Technik an der RWTH Aachen
Pauwelsstraße 30
52074 Aachen, FRG

Adress from 1st July 1993
Laboratoire de Physique Statistique
Ecole Normale Superieure
24 rue Lhomond
75231 Paris cedex 5, France

Clusters in dispersions of rod-like boehmite particles

J. Buitenhuis, A. P. Philipse, J. K. G. Dhont, and H. N. W. Lekkerkerker

Van't Hoff Laboratory for Physical and Colloid Chemistry, Utrecht, The Netherlands

Key words: Boehmite rods — clusters

Recently, sterically stabilized boehmite rods have been synthesized [1] which form stable dispersions in cyclohexane. To use these dispersions for model studies on rod-like particle systems, clusters must be avoided. An experimental analysis of the presence of permanent clusters in these boehmite dispersions is presented, using a combination of transmission electron microscopy, static light scattering and filtration.

In a number of cases, static light-scattering measurements on a dilute dispersion of the boehmite rods in cyclohexane clearly deviate from theoretical scattering curves calculated on the basis of the length and width distribution obtained from electron microscopy. The scattering calculations were performed with Rayleigh-Gans-Debye theory, which was checked for validity by comparison with coupled dipole method results [2]. If the samples are filtrated using the proper pore size, much better agreement is obtained between the measured and calculated curves. These results demonstrate the presence of permanent particle clusters in these otherwise stable dispersions.

References

1. Buining PA, Veldhuizen YSJ, Pathmamanoharan C, Lekkerkerker HNW (1992) Colloids Surfaces 64:47—55
2. Buitenhuis J, Dhont JKG, Lekkerkerker HNW (199—) Scattering of light from cylindrical particles, Coupled Dipole Method calculations and the range of validity of the Rayleigh-Gans-Debye approximation (accepted for publication in the J Colloid Interface Sci)

Authors' address:

J. Buitenhuis
Van't Hoff Laboratory for Physical
and Colloid Chemistry
Utrecht University
Padualaan 8
3584 CH Utrecht, The Netherlands

Influence of GM3 and GD3 glycolipids on the conductometric properties of a model membrane system

C. Cametti, F. De Luca, M. A. Macri[1]
B. Maraviglia, R. Misasi[2], and M. Sorice[3]

Dipartimento di Fisica, Università "La Sapienza", Roma, Italy
[1]) Dipartimento di Fisica Medica, Università di Chieti, Italy
[2]) Dipartimento di Medicina Sperimentale, Università "La Sapienza", Roma, Italy
[3]) Istituto di Malattie Tropicali, Università "La Sapienza", Roma, Italy

Abstract: The electrical conductivity of phospholipid bilayers, to which different gangliosides containing one or two sialic acid groups (GM3 or GD3 respectively) have been added, has been measured in the frequency range from 10 kHz to 100 MHz. — The observed alterations in the overall conductivity of the aqueous suspensions may be partially ascribed to a surface conductance contribution due to the polar head groups of the gangliosides.

Key words: GM3 ganglioside — GD3 ganglioside

Gangliosides, which are anionic glycolipids with a strong amphiphilic character, consist of a double-chain hydrocarbon portion (the hydrophobic part) and a polar head group with sialic acid containing oligosaccharide (hydrophilic part). Gangliosides are also constituents of the plasmamembrane of mammalian cells and have a considerable biological relevance for the role they play as biotransductors in the membrane-mediated information and/or in the cell recognition [1].

For these aspects, the activity of gangliosides seems also correlated to the modification they are able to induce on the structure of the cell membrane, which, in turn, can modulate the activity of the normal components of the bilayers such as membrane proteins and immunoreceptors.

Recently, we have measured by means of radiowave electrical conductivity technique, the alterations in the passive electrical properties of human resting lymphocyte membranes, induced by two different monosialogangliosides, GM1 and GM3, possessing the same hydrophobic part, but different hydrophilic head groups [2].

In the effort towards less complex systems as membrane model systems, we have recently undertaken a systematic investigation on the alteration of the electrical properties of phospholipid bilayers into which gangliosides of different structure have been incorporated.

In this note, we report on some preliminary electrical conductivity measurements on dipalmitoylphosphatidylcholine (DPPC) aqueous suspensions in presence of various amounts of two different gangliosides containing one sialic acid group (GM3) and two sialic acid groups (GD3), but with the same hydrophobic chains (ceramide).

The DPPC and the two different gangliosides, purchased from Sigma Chem. Co. (St. Louis Missouri, USA) were used without further purification. In all the samples investigated, the lipid concentration in water was kept constant at the value of 10% wt/wt. The DPPC and GM3 and GD3 molecules were dissolved in chloroform/methanol 2:1 vol/vol solutions and then dried under vacuum by rotaring evaporation. The samples were prepared by adding appropriate amounts of pure water (with electrical conductivity lower than $1 \times 10^{-6} \Omega^{-1} cm^{-1}$ at 20°C) at a temperature above the main phase transition temperature $T_c \cong 42$°C. After a vigorous shaking of 3 min., the suspensions were sonicated by using a Bransonic 220 PBI device for about 45 min at a temperature of 42°C in order to get clear, homogenous, low-viscosity solutions of multilamellar vesicles (liposomes). Then, the suspension was kept for 12 hours at a temperature of 35°C.

The electrical conductivity measuring technique has been previously described [3]. The conductivity spectra measured in the frequency range from 10 kHz to 100 MHz for DPPC-GM3 and DPPC-GD3 aqueous suspensions are shown in Figs. 1 and 2. At the lipid concentration employed, phospholipid molecules are organized in closed, concentric, multilamellar aggregates of roughly spherical shape, randomly dispersed in the bulk aqueous phase [4].

These vesicles should contain a water core surrounded by several phospholipid bilayers, whose structure is altered by the insertion of gangliosides, each separated by a water layer. From a dielectric point of view, this fact suggests that each aggregate behaves as a particle of an effective conductivity σ_p, dispersed in a continuous medium of conductivity σ_m.

The multilayer structure should, in principle, result in a more complex dielectric behavior consisting of different relaxation regions associated to the presence of the different layers which separate the various components of the aggregate. However,

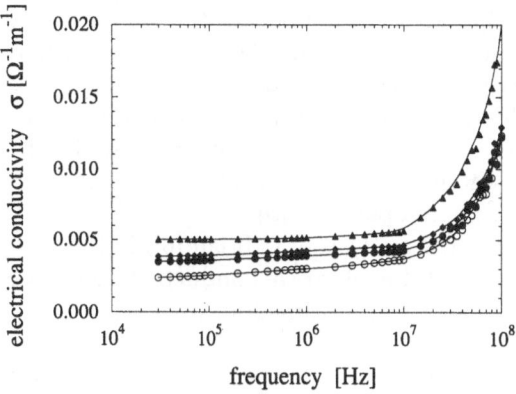

Fig. 1. Electrical conductivity of DDPC-GM3 aqueous suspensions as a function of frequency for various ganglioside concentrations. (○) $c = 0$ μg/ml; (●) $c = 50$ μg/ml; (◆) $c = 100$ μg/ml; (▲) $c = 200$ μg/ml. The lipid concentration is 10% wt/wt

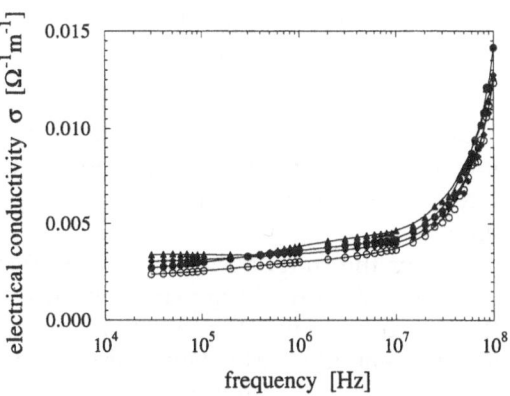

Fig. 2. Electrical conductivity of DPPC-GD3 aqueous suspensions as a function of frequency for various ganglioside concentrations. (○) $c = 0$ μg/ml; (●) $c = 50$ μg/ml; (◆) $c = 100$ μg/ml; (▲) $c = 200$ μg/ml. The lipid concentration is 10% wt/wt

the shielding effect of the exterior layer should make it difficult to detect the influence of the inner composite media which, in the radiofrequency region, behave as a homogeneous medium of mean conductivity σ_p and mean dielectric constant ε_p.

Further support for this picture comes from the behavior of the conductivity spectra which show the absence of a conductivity dispersion that, in systems where a less conductive layer separates two different high conductivity media, (like in biological cells), generally occurs in the frequency range from 100 kHz to 10 MHz.

Under the above assumptions, the simplest way to take into account the modifications of the external surface structure induced by different adsorbed molecules is to introduce a surface conductivity, similarly to what is generally done to describe the electrical double layer properties in solutions of charged colloidal particles [5].

Following the analysis of O'Brian [6], in the low-frequency limit, the electrical conductivity σ can be written, to a first approximation, as

$$\sigma = \sigma_m(1 - 3\phi f) , \qquad (1)$$

where σ_m is conductivity of the continuous medium, ϕ is the volume fraction of the dispersed phase, f is given by

$$f = \left(\sigma_m - \frac{2\Lambda s}{R}\right)\Big/\left(2\sigma_m + \frac{2\Lambda s}{R}\right)$$

and Λ_s is the surface conductance and R the average radius of the aggregates. Thus, the measurement of σ can, in principle, provide us to estimate Λ_s without any structural modification assumption.

Figure 3 shows the conductivity σ/σ_0 of the two samples investigated, normalized to the conductivity σ_0 of the control suspension, as a function of the ganglioside concentration. The figure also shows the upper and lower limits predicted by Eq. (1) for the surface conductance Λ_s.

It can be noticed that the introduction of a surface conductance term accounts for the observed conductivity behavior, at least in the low-concentration range of gangliosides.

The first indication one can draw from these preliminary results seems related to the role of different conformations of the GM3 and GD3 oligosaccharide chains play on the phospholipid bilayer structure. In fact, the surface conductivity of the lipid bilayer shows a marked alteration due to the insertion of gangliosides and, in such alteration, a detectable difference seems attributable to the two different oligosaccharide chains.

From the present knowledge of the conductivity changes introduced in a phospholipid bilayer by the presence of gangliosides it is not possible to give a full description of these effects. Further measurements over a wide frequency range and in a more extended ganglioside concentration are in progress and will provide new information on this problem.

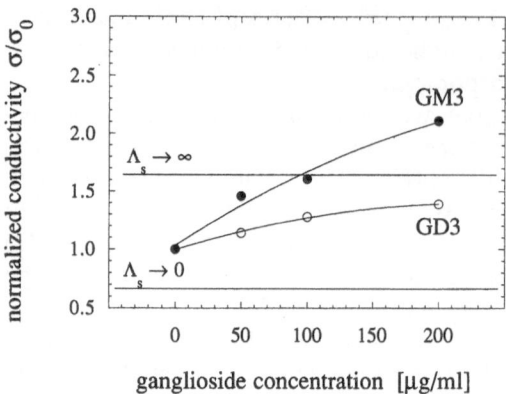

Fig. 3. The normalized conductivity σ/σ_0 as a function of ganglioside concentration (o) GD3; (●) GM3. Lines indicating the limit values corresponding to a surface conductance Λ_s equal to zero or to infinity respectively are also shown

References

1. Wiegandt H (1971) Adv Lipid Res 9:249
   Tettamanti G, Sonnino S, Ghidoni R, Masserini M, Venerando B (1985) in Physics of Amphiphiles: Micelles, vesicles and microemulsions, V De Giorgio and M Corti Eds, North-Holland
2. Cametti C, De Luca F, D'Ilario A, Macri' MA, Maraviglia B, Bordi F, Lenti L, Misasi R, Sorice M (1992) Biochim Biophys Acta 1111:197
3. Bottomley PA (1978) J Phys E: Sci Instrum 91:412
4. Corti M, Degiorgio V, Ghidoni R, Sonnino S, Tettamanti G (1980) Chem Phys Lipids 26:225
5. Takashima S (1989) Electrical properties of biopolymers and Membranes, Hilger-Bristol
6. O'Brien RW (1986) J Coll Interf Sci 113:81
   Midmore BR, Hunter RJ, O'Brien RW (1987) J Coll Interf Sci 120:210

Authors' address:

C. Cametti
Dipartimento di Fisica
Università "La Sapienza"
00185 Roma, Italia

Progress in Colloid & Polymer Science   Progr Colloid Polym Sci 93:191 (1993)

Electrical conductivity in the lamellar phase of a Water-in-oil microemulsion

A. Di Biasio[1]), C. Cametti[2]), P. Codastefano[2]),
P. Tartaglia[2]), J. Rouch[3]), and S. H. Chen[4])

[1]) Dipartimento di Matematica e Fisica, Università di Camerino, Italy
[2]) Dipartimento di Fisica, Università "La Sapienza", Roma, Italy
[3]) Centre de Physique Moléculaire Optique et Hertzienne (URA 283 du CNRS), Université Bordeaux I, Talence, France
[4]) Department of Nuclear Engineering and Center for Materials Science and Engineering, Massachusetts Institute of Technology, Cambridge, Massachusetts, USA

Abstract: The electrical conductivity of a three-component ionic microemulsion has been measured in the phase diagram region where the lamellar phase, made up of repeated layers of water, surfactant and oil, exists. The data have been analyzed on the basis of an effective medium theory expression, taking into account the ion distribution within the water layer according to the Poisson-Boltzmann equation. — PACS numbers: 64.60. Cn, 05.40. + j, 82.70. Dd

Key words: W/o microemulsion — Poisson-Boltzmann equation

The electrical conductivity behavior of water-in-oil microemulsions has been extensively studied over the past 10 years in order to obtain dynamic and structural information on a large variety of phenomena occurring in these systems, depending on the nature of the components and the composition of the system [1, 2]. Recently, the phase diagram of a three-component microemulsion made of water, decane and sodium-di-2 ethyl-hexyl sulfosuccinate (AOT) has been accurately determined by means of neutron [3] and light-scattering and conductometric measurements, and at high volume fractions well above the percolation temperature; a structure consisting of repeated leaflets (L_a phase) has been observed. In this region of the phase diagram, the electrical conductivity mechanism is completely different from that occurring in other molecular arrangements both below and above the percolation. In this case, the ionic transport is essentially due to the movement of the N_a^+ ions derived from the ionization of the surfactant molecules within the water layers and the electrical conduction reduces to a bulk phenomenon, the

contribution of the other mechanisms (i.e., hopping of the AOT anions from one droplet to another or opening of the surfactant layers to form extended water microchannels) being negligible.

Within this region, the properties of the water lamellar are constant, the thickness d_w of the water layer depending on the molar ratio of water to surfactant X = [water]/[AOT] only:

$$d_w = \frac{2dXM_w\rho_{AOT}}{M_{AOT}\rho_w} \ , \tag{1}$$

whereas the distance l depends on the composition through the fractional volume $\phi = \phi_w + \phi_{AOT}$, where ϕ_w and ϕ_{AOT} are the volume fractions of water and surfactant, respectively, according to the expression $l = \dfrac{d_w + 2d}{\phi}$. Here, d is the length of an AOT molecule and M and ρ are the molecular weights and densities of water and surfactant. Moreover, each polar group of the surfactant molecule carries an electronic charge due to its ionization and, consequently, contributes one sodium ion to the water layer.

In the present work we have analyzed the conductivity behavior of a three-component microemulsion in the temperature range 13° to 70°C and at fractional volume ϕ between 0.25 and 0.85 which completely covers the region where the lamellar phase exists. The molar ratio X was kept constant at the value X = 40.8 to which it corresponds to a water thickness of about d_w = 39.5 Å in the lamellar phase. If the microemulsion in this phase region is considered as a set of large, randomly oriented multilamellae, the overall conductivity of the system can be written, to a first approximation, as

$$\sigma = \frac{1}{3} \frac{\sigma_l\sigma_w}{\sigma_w(1 - \phi_w) + \sigma_l\phi_w}$$

$$+ \frac{2}{3} (\sigma_l(1 - \phi_w) + \sigma_w\phi_w) \ , \tag{2}$$

where σ_l and σ_w are the conductivities of the oil layer coated by the surfactant molecule and the water layer, respectively. The conductivity σ_w, depending on the orientation of the water layer in the applied external electric field, must be associated to the non-uniform spatial ion distribution. According to the model we adopted, this distribution as a function of distance χ, obtained from the Poisson-Boltzmann equation in planar geometry, can be written as

$$n(\chi) = \frac{zefn_0}{\cos^2(\sqrt{\frac{f}{2}}\kappa_D\chi)} \tag{3}$$

where κ_D^{-1} is the Debye screening length, ze the electronic charge, n_0 the unperturbed ion concentration, and f a constant derived from the electroneutrality condition in the whole water layer.

As we have noted, owing to the random distribution of the lamellar orientation, the conductivity must be evaluated in two extreme limits, when the external field is perpendicular and parallel to the lamellar. These conditions yield, respectively,

$$\sigma_\perp = zef\frac{n_0}{N_A}\ \lambda(T) \tag{4}$$

and

$$\sigma_{||} = ze\frac{\bar{n}}{N_A}\ \lambda(T), \tag{5}$$

where N_A is the Avogadro number and λ the equivalent conductance of the ions, and

$$\begin{aligned}
\bar{n} &= \frac{2}{\kappa_D d}\int_0^{\frac{\kappa_D d}{2}-\eta} d\xi\, n(\xi) \\
&= \frac{2\sqrt{2}fn_0}{\kappa_D d_w}\ \mathrm{tg}\left(\sqrt{\frac{f}{2}}\left(\frac{\kappa_D d}{2}-\eta\right)\right)
\end{aligned} \tag{6}$$

the average effective ion density that contributes to the conductivity. Here, we have introduced a parameter η to take into account the hindering effect at the charged surfactant-water interface on the transport of the ionic charge and, for the sake of simplicity, we assume that counterions lying in a layer adjacent to the charged surface are firmly bound, thus possessing a negligible mobility. In the limit $\sigma_l << \sigma_\perp$ and $\sigma_{||}$, Eq. (2) reduces to

$$\frac{\sigma(T)}{\phi_w} = \frac{8z^2e^2\sqrt{2f}}{3S_0d_wN_A}\ \mathrm{tg}\left[\frac{f}{2}\left(\frac{\kappa_D d_w}{2}-\eta\right)\right]\frac{\lambda(T)}{\kappa_D d_w}, \tag{7}$$

where $S_0 = 2/(n_0 d_w)$ is the average interfacial area occupied by each surfactant molecule, carrying the ionic SO_3^- group. The $T-\phi$ phase diagram of the microemulsion is shown in Fig. 1 and was completely determined from an extensive set of conductivity measurements at different volume fractions ϕ. The electrical conductivity of the microemulsions shows a complicated behavior reflecting the structural changes occurring in the system at constant ϕ as a function of temperature. A typical example is shown in Fig. 2, where the change in the slope

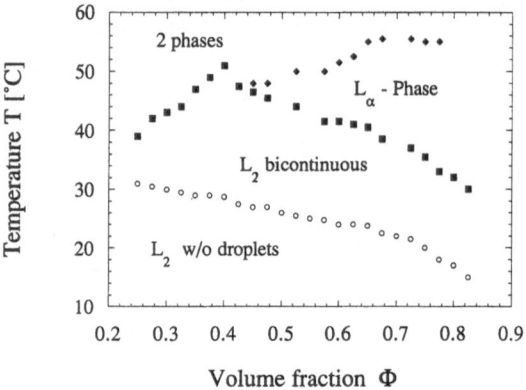

Fig. 1. The phase diagram of the water-decane-AOT microemulsion at $X = 40.8$. The L_a phase extends at higher temperature above L_2 bicontinuous region. The percolation loci in the L_2 phase, between the water/oil droplets and the bicontinuos structure is also shown

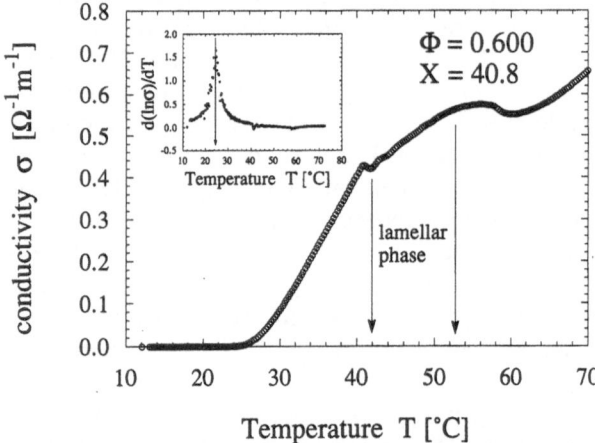

Fig. 2. Electrical conductivity of the microemulsion. The arrows mark the temperature interval where the lamellar phase exists. The inset shows the percolation temperature as the inflection point in the σ vs T plot

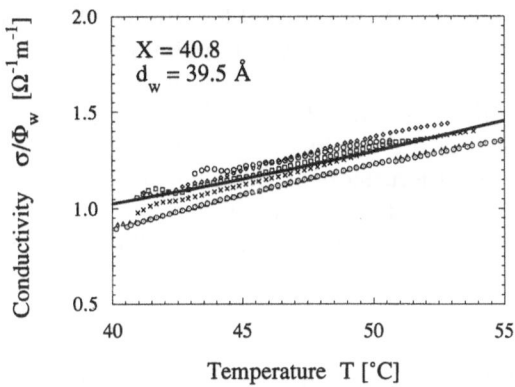

Fig. 3. The normalized conductivity for $X = 40.8$ in the lamellar region. The data refer to different volume fraction ϕ. The full line represents the calculated values according to Eq. (7). $\circ$: $\phi = 0.525$, $\diamond$: $\phi = 0.575$, $\square$: $\phi = 0.600$, $\times$: $\phi = 0.625$, $\odot$: $\phi = 0.675$, $\triangle$: $\phi = 0.750$

gives evidence for the phase boundaries between bicontinuous [4] and the L_a phases. The inset shows the maximum in $d(\ln\sigma)/dT$ as a function of temperature, which displays the percolation temperature. The experimental values of the electrical conductivity of the microemulsion at $X = 40.8$, normalized to the volume fraction of water ϕ_w, in the temperature range where the lamellar structure occurs, for ϕ between 0.45 and 0.70, are shown in Fig. 3. The solid line represents the calculated values according to Eq. (7), assuming for the equivalent conductance λ the value at infinite dilution. As it can be seen, a good agreement with the experimental data is found with the parameter η corresponding to a thickness of immobilized ions of the surfactant-water interface equal to 1.7 times the Debye screening length. In the present case, with $X = 40.8$, we obtain $\kappa_D^{-1} \approx 3.6$ Å. This means that when the conduction occurs in a narrow water layer only a fraction of counterions contributes to the ionic conductivity, whereas a large fraction is immobilized in the vicinity of the charged groups by means of strong electrical interactions. In the present case, approximately 30% of the total number of ions must be considered strongly bound to the charged surface.

These systems which exhibit a repeated bilayer-water structure are also of fundamental biophysical interest since they serve as simple model biomembranes where the ionic conduction transport may be investigated without the influence of proteins and other biomolecules which are usually embedded in the cell membrane.

Acknowledgement

This work has been supported by grants from the Ministero dell'Università e della Ricerca Scientifica e Tecnologica and the Gruppo Nazionale di Struttura della Materia del Consiglio Nazionale delle Ricerche, Italy. Work of SHC is supported by a grant from Materials Science Division of US DOE.

References

1. Cametti C, Codastefano P, Tartaglia P, Rouch J, Chen SH (1990) Phys Rev Lett 64:1461
2. Cametti C, Codastefano P, Tartaglia P, Rouch J, Chen SH (1992) Phys Rev A 45:R5358
3. Sheu EY, De Tar MM, Kotlarchyk M, Lin JS, Capel M, Storm DA (1992) in: Structure and Dynamics of Strongly Interacting Colloids and Supramolecular Aggregates in Solution edited by Chen SH, Huang JS, Tartaglia P (Kluwer Acad Pub vol 369)
4. Chen SH, Chang SL, Strey R (1990) Progr Colloid Polym Sci 81:31 and (1990) J Chem Phys 93:1907

Authors' address:

C. Cametti
Dipartimento di Fisica
Università "La Sapienza"
Piazzale Aldo Moro 2
00185 Roma, Italia

The effect of zeta potential on some dielectric properties of colloidal dispersions

F. Carrique[1]) and A. V. Delgado[2])

[1]) Departamento de Física Aplicada 1, Facultad de Ciencias Universidad de Málaga, Spain
[2]) Departamento de Física Aplicada, Facultad de Ciencias Universidad de Granada, Spain

Abstract: Some aspects of the theory of the dielectric response and AC conductivity of dilute suspensions of colloidal particles are analyzed in this work. The effect of the zeta potential, ζ, of the particles on such properties is considered. Using Kramers-Krönig relations, the contributions to the real part of the conductivity of the dispersions are separated, and their ζ-potential dependence is discussed. The energetic aspects of the response of the suspensions to alternating electric fields of different frequencies are also analyzed; the different contributions to the total energy loss in the systems are considered separately. Some theoretical results have been considered from the time domain point of view; thus the dependence of the total current density (and its components) on ζ is also studied. As a whole, our results indicate that the theory does predict the fundamental importance of the zeta potential to account for the dielectric parameters of colloidal suspensions. The relative importance of the individual contributions to quantities of interest, although implicit in the theory, is made explicitly clear.

Key words: Zeta potential — complex conductivity — dielectric constant — power dissipation

Introduction

The study of the conductivity and dielectric behavior of colloidal suspensions in the presence of alternating electric fields of frequency ω ($\vec{E} = \vec{E}_0 e^{i\omega t}$) can improve our understanding of the electrokinetic phenomena characteristic of the suspensions [1]. It has been shown that the strong dielectric dispersion that is observed at low frequencies in any colloidal system is related to polarization phenomena that occur in the electric double layer surrounding the particles [2—4].

In fact, when an electric field is applied to the suspension, the polarization of the solid/solution interface is, in general, out of phase with respect to the field, this giving rise to a phase difference between the field and the system's response as represented by the displacement vector $\vec{D}$. The dielectric constant of the system will hence be complex: $\varepsilon^*(\omega) = \varepsilon'(\omega) - i\varepsilon''(\omega)$, and the relation between $\vec{D}$ and $\vec{E}$ will read:

$$\vec{D}(\omega, t) = \varepsilon^*(\omega)\,\vec{E}(\omega, t)\,. \tag{1}$$

This phase difference will manifest as a relaxation effect in the dielectric parameters of the colloidal suspension. The exact nature of such relaxation will depend on the properties of the colloidal paticles, of the solution, and of their common interace.

Although some models have been proposed to explain the dielectric relaxation of colloidal suspensions [2—7], the most general treatment to our knowledge was elaborated by DeLacey and White [8]. This model, although only valid for dilute suspensions and low frequencies ($\omega << 10^{10}\,\text{s}^{-1}$), does not have any limitations concerning the values

of κa (κ^{-1} is the Debye-Hückel length, see [9], and a is the particle radius) and zeta potential, and the ionic characteristics of the medium. In this work we will consider how the zeta potential of the particles affect some of the results of this model. Before going into the discussion of those results, it seems useful to become familiar with the notation and interesting quantities in the field. This will be done in the following section.

Notation and dielectric quantities of interest

The conductivity of a dilute colloidal suspension in the presence of an alternating field of frequency ω can be written as:

$$K^*(\omega) = K_e^*(\omega) + \phi\Delta K^*(\omega)\,, \tag{2}$$

where $K_e^*(\omega)$ is the complex conductivity of the pure electrolyte solution, and ϕ is the volume fraction of solids in the suspension. The conductivity increment due to the presence of the particles in the medium, $\Delta K^*(\omega)$, has the following form [8]:

$$\Delta K^*(\omega) = \Delta K(\omega) + \omega\varepsilon_0\Delta\varepsilon''(\omega) + i\omega\varepsilon_0\Delta\varepsilon'(\omega)\,. \tag{3}$$

Here, ε_0 is the dielectric permittivity of a vacuum, and

$$\Delta K(\omega) = \frac{K(\omega) - K^\infty}{\phi} \tag{4}$$

$$\Delta\varepsilon'(\omega) = \frac{\varepsilon'(\omega) - \varepsilon_{rl}}{\phi} \tag{5}$$

$$\Delta\varepsilon''(\omega) = \frac{\varepsilon''(\omega)}{\phi}\,. \tag{6}$$

The term $K(\omega) + \omega\varepsilon_0\varepsilon''(\omega)$ is often called suspension conductivity, K^∞ is the conductivity (dc) of the electrolyte solution, and ε_{rl} is the dielectric constant of the latter.

In order to get a better insight into the contents of the theory, we have numerically integrated its differential equations, thus obtaining the frequency dependence of any of the dielectric quantities for different experimental conditions. Figure 1 shows the behavior of the quantity

$$\frac{\Delta K_{eff}(\omega)}{K^\infty} = \frac{Re[\Delta K^*(\omega)]}{K^\infty} \qquad (7)$$

as a function of ω for different values of the zeta potential. For these and subsequent results, spherical particles of 100 nm radius and dielectric constant $\varepsilon_l = 2$ will be assumed. KCl is the electrolyte, and its concentration has been adjusted to give $\kappa a = 10$ in all cases. The value of ε_{rd} has been fixed to 78.54, and the temperature is 298 K. As observed in Fig. 1, as ζ increases, the negative effect of the dielectric particles on the bulk conductivity of the system is compensated by the extra conductivity associated to the electric double layers. Thus, $\Delta K_{eff}(\omega)$ becomes less negative the higher the value of ζ. Concerning the increase in $\Delta K_{eff}(\omega)$ observed at high frequencies, it has been related [8] to the smaller distortion effect of the particles on the ion trajectories upon increasing frequency.

In Fig. 2 we have plotted the frequency dependence of $\Delta\varepsilon'(\omega)$. The sensitivity of this parameter to ζ variations is again clearly seen. The observed behavior can be explained by noting that $\Delta\varepsilon'(\omega)$ depends mainly [8] on the strength of the dipole moment that the electric field induces in the solution-particle system. For low zeta potentials we have a "negative" polarizability effect: the presence of

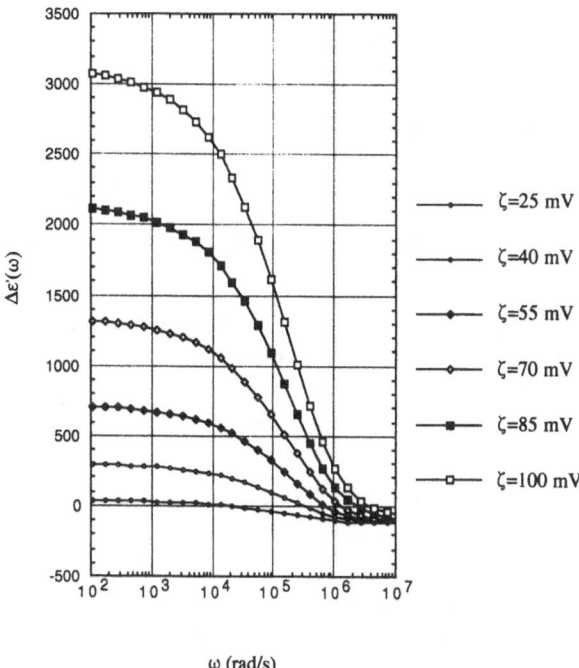

Fig. 2. Increment of the real part of the dielectric constant as a function of frequency for different zeta potentials

the particles means that high dielectric constant material -the solution- is being substituted by low dielectric constant material -the particles. When $|\xi|$ is increased and a consistent double layer develops around the particles, these become more polarizable, and we have a "positive" polarizability associated to the ion cloud.

The theory of DeLacey and White does not consider explicitly the separation of the two contributions to $Re[\Delta K^*(w)]$, i.e., $\Delta K(\omega)$ and $\omega\varepsilon_0\Delta\varepsilon''(\omega)$ (Eq. (3)). We have carried out such separation by means of the Kramers-Krönig relation [10, 11]:

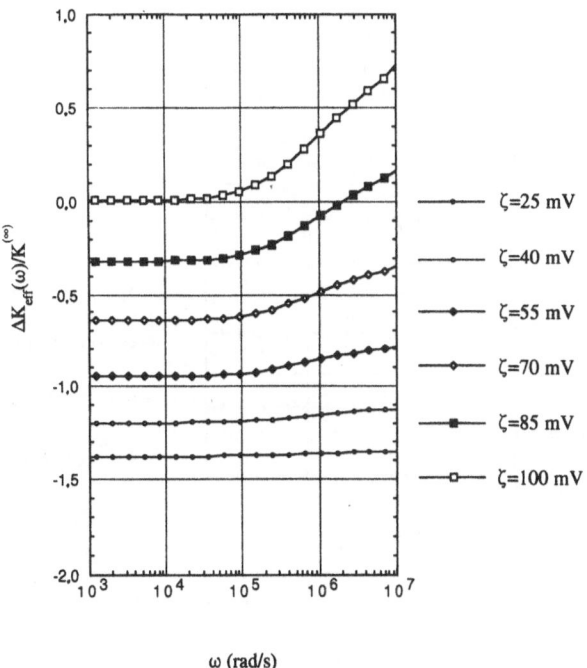

Fig. 1. Effective conductivity increment, normalized by the conductivity K^∞ of the solution, as a function of frequency for different ζ values

$$\varepsilon''(\omega) = -\frac{2\omega}{\pi} \int_0^\infty \frac{\varepsilon'(\omega') - \varepsilon_\infty}{\omega'^2 - \omega^2} \, d\omega', \qquad (8)$$

and using the $\varepsilon'(\omega)$ spectrum obtained by integration of DeLacey and White's equations. In Eq. [8] ε_∞ is the asymptotic high frequency value of $\varepsilon'(\omega)$. Using Eq. [8] one obtains $\Delta\varepsilon''(\omega)$ and, hence, that part of $Re[\Delta K^*(\omega)]$ due to dielectric loss, $\omega\varepsilon_0\Delta\varepsilon''(\omega)$. The difference between $Re[\Delta K^*(\omega)]$ and $\omega\varepsilon_0\Delta\varepsilon''(\omega)$ then yields the conductivity term not related to dielectric losses, $\Delta K(\omega)$.

With these data at hand, some quantities related to the energy loss in the system can be computed, namely the total energy dissipated per unit volume and unit time, $W_{dt}(\omega)$ given by:

$$W_{dt}(\omega) = \frac{1}{2} \; [K^\infty + \phi\Delta K(\omega)$$

$$+ \; \phi\omega\varepsilon_0\Delta\varepsilon''(\omega)] \; E_0^2 \;, \tag{9}$$

E_0 being the applied field amplitude. The power density dissipated as a consequence of dielectric loss can be written:

$$W_d(\omega) = \frac{1}{2} \; [\phi\omega\varepsilon_0\Delta\varepsilon''(\omega)] \; E_0^2 \;. \tag{10}$$

Furthermore, proper treatment of the results allows to study the system's response in the time domain instead of the frequency domain. A fundamental quantity is the dielectric response function $\Phi(t)$ [12]:

$$\Phi(t) = \frac{2}{\pi} \int\limits_0^\infty \frac{\varepsilon'(\omega') - \varepsilon_\infty}{\varepsilon_s - \varepsilon_\infty} \; \cos\omega'td\omega' \;, \tag{11}$$

ε_s being the low frequency limit of $\varepsilon'(\omega)$. $\Phi(t)$ is intimately related to the specific mechanisms giving rise to the phase lag between $\vec{D}$ and $\vec{E}$. When a time-independent electric field $\vec{E}_0$ is applied to the system at $t = 0$, a transient current density, $\vec{J}_t$, is generated in the suspension. The total current density $\vec{J}_t$ includes two contributions, $\vec{J}_c$ (conduction current density) and $\vec{J}_a$ (absorption current density):

$$\vec{J}_t(t) = \vec{J}_c + \vec{J}_a(t) \tag{12}$$

$$\vec{J}_c = [K^\infty + \phi\Delta K(0)] \; \vec{E}_0 \tag{13}$$

$$\vec{J}_a(t) = [\phi\varepsilon_0(\Delta\varepsilon_s' - \Delta\varepsilon_\infty')\Phi(t)] \; \vec{E}_0 \;. \tag{14}$$

Here, $\Delta\varepsilon_s'$ and $\Delta\varepsilon_\infty'$ mean, respectively, the low and high frequency asymptotic values of $\Delta\varepsilon'$ (Eq. 5). For the calculation of these currents, a constant field of 10 V/m, and a volume fraction $\phi = 0.05$ will be assumed.

Contributions to $Re[\Delta K^*(\omega)]$. Zeta potential effects

Figure 3 shows the frequency variation of the two contributions to $Re[\Delta K^*(\omega)]$: $\Delta K(\omega)$ and $\omega\varepsilon_0\Delta\varepsilon''(\omega)$. The first feature that must be noted is that ΔK is independent of frequency for any value of ξ; the fact that the value of ΔK increases with $|\zeta|$ can be explained using the reasoning mentioned above concerning the compensation of the negative effect that particles have on bulk conductivity by the extra conductance associated to the double layers [8].

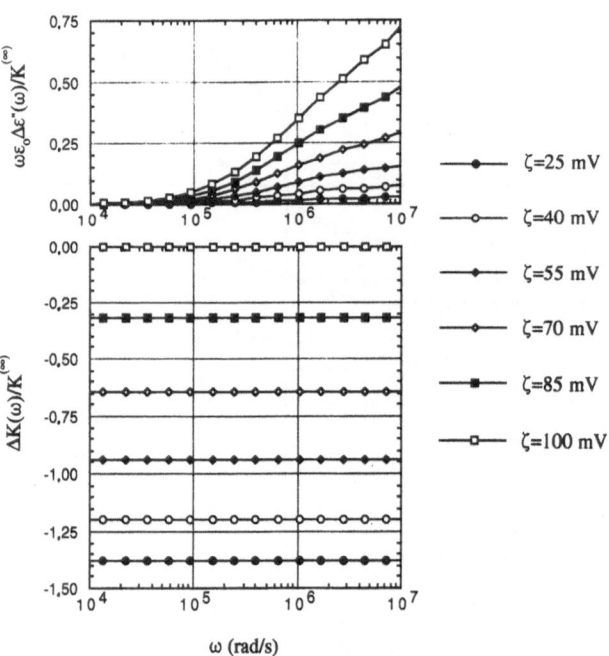

Fig. 3. Contributions to $Re[\Delta K^*(\omega)]$, normalized by the conductivity of the solution, plotted as as function of frequency for different ζ values

In Fig. 4 we have plotted the variation with ω of the quantity $\Delta\varepsilon''(\omega)$. Note the occurrence of maxima in all curves corresponding to relaxation frequencies of the system; these are a measure of the similarity between the period $2\pi/\omega$ of the applied field and the characteristic diffusion time of the ions in the double layer. The relaxation frequencies are essentially independent of the ζ value: DeLacey and White [8] justified that the positions of the maxima were related to the value of κa, which is maintained constant in our calculations. The increase in $\Delta\varepsilon''$ with ζ for a given frequency (Fig. 4) is due to the fact that $\Delta\varepsilon''$ depends on the imaginary part of the polarizability, which increases with $|\zeta|$ at constant κa [8].

After the individual analysis of the components of the complex dielectric increment $\Delta\varepsilon^* = \Delta\varepsilon' + i\Delta\varepsilon''$, it seems of interest to study (Fig. 5) the Cole-Cole diagram $\Delta\varepsilon''$ vs. $\Delta\varepsilon'$ for this quantity [10—13]. Taking into account the asymmetry of the $\Delta\varepsilon''$-ω curves about their maxima (Fig. 4), it is clear that the plots in Fig. 5 are not depressed symmetric semicircles, as suggested by other authors [14]. This means that the relaxation response of a colloidal

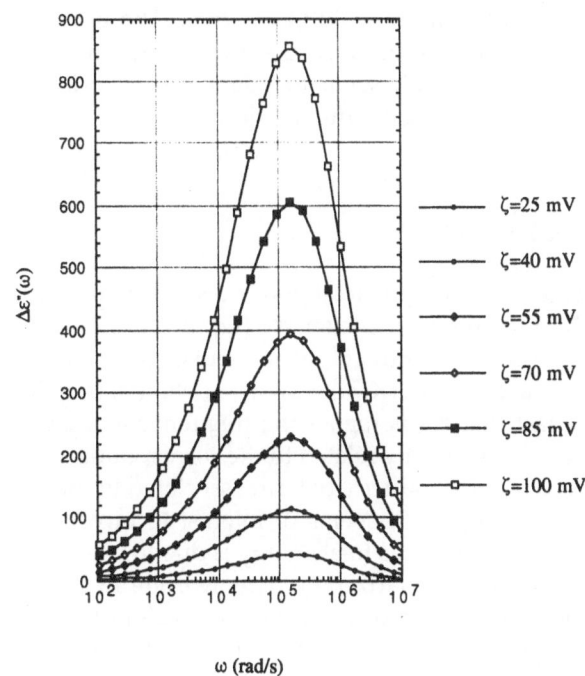

Fig. 4. Increment of the imaginary part of the dielectric constant as a function of frequency and zeta potential

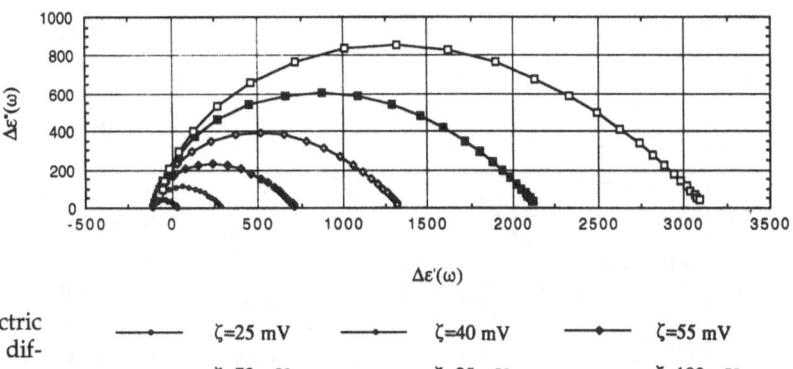

Fig. 5. Cole-Cole diagrams for the dielectric constant increment of suspensions with different zeta potentials

system must be explained on the basis of a distribution of relaxation times in the system, if the generalized Debye model (multiple relaxation times and non-interacting particles) is to be applied to the type of dielectric we are studying. Lim and Franses

[14] have proposed a symmetric distribution of relaxation times to explain their results, but such distribution does not seem to suit the dielectric theory of DeLacey and White. Our preliminary results indicate that some type of empirical asym-

metric distribution could fit the data. The problem is still open, the more so since most experimental results [1, 14—17] seem to adapt satisfactorily to symmetric Cole-Cole distributions [11].

Energy dissipation

Figure 6 shows the total power dissipated per unit volume of the suspension W_{dt} (Eq. (9)). As observed, it increases with $|\zeta|$, as it should, given the increasing trend of $\Delta K(\omega)$ with zeta potential (Fig. 3). The fact that W_{dt} is higher the frequency at a given value of ζ is directly related to the corresponding increase in the conductivity term associated to dielectric loss (Fig. 3). It seems of interest to investigate what fraction of this total energy is due to processes not related to polarization. This is done in Fig. 7, which shows that for low frequencies almost 100 % of the dissipated energy must be ascribed to conductive processes not related to bounded charge, this indicating that for such frequency range dielectric losses are both negligible and zeta potential independent.

$$\phi = 5 \cdot 10^{-2}$$

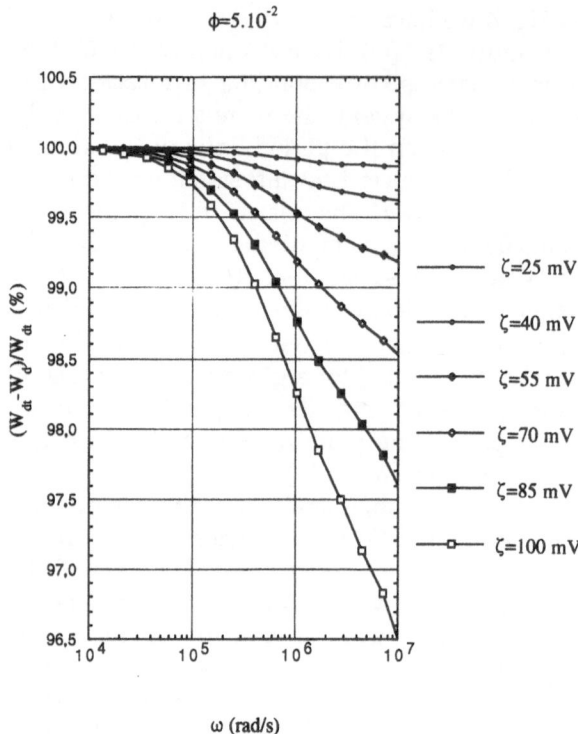

Fig. 7. Contribution of non-dielectric energy loss as compared to total energy loss plotted as a function of frequency for several values of ζ

$$\phi = 5 \cdot 10^{-2}$$

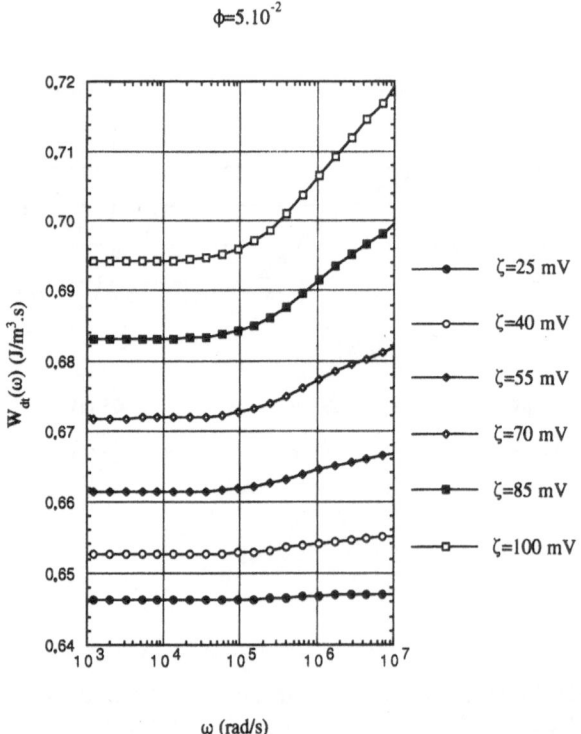

Fig. 6. Total dissipated power per unit volume of suspension as a function frequency for the zeta potentials indicated

When frequency is increased above $\sim 10^5$ s^{-1} dielectric loss becomes more significant due to the frequency variation of $\omega \varepsilon_0 \Delta \varepsilon''$ (Fig. 3). This contribution to the total energy dissipated is more important and increases more rapidly with ω the higher the zeta potential of the particles. Note, however, that even for the highest values of ζ and ω investigated (100 mV and 10^7 s^{-1}, respectively), the dielectric energy loss is at most $\sim$ 4% of the total dissipated energy. The conductive loss associated to the electrolyte is by far the most important contribution to the total power loss of a dilute colloidal suspension in the presence of ac fields.

Time domain analysis of the suspensions

From an experimental point of view, the analysis of the electric behavior of a colloidal system in the initial instants after the application of a step electric field can also characterize the dielectric response of the system. We will use here this alternative point of view to study the consequences of DeLacey and White's equations.

Let us begin with the analysis of the time dependence of the absorption current density, J_a which is related to the number of dipoles per unit volume of suspension: we expect it to decrease with time until reaching a zero value. Figure 8 shows the absorption current density increment ($\Delta J_a(t) = J_a(t)/\phi$). This quantity is independent of the volume fraction, since $\Phi(t)$ does not depend on ϕ (Eqs. (11) and (14)). Note how ΔJ_a does indeed tend to zero for times longer than $\sim 10^{-5}$ s. Also, the absorption current is larger the higher the value of $|\zeta|$, in agreement with the increase in the strenght of the dipole moment that the field induces in the system, as shown in Fig. 2. The increase with $|\zeta|$ of the polarizability of the suspension gives rise to a more important response shortly after the application of the field.

Figure 9 gives the variation with time of the total current density; the shape of the curves is typical of transient currents [12]. The attainment of a constant value (that increases with ζ due to the growing trend of conductivity with ζ) for times larger than $\sim 10^{-5}$ s is clearly observed. It is to be expected that after this time the conduction currents (Eq. (13.)) are predominant, and a steady current is reached. That this is so can be seen in Fig. 10, where the conduc-

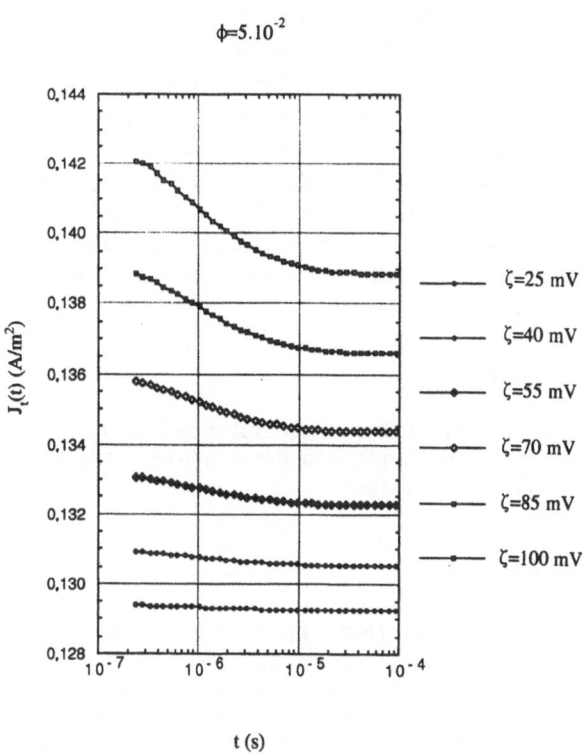

Fig. 9. Same as Fig. 8 for total current density

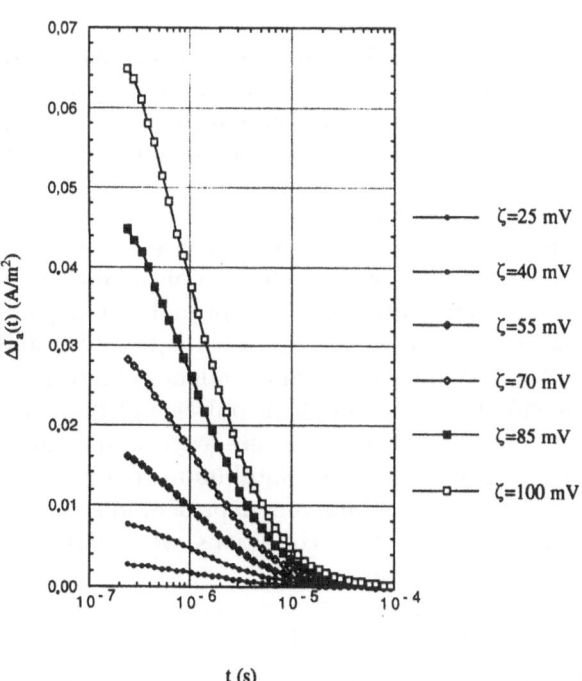

Fig. 8. Absorption current density as a function of time and zeta potential

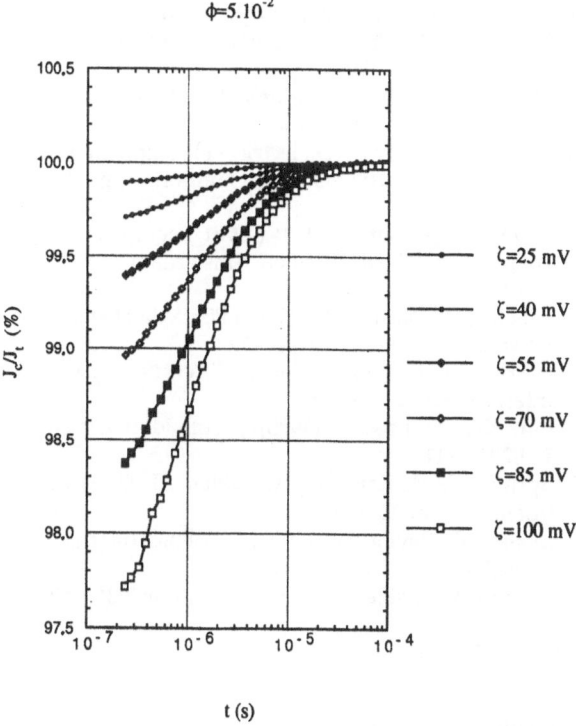

Fig. 10. Time variation of the conduction current density relative to total current density for several ζ values

tion current density is normalized by the total current density. Note how, for long times, the whole current is due to conduction, whereas at shorter times the absorption currents (more important the higher the zeta potential) reduce the significane of the conductive term. Although the contributions are small, they can be measured, and give an alternative (but physically equivalent) insight into the dielectric behavior of colloidal suspensions.

Acknowledgement

Financial support from DGICYT, Spain (Project No. PB89-0461) and Fundacio'n Ramo'n Areces, Spain, is gratefully acknowledged.

References

1. Springer MM (1979) Ph D Thesis. Wageningen University, The Netherlands, pp 49—61
2. Schwarz GJ (1962) J Phys Chem 66:2636—2642
3. Schurr JM (1964) J Phys Chem 68:2407—2413
4. Dukhin S, Shilov VN (1974) Dielectric Phenomena and the Double Layer in Disperse Systems and Polyelectrolytes. Wiley, New York, pp 42—52
5. Chew WC (1984) J Chem Phys 80:4541—4552
6. Fixman M (1983) J Chem Phys 78:1483—1491
7. Fixman M (1980) J Chem Phys 72:5177—5186
8. DeLacey EH, White LR (1981) J Chem Soc Faraday Trans 2 77:2007—2039
9. Hunter RJ (1987) Foundations of Colloid Science, Vol 1. Clarendon Press, Oxford, ch. 6
10. Jonscher AK (1983) Dielectric Relaxation in Solids. Chelsea Dielectric Group, London, pp 47—52, 96—101
11. Böttcher CJF, Bordewijk P (1978) Theory of Electric Polarization, Vol II. Elsevier, Amsterdam, pp 30—38, 62—67
12. Albella JM, Martínez JM (1984) Física de Dieléctricos. Boixareu, Barcelona, pp 75—122
13. Carrique F, Criado C, Delgado AV (1993) J Colloid Interface Sci 156:117—120
14. Lim KH, Franses EI (1986) J Colloid Interface Sci 110:201—211
15. Springer MM, Korteweg A, Lyklema J (1983) J Electroanal Chem 153:55—66
16. Myers DF, Saville DA (1991) J Colloid Interface Sci 131:448—460
17. Rosen LA, Saville DA (1991) Langmuir 7:36—42

Authors' address:

Dr. A. V. Delgado
Departamento de Física Aplicada
Facultad de Ciencias
Universidad de Granada,
18071 Granada, Spain

Ionic micellar-water system characterization by NMR diffusion measurements

E. Tettamanti and C. Casieri

Dipartimento di Fisica, Università di L'Aquila, Italy

Abstract: In order to study the phenomenon of water interacting with macro molecules, a ionic micellar system has been chosen. The dimensions of the micelle have been determined by diffusion measurements of the micelle in heavy water solution at different molar concentrations and vs. temperature by NMR gradient pulse technique. The results reported show that at low concentration (≤ 0.5 M), the micelle is a suitable system for the study of the water macro molecules interaction by NMR T_1 proton relaxation.

Key words: NMR — diffusion coefficient — ionic micelle

The micellar systems are suitable models for investigating the problem of water macro molecules interactions, which are very important from a biological point of view as well as for applied science.

We chose a surfactant p-octyloxybenzyl — trimetylammonium bromide that in water aggregates in micellar structure (1). Before using this system to study the water interacting behavior, we turned our attention to characterization of the micellar dimensions and how these depend on temperature and molar concentration. For this purpose we performed measurements of the diffusion of the micelle in heavy water solution at different molar concentration and at two different temperature (298.5 and 312 K), by NMR pulsed field gradient technique (2). The strength of the gradient pulses and the delay between 90° and 180° pulses were opportunely chose to eliminate the contribution to the NMR signal coming from monomers present in solution. In these experimental conditions we determined the diffusion of the micelle.

From the diffusion values computed by the exponential relations obtained by fitting the experimental diffusion data vs. molar concentration at two different temperatures, we derived the correspondent apparent micellar radius through the Stokes-Einstein relation, shown in Fig. 1.

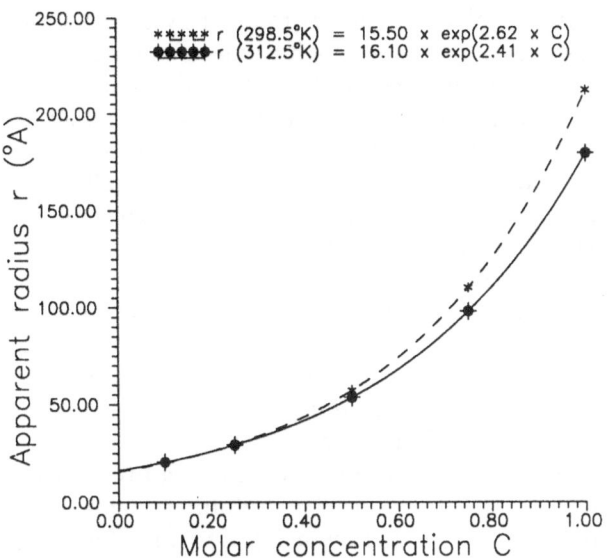

Fig. 1. Apparent radius vs. molar concentration at 298.5 K and 312 K. The diffusion data used in the Stokes-Einstein relation are computed by the exponential law obtained by fitting the experimental diffusion data

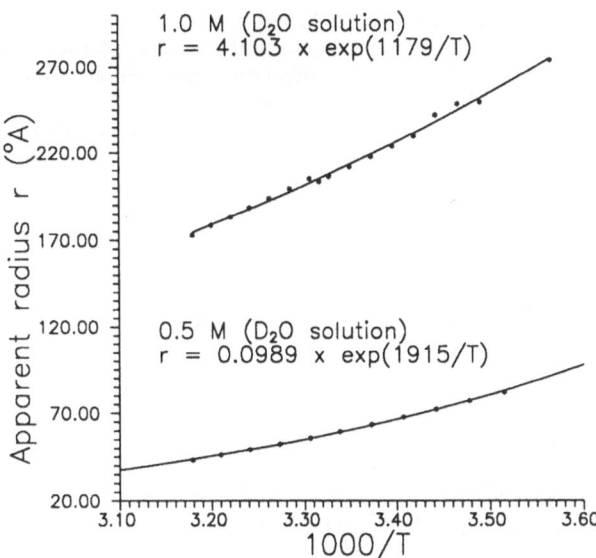

Fig. 2. Apparent radius vs. temperature for solutions at 1.0 M and 0.5 M

We note that at concentration below 0.5 M, the apparent radius is slowly dependent on the concentration and temperature. This suggests that, at concentration ≤ 0.5 M, the micellar structure is stable without appreciable variation of the dimensions vs. temperature. The values of apparent radius so obtained, extrapolated at low concentration, are in good agreement with that obtained for this system by light-scattering technique (3). The increase of radius with concentration suggests that micelles coalesce in a larger structure with a shape that is no longer spherical but, is probably ellipsoidal or rod-like (4, 5). In this situation the radius so determined should be considered as a mean radius.

The temperature dependence of the apparent radius at two different concentrations, 1.0 M and 0.5 M in the range 278 ÷ 314 K, shows an exponential behavior with two different activation energies (Fig. 2). The fact that no discontinuities are observed shows that, in the temperature range studied, no abrupt structural transitions are present. The radius variation may be attributed to the temperature variation of water layer bonded or structured, or counterion distribution around the micellar structure. These considerations are supported by the activation energy values obtained for the two molar concentrations. In fact, the activation energy relative to the 0.5 molar sample is higher than that of the 1.0 M. Then, it is possible to argue that at lower concentration the micelles are like isolated and stable systems surrounded by a strongly bound water layer. At higher concentration the overlapping of the counterion cloud may generate a coalescence of micelles (connected spheres) that is not spherical but probably ellipsoidal (connected spheres) and less stable and, thus easily modified by temperature variations.

The diffusion studies here seem to give information on the dimensions of micellar systems at high concentration that are generally not investigated by other techniques. The dimensions so obtained, extrapolated at lower concentration, agree well with those obtained at very low concentration by other techniques, such as light scattering. From the results presented it seems that the system investigated, at concentration ≤ 0.5 molar, may be a suitable system for investigating the problem of water-macro molecular interaction. Works, including nuclear proton relaxation T_1 studies, are in progress in order to investigate the dynamics of structured water interacting with micellar surface.

References

1. Fendler JH (1982) In: Membrane Mimetic Chemistry. Wiley Interscience, New York; 1a ibid Chapter II and references cited therein
2. Stejscal EO, Tanner JE (1965) J Chem Phys 42:288

Progress in Colloid & Polymer Science

Progr Colloid Polym Sci 93:202 (1993)

3. McKenzie DC, unpublished results
4. Raul Zana, in Surfactant Solutions New Methods of Investigation (1987), Surfactant Science Series Vol 22, Marcel Dekker, Inc. New York
5. Tadros F (1984) in Surfactant, Inc. Academic Press

Authors' address:

E. Tettamanti
Dipartimento di Fisica
Università di L'Aquila
67100 L'Aquila, Italy

Viscoelastic lecithin reverse micelles: A simple model system for equilibrium polymers?

C. Cavaco and P. Schurtenberger

Institut für Polymere, ETH Zentrum, Zürich, Switzerland

Abstract: We report on a detailed study of lecithin reverse micellar solutions with novel polymer-like properties using dynamic (QLS) and static (SLS) light scattering and rheological measurements. The QLS and SLS experiments show a pronounced water-induced micellar growth from small and almost spherical reverse micelles at $w_0 = 2.0$ to giant polymer-like reverse micelles at $w_0 \geq 8.0$ for lecithin/cyclohexane w/o-microemulsions. An excellent agreement between polymer theory for dilute and semidilute solutions and the light scattering results was found. At volume fractions $\Phi << \Phi^*$, where Φ^* is the overlap threshold, the scattering data is in good agreement with the worm-like chain model. At $\Phi > \Phi^*$ a viscoelastic network is formed, and properties such as the static (ξ_s) and hydrodynamic (ξ_h) correlation length or the osmotic compressibility exhibit universal behavior which is in agreement with scaling theory and experimental results from polymer literature.

While the dependence of the zero shear viscosity η_s on Φ can be explained by the 'living polymer' model, the frequency-dependent measurements of storage and loss moduli G' and G" are in clear disagreement with the single exponential stress relaxation predicted by this model. The frequency-dependent measurements could indicate the formation of crosslinks between the entangled micelles, a hypothesis further supported by the very large values for the "scission energy" found in a preliminary study of the temperature dependence of the micellar size distribution at low values of Φ.

Key words: Lecithin reverse micelles — sphere-to-coil transition — equilibrium polymers

In contrast to aqueous micellar systems, reverse micelles at moderately high values of surfactant concentration and molar ratio of water to surfactant, w_0, are generally believed to have a droplet-like structure. However, we recently discovered the formation of gel-like, viscoelastic reverse micellar solutions in the system lecithin/organic solvent/water. Their unusual polymerlike properties could be explained with a water-induced one-dimensional micellar growth into very long and flexible cylindrical reverse micelles [1—4], i.e., this system exhibits a rather unique behavior with a characteristic sphere-to-rod transition normally observed in aqueous solutions only.

Here, we report on a detailed study of these lecithin reverse micellar solutions with novel polymer-like properties using dynamic (QLS) and static (SLS) light scattering and rheological measurements. The QLS and SLS experiments show a pronounced water-induced micellar growth from small and almost spherical reverse micelles at $w_0 = 2.0$ to giant polymer-like reverse micelles at $w_0 \geq 8.0$ for lecithin/cyclohexane w/o-microemulsions. An excellent agreement between polymer theory and the light scattering results was found. At volume fractions $\Phi << \Phi^*$, where Φ^* is the overlap threshold, the scattering experiments probe structural properties of the individual micelles, and the data is in good agreement with the worm-like chain model. At higher lecithin volume fractions ($\Phi > \Phi^*$) a viscoelastic network is formed, and properties such as the static (ξ_s) and hydrodynamic (ξ_h) correlation length or the osmotic compressibility exhibit scaling laws which are in excellent agreement with scaling theory and experimental results from polymer literature.

While the dependence of the zero shear viscosity η_s on Φ can be explained by the "living polymer" model [5], the frequency dependent measurements of storage and loss moduli G' and G" are in clear disagreement with the single exponential stress relaxation predicted by this model. The frequency dependent measurements could indicate the formation of crosslinks between the entangled micelles, a hypothesis further supported by the very large values for the "scission energy" found in a preliminary study of the temperature dependence of the micellar size distribution at low values of Φ. A different behavior is observed for the frequency dependence of G' and G" in lecithin/isooctane w/o-microemulsions. Here, the data appears to be in agreement with the theoretical model of "living polymers", but it indicates that breathing and/or Rouse motion [6] become important at higher values of w_0. This fact could explain the low exponent for the dependence of the zero shear viscosity on volume fraction previously reported for $w_0 \geq 2.5$ [2].

References

1. Scartazzini R, Luisi PL (1988) J Phys Chem 92:829
2. Schurtenberger P, Scartazzini R, Luisi PL (1989) Rheol Acta 28:372
3. Schurtenberger P, Scartazzini R, Magid LJ, Leser ME, Luisi PL (1990) J Phys Chem 94:3695
4. Schurtenberger P, Magid LJ, King SM, Lindner P (1991) J Phys Chem 95:4173
5. Cates ME, Candau SJ (1990) J Phys Condens Matter, and references therein 2:6869
6. Granek R, Cates ME (1992) J Chem Phys 96:4758

Authors' address:

Dr. Peter Schurtenberger
Institut für Polymere
ETH Zentrum
8092 Zürich, Switzerland

The structure of biopolymer-containing microemulsions: A light-scattering study at the optical match point

S. Christ, P. Schurtenberger, and P. L. Luisi

Institut für Polymere, ETH Zentrum, Zürich, Switzerland

Abstract: In this study, we focus on the size and structure perturbations induced in the reverse micellar system upon solubilization of the enzyme a-chymotrypsin by using light-scattering techniques. The knowledge of the geometrical parameters is very important in order to develop a good theoretical model of the driving forces responsible for polymer and biopolymer solubilization in microemulsions. Since the size difference between the empty reverse micelles and the average size measured in
the biopolymer-containing solution is often very small, the size distribution changes induced by the presence of biopolymers cannot be determined unambiguously by normal light-scattering experiments.

However, we can overcome this problem by suppressing the scattering contributions from the empty reversed micelles close to the so-called optical match point of the microemulsion droplets. Light-scattering experiments at the optical match point thus provide a very sensitive method to quantitatively test the currently existing theoretical models for protein solubilization in microemulsions. Using the size distributions predicted by these models, we can calculate both the total scattering intensity as well as the z-average diffusion coefficient unambiguously from the known dielectric constants of the components following established procedures.

Key words: Biopolymer-containing microemulsions — optical match point — light scattering

Reverse micelles and water-in-oil microemulsions can solubilize guest molecules such as ions, enzymes or synthetic polymers which would otherwise have only limited or no solubility at all in the organic solvent. This property makes microemulsions very interesting model systems for both basic research as well as an increasing number of applications such as enzymatically catalyzed organic-phase synthesis, chromatographic separation or extraction of proteins through phase transfer methods. In this study, we focus on the size and structure perturbations induced in the reverse micellar system upon solubilization of the enzyme a-chymotrypsin by using light-scattering techniques. The knowledge of the geometrical parameters is very important to correctly interpret the observed phenomena and to develop a good theoretical model of the driving forces responsible for polymer and biopolymer solubilization in microemulsions.

In recent years, this proplem was often discussed and a number of experimental studies and theoretical models have been presented [1]. However, no unambiguous solution to this problem has been found, and the situation still remains controversial. Since the size difference between the empty reverse micelles and the average size measured in the biopolymer-containing solution is often very small, the size distribution changes induced by the presence of biopolymers cannot be determined unambiguously by normal light-scattering experiments. However, we can overcome this problem by suppressing the scattering contributions from the empty reversed micelles close to the so-called optitcal match point of the microemulsion droplets [2].

Progress in Colloid & Polymer Science   Progr Colloid Polym Sci 93:204 (1993)

It is the aim of this study to show that we can use light-scattering experiments close to the optical match point as a very sensitive method to quantitatively test the currently existing theoretical models for protein solubilization in microemulsions. Our light-scattering experiments clearly indicate that incorporation of a-chymotrypsin results in a small alteration of the micellar radius only, although the scattering intensity significantly changes due to the strong modification of the scattering profile of the reverse micelles. We are able to show that we can use light-scattering experiments close to the optical match-point as a very sensitive method in order to quantitatively test the currently existing theoretical models for protein solubilization in microemulsions. Using the size distributions predicted by these models, we can calculate both the total scattering intensity as well as the z-average diffusion coefficient unambiguously from the known dielectric constants of the components. As a first test, we have performed calculations on the basis of the model presented by Rahaman and Hatton, which would predict much higher effects on both the size and the scattering intensity of the reverse micelles after solubilization of a-chymotrypsin than experimentally observed. However, additional experiments will be required in order to quantitatively assess the structure and size distribution of the protein-containing microemulsion.

References

1. Rahaman RS, Hatton TA (1991) J Phys Chem (and refs. therein) 95:1799
2. Ricka J, Borkovec M, Hofmeier U (1991) J Chem Phys 94:8503

Authors' address:

Dr. Peter Schurtenberger
Institut für Polymere
ETH Zentrum
8092 Zürich, Switzerland

Ultrasonic study of phospholipid multibilayer systems. Thermotropic transitions and melittin-induced transitions

A. Colotto[1]), D. P. Kharakoz[2]), K. Lohner[1]), and P. Laggner[1])

[1]) Institute of Biophysics and X-Ray Structure Research. Austrian Academy of Sciences, Graz, Austria
[2]) Institute of Biological Physics. Russian Academy of Sciences, Pushchino, Moscow Region, Russia

Abstract: Temperature dependencies of sound velocity and absorption in dilute suspensions (2.3 wt%) of 1.2-dipalmitoyl-*sn*-glycero-3-phosphatidylcholine (DPPC) multilamellar systems, with and without melittin, have been measured at 7.2 MHz. Relaxational characteristics of heterophase fluctuations of density in the vicinity of the main phase transition have been determined for the pure lipid system, allowing for the first time the estimation of the interphase line tension, in the fluid bilayer, through acoustic data. Significant changes in the acoustic data of the melittin containing systems, are indicative that the peptide effects on the bilayers are related to relaxational processes connected to dynamic properties of the gel phases of the system.

Key words: Ultrasound velocimetry — lipid phase transition — line tension — melittin

Measurement of ultrasound velocity and absorption by a differential fixed-path interferometer [1] have been used to investigate phase transitions in lipids and lipid-protein complexes.

An attempt has been made to gain deeper insight into the properties of the domains involved in the mechanisms of phospholipid phase transitions. A comparison is made between transitions induced by temperature change and by the interaction with melittin (the main peptide component of bee venom, composed of 26 amino acids [2]), respectively.

For pure DPPC multilamellar liposomes, the results were analyzed in terms of a model of density fluctuations based on i) the Frenkel thermodynamic theory of heterophase fluctuations, with the Gibbs chemical potential μ_n of a cluster of size n (n pairs of lipid molecules in the cluster), given by:

$$\mu_n = n \times \Delta\mu + \gamma \times 2\pi \times [\sqrt{(n/\pi)} - 1/2] , \quad (1)$$

where γ is the interphase line tension coefficient and $\Delta\mu$, the chemical potential difference between two phases; and ii) on a chain kinetic scheme of creation and growth of clusters of the new phase inside the parent phase [3]:

$$C_0 \underset{b_7}{\overset{f_7}{\rightleftarrows}} C_7 \underset{b_8}{\overset{f_8}{\rightleftarrows}} C_8 \underset{b_9}{\overset{f_9}{\rightleftarrows}} \cdots \underset{b_{n-1}}{\overset{f_{n-1}}{\rightleftarrows}} C_{n-1} \underset{b_n}{\overset{f_n}{\rightleftarrows}} C_n \rightleftarrows \cdots ,$$

$$(2)$$

where the backward rate constants are proportional to the number of boundary molecules contained in a cluster,

$$b_n = b_0 \times 2\pi \times \sqrt{(n/\pi - 1/2)} , \qquad (3)$$

and the forward rate constants are given by the equilibrium condition:

$$f_n = b_n \times [-kT \times \ln (\mu_n - \mu_{n-1})] . \qquad (4)$$

Using the system of equations which describes this linear kinetic chain scheme and applying the theory of linear acoustic relaxation the measured sound velocity and absorption are obtained as functions of b_0 and γ which serve as adjustable parameters and which values are obtained by means of a fitting procedure. The final values being: $\gamma = 6 \times 10^{-7}$ erg/cm and $b_0 = 40\ \mu s^{-1}$.

Remarkable anomalies were observed when melittin was added at already very low doses: at the pre-transition, melittin was found to lead to a strong increase in absorption at molar ratios $R < 5 \times 10^{-3}$ (peptide to lipid), while at the main transition the absorption decreased. Qualitatively, this indicates that melittin induces the formation of domains differing strongly in density and/or compressibility from the pure ones. In combination with recent x-ray data (A. Colotto et al. 1993, Biophys. J.), a model is suggested whereby melittin induces and stabilizes domains with hexagonally ordered hydrocarbon chains. The melittin-induced transition is comparable to the thermotropic ones in that it also carries a cooperative nature [4] and, therefore, may also be modeled in terms of domain density fluctuations.

References

1. Sarvazian AP (1982) Ultrasonics 20:152—154
2. Habermann E, Jentsch E (1967) Hoppe Seyler's Physiol Chem 348:37—50
3. Tsong Y, Kanehisa I (1977) Biochemistry 16:2674—2680
4. Posch M, Rakusch U, Mollay C, Laggner P (1983) J of Biol Chem 258:1761—1766

Authors' address:

Colotto Adriana
Institute of Biophysics and X-ray Structure Research, Austian Academy of Sciences
Steyrergasse 17/6
8010 Graz, Austria

X-ray and neutron-scattering measurements on concentrated non-ionic amphiphile solutions

I. S. Barnes[1]), M. Corti, V. Degiorgio, and T. Zemb[2])

Department of Electronics, University of Pavia, Italy
[1]) Department of Applied Mathematics, Australian University, Camberra, Italy
[2]) S C M Bat 125, CEN Saclay, Gif sur Yvette, France

Abstract: The structure of aqueous solutions of the non-ionic amphiphile $C_{12}E_8$ is studied by small-angle x-ray and neutron scattering along the isothermal path across the single-phase region from 0% to 100% amphiphile volume fraction. Solution structuring is present even in the pure amphiphile, as shown by x-ray data.

Key words: Non-ionic micelles — x-ray scattering

Aqueous solutions of the non-ionic polyoxyethylene amphiphiles $C_{12}E_8$, $C_{10}E_5$ and C_8E_4, whose chemical formula is $C_iH_{2i+1} (OCH_2CH_2) j$, have been studied by neutron scattering along an isothermal path across the single-phase region, which is below the cloud point curve and above the mesophase region [1, 2]. It has been shown that the solution is structured for all concentrations ranging from the micellar region to the pure liquid amphiphile. As the amphiphile volume fraction increases, the micellar structure becomes less and less sharp, but some orientational correlations between neighboring amphiphile molecules are preserved even at high concentrations. A structure peak is clearly observable until 95% amphiphile volume fraction, but the pure amphiphile neutron spectrum is absolutely flat. This is due to the fact that the neutron scattered intensity is mainly determined by the large contrast between the deuterated water and the hydrogenated amphiphile. Therefore, orientational correlations in the pure liquid, which could be visible only through the very small contrast between the hydrophobic and hydrophilic part of the amphiphile, do not give enough neutron counts above incoherent scattering. In order to overcome this problem, measurements were repeated with small-angle x-rays, where electron-densities of the hydrophilic and the hydrophobic part of the amphiphile and water are of the same order of magnitude.

Figure 1 shows the x-ray spectra for the system $C_{12}E_8$ in deuterated water at 60°C, made at the LURE synchrotron in Orsay, France, with the high-resolution, small-angle camera D22. As expected, the structure peak is also visible at 100% volume fraction. This definitely confirms the statement of [2] of orientational correlations in the amphiphile. In analogy with experiments on block-copolymer melts [3], the existence of a peak is a direct result of the block structure of the amphiphile monomer with attractive head-head and tail-tail interactions. This is called the correlation-hole effect [4, 5].

An interesting difference between neutron and x-ray spectra is also found at volume fractions below 70%. X-ray spectra are sharper and have a small second peak at higher q's, q being the scattering vector. For these concentrations, the reminiscence of a micellar structure affects the spectra by means of the micelle form factor $P(q)$, which is quite different for neutrons and x-rays. If micelles are modeled as inhomogeneous spheres with two different contrasts for the hydrophobic and the hydrophilic parts, the x-ray form factor is double peaked with lower amplitude at $q = 0$, while the neutron form factor is monotonically decreasing with q. This explains the differences.

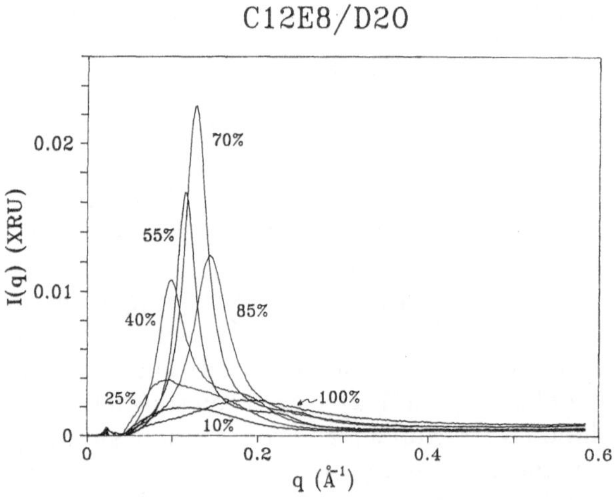

References

1. Degiorgio V, Corti M (1988) Chem Phys Lett 151:349
2. Degiorgio V, Corti M, Piazza R, Cantù L, Rennie A. R. (1991) Colloid Polym Sci 269:501
3. Stühn B, Rennie A (1989) Macromolecules 22:2460
4. de Gennes PG (1969) Scaling concepts in polymer physics. Cornell University Press, Ithaca, p 65
5. Leibler L (1980) Macromolecules 13:1602

Authors' address:

Prof. Mario Corti
Dipartimento di Elettronica
Università di Pavia
via Abbiategrasso 209
27100 Pavia, Italy

Fig. 1. X-ray scattering intensity vs q for different amphiphile volume fractions

Surface forces between adsorbed polyelectrolytes in salt solution

M. A. G. Dahlgren and P. M. Claesson

Laboratory for Chemical Surface Science,
Dept. of Chemistry,
Royal Institute of Technology, Stockholm, Sweden; and Institute for Surface Chemistry, Stockholm, Sweden

Abstract: Highly charged cationic polyelectrolyte was adsorbed onto mica from 10^{-4} M KBr and 10^{-4} M K_2SO_4 solutions. The results show that the structure of the adsorbed layer and the total amount adsorbed is dependent on the anion valency. Both the layer thickness and the adsorbed amount increase when the salt anion is of higher valency. — Desorption over several days into 10^{-4} M K_2SO_4 solution was followed. This indicates the adsorbed layer initially overcompensates the mica lattice charge. After some time, the sign of the charge is re-reversed, due to desorption. Adhesion between the adsorbed layers increases during the desorption process. This adhesion is attributed to bridging of entropic origin.

Key words: Surface force — polyelectrolyte — adsorption — desorption — bridging

Introduction

The aim of this study was to investigate how the counterion valency affects the adsorption of cationic polyelectrolytes onto negatively charged surfaces. The polyelectrolyte used is poly ((*N, N, N,*)-trimethyl-aminochloride-propyl-methacrylamide) (called MAPTAC, Fig. 1), with a charge density of 1.0 and molecular weight of 1.0×10^5 g/mol. MAPTAC was adsorbed from salt solutions containing 10^{-4} M KBr and 10^{-4} M K_2SO_4, respectively. The MAPTAC content of the solutions was 10 ppm.

Experimental technique

The force between MAPTAC-coated surfaces as a function of surface separation was determined by using a Surface Force Apparatus (SFA). The present study was conducted using an SFA developed by Parker et al. [1] In this apparatus two negatively charged molecularly smooth muscovite mica surfaces are mounted in a crossed cylinder configuration.

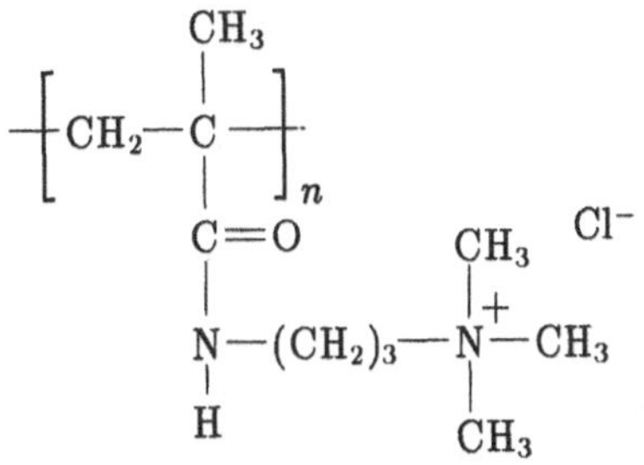

Results

After adsorption overnight, the force — distance profiles were measured. The results for both 10^{-4} M KBr and 10^{-4} M K_2SO_4 are shown in Fig. 2. When the inert salt is KBr, the MAPTAC neutralizes the mica lattice charge, and adsorbs in a flat conformation. The position of the attractive force minimum is located at 1 nm mica — mica separation. When the solution contains K_2SO_4 as the inert salt, a repulsive barrier is present before the surfaces are brought into adhesive contact. Also, the adsorbed layer thickness is larger, giving a final surface separation of 2.5 nm, still a very thin layer. The decay-length of the repulsive force is 17 nm, which is as expected for a double-layer force in a 2:1 electrolyte at a concentration of 10^{-4} M.

After the adsorption of MAPTAC from K_2SO_4 solution, the salt and polyelectrolyte solution was replaced with a polyelectrolyte-fee 10^{-4} M K_2SO_4 solution. Then the desorption of polyelectrolyte from the mica surface was followed during several

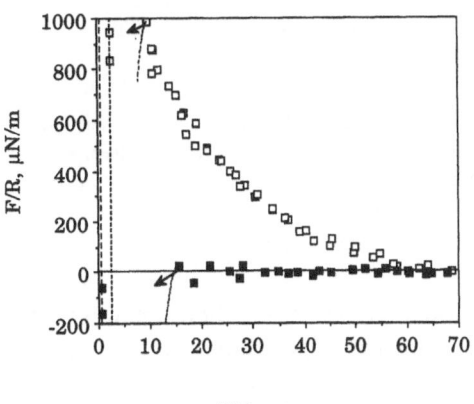

Fig. 1. Molecular structure of MAPTAC

Fig. 2. Force-distance profiles for mica surfaces with adsorbed MAPTAC in the presence of 10 ppm polyelectrolyte ■ in 10^{-4} M KBr, □ in 10^{-4} M K_2SO_4. Arrows indicate inward jumps

days (Fig. 3). Notably, the repulsive barrier is lower after 1 day of desorption than previously seen (Fig. 2.). After 3 days of desorption the force is purely attrative. An even longer desorption time causes the build-up of a repulsive force. The position of the attractive force minimum is located at 2.5 nm mica — mica separation during the entire desorption experiment (one week), and is also the same as when there was MAPTAC in the solution. We interpret this change in magnitude of the repulsive barrier as an indication that the original double-layer repulsion was due to an overcompensation of the mica lattice charge by adsorption of polyelectrolyte. During the desorption the net charge of the system is changing to negative. Nevertheless, the polyelectrolyte molecules start to extend further out from the surfaces. This is supported also by the change in decay-length. After the fourth day in the desorption experiment, the decay length is no longer consistent with what is expected for a double-layer force in a 2:1 electrolyte at a concentration of 10^{-4} M.

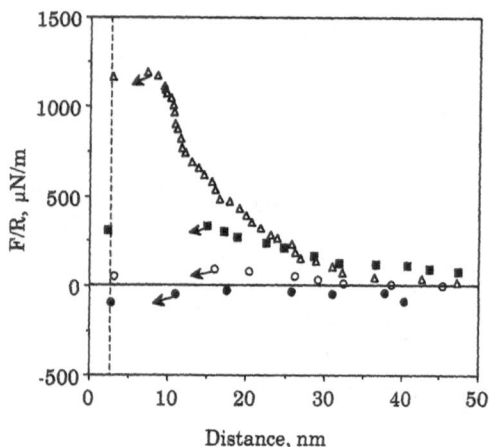

Fig. 3. Force-distance profiles for mica coated with MAP-TAC during desorption in 10^{-4} M K_2SO_4 solution. ■ after 1 day; ● after 3 days; ○ after 4 days; △ after 5 days. Arrows indicate inward jumps

The adhesion force observed when separating the surfaces was measured. The results are given in Fig. 4. After the adsorption, the adhesion measured was 35 mN/m (compared to 180 mN/m when adsorbed from 10^{-4} M KBr solution). After the first day in the desorption experiment, the adhesion has increased slightly to 37 mN/m. However, during the next few days, the adhesion increases to reach a maximum of 260 mN/m after 6 days. We attribute this to bridging forces of entropic origin [2] which should be facilitated by a more extended conforma-

tion. The increased adhesion lends support for the idea that the adsorbed molecules starts to extend further out. This change in adhesion when going from an over-compensated system to an under-compensated one is in accordance with recent theoretical findings [3].

References

1. Parker JL, Christenson HK, Ninham BW (1989) Rev Sci Instrum 60:3135—3138
2. Åkesson T, Woodward C, Jönsson B (1989) J Chem Phys 91:2461—2469
3. Dahlgren MAG, Waltermo A, Blomberg E, Claesson PM, Sjöström L, Åkesson T, Jönsson B J Phys Chem

Authors' address:

Mats A. G. Dahlgren
Laboratory for Chemical
Surface Science, Dept. of Chemistry
Royal Institute of Technology
100 44 Stockholm, Sweden

Film formation between two emulsion drops in Brownian flocculation and coalescence

K. D. Danov, N. D. Denkov, D. N. Petsev, I. B. Ivanov, and R.P. Borwankar*)

Laboratory of Thermodynamics and Physico-chemical Hydrodynamics, University of Sofia, Faculty of Chemistry, Sofia, Bulgaria
*) KRAFT General Foods, Inc., Technology Center, Glenview, Illinois, USA

An original kinetic scheme for emulsion drop diffusion toward a central drop in the presence of potential interactions is suggested. Four basic stages are defined: flocculation of nondeformed drops, deformation and film formation, film thinning, and coalescence. The diffusion flux, effective thermodynamic force and pair probability are determined. A transcendental equation for the critical distance of film formation is derived. Original relationships for the energy of deformation, including the Gibbs elasticity, van der Waals, and electrostatic interactions are obtained. The hydrodynamic resistance, including edge effects, is also taken into

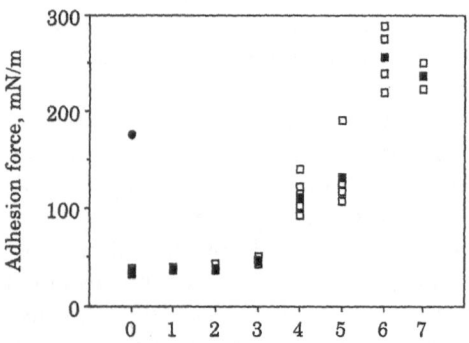

Fig. 4. Variation of the adhesion force during the desorption experiment. □ different measurements in K_2SO_4; ■ mean value for 1 day's measurements in K_2SO_4; ● mean value for measurements in KBr

account. The equation for the distance of film formation is investigated numerically for some real
systems. It is shown that the coalescence may occur
after a stage of film thinning or without such stage,
depending on the system parameters.

Authors' address:

K. D. Danov
Laboratory of Thermodynamics and
Physico Chemical Hydrodynamics
University of Sofia
J. Boucher Ave 1
1126 Sofia, Bulgaria

Fractal structures of soil particles as revealed by small-angle x-ray scattering studies

G. Degovics[1]), P. Laggner[1]), M. Borkovec[2]),
Q. Wu[2]), and H. Sticher[2])

[1]) Institute of Biophysics and X-Ray Structure Research
Austrian Academy of Sciences, Graz, Austria
[2]) Institute of Terrestrial Ecology, Federal Institute of
Technology, Zürich, Switzerland

Key words: Soil particles — soil-fine-structure — small-
angle x-ray scattering — surface fractals

Sorption properties and leachability of soils are
largely determined by their contents of fine particles. Typically, 1 cm^3 of soil contains $10^{17} - 10^{19}$
particles. From this, it is clear that the physico-
chemical characterization of soils requires quantitative knowledge of the particles in the nanometer
to micrometer range. While the classcal methods
such as sieve fractionation and sedimentation are
patent tools for the micron size range and above, little direct information exists on the submicron range
down to atomic or molecular dimensions.

Small-angle x-ray scattering (SAXS) is capable of
entering these domains. In the present work a
systematic application on problems of soil-
fine-structure is reported. For size fractions, SAXS
allows to distinguish unambiguously between
smooth surfaces and surface fractals. In the case of
smooth surfaces the scattering intensity for large
scattering vectors follows the Porod law [1]

$$I(q) \sim q^{-a}, \tag{1}$$

where $a = 4$ in the case of point collimation
geometry. For surface fractals we observed also a
power law Eq. (1), but with a smaller exponent,
from which the surface fractal dimension D_s can be
directly estimated. In the present case of infinite
slit-length collimation geometry the exponent is
related to the surface fractal dimension by

$$a = 5 - D_s. \tag{2}$$

We have shown that soil particles have a rough surface and their surface area scales as $A \sim r^{Ds}$,
where $D_s = 2.4 \pm 0.1$. This value has been determined by independent experiments such as small-
angle x-ray scattering, nitrogen gas adsorption and
methylene blue adsorption from solution. Our
estimate agrees perfectly with the previous result of
$D_s = 2.35 \pm 0.05$ for soil particles based on
nitrogen gas and water vapor adsorption data [2].
These results have been further interpreted using
recent soil particle size distribution data in the size
range from 20 nm to 100 µm particle radius. The
number of particles per unit volume with a radius
larger than r is found to follow a power law $N(r) \sim r^{-D}$ with the exponent $D = 2.8 \pm 0.1$ in all sols investigated. The power law is typically valid between
two cut-off radii $r_1 \ll r \ll r_2$ with values around $r_1 \simeq$
$10 - 100$ nm and $r_2 \simeq 10 - 500$ µm. Using the
known specific surface area of individual particle
size fractions and together with the size distribution
one can accurately predict the specific surface area
of the unfractionated sample. The major part of the
surface area originates from particles with radii between 20 nm and 1 µm.

References

1. Porod G (1982) In: Glatter O, Kratky O (eds) Small-
   angle x-ray scattering. Acad Press London, pp 17—51
2. Sokolowska Z (1989) Geoderma 45:251—265

Authors' address:

Gabor Degovics
Institute of Biophysics and X-Ray
Structure Research
Austrian Academy of Sciences
Steyrergasse 17/VI
8010 Graz, Austria

Progress in Colloid & Polymer Science

Progr Colloid Polym Sci 93:210 (1993)

Effect of some inhibitors on the zeta-potential of calcium oxalate monohydrate particles

J. Callejas-Fernández, R. Martínez-García, R. Hidalgo-Alvarez, and F. J. de las Nieves

Biocolloids and Fluid Physics Group, Department of Applied Physics, University of Granada, Spain

Abstract: The ζ-potential of dilute dispersions of calcium oxalate monohydrate (COM) in the presence of several additives has been studied. The experimental ζ-potential values were obtained from electrophoretic mobility measurements by using the Smoluchowski equation. The results were compared with the theoretical data obtained from a simple electric double layer (e.d.l.) model, considering the calcium concentration (pCa) changes. For urea, creatinine, D-glutamic acid, and methylene blue the slight ζ-potential variations were explained via small changes in the pCa values of the dispersions. However, for D-tartaric, chondroitin sulfate, heparin, and sodium dodecyl sulfate the results can only be explained via specific adsorption of these molecules. In these latter cases the ζ-potentials were negative and relatively high, which is an indication of the inhibitory effect of these molecules on the aggregation of COM particles.

Key words: Calcium oxalate monohydrate — zeta-potential, inhibitors

Introduction

An important aspect during the formation process of kidney stones is the aggregation of calcium oxalate crystals, which can appear in different crystalline forms: mono-, di-, and trihydrate. The electric state of the microparticle surface has a great influence on the tendency of the crystal to be in a separate or aggregate forms [1, 2]. This is the reason why the surface charge of the calcium oxalate particles is of prime interest; particularly, that of calcium oxalate monohydrate (COM), which is the main component of the kidney stone.

In previous papers, we have studied the electrokinetic behavior of COM particles in the presence of some electrolytes in solution [1] or when some substances were added during the nucleation process of the COM crystals [3]. In this work, we will study the effect of some substances, called inhibitors or promotors, on the electric state of the

COM particles surface. Substances such as urea, creatinine, D-glutamic, D-tartaric, glucosaminoglycans, methylene bleu, and sodium dodecyl sulfate (SDS) were used throughout this work. Some of these additives are present in human urine and others have been considered to be potential inhibitors of aggregation [4, 5]. To determine the electrokinetic behavior of the COM particles, electrophoretic mobility measurements were carried out and the experimental results were converted into zeta-potential (ζ) data by the simple Smoluchowski's equation.

Materials and methods

Calcium oxalate monohydrate (COM) crystals were obtained by precipitation in a supersaturated solution of sodium oxalate and calcium chloride. The ratio between calcium and oxalate ions was 5:1, because the resulting COM crystals were most similar to commercial COM samples [3]. Details about the amounts and experimental conditions of the precipitation reaction can be found in [1, 3].

The crystals were analyzed by x-ray diffraction and thermogravimetric analysis, as described in [6, 7]. After storage of the calcium oxalate precipitates for at least 1 month the content in COM crystals was more than 90%.

Electrophoretic measurements were carried out with a Zeta-Sizer IIc (Malvern Inst., England) at a temperature of 298 K in a cylindrical cell.

All chemicals were of analytical grade from Merck. The sodium heparin and D-glutamic acid were obtained from Sigma Co. The calcium concentrations in solution were determined with a Ca^{2+}-specific electrode (Ingold, Switzerland).

Results and discussion

The calcium (Ca^{2+}) and oxalate ($C_2O_4^{2-}$) ions are potential determining ions (p.d.i.) and their concentrations in the dispersions are decisive on the ζ-potential of the COM particles. Using the simple Gouy-Stern model of the e.d.l., it is possible to relate the surface potential, the electric potential at the slipping plane (identified as the ζ-potential) and the electrical surface charge density [6, 8]. The result should be an exponential relationship between the surface potential and the calcium concentration (pCa). However, if we plot the ζ-potential data versus the pCa of the liquid, the result can be approximated by a straight line near the isoelectric point (i.e.p.) of the calcium oxalate particles [1, 6, 8]. To understand the results presented in this paper, we have used the linear dependence previously found [1] between the ζ-potential of COM particles and the calcium concentration (pCa), at pH $\approx$ 6:

$$\zeta \text{ (mV)} = (54 \pm 5) - (10 \pm 1) \text{ pCa} . \qquad (1)$$

Thus, at pCa $\approx$ 5.2 the COM particles are electrokinetically uncharged. Below that pCa the ζ-potential will be positive and above that pCa the particles will acquire a negative electrokinetic charge.

Figure 1 shows the ζ-potential of COM particles versus urea concentration (curve 1). ζ is positive within the concentration range used, although it diminishes as urea content increases. The pCa and pH of the solution were practically constant, with a value of 4.0 and 5.8, respectively. Using Eq. (1), pCa gives a constant ζ-potential of 14 mV, which is in agreement with the experimental value at low urea concentration, but does not explain the ζ/pCa curve found. Robertson et al. [9] has found that within the concentration range of urea in human urine (11—20 g/l), this molecule yields a small but significant inhibitor effect, which could be due to its adsorption on the active COM crystals or the formation of a hydration layer which constitutes a barrier for the diffusion of Ca^{2+} and $C_2O_4^{2-}$ ions from the solution to the surface of the crystals. For urea concentration equal or lower than 10^{-4} M the p.d. ions are responsible of the ζ values, but at high concentrations the diminution of ζ could be due to the compression of the e.d.l. or the progresive formation of the layer suggested by Robertson et al. [9].

Creatinine is a substance removed by urine in quantities almost proportional to the volume of muscular mass of the person. Coe [10] found an amount of 1675 mg of creatinine in urine collected for 24 h. In Fig. 1, we also show the ζ-potential of COM particles versus creatinine concentration (curve 2). The ζ-potential is constant as the concentration of that molecule increases. The pH of the dispersions changes from 5.2 to 7.0 with the amount of creatinine, while the pCa remains constant (pCa = 4.1). For this pH change ζ is constant [1, 3] and introducing the pCa in Eq. (1), we obtain a ζ-potential value of 12—13 mV, which is quite similar to that experimentally found. Thus, for the system

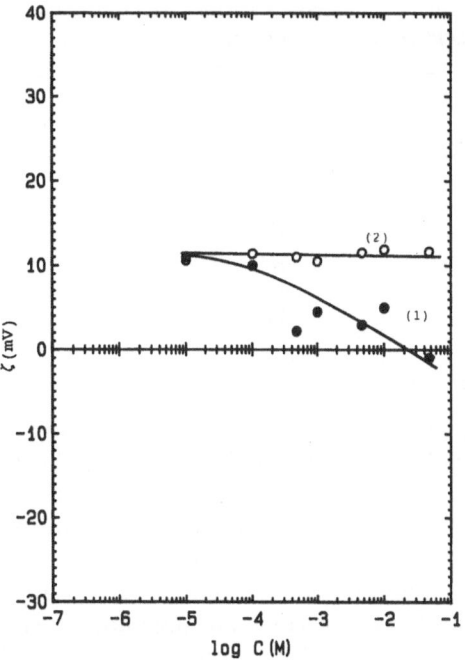

Fig. 1. ζ-potential of COM particles versus urea (curve 1, ●) and creatinine (curve 2, ○) concentrations

COM/water/creatinine the electric potential in the e.d.l. is determined by Ca^{2+} and $C_2O_4^{2-}$ concentrations and the creatinine seems to be an electrokinetically indifferent molecule.

D-glutamic acid (D-G) plays an important role in the aminoacid metabolism, because it can act like an element to captivate aminoacids that participate in the urea cycle. Azoury et al. [11] have indicated that D-G can influence the crystallyzation process of COM. Our interest is to know if this substance acts on the electrokinetic properties of COM/water interface. Figure 2 shows the ζ-potential values versus D-G concentration (curve 1). A slight decrease at a concentration between 10^{-5} and 10^{-2} M can be observed, as well as a similar ζ values at extreme concentrations. Table 1 shows the pH and pCa values at different D-G concentrations. As the amount of D-G increases the pH decreases, which provokes a decrease in pCa. According to Eq. (1), this means a continuous increase in the ζ-potential (curve 2). This result is different from that experimentally found. This is a consequence of the special properties of this component. D-Glutamic acid has pK = 2.19 and 4.15, which means a positively charged molecule at pH < 2.19, uncharged molecule at pH between 2.19 and 4.15, and negatively charged molecule at pH > 4.15. Thus, at

Table 1. pH and pCa values of COM dispersions versus D-glutamic acid (D-G) concentration

C(M)	pH	pCa
10^{-6}	5.0	4.1
10^{-5}	4.9	4.0
$5 \cdot 10^{-5}$	4.5	3.9
10^{-4}	4.3	3.9
$5 \cdot 10^{-4}$	3.8	3.7
10^{-3}	3.6	3.6
10^{-2}	3.2	3.5

D-G concentrations lower than 10^{-4} M, the pH is higher than 4.2, and the molecule is negatively charged. The opposite charges probably produce a specific adsorption of D-G and a diminution in the ζ-potential. At D-G concentrations higher than 10^{-4} M the pH remains between 2.2 and 4.2, and, therefore, the D-G molecules are uncharged. The decrease in the adsorption improves the flow of potential determining ions (specially calcium) toward the surface and they will determine the electrokinetic potential, producing a new increase (as can be seen in Fig. 2 for a D-G concentration of 10^{-2} M).

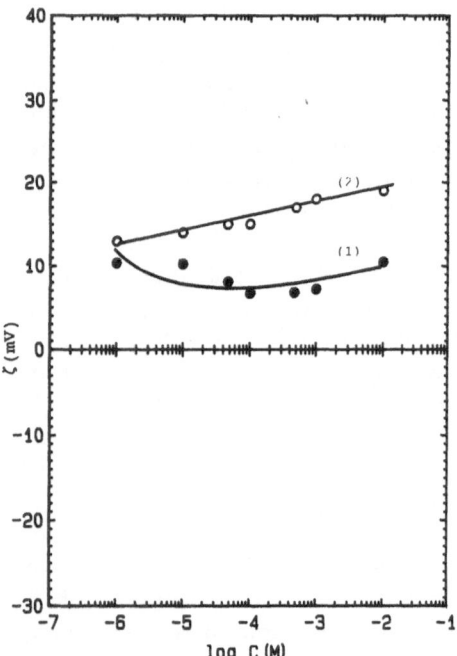

Fig. 2. ζ-potential of COM particles versus D-glutamic acid concentration: experimental, curve 1 (●); theoretical, curve 2 (○) (according to Eq. (1))

Another important aminoacid is D-tartaric acid (D-T). According to Sur et al. [12], D-tartaric acid inhibits calcium oxalate crystallization because it can form complexes with the Ca^{2+} ion. The influence of D-T on the ζ-potential can be seen in Fig. 3 (curve 1). This curve was obtained without pH control. If we maintain the pH of the dispersion at a value of around 7, the ζ-potential varies, as curve 3 shows. Table 2 shows the pCa and pH values as D-T concentration increases. These pCa values can explain the results of curve 1: using Eq. (1) and the pCa values from Table 2, we can obtain curve 2 for ζ-potential, which is in coincidence with the experimental curve 1 (when there is not pH control). This result suggests that the anion tartrate ($H_5C_4O_4^-$) does not interact with the COM surface, because the calcium concentration determines the ζ value. However, when we maintain constant pH (around 7) the ζ-potential decreases (Fig. 3, curve 3). Under these conditions, the pCa increases from 3.8 to 4.6, which means a decrease in calcium ions as a consequence of the formation of complexes between Ca^{2+} and D-T. These pCa values would explain a diminution in ζ, but not negative values of the electrokinetic potential. Thus, at pH $\approx$ 7 the adsorption of $H_5C_4O_4^-$ species on COM particles could be responsible for the negative ζ-potential values at high concentration of D-tartaric acid.

The glycosaminoglycans appear in the biological liquids of mammalians. Particularly, they are present in human urine and can act as inhibitors during the formation process of kidney stones [5, 9, 13, 14]. The main molecules are chondroitin sulfate (C-S) and sodium heparin (S-H). Gjaldbaeck [5] indicates that C-S does not influence the crystal growth of COM, but can influence the aggregation of COM particles. Figure 4 (curve 1) shows the ζ-potential values of COM particles versus the C-S concentration. ζ is positive at low content of C-S, decreases as C-S concentration increases, and is negative at C-S concentrations higher than 10^{-4} g/l. When the C-S concentration is higher than 10^{-3} g/l, the ζ-potential seems to reach a saturation plateau of around —22 mV. For the concentration range of C-S studied, pH and pCa remained constant with values of 5.4 and 4.0, respectively. Using Eq. (1) and that pCa, we obtain a constant ζ-potential value of 14 mV, for all the C-S concentrations, which is completely different from the experimental result. Thus, the specific adsorption of C-S molecules seems to be the explanation for that result. The negative charge of the chondroitin sulfate

Table 2. pH and pCa values of COM dispersions versus D-tartaric acid (D-T) concentration

C(M)	pH	pCa
10^{-5}	5.3	4.1
10^{-4}	4.1	3.8
$5 \cdot 10^{-4}$	3.4	3.5
10^{-3}	3.2	3.4
$5 \cdot 10^{-3}$	2.8	3.1
10^{-2}	2.7	3.0
$5 \cdot 10^{-2}$	2.4	2.8

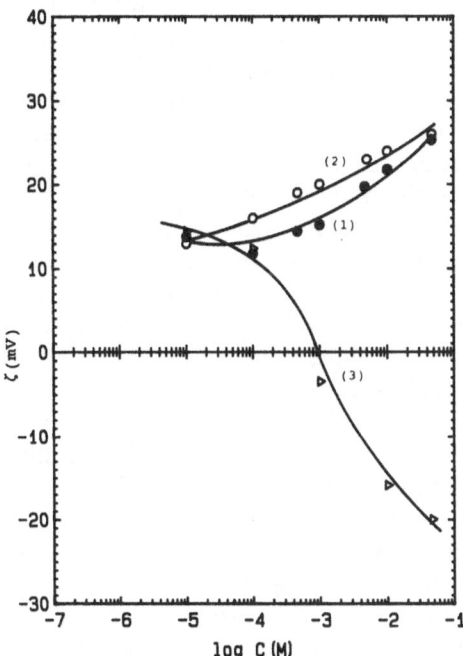

Fig. 3. ζ-potential of COM particles versus D-tartaric acid concentration: experimental, curve 1 (●); theoretical, curve 2 (○) (according to Eq. (1)); experimental with constant pH (pH = 7), curve 3 (△)

counterbalance the positive charge of the surface and ζ becomes more negative as the amount of adsorbed C-S molecules increases. At a C-S concentration of 10^{-3} g/l the surface of COM particles seems to be saturated and there is no more adsorption of C-S molecules; as a consequence, the ζ-potential reaches a plateau. Robertson et al. [9] suggested two mechanisms to explain this type of result: formation of complexes with the p.d. ions, or adsorption on the crystal surface. Due to the

plateau reached by ζ at higher C-S concentrations, we are in agreement with Curreri et al. [15] about the adsorption mechanism.

The effect of sodium heparin (S-H) on the ζ-potential of COM particles can be observed in Fig. 4 (curve 2). The trend of the results is similar to that shown for C-S, although the ζ values are higher in all cases. Again, the pH and pCa values were constant (5.4 and 4.1, respectively) and, therefore, the results cannot be explained by changes in the concentration of p.d. ions (calcium or oxalate). Thus, as several authors have suggested [5, 14], the specific adsorption of S-H on the surface of COM particles seems to again be the explanation for those results. However, with S-H the ζ values did not show a plateau within the concentration range studied, because the saturation of the COM particle surface with S-H molecules was probably not reached.

Some authors [9, 16] have indicated that methylene blue (M-B) can interact with the surface of COM particles and, therefore, modify the electric state of the COM/water interface. Figure 5 (curve 1) shows the ζ-potential values of COM particles versus the M-B concentration. For concentrations higher than 5.10^{-5} M it was not possible to measure electrophoretic mobilities, because the

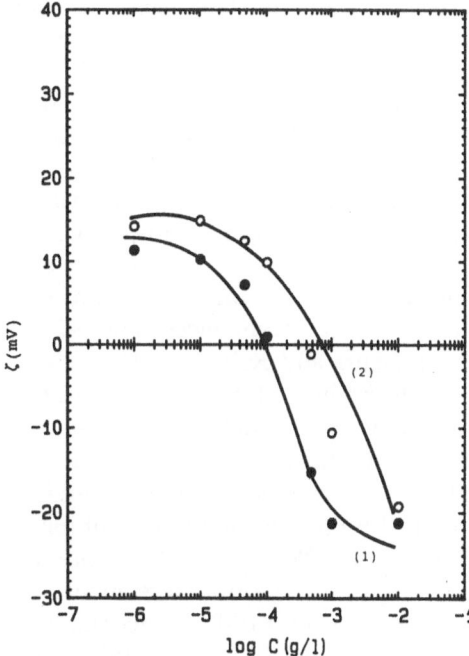

Fig. 4. ζ-potential of COM dispersions versus chondroitin sulfate (curve 1, ●) and sodium heparin (curve 2, ○) concentrations

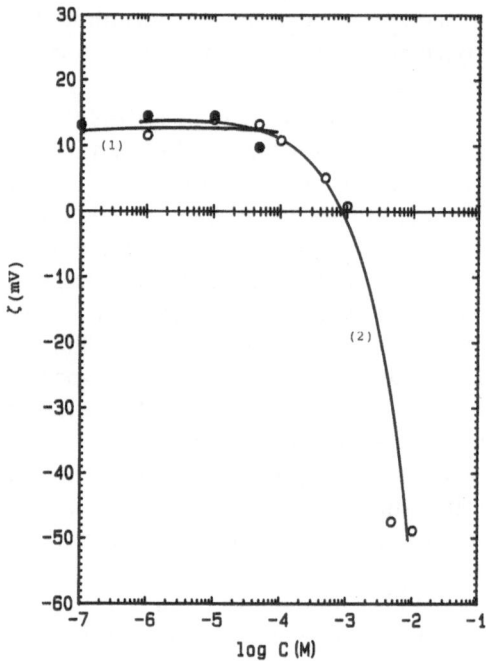

Fig. 5. ζ-potential of COM dispersions versus methylene blue (curve 1, ●) and sodium dodecyl sulfate (curve 2, ○) concentrations

Zetasizer IIc did not detect the light dispersed by the particles. Within the concentration range studied, ζ was almost constant, with a slight decrease at a concentration of 5.10^{-5} M. In this range the pH was constant (≈ 6.6) and the pCa changes were not significant ($\approx 4.1 - 3.8$), except for a concentration of 5.10^{-5} M where the pCa was 2.5. Using Eq. (1) and a pCa value of 4.0, a ζ-potential of 14 mV can be obtained, which is quite similar to that experimentally found. Thus, the M-B molecule seems to be an indifferent substance for the ζ-potential of COM/water interface and will not change the e.d.l. structure (for concentrations lower than 5.10^{-5} M). Scurr and Robertson [4], using a light-scattering technique to measure the size of COM particles, have found that in the presence of several M-B concentrations (between 10^{-7} and 10^{-4} M) the particle size remained constant. Therefore, the presence of M-B did not modify the inhibition of the aggregation of COM crystals. Therefore, our results (see Fig. 5, curve 1) are in agreement with those obtained in [4] with a different technique.

Sodium dodecyl sulfate (SDS) is an ionic surfactant widely used in the preparation of emulsions and suspensions of different products. Therefore, it could be interesting to study its influence on the electrokinetic behavior of COM/water interface. Figure 5 (curve 2) shows the ζ-potential of COM particles versus SDS concentration. As SDS concentration increases the ζ-potential decreases and becomes negative for concentrations higher than 10^{-3} M. The pH and pCa values of the COM dispersions are shown in Table 3. The pH was almost constant, while the pCa increased. Skrtic et al. [17] have shown that the increasing pCa is a consequence of the precipitation of $Ca(DS)_2$, for a specific concentration of SDS. For SDS concentration lower than 10^{-3} M the pCa variation explains the decrease in the ζ-potential. But the pCa cannot explain the negative ζ-potential values. The specific adsorption of SDS seems to be the reason for the negative character of the microcrystal surface. This adsorption confers a negative and high ($\zeta \approx -50$ mV) surface charge which colloidally stabilizes the dispersion. Under these conditions the electrokinetic properties of COM particles are completely determined by the presence of SDS molecules. This result is similar to that found with other cationic system (polystyrene latex) and the same surfactant [18]: the electrokinetic behavior was completely controlled by the amount of adsorbed SDS molecules. Thus, the adsorption of SDS on the COM particle surface should be the reason for the negative ζ-potential values, although the COM surface could not be saturated with SDS molecules because a plateau was not found within the concentration range studied.

In conclusion, the presence of some substances in solution can modify in a significant way the e.d.l. structure of the water/COM particle interface. This modification can be explained via pCa changes in the dispersion or specific adsorption on the COM particle surface. Thus, urea, creatinine, D-glutamic

Table 3. pH and pCa values of COM dispersions versus sodium dodecyl sulfate (SDS) concentration

C(M)	pH	pCa
10^{-6}	5.2	4.2
10^{-5}	5.3	4.1
$5 \cdot 10^{-5}$	5.2	4.3
10^{-4}	5.1	4.4
$5 \cdot 10^{-4}$	5.2	5.0
10^{-3}	5.4	5.5
$5 \cdot 10^{-3}$	5.3	6.5
10^{-2}	5.8	7.0

and methylene blue slightly affected the ζ-potential of COM particles, via small pCa changes. However, D-tartaric, chondroitin sulfate, heparin and SDS strongly affected the e.d.l. of water/COM particle interface, via specific adsorption of those molecules. In these cases, the high ζ-potential, found at relatively high concentrations of these additives, indicates stable dispersions, from the colloidal point of view. Therefore, these molecules can be considered as inhibitors of the aggregation of COM particles.

Acknowledgement

This work was supported by Comisión Interministerial de Ciencia y Tecnología (CICYT) (projects numbers PA 86-0235 and MAT 90-0695-C02-01) and by Consejería de Educación y Ciencia de la Junta de Andalucía (Ayuda Consolidación de Grupos).

References

1. Callejas-Fernández J, de las Nieves FJ, Salcedo J, Hidalgo Alvarez R (1990) J Colloid Interface Sci 135:154—164
2. Norman RW, Scurr DS, Robertson WG, Peacok M (1984) Brit J Urol 56:594—598
3. Callejas-Fernández J, de las Nieves FJ, Martínez-García R, Hidalgo-Alvarez R (1991) Progr Colloid Polym Sci 84:327-333
4. Scurr DS, Robertson WG (1980) J Urol 136:128—131
5. Gjalbaeck JC (1982) Clin Chim Acta 120:363—365
6. Callejas-Fernández J, Martínez-García R, Hidalgo-Alvarez R, de las Nieves FJ (1992) Colloids Surfaces (in press)
7. Callejas-Fernández J, de las Nieves FJ, Martínez-García R, Hidalgo-Alvarez R (1992) J Surface Sci Tech (in press)
8. Callejas-Fernández J, de las Nieves FJ, Martínez-García R, Hidalgo-Alvarez R (1991) Colloids Surfaces 61:123—135
9. Robertson WG, Peacok M, Nordin BEC (1973) Clin Chim Acta 43:31—37
10. Coe FL (1983) Kidney Int 24:392—403
11. Azoury R, Randolph AD, Drach GW, Perlberg S, Garti N, Sarig S (1983) J Crystal Growth 64:389—392
12. Sur BK, Pandey HN, Deshpande S, Pahwa R, Sing RK, Taraclandra Y (1981) In: Smith L, Robertson WG, Finlayson (eds) "Urolithiasis: Clinical and Basic Research", Plenum Press, New York
13. Ryall RL, Harnett RH, Marschall VR (1981) Clin Chim Acta 112:349—356
14. Leal J, Finlayson B (1977) Invest Urology 14:278—283
15. Curreri PA, Onoda GY, Finlayson B (1987) In: Brash JL, Horbett TA (eds) "Protein at Interfaces. Physicochemical and Biochemical Studies", ACS Symp Ser 34:278—287, Washington DC
16. Rollins R, Finlayson B (1973) J Urology 110:459—463
17. Skrtic D, Filipovic-Vincekovic N (1988) J Crystal Growth 88:313-320
18. Galisteo-González F, Cabrerizo-Vilchez MA, Hidalgo-Alvarez R (1991) Colloid Polym Sci 269:406—411

Authors' address:

Dr. F. Javier de las Nieves
Biocolloids and Fluid Physics Group
Department of Applied Physics
University of Granada
18071 Granada, Spain

Electrokinetic properties of monodisperse colloidal dispersions of zinc sulfide

J. D. G. Durán, M. C. Guindo, and A. V. Delgado

Departamento de Física Aplicada, Facultad de Ciencias Universidad de Granada, Spain

Abstract: The electrophoretic behavior of monodisperse spherical zinc sulfide particles is studied in this work in order to contribute new data to the electrical characterization of the surface of this technologically important material. The electrophoretic mobility of ZnS suspensions is analyzed as a function of pH for different concentrations of inorganic electrolytes. The effect of NaCl, $NaNO_3$, KCl and KNO_3 is typical of indifferent electrolytes, the pH corresponding to the isoelectric point of the particles ($\sim$ 5.5) remaining the same for whatever concentration of the electrolytes. $CaCl_2$, on the contrary, seems to have some specific interaction with the ZnS surface. The nature of that interaction remains unclear. The cations Ag^+, Mn^{2+}, and La^{3+} can be considered as activating cations for the zinc sulfide/solution interface. Their hydrolysis products adsorb onto the ZnS particles and provoke up to three charge reversals of the latter when the pH is varied between 2 and 10. The effect is most important for Ag^+ and La^{3+}: when the electrophoretic mobility of colloidal zinc sulfide is measured as a function of the concentration of $AgNO_3$ and $La(NO_3)_3$ at pH close to 8, the charge of the particles is changed from negative to positive, even for initial concentrations as low as 2.5×10^{-4} M ($AgNO_3$) and 2.5×10^{-5} M ($La(NO_3)_3$). Lanthanum species give rise to the highest positive mobilities found for ZnS in this work.

Progress in Colloid & Polymer Science   Progr Colloid Polym Sci 93:216 (1993)

Equilibrium diagrams of the concentration of the different species appearing in solution or precipitating onto the particles are considered to explain the results.

Key words: Colloidal zinc sulfide — electrophoresis — activating ions — hydrolysis

phenomenon takes place provides information on the surface chemistry of metal sulfides. The effect of the adsorption of Cu(II) cation on ZnS has already been studied [9, 13]. In this work, after characterizing the surface of our zinc sulfide particles, the effect of the cations Ag^+, Mn^{2+}, and La^{3+} will be considered.

Introduction

There has recently been a renewed interest in the investigation of inorganic colloids, both from the fundamental and applied points of view. Thus, they show very versatile surface properties, their surface charges being controllable by addition of proper electrolytes [1]. The colloid chosen for this study is ZnS (sphalerite); this material has found many technical applications: it is a semiconductor that can also be used in the preparation of phosphors for cathode ray tubes, in water purification, in mineral flotation or in pigment manufacture [2—6].

A number of works have been published on the surface properties of zinc sulfide. Most of those works have used natural samples of mineral sphalerite, which explains the considerable variety of results obtained, since the mineral sources, the amount of impurities, etc., can be different for different authors [7—9]. Hence, some works have dealt with synthetic ZnS [6, 9, 10], but it was rather recently that this sulfide was first prepared and studied in the form of monodisperse colloidal spheres [11], as will be done in the present work. The interest of using such monodisperse, chemically pure, samples is clear: we will not face any problems concerning the homogeneity and purity of the material, and will be closest to the colloidal model assumed in most theories.

In this work, we will focus on the so-called "surface activation" resulting from adsorption of hydroxylated ions on spherical zinc sulfide particles. The interaction between such ions and the colloid will be characterized by electrophoresis: the study of the electrophoretic mobility as a function of pH in the presence of different activating metallic cations shows substantial modifications of the surface potential of zinc sulfide, as compared to its value when only indifferent electrolytes are present. Depending on the activating cation used, several charge reversal (CR) pH values can be found when the pH is varied at constant concentration of electrolyte [12]. The pH value(s) at which the CR

Experimental

The zinc sulfide particles were prepared following the method described by Wilhelmy and Matijević [11]. TEM micrographs showed that they were spherical and considerably monodisperse (average diameter 320 ± 20 nm). The x-ray powder diffractogram was found to correspond to pure sphalerite. The specific surface area of the solid obtained was 43.7 m^2/g as deduced from the BET method using a Quantasorb Jr apparatus (Quantachrome).

It must be mentioned that, when working with metal sulfides, the possible surface oxidation of the particles must be kept in mind, and for this reason many authors proposed to work, in as much as possible, under nitrogen atmosphere [8—10]. Our ZnS spheres were prepared in the presence of a (strongly oxidizing) nitric acid solution. We have not observed (even working in air) any subsequent modification in our samples, and some experiments were run in parallel in the presence of N_2 and in ambient conditions without finding significantly different results.

Chemicals used in the preparation of the particles and of the suspensions were supplied by Merck or Carlo Erba with analytical quality. They were used as received, without further purification. Only thioacetamide used in the synthesis of the particles was recrystalized from spectroscopic quality benzene.

For the preparation of the suspensions, water was twice distilled, deionized and filtered through 0.2 μm membranes (Milli-Q Reagent Water System, Millipore). The mother dispersions were kept in the dark in refrigerated polyethylene bottles. Their pH was 5—5.5.

The electrophoretic mobilities were measured by means of a Malvern Zeta-Sizer IIc apparatus at 25°C. The measurements were carried out 24 h after the preparation of the suspensions, the solids concentration of which was ~ 0.05 g/l. A relative error ≤5% can be ascribed to the mobility data presented in this work.

Results and discussion

Indifferent electrolytes

Figure 1 shows the variation of the electrophoretic mobility of ZnS with pH in three different conditions: 10^{-2} M constant concentrations of KCl and KNO_3, and no electrolyte added except the amounts of acid or base necessary to adjust the pH. The isoelectric point (i.e.p.) found was ~ 5.5 in all cases. Thus, no specific effects of either KCl or KNO_3 (and, in particular, of the anions Cl^- or NO_3^-) seems to exist on the surface of zinc sulfide. It must also be noted that, contrary to claims by others authors [10], very reproducible results were obtained when nitrate ion was present in the suspensions, so that any further oxidation of the surface of the colloidal particles can be neglected. Results (not shown here for brevity; see, however, Fig. 2) corresponding to experiments carried out in the presence of 10^{-2} M NaCl and $NaNO_3$ are essentially not distinguishable from those shown in Fig. 1. It can be said that electrolites such as NaCl, KCl, $NaNO_3$ or KNO_3 are indifferent for the ZnS/solution interface.

The pH corresponding to the i.e.p. found in this work is in good agreement with values given by other authors [7—9], both for mineral and synthetic sphalerite. However, several authors [6, 10, 11] report i.e.p. values close to 3 for the same compound. According to Healy et al. [8], the preequilibrium pH of the zinc sulfide particles considerabl affects their surface chemistry: at a pH close to 5, elemental sulfur (S^0) originally present on the particles (as a result of S^{2-} oxidation in the acidic medium) is gradually replaced by other surface phases such as ZnHSOH(s), $Zn_2(OH)_2SO_4$(s) and $Zn(OH)_2$(s). The shifts in the i.e.p. reported by different authors are probably related ro the extent to which such surface reactions have proceeded.

The pH dependence of the electrophoretic mobility when the background electrolyte is $CaCl_2$ is shown in Fig. 2. Data corresponding to 10^{-2} M NaCl are included for comparison. It can be seen that both at low and high pH values, the electrophoretic mobility values obtained with $CaCl_2$ in the medium are lower in absolute value than those corresponding to NaCl, as a consecuence of double layer compression. The most noticeable fact in Fig. 2 is the shift of the i.e.p. to a pH value of ~ 4.5, although no effect of $CaCl_2$ concentration on

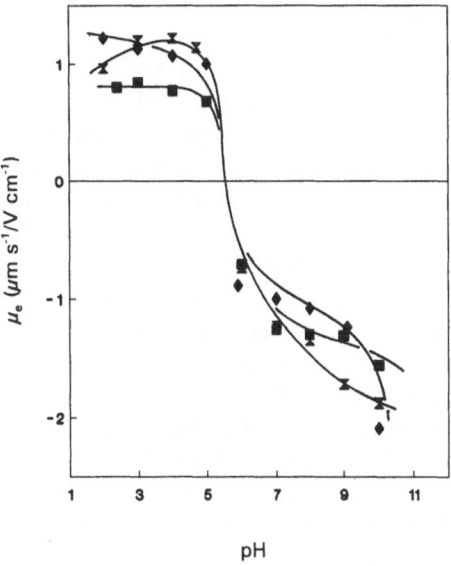

Fig. 1. Electrophoretic mobility of ZnS as a function pH in the presence of 10^{-2} M KCl (■), KNO_3 (◆), and with no salt added (⋎)

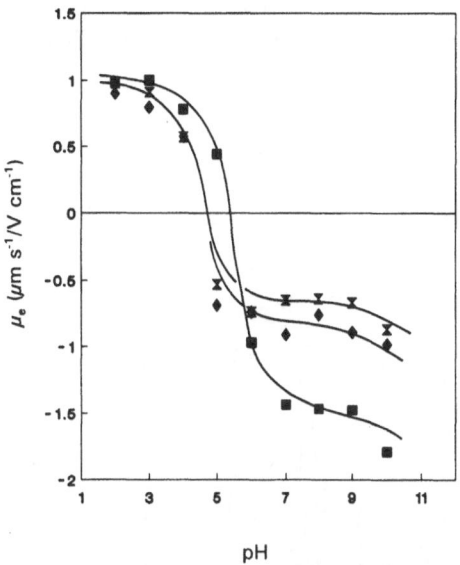

Fig. 2. Electrophoretic mobility of ZnS spheres as a function of pH in the presence of 10^{-2} M NaCl (■), 10^{-2} M $CaCl_2$ (⋎), and 10^{-3} M $CaCl_2$ + 10^{-2} M NaCl (◆)

the i.e.p. can be observed. This shift cannot be justified by experimental errors: note, for instance, how at pH = 5 electrophoretic mobility is positive when there is only NaCl in the dispersion medium, but changes to negative when $CaCl_2$ is added. We do not have an explanation for this unexpected phenomenon.

Activating electrolytes for ZnS

The activating effect of Cu^{2+} on ZnS has been already reported by other authors [9], who have shown that this cation may change the sign of the zeta potential of zinc sulfide particles. In this work we will evaluate to what extent other hydrolizable cations can have similar effects on this material. We will check the effect of the ionic charge by using Ag^+, Mn^{2+} and La^{3+} salts. The different solubilities of the sulfides of these cations will allow us to study different mechanisms of surface activation of synthetic monodisperse sphalerite.

Thus, Fig. 3 shows the electrophoretic mobilities of ZnS as a function of pH for three different concentrations of silver nitrate with 10^{-2} M KNO_3 as background electrolyte. The general trends observed in this figure follow closely those described by James and Healy [12] for SiO_2. We will follow their notation and denote CR_1, CR_2 and CR_3 the pH values corresponding to successive charge reversals of ZnS particles as the pH is increased.

Note in Fig. 3 how the CR_1 value is about 3 pH units below the i.e.p. found in Fig. 1. Such a pH shift could be justified on the basis on the following process [9, 13]:

$$2Ag^+ + (ZnS)^{Zn^{2+}} \rightleftharpoons (ZnS)^{2Ag^+} + Zn^{2+}$$

(equilibrium constant $K = 6.3 \times 10^{25}$, from pK_s data in [14]), which favors the surface substitution of Ag^+ for Zn^{2+}; the i.e.p. (CR_1) would correspond to that of $Ag_2S(s)$.

The most significant characteristic of Fig. 3 is the abrupt change that the electrophoretic mobility undergoes from negative to positive (CR_2) at a pH $\simeq 8$ for the highest concentrations of $AgNO_3$. This charge reversal has been explained by surface precipitation of $Ag_2O(s)$, which occurs at a pH value slightly lower than that corresponding to the bulk precipitation of silver oxide [12]. This can be confirmed if a calculation is carried out of the concentrations of the different species as function of pH, as shown in Fig. 4 (data from [14] were used). It can be observed that the increase in initial $AgNO_3$ concentration shifts the pH of bulk precipitation to lower values, in agreement with the mobility data plotted in Fig. 3.

When the pH is increased above CR_2, a third charge reversal is detected (CR_3) (Fig. 3). Apparently, the zinc sulfide particles are then completely covered by surface-precipitated silver oxide: the pH value corresponding to CR_3 would hence be the i.e.p. of Ag_2O, and this compound would dominate the electrokinetics of our colloidal system.

The effect of $[AgNO_3]$ changes on the electrophoretic mobility of monodisperse ZnS spheres is depicted in fig. 5 for pH 8, when the maximum variations in mobility are expected. When the concentration of initial silver nitrate is $\sim 2.5 \times 10^{-4}$ M the negative charge that the particles have at pH 8 is changed to positive: apparently, for such concentration the amount of surface Ag_2O is sufficient to alter the electrophoretic behavior of ZnS particles. Although these experiments were carried out in the presence of 10^{-2} M KNO_3, the mobility decreases slightly when $AgNO_3$ concentration is above 10^{-3} M as a consequence of double-layer compression by the added silver nitrate.

Figure 6 shows the mobility variation as a function of pH for three concentrations of $MnCl_2$ (10^{-2} M NaCl was also added to maintain an approximately constant ionic strength). Note that Mn^{2+} can be considered an activating cation for the ZnS/solution interface. However, contrary to findings with Ag^+ (Fig. 3), the CR_1 value changes very little with respect to the i.e.p. found in absence of any activating cation. The reason may be that the reaction:

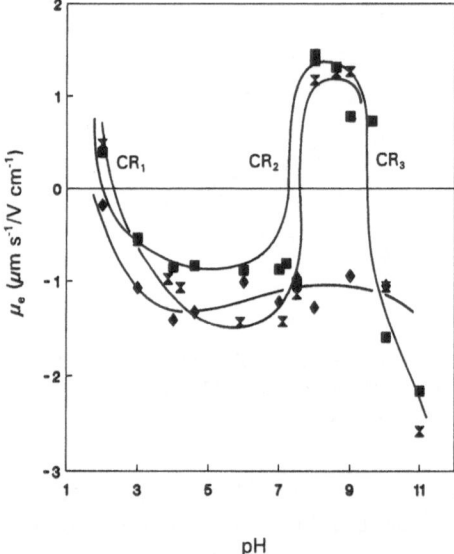

Fig. 3. Electrophoretic mobility of ZnS vs. pH in the presence of 10^{-2} M KNO_3, for the following molar concentrations of $AgNO_3$: 10^{-4} (◆), 5×10^{-4} (✗), and 10^{-3} (■)

(a)

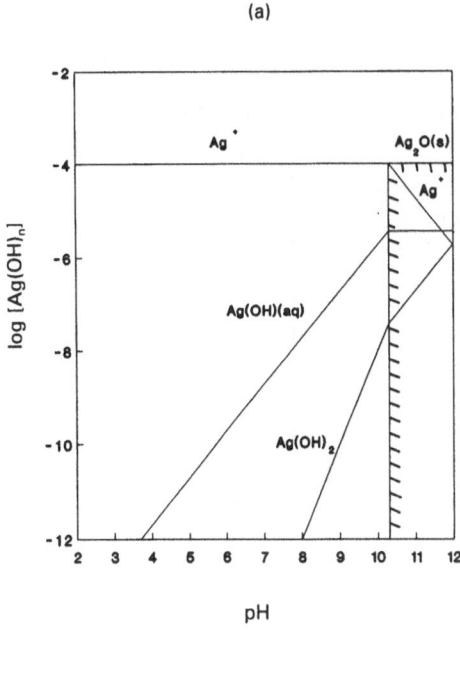

(b)

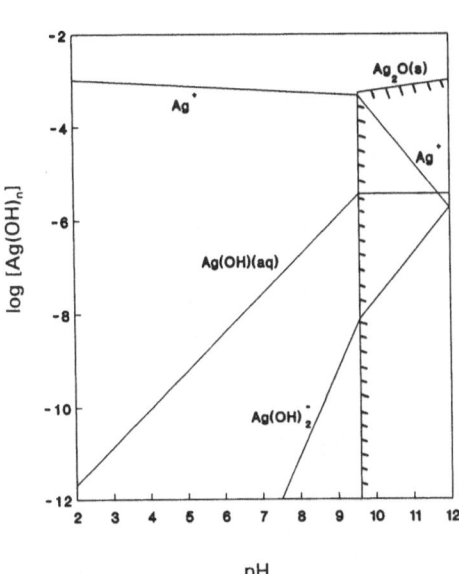

Fig. 4. Equilibrium diagrams for the concentration of $Ag(OH)_n$ complexes as a function of pH. Initial $[AgNO_3]$: 10^{-4} M (a); 10^{-3} M (b)

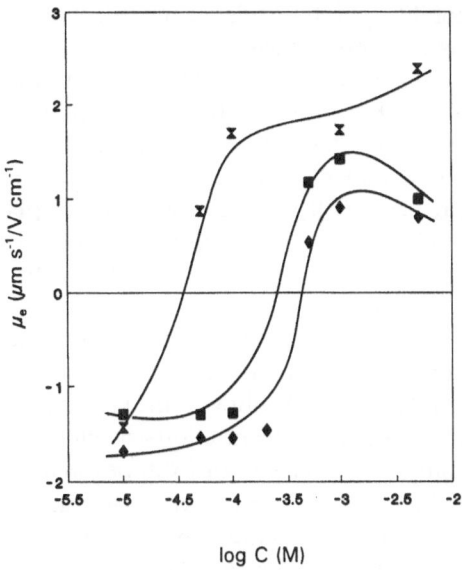

Fig. 5. Electrophoretic mobility of ZnS as a function of the initial concentration of activating salt at a constant ionic strength of 10^{-2} M. (■): $AgNO_3$, pH 8; (◆): $MnCl_2$, pH 8.5; (✗): $La(NO_3)_3$, pH 8

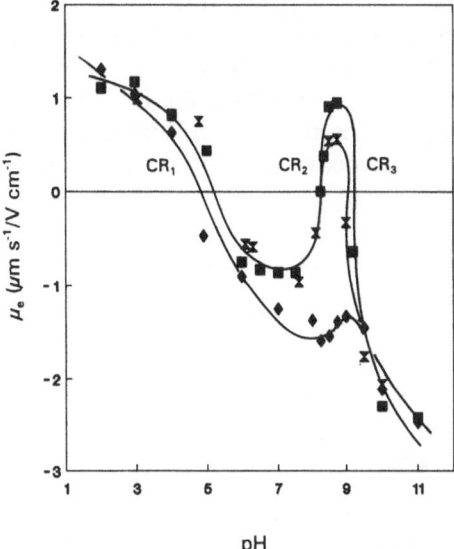

Fig. 6. Electrophoretic mobility of ZnS vs. pH in the presence of 10^{-2} M NaCl, for the following molar concentrations of $MnCl_2$: 10^{-4} (◆), 5×10^{-4} (✗), and 10^{-3} (■)

$$Mn^{2+} + (ZnS)^{Zn^{2+}} \rightleftharpoons (ZnS)^{Mn^{2+}} + Zn^{2+}$$

is not thermodynamically favored, since its equilibrium pK is -11.2 (pK_s data from [14]). Hence, the mobility curve remains unaltered (as compared to that found in the absence of Mn^{2+}) up to the pH at

which the formation of $Mn(OH)_2(s)$ takes place on the particles. The electrophoretic mobility thus changes from negative to positive (CR_2) if the concentration of $MnCl_2$ in the dispersion medium is high enough ($\geq 5 \times 10^{-4}$ M; see Fig. 5). For lower concentrations, only a slight maximum in the (still negative) mobility-pH trend is observed. Equilibrium diagrams of the complex Mn species in solution, not shown in this work, demonstrate that CR_2 takes place at a pH slightly lower than that corresponding to the bulk precipitation of $Mn(OH)_2$. When that precipitation takes places, a third charge reversal (CR_3) is again observed at pH $\simeq$ 9. The precipitate was even observable by the naked eye as a brown solid appearing in the suspensions.

The effect of the concentration of manganese chloride on the mobility at constant pH (8.5) and ionic strength (10^{-2} M) is shown in Fig. 5. As observed, the $MnCl_2$ concentration needed to change the charge of ZnS particles from negative to positive is slightly higher than that of silver nitrate, this confirming that Ag^+ cations are stronger activating agents than Mn^{2+} for the ZnS surface.

In Fig. 7, we have plotted the variation of electrophoretic mobility with pH for two initial concentrations of the last cation studied, La^{3+}, in the presence of 10^{-2} M KNO_3 as background electrolyte. For the highest concentration of $La(NO_3)_3$ investigated (10^{-3} M), the pH corresponding to CR_1 is again lower than the i.e.p. found in the absence of any activating cation. The effect is not, however, as pronounced as it was when $AgNO_3$ was the electrolyte (Fig. 3). The surface substitution of Zn^{2+} by La^{3+} is not thermodynamically favored (pK =

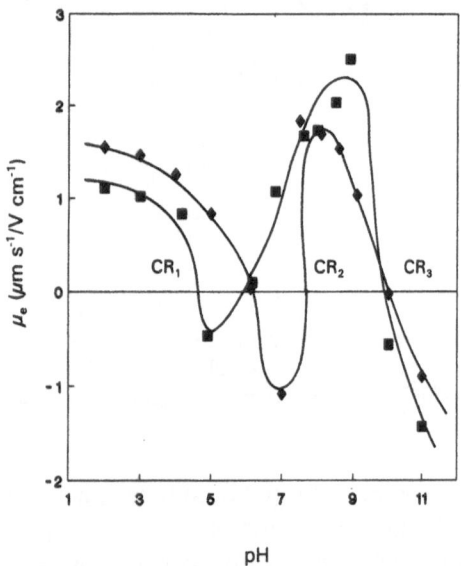

Fig. 7. Electrophoretic mobility of ZnS as a function of pH in the presence of 10^{-2} M KNO_3, for the following molar concentrations of $La(NO_3)$: 10^{-4} (♦), and 10^{-3} (■)

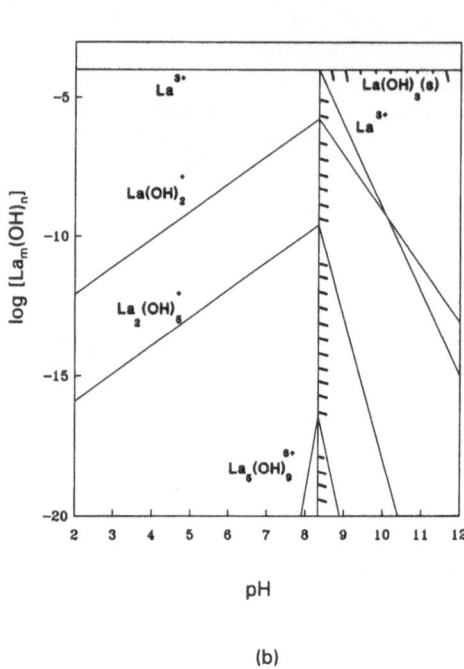

(a)

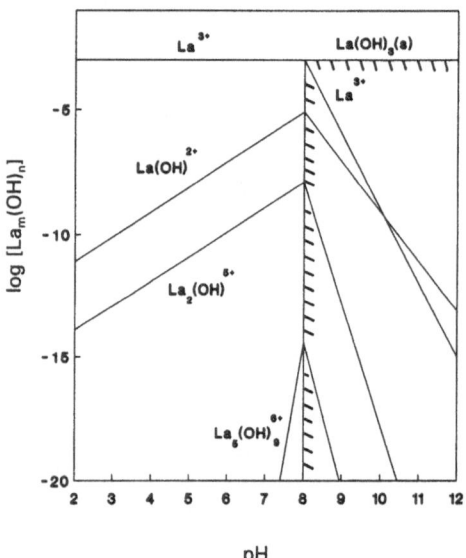

(b)

Fig. 8. Equilibrium diagrams for the concentration of $La_m(OH)_n$ complexes as a function of pH. Initial [$La(NO_3)_3$]: 10^{-4} M (a); 10^{-3} M (b)

12.6), (pK_s data from [14, 15]). The mobility maximum and the third charge reversal pH is again found in the presence of $La(NO_3)_3$; the positive mobility attained by the particles is the highest in the case. The existence of highly charged species, demonstrated by the diagrams shown in Fig. 8 (data from [16, 17]) can account for this fact. The analysis of Figs. 7 and 8 also shows that CR_2 occurs at a pH somewhat lower than that corresponding to bulk precipitation of $La(OH)_3$, a result which is common to all the activating cations studied.

The behavior of electrophoretic mobility vs. initial $[La(NO_3)_3]$ at pH 8 (and a ionic strength of 10^{-2} M) is plotted in Fig. 5. Note how the activating effect of the lanthanum species appears at much lower concentrations than those of Ag^+ and Mn^{2+}. It is clear that the cation charge is, together with the solubility of the cation sulfide and hydroxide, an essential factor in the surface activation of zinc sulfide.

Acknowlededgement

Financial support from DGICYT, Spain, under Project No. PB89—0461, and Fundación Ramón Areces, Spain, is gratefully acknowledged. The x-ray analysis of the samples was kindly carried out by Dr. D. Martín, Department of Mineralogy and Petrology, University of Granada.

References

1. Johnson JE, Matijević E (1992) Colloid Polym Sci 270:353—363
2. Schlam E (1973) Proc IEEE 61:894—901
3. Henglein A (1984) Pure & Appl Chem 56:1215—1224
4. Healy TW, Moignard MJ (1976) In: Fuerstenau MC (ed) Flotation. AIME, New York, vol. 1
5. Park SW, Huang CP (1987) J Colloid Interface Sci 117:431—441
6. Williams R, Labib ME (1985) J Colloid Interface Sci 106:251—254
7. Fuerstenau MC, Clifford KL, Kuhn MC (1974) Int J Min Proc 1:307—318
8. Moignard MJ, Dixon DR, Healy TW (1977) Proc Australas Inst Min Metall 263:31—38
9. Pugh RJ, Tjus K (1987) J Colloid Interface Sci 117:231—241
10. Nicolau YF, Menrad JC (1992) J Colloid Interface Sci 148:551—570
11. Wilhelmy DM, Matijević E (1984) J Chem Soc Faraday Trans 1 80:563—570
12. James RO, Healy TW (1972) J Colloid Interface Sci 40:53—64
13. Jain S, Fuerstenau DW (1985) In: Forsberg KSE (ed) Sulfide Minerals. Elsevier, Amsterdam/New York
14. Butler JN (1964) Ionic equilibrium. A mathematical approach. Addison-Wesley, London
15. Waggoner WH (1985) J Chem Ed 35:339—342
16. Ringbom A (1975) Formación de complejos en Química Analítica. Alhambra, Madrid
17. Kolthoff IM (1976) Análisis químico cuantitativo. Nigar, Buenos Aires

Authors' address:

Dr. Angel V. Delgado
Departamento de Física Aplicada
Facultad de Ciencias
Universidad de Granada
18071 Granada, Spain

X-Ray diffraction analysis of mixed colloids

R. Despotović, B. Fraj, and D. Mioč

Department of Colloid Chemistry,
Ruder Bošković Institute, Zagreb, Croatia

Abstract: The main aim of this study was to find out to what extent the concentration of cationic n-dodecylamine nitrate DAN and/or anionic sodium n-dodecyl sulphate SDS surfactants influence the ratio between hexagonal H and cubic C type of negative colloid AgI particles prepared by dilution of AgI complex in NaI solution by pouring into DAN and/or SDS solutions of different concentrations. Using x-ray diffraction data it has been shown that the cubic modification of negative colloid silver iodide sol particles are predominantly formed with $C = 89 \pm 2\%$. Concluded was that SDS could be effective only as a neutral stabilization ion. Meanwhile, in an aqueous solution containing cationic surfactant DAN the part of cubic AgI modification decreases with DAN concentrations increasing: 87% C at 0.00001 mol DAN/dm^3, 17% C at 0.0001 mol DAN/dm^3, and 15% C at 0.001 mol DAN/dm^3. The x-ray diffraction data direct us to very sensitive mutual interactions between primary AgI particles and different DAN associate structure present.

Key words: Colloids — x-ray diffraction analysis — surfactant

Introduction

Silver iodide sol in an aqueous suspension is polymorphic and the fractions cf cubic (sphalerite type) and hexagonal (wurtzite type) modifications of AgI vary with experimental conditions [1]. As a rule in an inorganic sol + surfactant system, the surfactant dictates the colloid properties of the whole system [2]: the main aim of this study represents an attempt to find out to what extent the concentration of cationic and/or anionic dodecyl derivatives influence the ratio between hexagonal and cubic type of negative colloid AgI sol.

Experimental

5.87 g of dried AgI were placed into a 25.0 ml measuring flask. A fresh solution of 4.00 mol NaI/dm^3 was added and when the AgI was completely dissolved in the added NaI solution, NaI solution up to 25.0 ml was added into the flask. The sols were prepared by adding of 20.00 µl complex solution into 8.00 ml of surfactant solution: 0.00001, 0.0001 and/or 0.001 mol n-dodecylamine nitrate (DAN) or 0.00001, 0.0001, 0.001, and/or 0.01 mol sodium n-dodecyl sulphate $(SDS)/dm^3$. The hexagonal diffraction line (100) at the Bragg angle $\Theta = 11.2°$, and overlapping diffraction lines (111) and (002) at $\Theta = 11.9°$ of the cubic C and hexagonal H modification were recorded. Sometimes we also used hexagonal line (101) at $\Theta = 12.7°$.

Results and Discussion

The employed surfactant affects the crystallographic and colloid properties of AgI in various ways. The part of cubic AgI is approximately the same at varied amounts SDS present in systems and it is approx. 89 ± 2%. Being a negative ion, SDS cannot cause the coulombic coagulation of the negative AgI sol: so that SDS could be effective in the silver iodide suspensions only as a neutral stabilization ion. The suspensions prepared in the supernatant liquor containing cationic surfactant DAN at very low surfactant concentrations (0.0000001 or 0.00001 mol DAN/dm^3) DAN does not affect the crystallographic modification of AgI, contrary to the results at the higher DAN concentration: 87% C at 0.00001 mol DAN/dm^3, 17% C at 0.0001 mol DAN/dm^3, and 15% C at 0.001 mol DAN/dm^3, indicating that the DAN associates grow and appear as a different colloid specia affecting on different way the formation of AgI solid phase. Each surfactant composition has a different surface tension at the crystal/solution interface, causing different properties of the interface layer, e.g., the crystallographic plane characteristics in good agreement with Težak's concepts of interrelation between the ions in the liquid phase and the chemical and physical properties of solid phase [3]. Conclusively, the x-ray diffraction data direct us to very sensitive mutual interactions between the present AgI particles and DAN associate structure [4].

References

1. Despotović R (1992) Jorn Com Esp Deterg 23:295—306
2. Despotović, Filipović Vinceković N, Subotić B (1976) In: Kerker M (ed) Colloid and Interface Science, Academic Press Inc, New York, San Francisco, London 4:297—313
3. Težak B (1966) Discussion Faraday Soc 42:175—186. Težak B (1968) Croat Chem Acta 40:63—78
4. Despotović R (1991) Progr Colloid Polym Sci 84:66—68

Authors' address:

Prof. Dr. Radoslav Despotović
Ruder Bošković Institute
P.O. Box 1016
Zagreb, Croatia

Structural changes of citrate synthase upon ligand binding and upon denaturation

H. Durchschlag[1]), G. Purr[1]), R. Jaenicke[1]), and P. Zipper[2])

[1]) Institute of Biophysics and Physical Biochemistry, University of Regensburg, FRG
[2]) Institute of Physical Chemistry, University of Graz, Austria

Abstract: Conformational changes of citrate synthase upon ligand binding and upon denaturation were monitored in solution by spectroscopic techniques, analytical ultracentrifugation and small angle x-ray scattering. Mafor contributions to structural alternations and to the stabilization against denaturation are attributable to the substrate oxaloacetate

Key words: Citrate synthase — ligand binding — oxaloacetate complex — denaturation — structural changes

Citrate synthase, the primary pacemaker enzyme of the tricarboxylic acid cycle, catalyzes the condensation of oxaloacetate with acetyl-coenzyme A to form citrate and coenzyme A. As taken from x-ray crystallography, there are drastic differences between open and substrate-induced closed forms of the enzyme [1, 2]. In solution, it is difficult to detect the corresponding structural changes. Using a series of solution techniques and evaluation procedures [3, 4], we succeeded in quantifying significant changes of molecular parameters, in close agreement with the crystal data. The effects differed markedly for different ligands (substrates, products, and/or analogues), with oxaloacetate representing the main trigger for the conformational change. UV absorption, fluorescence, and circular dichroism spectroscopy were found to exhibit a series of spectral changes, due to alterations in the neighborhood of intrinsic chromophores and fluorophores (Trp and/or Tyr); analytical ultracentrifugation and small-angle x-ray scattering allowed to monitor significant changes of the overall structure, reflected by changes of the sedimentation coefficient, distance distribution function $p(r)$, and radius of gyration (Table 1).

Denaturation of the enzyme was followed by spectroscopic and ultracentrifugal analyses. While acid denaturation indicated only incomplete unfolding ($\Delta s_{20,w}$ about —40% at pH 2), denaturation

Table 1. Structural changes of citrate synthase upon oxaloacetate binding, as revealed by different techniques

Characteristics	Results
λ_{max} of UV absorption	—(3.0 ± 0.5) nm
λ_{max} of fluorescence emission	+(1.7 ± 0.5) nm
λ_{max} of fluorescence excitation	+(7.3 ± 0.5) nm
Near-UV CD	increase of θ_{260}
Far-UV CD	no change
Sedimentation coefficient $s_{20,w}$	+(1.0 ± 0.5) %
Radius of gyration	—(3.8 ± 0.7) %
Position of maximum of $p(r)$ function	—(2.9 ± 0.3) %

by 8 M urea or by 6 M guanidinium chloride was more effective ($\Delta s_{20,w}$ about —70%). The process of unfolding was monitored by registration of transition curves created at different concentrations of the denaturing agents used. Oxaloacetate was found to stabilize the enzyme to some extent against denaturation, as follows from a pronounced shift of the transition point towards elevated denaturant concentration (2.5 ⇒ 3.0 M guanidinium chloride).

References

1. Remington S, Wiegand G, Huber R (1982) J Mol Biol 158:111—152
2. Wiegand G, Remington S, Deisenhofer J, Huber R (1984) J Mol Biol 174:205—219
3. Purr G (1989) Diplomarbeit, University of Regensburg, FRG
4. Durchschlag H, Zipper P, Wilfing R, Purr G (1991) J Appl Cryst 24:822—831

Authors' address:

Dr. Helmut Durchschlag
Institut für Biophysik und Physikalische Biochemie
Universität Regensburg
Universitätsstraße 31
93040 Regensburg, FRG

Radiation damage and modification of radiation action of x-rays and UV-light on enzymes

H. Durchschlag[1], B. Feser[1], C. Fochler[1], T. Seroneit[1], E. Swoboda[1], C. Wlček[1], and P. Zipper[2]

[1] Institute of Biophysics and Physical Biochemistry, University of Regensburg, FRG
[2] Institute of Physical Chemistry, University of Graz, Austria

Abstract: Several enzymes have been irradiated with x-rays and UV-light in aqueous solution. Various aspects of radiation damage have been studied systematically by a series of solution techniques. The radiation action may be modified effectively by many protectants.

Key words: Enzymes — radiation damage — radioprotective substances — ultraviolett radiation — x-radiation

The radiation damage (x-ray, UV-C) of various enzymes has been investigated in aqueous solution. Enzymes with essential sulfhydryls (glyceraldehyde-3-phosphate dehydrogenase, malate synthase), sulfhydryls (citrate synthase, lactate dehydrogenase), disulfides (lysozyme, ribonuclease A), and the antioxienzyme catalase have been the subject of intense studies.

A series of biochemical and physico-chemical techniques has been applied (e.g., [1—4]), in order to differentiate between different radiation effects. Both functional and structural changes of several molecular parameters were monitored; the effects still intensified in the post-irradiation period. Enzymic tests registered changes of enzyme activity; UV absorption, fluorescence and near- and far-UV circular dichroism spectroscopy revealed statements on the integrity of aromatic amino acids, on enzyme aggregation and helix content; electrophoretic analyses monitored fragmentation, unfolding and cross-linking of subunits; densimetry indicated volume changes; analytical ultracentrifugation, small-angle x-ray scattering, light scattering, size exclusion chromatography and viscometry yielded detailed statements on changes of the overall structure and the progress of enzyme aggregation. Determinations of SH and SS groups rendered changes of cysteins and cystins.

A modification of the radiation action was achieved by various measures and additives. A considerable influence was found by variation of dose, enzyme concentration, and gassing conditions. A plethora of additives has been tested, looking for their action as protectants or their repair ability. The additives screened included many antioxidants, reductants, antioxienzymes, scavengers, thiols, specific ligands, salts, buffer components, metal ions, chelating agents. For example, many thiols, dithionite, formate, mannitol, EDTA, NADH, ascorbate, and substrates turned out to be very potent protectives against primary changes induced by x-rays; catalase suppressed p.r. inactivation effectively. Addition of superoxide dismutase may protect enzymes against UV-irradiation. Results were summarized in dose-effect curves, stability and correlation plots. Normalization of parameters allowed a quantitative comparison of the protective efficiency of the additives. The findings indicate that with x-rays primary changes are caused mainly by $\cdot OH$, secondary damages by H_2O_2; in the case of UV-C light production of O_2^- may be responsible for primary changes. The results are of interest for many fields of research using x-rays and UV-light.

References

1. Durchschlag H, Wlček C, Zipper P, Wilfing R (1991) Proc 5th Working Meeting on Radiation Interaction (Mai H, Brede O, Mehnert R, eds). ZfI, Leipzig, pp 123—130
2. Zipper P, Durchschlag H (1991) Proc 7th Tihany Symposium on Radiation Chemistry (Dobó J, Nyikos L, Schiller R, eds). Hungarian Chemical Society, Budapest, pp 437—447
3. Durchschlag H, Zipper P (1991) Proc 7th Tihany Symposium on Radiation Chemistry (Dobó J, Nyikos L, Schiller R, eds). Hungarian Chemical Society, Budapest, pp 449—458
4. Durchschlag H, Zipper P (1991) In: Anticarcinogenesis and Radiation Protection 2 (Nygaard OF, Upton AC, eds). Plenum Press, New York, pp 269—274

Authors' address:

Dr. Helmut Durchschlag
Institut für Biophysik und Physikalische Biochemie
Universität Regensburg
Universitätsstraße 31
93040 Regensburg, FRG

Murphy's law of particle sizing: comparison of the size distributions of lecithin liposomes obtained by dynamic light scattering and electron microscopy

S. Egelhaaf[1]), P. Schurtenberger[2]), E. Wehrli[1]), M. Adrian[3]), and P. L. Luisi[2])

[1]) Laboratorium für Elektronenmikroskopie, Institut für Zellbiologie, ETH Zürich, Switzerland
[2]) Institut für Polymere, ETH Zürich, Switzerland
[3]) Laboratoire LAU, Université de Lausanne, Dorigny, Switzerland

Abstract: Frequently, a mean radius is used to characterize colloidal systems. As most real systems possess a broad size distribution, such a mean size will hardly be sufficient to describe the studied system, and more detailed information on the size distribution is desired. Because different techniques imply a different weighting of the individual sizes, the obtained size distributions depend on the method used. On the other hand, a number or mass distribution is the most desirable result for many experiments, and therefore the measured distribution must be transformed. Although such a transformation is often possible, the limitations of the measuring technique and, hence, of the transformed distribution must be considered. — Lecithin lipsomes were used as model colloids. Their size distributions were investigated with dynamic light scattering and both classical freeze-fracture electron microscopy and cryo transmission electron microscopy. Algorithms were developed in order to convert the thus obtained size distributions into "true" number distributions.

Key words: Size distribution — liposome — light scattering — cryo electron microscopy — freeze-fracture electron microscopy

Frequently, a mean radius is used to characterize colloidal systems. As most real systems possess a broad size distribution, such a mean size will hardly be sufficient to describe the studied system, and more detailed information on the size distribution is desired. Because different techniques imply a different weighting of the individual sizes, the obtained size distributions depend on the method used. On the other hand, a number or mass distribution is the most desirable result for many experiments and, therefore, the measured distribution must be transformed. Although such a transformation is often possible, the limitations of the measuring technique and, hence, of the transformed distribution must be considered.

Lecithin liposomes were used as model colloids. Their size distributions were investigated with dynamic light scattering and both classical freeze-fracture electron microscopy and cryo transmission electron microscopy. Algorithms were developed and tested in order to convert freeze-fracture weighted size distributions into "true" number weighted size distributions by taking into account capping effects and cutting probability. Furthermore, they permit a reliable and numerically stable conversion of dynamic light scattering data into "true" number weighted size distributions. Our results show that the thus obtained size distributions quantitatively agree with each other and they demonstrate the weaknesses and the strengths of each technique. We can now aim for a self-consistent description of the number, mass or intensity weighted size distribution of a colloidal suspension using a combination of complementary experimental techniques.

Authors' address:

PD Dr. Peter Schurtenberger
Institut für Polymere, ETH Zürich
Universitätsstrasse 6
8092 Zürich, Switzerland

Size distributions of equilibrated surfactant vesicles

J. C. Eriksson, M Bergström, and S. Ljunggren

Department of Physical Chemistry, Royal Institute of Technology, Stockholm, Sweden

Abstract: Our model calculations show that the formation of single-shell, spherical surfactant vesicles which are in full equilibrium with the surrounding solution is geared primarily by the residual bilayer tension ($10^{-6} - 10^{-8}$ Nm^{-1}) and a statistical factor accounting for composition and chain packing density fluctuations and, moreover (which is in line with what Helfrich [4] has derived earlier), that the (size-independent) bilayer bending properties do not influence the size distributions appreciably. The relative width of a vesicle size distribution, $\sigma_R / < R >$ is typically found to be about 0.35—0.5 which agrees with recent experimental observations due to Kaler et al. [1].

Key words: Vesicles, formation from surfactant solution — vesicles, calculation of size distributions — bilayer vesicles, equilibrium formation of — surfactant vesicles and curvature elasticity — aggregation of surfactant monomers to vesicles

Background

Hitherto, it has commonly been assumed that, by and large, the bilayer bending properties are crucial for the formation (and shape) of vesicles [2—5]. However, our recent model calculations for SDS/dodecanol(DOH) surfactant vesicles carried out at constant chemical potentials indicate that the bending properties are, in fact, only of a secondary importance. Instead, the residual bilayer tension ($10^{-6} - 10^{-8}$ Nm^{-1}) and a statistical factor related to the fluctuations in chain packing density, composition and shape play major roles. Here, we restrict the treatment to spherical vesicles which are formed in a dynamic aggregation equilibrium with the surrounding surfactant solution. In analogy with the case of droplet microemulsions [6], the influence of shape fluctuations on the vesicle size distribution is likely to be minor. This question will be considered in detail, however, in a forthcoming paper.

Theory based on curvature-dependent monolayer tensions

For each of the two monolayers composing the closed vesicle bilayer we may make the following (extended) Helfrich ansatz

$$\gamma = \gamma_0 + 2k_c (H - H_0)^2 + \bar{k}_c K + k_c' (H - H_0)^3$$

$$+ \bar{k}_c' (H - H_0) K , \qquad (1)$$

where H and K denote the mean and Gaussian curvatures, respectively, of the water/hydrocarbon dividing surface (located somewhat underneath the polar head groups), and where H_0 is the spontaneous curvature and k_c, $\bar{k}_c$, k_c', $\bar{k}_c'$ denote bending constants related to the second-and third-order terms in curvature. For a spherical bilayer of thickness 2ζ this yields for the bilayer tension γ_b:

$$\gamma_b/2 = \gamma_\infty + (k_1 \zeta + k_2)/R^2 - k_3 \zeta/R^4 \qquad (2)$$

and, hence,

$$\gamma_b R^2/2 = \gamma_\infty R^2 + (k_1 \zeta + k_2) - k_3 \zeta/R^2 , \qquad (3)$$

where R is the radius of the middle surface, and where

$$\gamma_\infty = \gamma_0 + 2k_c H_0^2 - k_c' H_0^3 \qquad (4)$$

$$k_1 = -4k_c H_0 + 3k_c' H_0^2 \qquad (5)$$

$$k_2 = 2k_c + \bar{k}_c - 3k_c' H_0 - \bar{k}_c' H_0 \qquad (6)$$

$$k_3 = k_c' + \bar{k}_c' . \qquad (7)$$

Our model calculations, to be discussed below, indicate that the chain packing constraints together with the curvature-dependence of the hydrocarbon/water contact free energy are chiefly responsible for obtaining $k_3 < 0$ and, hence, a resulting vesicle free energy minimum for a particular radius $R = (-k_3 \zeta/\gamma_\infty)^{0.25}$ in accordance with Eq. (3). The corresponding bilayer tension is an even fourth-order function of the curvature, having a minimum equal to $2\gamma_\infty$ for $1/R = 0$. It increases more rapidly with curvature than the quadratic function corresponding to the usual Helfrich expression with $k_c' \; \bar{k}_c' = 0$.

Fluctuations about the equilibrium DS/DOH composition occur for the two monolayers in approximately parabolic free energy wells which, however, flatten out with the vesicle size similarly as we have investigated earlier for rodshaped micelles [7]. In particular, by assuming that the free energy per chain is proportional to the square of $\Delta x = x - x_{eq}$ (where χ is the monolayer mole fraction

of DOH), one can define a "force constant", C_x (in kT units) which accounts for how easily additional composition states can be reached by fluctuations. In a corresponding manner one can introduce an analogue constant C_a related to the fluctuations in chain packing density. By integrating over the various compositional states, one finds that the vesicle size distribution, $\phi(R)$, where ϕ stands for volume fraction, can be written

$$\phi(R) = \frac{64\pi^5 R^6}{\bar{a}^2 C_x C_a V_w} \times \exp[-4\pi R^2 \gamma_b/kT] \qquad (8)$$

for mixed vesicles, and

$$\phi(R) = \frac{16\pi^3 R^4}{\bar{a} C_a V_w} \times \exp[-4\pi R^2 \gamma_b/kT] \qquad (9)$$

for pure vesicles, where $\bar{a}$ is the average area per hydrocarbon chain V_w. We note that since C_x is of the order of 4 and C_a of the order of 10, the statistical (pre-exponential) factor may become very large for large vesicles.

Model calculations

We have carried out model calculations for two matching, SDS/DOH monolayers of spherical shape which together compose a bilayer vesicle. These calculations have been directed towards quantifying the overall excess free energy of a vesicle, ε, in the equilibrium case when the inner (i) as well as the outer monolayer (e) are in dynamic equilibrium with the surrounding surfactant solution, as a function of the vesicle radius R and for all possible compositional states. The various free energy contributions included are:

i) the Tanford hydrocarbon chain/water (hydrophobic) excess free energy;
ii) the electrostatic free energy as obtained from Poisson-Boltzmann electrostatics to second order in curvature;
iii) a curvature-dependent (Gruen) hydrocarbon chain configurational free energy;
iv) a curvature-dependent hydrocarbon/water contact free energy;
v) an ideal mixing free energy contribution (in the SDS/DOH Cease);
iv) a constant head group term.

The size distribution, $\phi(R)$, is obtained by means of the partition-function-like expression

$$\phi(R) = \sum e^{-\varepsilon \left(N_w,\, N^i_{DS'},\, N^i_{OH'},\, N^e_{DS'},\, N^e_{OH'}\right)/kT} , \qquad (10)$$

where the sum extends over all the different compositional state at a fixed radius R which arise due to fluctuations.

Vesicle size distributions

From the present work, it follows that the vesicle size distribution reflects a competition between a rapidly growing statistical factor and the diminishing exponential factor $\exp[-4\pi R^2 \gamma_b/kT]$. Except for very small vesicles, $\gamma_b R^2$ is simply given by (cf. Eq. (3)) $\gamma_b R^2 = 2\gamma_\infty$ (monolayer) $\times R^2 + 2\,(k_1\xi + k_2)$ and, accordingly, the bending constants of the monolayer (i.e., $\bar{k}_c$, $\bar{k}_c$, $\bar{k}'_c$, $\bar{k}'_c$) do not affect the shape of the distribution curves but only the overall volume fraction.

Referring to the pure case and by invoking vesicle shape fluctuations (that we have neglected so far), Helfrich [4] earlier derived a vesicle size distribution of similar mathematical form as Eq. (9), $\sim R^3 \times \exp[-4\pi R^2 \gamma/kt]$. These shape fluctuations were found to yield a pre-exponential factor that is proportional to the number of amphiphile molecules in the bilayers. In our theoretical approach, however, where a vesicle is considered basically as a chemical

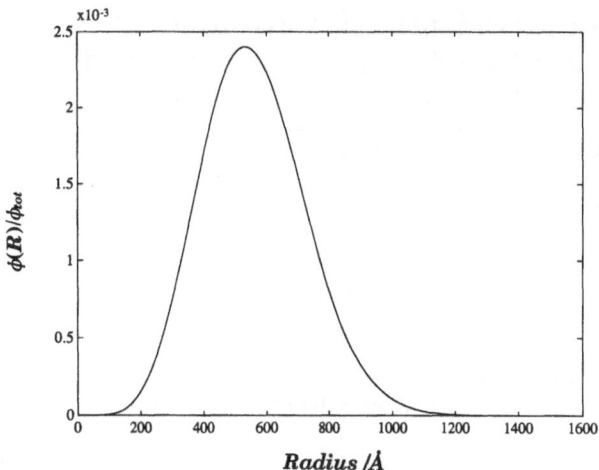

Fig. 2. Normalized vesicle size distribution for the mixed SDS/DOH case obtained at the following solution state. Added salt, $c_s = 1.0$ M, SDS monomer concentration, $c_2 = 0.1965$ mM. The monolayer tension γ_∞ was equal to 1.44×10^{-7} N/m

complex, the pre-exponential factor has quite a different, purely statistical origin.

Examples of vesicle size distribution that we have generated are shown in Figs. 1 (pure SDS vesicles) and 2 (mixed SDS/DOH vesicles). They are rather broad with relative standard deviations in the range 0.35—0.5. Furthermore, there is a certain skewness due to the circumstance that the statistical factor favors the formation of large vesicles. These features are all in line with recent experimental findings of Kaler et al. [1].

References

1. Kaler EW, Herrington KL, Murthy AK, Zasadzinski JAN (1992) J Phys Chem 96:6698—6707
2. Safran SA, Pincus P, Andelman D (1990) Science 248:354
3. Safran SA, Pincus P, Andelman D, Mackintosh FC, (1991) Phys Rev A 43:1071
4. Helfrich W, (1986) J Physique 47:321—329
5. Ou-Yang Z, Helfrich W, (1989) Phys Rev A 39:5280
6. Eriksson JC, Ljunggren S (1990) Prog Colloid Polym Sci 81:41—53
7. Bergström M, Eriksson JC (1991) Langmuir 8:36—42

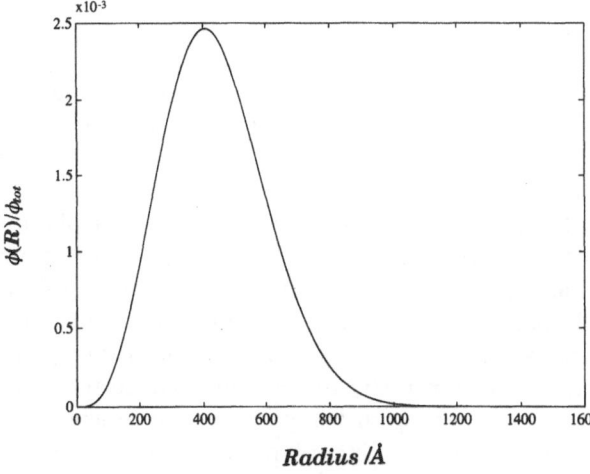

Fig. 1. Normalized vesicle size distribution for the pure SDS case generated at the following solution state. Added salt, $c_s = 0.9$ M, SDS monomer concentration, $c_2 = 0.3945$ mM. The monolayer tension γ_∞ was equal to 1.48×10^{-7} N/m

Authors' address:

Professor Jan Christer Eriksson
Department of Physical Chemistry
Royal Institute of Technology
100 44 Stockholm, Sweden

Reversed micellar approach as a new tool for the formation and structural studies of protein Langmuir-Blodgett films

V. Erokhin, S. Vakula, and C. Nicolini

Consorzio Technobiochip, Institute of Biophysics, University of Genova, Italy

Key words: LB films — reversed micelles — cytochrome c — circular dichroism — monolayers

Reversed micelles of AOT with cytochrome c were used as spreading solutions for Langmuir-Blodgett film formation. A condensed monolayer with closely-packed surfactant molecules in the air and closely packed protein molecules under the water surface was formed after several repetitions of the compression-expansion cycle. The resulting film was transferred onto the solid substrate. The films were studied by circular dichroism and by small-angle x-ray scattering. It was shown that the protein in the film is not denatured and the film is more ordered in respect to monocomponent protein Langmuir-Blodgett films.

Authors' address:

C. Nicolini
Institute of Biophysics
University of Genova
Via Giotto 2
16153 Genova Sestri Ponente, Italia

Thermal stability of photosynthetic reaction centers from rhodobacter spheroids in Langmuir-Blodgett films, studied by circular dichroism measurements

V. Erokhin, F. Antolini, P. Facci, and C. Nicolini

Institute of Biophysics, University of Genova, Italy

Key words: Langmuir-Blodgett films — photosynthetic reaction centers — circular dichroism denaturation — monolayers

The change in the secondary structure of photosynthetic reaction centers from rhodobacter spheroids in Langmuir-Blodgett film was studied by circular dichroism measurements after heating to different temperatures. It was shown that the secondary structure of the protein was not affected in Langmuir-Blodgett film with heating up to 200°C, while protein in solution denatured already at 55°C.

Authors' address:

C. Nicolini
Insitute of Biophysics
University of Genova
Via Giotto 2
16153 Genova, Italia

Properties of silica filler produced by water glass carbonation

Z. Fekete and I. Szabó

Research Institute of Chemical Engineering of the HAS
Veszprém, Hungary

Key words: Silica filler — rubber industry — pore characteristics

A special kind of amorphous silica is a good filler used worldwide by rubber industry. It diminishes not only the price of natural and synthetic base rubbers, but increases their resistance against dynamic stresses; this material also has excellent reinforcing properties.

The main parameters on which the degree of reinforcing depends are, in order of importance, pore characteristics, specific surface area and size of ultimate particles, which are considered dense, inside anhydrous SiO_2 spheres. These three parameters are obviously in relation because the pore area- and volume distribution are the functions of particle size and kind of packing (i.e., "average" coordination number of ultimate particles) and, on the other hand, the specific surface area, which is the whole area of spaces among the touching spheres, is inversely proportional with particle diameter.

This type of filler is generally produced with wet technology by acidifying water glass solutions with mineral acids, but the technology we developed uses low-concentration CO_2 gas as a Lewis acid to decrease the pH of starting water glass solution, thus forcing SiO_2 to polymerize. The dried product is useful provided it is a fine powder of loose aggregates of silica particles constituting ultimate particles of 15—20 nm and its specific surface area is at least 150 m^2/g.

In the temperature and water glass concentration conditions used by us, basically two types of material form if we consider pore characteristics (Fig. 1). One type has mainly pores that are smaller than 10 nm; the other, however, shows a firm maximum in the pore diameter range of 30—40 nm. Only the latter type proves a good filler, although the specific surface area of both may be similarly high enough. This is why only the area of pores larger than 10 nm are thought to contribute the "real" specific surface area. The volume of all pores is an important parameter from a practical point of view, and this is also proportional with the area of pores larger than 10 nm (Fig. 2).

The key to favorable effect of the filler is how it can be kneaded into the raw elastomer mixture and this can be characterized by its looseness, which depends on the coordination number of ultimate particles. If the area of pores that are larger than 10 nm is plotted against the counted coordination number (Fig. 3.), one can see that the fillers with low coordination number have larger pore areas which are in the desired range.

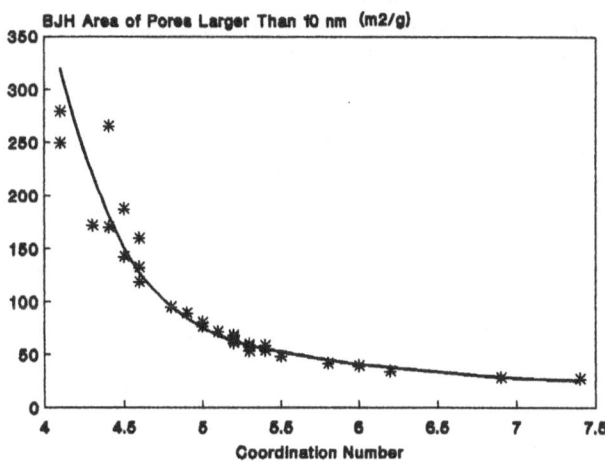

Fig. 3

Of the main steps of technology, i.e., precipitation, filtration, further acidification, and drying, only the first seems to significantly influence the structure of silica produced; all the others only modify it. Thus, starting SiO_2 and Na^+ concentrations, duration of precipitation and magnitude of shear stresses during agitation are the most important technological parameters.

The investigated silica samples were produced in a 20 l jacketed vessel equipped with an agitator and they were dried in a laboratory scale spray drier. The pore size distribution and the Barett Joyner Halenda's pore areas and volumes were determined with an ASAP 2000 instrument purchased from Micromeritics.

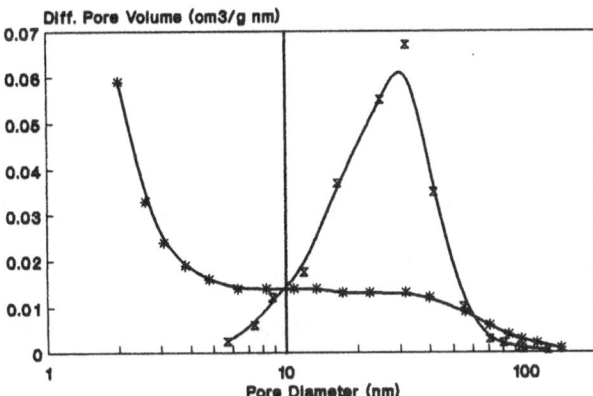

Fig. 1

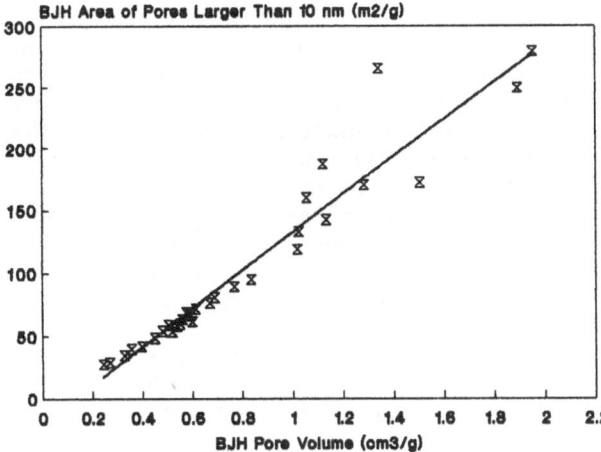

Fig. 2

Authors' address:

Z. Fekete
Reasearch Institute of
Chemical Engineering of the HAS
Veszprem
Egyetem u. 2
8200 Hungary

Fiber-optic dynamic light scattering; neither homodyne nor heterodyne

L. Bremer[1, 2]), L. Deriemaeker[1]), R. Finsy[1]),
E. Geladé[2]), and J. Joosten[2])

[1]) Theoretical Physical Chemistry, Vrije Universiteit Brussel , Belgium
[2]) Department for Physical and Analytical Chemistry, DSM Research Geleen, The Netherlands

Key words: Dynamic light scattering — concentrated dispersions — photon correlation spectroscopy — particle sizing — on-line measurement

In recent years, single-mode optical fibers have been used to study the dynamics of concentrated dispersions of sub-micron particles. Both heterodyne and homodyne detection modes have been reported [1]. Essentially, homodyne experiments probe changes in the position of the particles with respect to each other and heterodyne measurements changes with respect to the local oscillator, i.e., the exit face of the optode. Heterodyne measurements are performed with an optode with a perpendicular exit face. The reflected light acts as local oscillator. In order to obtain "full" heterodyning the intensity of the reflected light must be much higher than the scattered intensity. This implies a low signal-to-noise ratio and the recovery of the dynamic information out of the measured signal becomes less reliable. Homodyne measurements may be accomplished by using a slanted exit face optode. In this optode the back-reflected light is largely attenuated due to the predominant reflection into the cladding of the fiber [2]. However, a small amount of back-reflected light has a large effect on the measurement and minor contaminations of the optode may cause back-reflected or scattered local oscillator light.

Fiber Optics Dynamic Light Scattering (FOQELS) experiments have been performed with several probes, including a slanted optode. Analysis of measurements on monodisperse latices with the singular value decomposition and reconstruction (SVR) method [3] shows that a homodyne and a heterodyne term is contained in the intensity autocorrelation function. The proportion of both contributions can be calculated from the signal-to-noise ratio. If the ratio between the homodyne and heterodyne contribution is known, the field correlation function can be calculated from the data [4]. This method is used to analyze measurements of monodisperse latices and of binary mixtures of monodisperse latices. In binary mixtures with particle diameters in a ratio down to about 2 to 1 the two size classes can be separated.

References

1. Thomas JC (1989) Langmuir 5:1350
2. Wiese H, Horn D (1991) J Chem Phys 94—10:6429
3. Finsy R, de Groen P, Deriemaeker L, Van Laethem M (1989) J Chem Phys 91(12):7374
4. Joosten JGH, McCarthy JL, Pusey (1991) Macromolecules 24:6690

Authors' address:

Prof. Dr. Robert Finsy
Theoretical Physical Chemistry
Vrije Universiteit Brussel
Pleinlaan 2
1050 Brussel, Belgium

Interfacial properties of surfactin

A. Gilardoni and G. Gabrielli

Department of Chemistry, University of Florence, Italy

Abstract: Surfactin, a lipopeptide produced by *Bacillus subtilis*, is characterized by a lactone bond. — Interfacial properties of surfactin were studied at w/a interface: spreading isotherms at 17.5° and 20.5°C were recorded as a function of subphase composition (water at two different acid pH, mono and divalent ions solutions at the same pH and ionic strength). — The isotherms were affected by temperature and subphase composition, both in the shape and in the limiting area, although the films remain always in the liquid condensed phase. — The monolayers were then transferred onto a solid substrate according to the Langmuir-Blodgett technique: the ellipsometric measurements confirm the planar disposition at the w/a interface and the FTIR spectra show the presence of the lactone ring also in the bidimensional state of molecule. — Furthermore, mixed monolayers of surfactin with methylstearate, which is very stable in bidimensional state, were studied at the w/a interface.

Key words: Surfactin — monolayers — spreading isotherms — bidimensional mixtures

Introduction

Surfactin (PM 1050) is considered a very interesting surfactant because of its surface and biodegradability properties [1].

It is constituted by a sequence of seven aminoacids and an hydroxyacid forming a lactone bond [2].

The aim of this work is to investigate the interfacial behavior of surfactin alone and in mixture with a component very stable in bidimensional state, methylstearate, at the water/air interface.

Materials and methods

Surfactin (SUR) was supplied by Eniricerche, methylstearate (MS) by Merck. Substances employed for subphase: NaCl, $CaCl_2 2H_2O$, HCl (supplied by Merck), water bidistilled and free from organic impurities.

Spreading solvent: chloroform (Merck).

Spreading isotherms of surfactin were recorded using a Lauda FW2 film balance (continuous compression at 7.725 cm^2/min).

A KSV-LB5000 [3] was employed to transfer monolayer according to the Langmuir-Blodgett technique (dipping rate 1 mm/min at π = 35 mN/m) and to study the bidimensional behavior of the mixture.

The ellipsometric measurements were performed by an ellipsometer by Rudolph Research USA [4] and the FTIR spectra were recorded using a Perkin Elmer instrument.

Spreading isotherms of the mixtures of surfactin with methylstearate in different weight fractions were recorded using the KSV-LB5000 balance.

Results and discussion

Spreading isotherms of surfactin

The isotherm shape seems to be affected by the temperature and subphase composition. The presence of ions in the subphase solutions produces an increase in the limiting area, although the films always remain in the liquid-condensed phase. This expansion effect is likely due to a modification of the electric double layer at the interface and not to the interaction with carboxylic groups: in fact, in this latter case, the monolayers on divalent salt solution subphases would be condensed [5].

Ellipsometric thickness of the monolayer transferred onto a solid substrate, which isn't reproducible and below 10 A, together with the value of the limiting area (140—180 A^2/mol), suggests a planar disposition of surfactin in monolayer.

The plateau around π = 49 mN/m can be assumed as the result of a collapse, which appears comparable with those occurring by sliding of a monomolecular plane onto another [6].

This attribution is confirmed by these observations: the π of the plateau depends on compression methods [7]; the curves relative to different times in discontinuous compression separate just before the plateau; the surface entropy at plateau is negative and of the same order of magnitude as in polipeptides [8]; the limiting area after plateau is too small to be compatible with surfactin structure.

FTIR Spectra

Multilayers of surfactin were transferred according to the Langmuir-Blodgett technique onto germanium support from subphase at pH = 1 and pH = 4 and FTIR-ATR spectra were recorded using a Perkin Elmer apparatus. Spectrum of solid surfactin was recorded using KBr plate.

Bands related to the presence of aminoacidic and acidic groups are evident in all the spectra: v(C = O) at 1725 cm^{-1} due to the acid and v(C = O) at 1655 cm^{-1} due to the amide.

The spectra are very similar both in shape and peak frequencies, so that we can assume that the structure of solid surfactin is retained in bidimensional state at w/a interface.

Mixtures of SUR and MS at the water/air interface

Spreading isotherms of mixtures of SUR and MS in different weight fractions at the w/a interface (subphase pH = 4) were recorded (Fig. 1).

MS is known as a very stable component in bidimensional state and as a very simple lipidic material; furthermore, it is not sensitively affected by either the pH and the presence of ions in the subphase. Monolayers of MS, alone and in mixture with other components, were already characterized [4, 9].

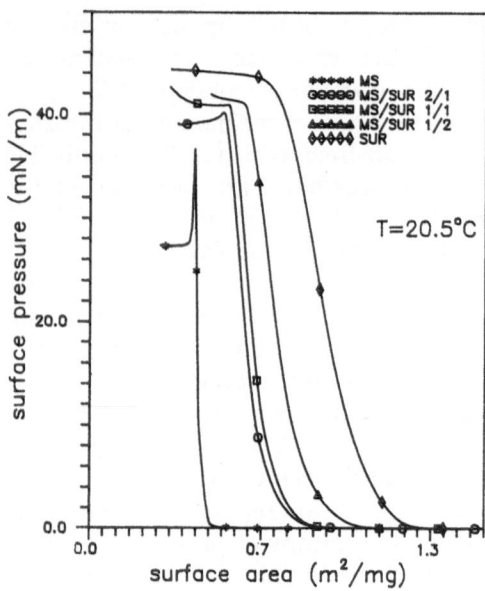

Fig. 1. Isotherms of surfactin, methylstearate and their mixtures on pH = 4 subphase at 20.5°C

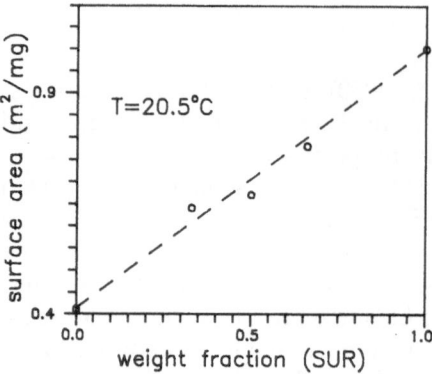

Fig. 2. Surface area at π = 8 mN/m in function of weight fractions of the MS-SUR mixtures at 20.5°C

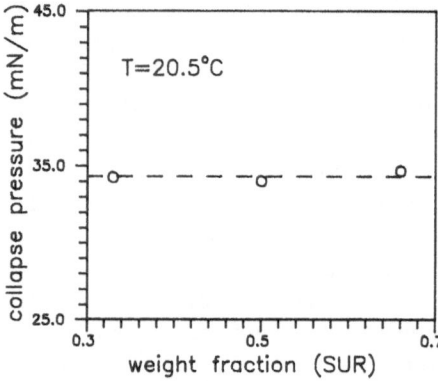

Fig. 3. Collapse pressure (graphically estimated) in function of weight fractions of the MS-SUR mixtures at 20.5°C

The evaluation of thermodynamic parameters allows us to deduce (or not) the miscibility of the two component [10].

Surface areas (at π = 8 mN/m) as a function of the weight fractions of the mixtures substantially follow an ideal behavior (dashed line): this can point out ideal miscibility or immiscibility (Fig. 2).

Collapse pressures (graphically estimated) do not vary significantly with the weight fractions of the mixtures: according to the Crisp rule [11], this points out the immiscibility of the two component at the w/a interface (Fig. 3).

This result is not surprising because of the very different structure of SUR and MS. It is largely demonstrated that similar orientation of the two components at the interface is very important to obtain bidimensional miscibility [12], and it is reasonable that the short chain of SUR and the chain of the fatty ester do not interact among them because of their different disposition with respect to the interfacial plane: the former one almost horizontal and the latter one vertical.

Conclusions

Spreading isotherms of SUR at the w/a interface were investigated.

From limiting area evaluation and ellipsometric measurements we can deduce that SUR is displaced quite horizontal in monolayer.

The immiscibility of SUR and MS in bidimensional state can be mostly attributed to the different orientation of the two components at the interface.

Acknowledgement

One of the authors (A. G.) thanks Eniricerche S. p. A. for a fellowship.

References

1. Arima K, Kakinuma A, Tamura G (1968) Biochem Biophys Res Comm 31:488

2. Kakinuma A, Hori M et al. (1969) Agr Biol Chem 33:1523
3. Bonosi F, Gabrielii G (1991) Colloids and Surfaces 52:277
4. Gabrielli G, Niccolai A, Dei L (1986) Colloid Polym Sci 264:972
5. Gabrielli G et al. (1989) Colloids and Surfaces 41:1
6. Dei L, Baglioni P, Gabrielli G (1987) in: "Surfactant in solution" Ed Mittal KL and Bothorel P Vol 5, pp 979
7. Gaines GL Jr (1966) "Insoluble Monolayers at liquid gas interfaces" Interscience Publ., NY
8. Gabrielli G et al. (1975) in: Adv Chem Series n 144, pp 347
9. Gabrielli G, Puggelli M, Gilardoni A (1992) Progr Coll Surfaces Sci 89:227
10. Adamson AW (1976) "Physical Chemistry of Surfaces" John Wiley & Sons
    Bacon KJ, Barnes GT (1987) J Coll Interf Sci 67:70
11. Crisp DJ (1949) "Surface Chemistry Suppl Research" London
12. Gabrielli G, Puggelli M, Baglioni P (1982) J Coll Interf Sci 86:485, Gabrielli G et al. (1988) Colloid Polym Sci 266:429

Authors' address:

G. Gabrielli
Department of Chemistry
University of Florence
Via G Capponi 9
50121 Florence, Italy

Building ion-selective membranes by insertion of ionophore in both lipid and polymer matrices

A. Gilardoni, E. Margheri, and G. Gabrielli

Dept. of Chemistry, University of Florence, Italy

Abstract: Building up of ion selective membranes is interesting both from a theoretical and technological point of view. — Actually, we built K$^+$ — ion selective membranes employing the Langmuir-Blodgett technique. — The selectivity was obtained using mixtures between the ionophore valinomycin (VAL) and methylstearate (MS) or polyoctadecylmetharylate (POMA). Such mixtures were studied at the air-water interface at 25 °C, recording spreading isotherms and determining the bidimensional phases of the films in the experimental conditions. — The behavior of surface parameters, such as collapse pressure as a function of the mixture composition, allowed us to determine bidimensional miscibilty of VAL with both MS and POMA. — The bidimensional films constituted by lipid or polymer molecules alone and in mixtures with VAL molecules were then transferred onto porous hydrophobic filters. — The transfer was obtained by dipping and drawing up the filter from the water subphase. The membranes were at last characterized by measuring their electrical resistance in presence of NaCl and KCl solutions in the $10^{-3} - 10^{-1}$ M concentration range. It was verified that K$^+$ — ion selectivity was always retained when the amphiphilic matrix contained VAL molecules.

Key words: Spreading monolayers — bidimensional mixtures — LB films — ion-selective membranes — electrical resistance

Introduction

Biological membranes are constituted by a lipid bilayer containing different kinds of components devoted to peculiar functions which confer specific properties to membranes [1]. The most commonly used models of natural membranes are represented by monolayers, black lipid membranes, vesicles and LB films [2].

In particular, ion-selective transport across the hydrophobic phase from the inner to the outer compartment of the bilayer and vice-versa is one of the most interesting properties investigated in both natural and model systems.

Aim of this work was to obtain membranes able to selectively transport one ion using both spreading monolayers at the water-air interface and Langmuir-Blodgett films as models of natural membranes.

We inserted the antibiotic valinomycin in a matrix reproducing the lipid matrix of biological bilayers, constituted by classical lipids, as methylstearate, or by a polymer, because polymers are known to give stable LB films [3].

It is well known that valinomycin selectively complexes and transports K$^+$ ions in biological systems, so we tried to build a K$^+$ selective membrane by transferring multilayers containing Val molecules onto solid supports, according to the Langmuir-Blodgett technique.

We tested the selectivity comparing the electrical resistance obtained in K$^+$ and Na$^+$ solutions at the same concentration.

Progress in Colloid & Polymer Science　　　　　　　Progr Colloid Polym Sci 93:234 (1993)

Materials and methods

Valinomycin (VAL) and methylstearate (MS) were supplied by Sigma Chemie. Polyoctadecylmethacrylate (POMA) as toluene solution at 23.4% w/w was supplied by Aldrich. Chloroform (purity $\approx$ 99%, Merck) was used as spreading solvent. 10^{-3} M to 10^{-1} M solutions of NaCl and KCl (purity $\approx$ 99.5, Merck) were used for electrical measurements. All the solutions were prepared using doubly distilled water freed from colloidal and ionic impurities by a Milli-Q (Millipore) ultrafiltration apparatus; water thus obtained had a resistivity greater than 18 M$\Omega \cdot$ cm.

Porous filters HVHP constituted by polyvinylidene difluoride having pore size of 45 μm and thickness of 125 μm, used as solid supports, were supplied by Millipore.

Spreading isotherms were recorded by a Lauda Film Balance FW2 at 25 °C. LB multilayers were prepared using a KSV balance LB5000 as reported elsewhere [4].

Electrical resistance measurements were performed using a teflon cell connected to an apparatus described in [4, 5].

Results and discussion

Spreading monolayers

We studied the bidimensional behavior of VAL/MS and VAL/POMA mixtures recording spreading isotherms of both pure components and mixtures at 25 °C using pure water as subphase. The investigated mixtures were in the following weight ratios: VAL/MS = 1/1; VAL/MS = 4/1; VAL/POMA = 1/1; VAL/POMA = 4/1.

Figure 1 shows the spreading isotherms obtained for VAL/MS and VAL/POMA systems.

We evaluated the bidimensional thermodynamic parameters in order to establish the miscibility between the components at the water-air interface.

The deviation from ideality of the surface areas and of the surface compressional modulus C_s^{-1} as a function of the mixture composition for both the systems and above all the variation of the collapse pressure values in the various mixtures (see Fig. 2) allowed us to state that VAL was miscible both with MS and POMA.

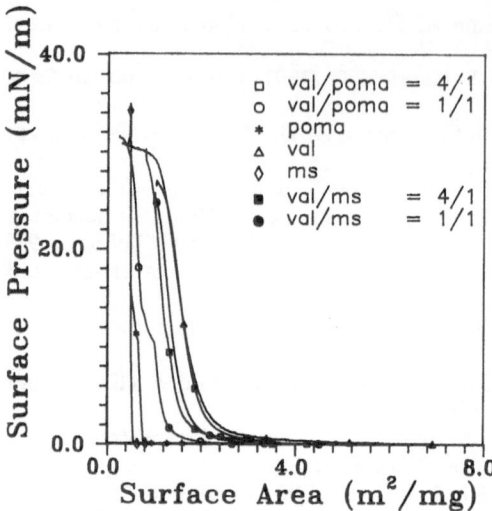

Fig. 1. Spreading isotherms at 25 °C on pure water of both pure components and mixtures in different weight ratios

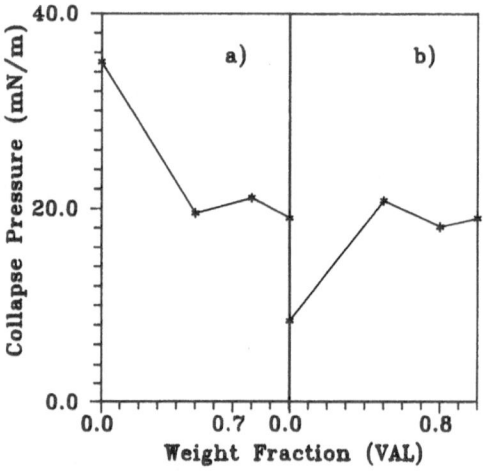

Fig. 2. Collapse pressure values versus mixture composition for a) VAL/MS system; b) VAL/POMA system

Miscibility of the monolayer components indicates the presence of a homogeneous monolayer at the air-water interface.

Once the bidimensional miscibility was stated, we transferred multilayers of the two systems onto hydrophobic filters.

Electrical measurements

Electrical resistance measurements were performed in order to deduce if VAL retains the K$^+$ ion selectivity also in the supramolecular systems and if there is a relationship between the kind of matrix in which VAL is inserted and selectivity.

The experimental parameter chosen for this purpose was $\Delta R\%$, defined as

$$\Delta R\% = \frac{R_{K^+} - R_{Na^+} \; 100}{R_{max}} \;, \qquad (1)$$

where R_{K^+} and R_{Na^+} were the electrical resistance values measured for the same membrane once in presence of KCl and one in presence of NaCl solutions at the same concentration. R_{max} was the higher resistance value measured between R_{K^+} and R_{Na^+} for a certain concentration. Thus, $\Delta R\%$ values lower than 0 were related to K^+ ion selective membranes.

In Table 1 the results obtained for VAL/MS system are shown. In presence of KCl solutions we always obtained resistance values lower than those obtained in presence of NaCl solutions at the same concentration, while at the same concentration there was not a similar behavior for a membrane constituted by MS alone, thus indicating that the presence of VAL on the filters conferred ion selectivity to the membranes.

As regards VAL/POMA system (see Table 1) the selectivity increased with increasing the number of layers transferred onto hydrophobic filters, thus indicating that, if the membrane is ion selective, an increase in the amount of VAL molecules on the filter makes possible an improvement in the selective behavior.

Table 1. Electrical Resistance values for Langmuir-Blodgett membranes

System	Conc. (m)	R_{Na^+} (Ω)	R_{K^+} (Ω)	$\Delta R\%$
Bilayer	$5 \cdot 10^{-4}$	$1.35 \cdot 10^6$	$1.42 \cdot 10^6$	47
MS	10^{-3}	$5.50 \cdot 10^5$	$2.22 \cdot 10^5$	−60
	$5 \cdot 10^{-3}$	$8.65 \cdot 10^4$	$7.95 \cdot 10^4$	−8
	$5 \cdot 10^{-2}$	$6.65 \cdot 10^4$	$2.79 \cdot 10^4$	−58
	$5 \cdot 10^{-4}$	$1.32 \cdot 10^6$	$1.14 \cdot 10^6$	−14
Bilayer	10^{-3}	$4.99 \cdot 10^5$	$1.40 \cdot 10^5$	−72
VAL/MS	$5 \cdot 10^{-3}$	$7.29 \cdot 10^4$	$6.44 \cdot 10^4$	−12
1/1	$5 \cdot 10^{-2}$	$1.46 \cdot 10^4$	$9.10 \cdot 10^3$	−35
	$5 \cdot 10^{-4}$	$1.22 \cdot 10^6$	$8.85 \cdot 10^5$	−27
Bilayer	10^{-3}	$2.86 \cdot 10^5$	$1.27 \cdot 10^5$	−55
VAL/MS	$5 \cdot 10^{-3}$	$1.01 \cdot 10^5$	$6.74 \cdot 10^4$	−33
4/1	$5 \cdot 10^{-2}$	$7.4 \cdot 10^3$	$6.00 \cdot 10^3$	19
Bilayer	$5 \cdot 10^{-4}$	$4.47 \cdot 10^6$	$1.32 \cdot 10^6$	70
POMA	10^{-3}	$2.47 \cdot 10^5$	$2.52 \cdot 10^5$	−2
	$5 \cdot 10^{-3}$	$6.27 \cdot 10^4$	$5.79 \cdot 10^4$	7
	$5 \cdot 10^{-2}$	$8.97 \cdot 10^3$	$6.023 \cdot 10^3$	−33
	$5 \cdot 10^{-4}$	$7.70 \cdot 10^5$	$5.57 \cdot 10^5$	−28
Bilayer	10^{-3}	$3.68 \cdot 10^5$	$3.12 \cdot 10^5$	−15
VAL/POMA	$5 \cdot 10^{-3}$	$7.67 \cdot 10^4$	$6.15 \cdot 10^4$	−20
4/1	$5 \cdot 10^{-2}$	$1.18 \cdot 10^4$	$9.01 \cdot 10^3$	−24
	$5 \cdot 10^{-4}$	$1.39 \cdot 10^6$	$1.21 \cdot 10^6$	−13
Four layers	10^{-3}	$3.55 \cdot 10^5$	$2.48 \cdot 10^5$	−30
VAL/POMA	$5 \cdot 10^{-3}$	$1.04 \cdot 10^5$	$6.39 \cdot 10^4$	−39
4/1	$5 \cdot 10^{-2}$	$3.70 \cdot 10^4$	$1.91 \cdot 10^4$	−49

Acknowledgement

Thanks are due to the Italian Ministero dell'Università e della Ricerca Scientifica e Tecnologica (MURST) and to Consiglio Nazionale delle Ricerche (CNR). One of the authors (A. G.) also thanks Eniricerche S. p. A. for a fellowship.

Conclusion

The obtained results showed that VAL is miscible at the water-air interface with MS and with polymer molecules having a skeleton very similar to the one of MS.

Furthermore, we demonstrated that if VAL is transferred onto porous filters, they behave like selective membranes, that is to say that the K^+ -ion selectivity is retained also in these supramolecular systems and that it was directly related to the content of VAL, that is to say to the number of layers.

References

1. Singer SJ and Nicolson GL (1972) Science 175:720
2. Fendler JH (1982) Membrane Mimetic Chemistry J. Wiley and Sons Roberts GG (1975) Adv Phys 34:475
3. Ulman A ed (1991) An Introduction to Ultrathin Films. From Langmuir-Blodgett to Self-Assembly, Academic Press
4. Gilardoni A, Margheri E and Gabrielli G (1992) Coll Surf 68:235

5. Margheri E, Niccolai A, Gabrielli G and Ferroni E (1991) Coll Surf 53:135

Authors' address:

Prof. G. Gabrielli
Department of Chemistry
University of Florence
Via G. Capponi
50121 Florence, Italy

Photophysical study of mixed monolayers of pyrene-labeled phospholipids

G. Caminati[1]), G. Gabrielli[1]), R. C. Ahuja[2]), and D. Möbius[2])

[1]) Dipartimento di Chimica, Italy
[2]) Max-Planck-Institut für Biophysikalische Chemie, Göttingen, FRG

Abstract: Mixed monolayers of dipalmitoylphosphatidic acid and two fluorescent analogs of dipalmitoylphosphatidylcholine were studied at water-air interface. Phosphatidylcholines with pyrene covalently bound in two different positions were used in order to study selectively the hydrophobic region of the monolayers as well as the polar groups/water interface. Monolayers were studied measuring surface pressure and surface potential-area isotherms, and recording fluorescence and reflection spectra as a function of compression as well as of monolayer composition. Information on the structure and the dynamic properties of the monolayer were extracted from the behavior of the ratio of the intensities of the monomer and excimer emission bands. Accessibility of the pyrene moiety in the phospholipid monolayer to substances present in the subphase was studied using methylviologen as a water-soluble quencher. Fluorescence and reflection data have shown that when pyrene is located in the hydrophobic portion of the monolayer, quenching of the excited state of pyrene occurs as a result of photoinduced electron transfer between pyrene and methylviologen. When the fluorescent probe is located in the polar region of the monolayer the decrease observed in fluorescence intensity derives also from the formation of a ground state complex between pyrene and the methylviologen quencher.

Key words: Monolayers — pyrene-labeled phospholipids — fluorescence and reflection spectroscopy

Introduction

Fluorescent lipid probes are widely used in research on biomembranes for the investigation of lateral diffusion and phase transition [1]. On the other hand, monolayers at air/water interface provide a simple mimetic model for lipid membranes where the effect of molecular organization can be elegantly investigated and systematically varied [2]. Information on the packing density, the orientation and the distribution of the interfacial film were extracted from the study of mixed lipid monolayers of dipalmitoylphosphatidic acid and two fluorescent phospholipids at water-air interface. We chose to investigate the properties of two phospholipid analogs with pyrene covalently bound to the molecule in two different positions, i.e., at the end of the hydrophobic chain and in the polar head group, in order to study selectively the hydrophobic region of the monolayers and the polar group/water interface. Information on the structure and the dynamic properties of the microenvironment of the probe in disperse systems may be obtained from the photophysical behavior of the chromophore [3, 4]. In this paper, the same procedure was applied to the study of monomolecular films at water-air interface as a function of monolayer composition and available surface area. Accessibility of the pyrene moiety in the mixed phospholipid monolayer to substances present in the subphase was studied using a water soluble quencher. Methylviologen was dissolved in the aqueous subphase and the resulting decrease of pyrene emission was studied as a function of quencher concentration.

Experimental

Dipalmitoylphosphatidic acid (DPPA), supplied by Sigma, was used as the matrix molecule; the pyrene-labeled phospholipids: 1-palmitoyl-2-(1-pyrenedecanoyl)-sn- glycero-3-phosphocoline (PYDD-PC) and N-(1-pyrensulfonyl)-1,2-hexadecanoyl-sn-glycero-3-phospho-ethanolammine, triethylammmonium salt (DPPEPY), were both supplied by Molecular Probes Inc. Methylviologen (MV) was purchased from Sigma. Surface pressure-area isotherms were recorded using a procedure reported elsewhere [5]. Reflection isotherms at constant wavelength and reflection spectra at constant surface area were performed using an apparatus previously described [6], Fluorescence isotherms at constant wavelength and fluorescence spectra at constant surface area were collected using photon counting techniques directly at water-air interface using fiber optics [7].

Results and discussion

Single-component monolayers: DPPEPY and PYDPPC

The two pyrene-labeled phospholipids in monolayer give different reflection spectra, the spectra of PYDPPC are similar to the absorption spectrum of pyrene monomer in solution [3], moreover, the bands are not shifted with compression of the monolayer and the relative intensities of the y-polarized (280 nm) band and the z-polarized (346 nm) band are similar in the whole surface pressure range. For DPPEPY monolayer the reflection bands are red shifted with compression, indicating the formation of weakly interacting ground state dimers. Furthermore, the y-polarized band increases in intensity much more than the z-polarized band when surface pressure increases; these results suggest a change in the orientation of the pyrene chromophore in the monolayer permitting the formation of loosely bound dimers.

Two-component monolayers:

DPPEPY/DPPA and PYDPPC/DPPA mixtures

The detailed study of the mixtures of the two phospholipids with DPPA has been reported in a previous paper [8]; the results showed that both DPPEPY and PYDPPC form homogeneous mixed monolayers at high matrix molar fractions. Fluorescence experiments in the DPPEPY/DPPA system showed that the excimer band appears at higher surface pressure than the monomer, but the intensity of the excimer band rapidly increases with compression [8]. On the contrary, for the PYDPPC/DPPA system, the monomer band appears together with the excimer band, and shows a higher intensity than the excimer. In Fig. 1, we report typical emission spectra for the 10:1 mixtures of the two mixed systems at 20 mN/m. The analysis of the monomer band in the fluorescence spectra of the 1:10 mixture showed that the pyrene group of DPPEPY is in a polar environment up to a surface area of 70 nm^2/molec, whereas in the case of PYDPPC the chromophore probes a hydrophobic moiety starting from surface pressure higher than 0.5 mN/m.

The values of the emission intensity of the monomer (I_m) and the excimer (I_e) were extracted

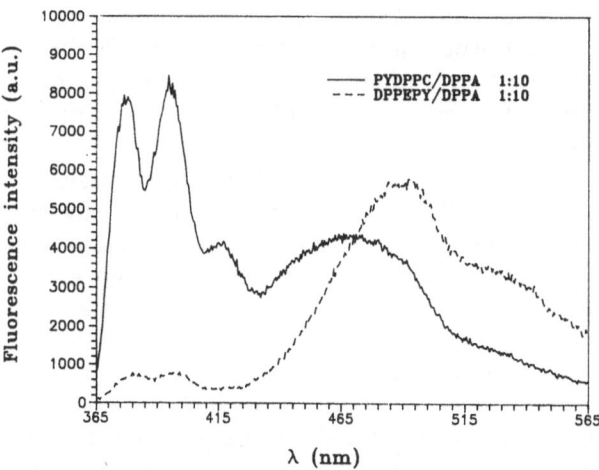

Fig. 1. Fluorescence emission spectra at water-air interface at 20 mN/m

from the spectra and correlated to diffusion of the labeled phospholipid in the monolayer (not reported here). For the system DPPEPY/DPPA, I_e/I_m values are linear with DPPEPY molar fraction only at low π, this shows that excimer formation is controlled by diffusion in the monolayer only at low DPPEPY surface density, whereas at higher π a different mechanism occurs (formation of loosely bound dimers in the ground state). In the system PYDPPC/DPPA, the ratio I_e/I_m is linear in the whole π range, suggesting that excimer formation is dynamic in nature and that it is controlled by diffusion in the monolayer.

Fluorescence quenching in mixed monolayers

Previous experiments showed that MV is electrostatically attracted to the monolayer from the subphase only for DPPA molar fraction higher than 0.5 [8] and that addition of methylviologen (MV) in the subphase has a strong condensing effect on the monolayers of both systems. When MV is added in the subphase, fluorescence emission decreases, evidencing a quenching process normally ascribed to photoinduced electron transfer [8].

For the DPPEPY/DPPA system excitation spectra recorded at excimer emission are shifted compared to the excitation spectra recorded at monomer emission in the presence of MV. Reflection spectra in the presence of MV were also recorded and it was found that in the case of DPPEPY mixtures a new band around 260 appears in addition to the characteristic band of MV at 280 nm. The same behavior in excitation and reflection spectra is found for the PYDPPC/DPPA system, but only at low surface pressure.

In Table 1 we report the ratio of the emission intensities with and without quencher in the subphase. The monomer and the excimer emissions are differently quenched in the low and high surface · pressure regime: for DPPEPY/DPPA system the maximum quenching is obtained for the excimer at high surface pressure, whereas for the PYDPPC/DPPA system the maximum quenching is obtained for the monomer at high surface pressure. This difference is ascribed to the formation of a ground state complex between pyrene and MV that superimposes to the electron transfer process as confirmed by reflection and excitation spectra.

Conclusions

These two labeled phospholipids can be successfully used as probes to investigate selectively the polar head group/water interface (DPPEPY)

or the hydrophobic region of the monolayer (PYDDPC) since they show completely different photophysical behavior in spreading monolayer. In the case of the PYDDPC/DPPA system, excimer formation is dynamic and it depends on the concentration of the probe in the monolayer. For the DPPEPY/DPPA system the pyrene groups interact also in the ground state, as shown by reflection and excitation spectra. The quenching results suggest the formation of a ground state complex between MV and pyrene when the chromophore is close to the water-monolayer interface. This occurs always for the DPPEPY/DPPA system where pyrene is located near the polar head group, whereas in the case of PYDPPC/DPPA system the complex forms only at low surface pressures when the alkyl chains are mobile, allowing the pyrene groups to approach the interface.

Acknowledgement

This work was supported by Italian CNR (Consiglio Nazionale delle Ricerche) and by MURST (Ministero dell' Università e della Ricerca Scientifica e Tecnologica).

Table 1. Fluorescence quenching data

DPPEPY/DPPA 1:10 monolayer

[MV] (μM)	π = 1.5 mN/m		π = 20 mN/m	
	monomer	excimer	monomer	excimer
0	1	1	1	1
0.025	1.04	1.11	1.11	0.37
0.05	1.48	2.16	1.07	3.25
10	4.66	3.18	7.8	39

PYDPPC/DPPA 1:10 monolayer

[MV] (μM)	π = 1.5 mN/m		π = 20 mN/m	
	monomer	excimer	monomer	excimer
0	1	1	1	1
0.025	1.5	0.30	2.03	0.62
0.05	1.70	0.27	2.27	0.60
10	21.9	1.50	1.72	3.75

References

1. Subramanian R, Patterson LK (1985) J Am Chem Soc 107:5820
2. Gaines G (1970) "Insoluble monolayers at liquid-air interface" Wiley-Interscience, New York
3. Birks JB (1970) "Photophysics of aromatic molecules", Wiley-Interscience, New York
4. Kalyanasundaram K, Thomas JK (1977) J Am Chem Soc 99:2039
5. Caminati G, Ahuja R, Möbius D (1991) Thin Solid Film 210/211:335
6. Gruniger D, Möbius D, Meyer H (1983) J Chem Phys, 79:3701
7. Ahuja R, Möbius D (1992) Langmuir 8:1136
8. Caminati G, Ahuja RC, Möbius D (in preparation)

Authors' address:

Dr. Gabriella Caminati
Dipartimento di Chimica
Via G. Capponi 9
50123 Firenze, Italy

Small angle X-ray scattering by TiO$_2$/ZrO$_2$ mixed oxide particles and a synroc precursor

D. Gazeau[1]), R. Amal[2]), J. Bartlett[3]), and Th. Zemb[1])

[1]) Centre d'Etudes Nucléaires de Saclay, Service de Chimie Moléculaire, Gif sur Yvette Cedex, France
[2]) School of Chemical Engineering, University of New south Wales, Kensington, Australia
[3]) Australian Nuclear Science and Technology Organization, Advanced Materials, Menai, New South Wales, Australia

Abstract: This high resolution small angle X-ray scattering study of a concentrated oxide sol, precursor of the SYNROC matrix for the storage of the high level radioactive waste, evidences a locally cylindrical microstructure. Locally, nanometric cylinders show disordered axis with some concentration dependent connections. This microstructure explains the paradoxical stability of this oxide dispersions. This stability has a steric origin and electrostatic repulsions are not needed. The addition of aluminium to the initial titanium-zirconium mixture enhances branching on the locally cylindrical microstructure. Finally, we show that the solid powder obtained after drying of the sol has the same specific area ($\sim$ 1000 m^2/g) as the sol.

Key words: Small angle X-ray scattering — high resolution, oxide dispersions — cylinders

Introduction

A typical nuclear power station generating 1000 megawatts of electrical power produces about 30 tons of used fuel per year. After uranium and plutonium have been recovered for recycling in reactors, about one ton of high level waste (HLW) is left behind. Almost all the radioactive species originally in the spent fuel ends up in this HLW. The radioactivity from fission product elements like caesium and strontium decays in about 1000 years. Thereafter, most of the radioactivity comes from the actinide elements. Their radioactivity takes at least of the order of 10^5 years to decay to acceptably low levels.

According to current waste management strategies, the HLW should be converted into a compact solid form and be buried as deep as possible underground while the radioactivity decays. Protection of the biosphere on a time scale exceeding several thousands years relies primarily upon the capacity of the geological barrier to minimize access of groundwaters to the HLW and to retard the migration of dissolved species to the biosphere. In addition, the waste form itself should be very stable in the geological environment.

The barriers may be divided into two categories. The first is the immobilization barrier. Primarily, this refers to the material which has been produced by solidification of the liquid waste. This matrix should be in a form that is mechanically sound and exhibits stability in the presence of groundwater. The barrier may include a corrosion resistant canister in which solidified waste has been placed. This package is then deposited in a repository which can take the form of a salt dome, a disused mine or a deep borehole. These and their surroundings constitute the geological barrier.

SYNROC was first developed and characterized in 1978 by Professor A. E. Ringwood and his colleagues at the Australian National University in Canberra. SYNROC is a synthetic rock composed of four titanate minerals: hollandite (BaAl$_2$Ti$_6$O$_{16}$), zirconolite (CaZrTi$_2$O$_7$), perovskite (CaTiO$_3$) and rutile (TiO$_2$), together with a small amount of metal alloy [1, 2]. Almost all the elements in the high level nuclear reactor wastes can form an integral part of the crystal lattices of these very stable SYNROC phases. Moreover, close analogues of the titanate phases of SYNROC also occur in nature. These ancient natural minerals have survived extreme geological conditions for millions of years, yet they still contain their original content of radioactive elements and daughters. Such evidence, provided by nature, strongly suggests that SYNROC can immobilize high level waste species for the geologically long periods needed for their radioactivity to decay to safe levels.

Synroc precursor powders have been produced by several advanced chemical routes involving alkoxide hydrolysis [3, 4]. The hydroxide-route [3], which involves the base-catalysed hydrolysis of a mixture of titanium, zirconium and aluminium alkoxides in ethanol, and subsequent sorption of barium and calcium cations under alkaline conditions, has been used to produce in excess of 3000 kg of powder. An alternative method, involving the spray-drying of a concentrated mixed-oxide sol containing titania, zirconia and alumina, has also been developed [4]. This latter method, which has been used to produce more than 100 kg of material, yields free-flowing, dust-free powders, with a high sorption capacity for high-level nuclear waste.

This paper reports a small-angle x-ray scattering study of the specific surface area and local microstructure of concentrated, mixed-oxide sols, containing titania and zirconia. These two oxides represent more than 75 wt% of the Synroc precursor powders. The small-angle scattering experiments, conducted in strongly absorbing materials over three decades of scattering wave-vector, allow direct, *in-situ* investigation of the microstructure of the viscous sols. Such data provide an insight into the anomalous stability of the sols and of inter-particle surface interactions. Small-angle scattering data obtained from dried Synroc precursor powders, produced by the hydroxide-route [3], are also reported.

Experimental

We studied small angle X-ray scattering by TiO_2/ZrO_2 acidic sols, prepared in ANSTO (Lucas Heights Laboratories, New South Wales) via the alkoxide route, hydrolyzing a mixture of iso-propyl titanate $(Ti(OC_3H_7)_4)$ and iso-butyl zirconate $(Zr(OC_4H_9)_4)$. The concentrated stock solution contains 15.0 moles TiO_2 per mole of ZrO_2, the ratio in SYNROC. The concentration of solid particles is 1070 g per liter. The concentrated sol, with a volume fraction of solid particles $\phi_1 = 0.20$, was further diluted with water to prepare samples with solid volume fractions $\phi_2 = 0.06$ and $\phi_3 = 0.03$. We also studied X-ray scattering by a full SYNROC precursor before and after drying. Details on samples compositions are given in table I.

Two small-angle X-ray cameras were used to cover a q range from 1.4×10^{-3} Å^{-1} to 0.6 Å^{-1}. A high resolution Bonse & Hart camera, coupled to a copper K_a 18 kW rotating anode ($\lambda = 1.54$ Å), allows intensity measurements at extremely small angles. The possibilities of this experimental arrangement have been extensively described [6, 7, 8]. The corresponding q range, from 1.4×10^{-3} Å^{-1} to 0.4 Å^{-1}, overlaps with the one of a Guinier-Mering type camera coupled to a molybdenum X-ray source. A curved quartz monochromator selects and focuses the K_a radiation ($\lambda = 0.71$ Å), allowing absolutely scaled scattered intensity measurements ($I(q)$ in cm^{-1}) in a q range from 0.05 Å^{-1} to 0.8 Å^{-1}. The use of such a molybdenum source, which emits photons with an energy of 17 keV, allows small angle X-ray scattering experiments with relatively

strongly absorbing materials like titanium and zirconium. Two dimensional scattering raw data have been obtained using a new "Image Plate" photosensitive detector, in which X-ray photons interact with europium sites homogeneously located on the plate [9]. Excited europium sites energy level is a triplet state, which makes possible temporary data storage. Further excitation with He-Ne laser radiation ($\lambda = 633$ nm) changes the excited europium sites into singlet ones. Then instantaneous fluorescence occurs ($\lambda = 490$ nm). The important point to notice is that the number of fluorescence photons for a given site is proportional to the number of incoming X-ray photons on that site. This type of detector exhibits excellent linearity over more than four decades in intensity. The detailed procedure allowing determination of the scattering power $I(q)$ from image processing is described in reference [10].

Intensities are measured in absolute units [11] which allows for direct comparison between the experimental scattered intensity and calculated scattered intensities for different models of the sol microstructure. Water, as well as Lupolen™ (BASF) [12], has been used to calculate small angle scattering cross-sections. Converted into real space, the three decades of scattering angles covered correspond to explored Bragg spacings between 1 micrometer and 1 nanometer. In order to be accepted as a good representation of a microstructure, the scattering power $I(q)$ of a physical model has to be close to the observed one over three decades of magnitude.

Results and discussion

1) Concentrated colloidal dispersions of mixed TiO_2/ZrO_2

As stated above, the scattered intensity was recorded in the q range from 1.4×10^{-3} Å^{-1} to 0.6 Å^{-1} in order to obtain structural information on length scales varying continuously from 10 Å to 4500 Å. Figure 1 shows the scattering patterns obtained for the three different volume fractions of the mixed oxides studied. The large linear behaviour measured at the low-q part of the scattering pattern is not accessible with classical small angle X-ray scattering apparatus and can only be obtained using a high resolution Bonse & Hart camera. This part of the scattering pattern reflects the morphology of particles for a typical size of 100 nm.

Table 1

TiO$_2$/ZrO$_2$ sols

ϕ	ρ g/cm³	Σ Å²/Å³	S m²/g	I_c(Å)	ξ(Å)	R_s(Å)	R_c(Å)	$t/2$(Å)
0.03	1.09	0.129	1180	85	85	23	16	7.8
0.06	1.20	0.108	897	83	81	28	19	9.3
0.20	1.67	0.056	335	59	50	54	36	18

ϕ	T (Cu/Mo)	e (Cu/Mo, cm)
0.03	0.452/0.708	0.032/0.120
0.06	0.364/0.848	0.024/0.024
0.20*)	0.217/0.388	0.013/0.069

*) stock sol, total oxide mass per liter: 1070 g

Synroc precursor

oxide particles	wt%	precursor solid content
TiO$_2$	70.74	320 g/l
ZrO$_2$	6.84	pH = 1.45
Al$_2$O$_3$	5.37	electrical conductivity
BaO	5.44	62 mS/cm
CaO	11.61	

ϕ, volume fraction; ρ, density; Σ, specific surface per unit volume; S, specific surface per unit of mass. I_c, average chord length. ξ, correlation length associated to the graphically determined cross-over. Rs, Rc and t/2, respectively the sphere and cylinder radii, and the half thickness of a lamella calculated from the Porod's limit. T is the transmission for copper and molybdenum K_a radiation and e is the sample thickness

For high q values, the scattered intensity exhibits a q^{-4} dependence characteristic of the Porod's law [13], which reflects the total specific interface Σ (in cm²/cm³ of solution) between the oxide and the solvent. The Porod's law gives the q dependence of the scattered intensity by an abrupt interface which separates two media having different electronic densities ρ_p and ρ_s

$$I(q) = 2 \cdot \pi \cdot (\rho_p - \rho_s)^2 \cdot \Sigma \cdot q^{-4} \tag{1}$$

In order to obtain the value of the total oxide/water specific interface we have to assume an homogeneous mixing of the titanium and zirconium oxides inside the solid particles. In fact, it

may be that the ZrO$_2$ component is coating the TiO$_2$ colloids. This effect would add a q^{-2} excess term to the observed pure Porod's limit. Our experiment rules out any significant separation of oxides at a scale larger than 10 Å. This "ideal mixing hypothesis" is used to calculate the density (d_p) of the TiO$_2$/ZrO$_2$ particles and their electronic density (ρ_p). The term Δ_p is the difference in the electronic densities of the scattering particles ρ_p and that of the solvent (water) ρ_s, expressed in cm⁻², as each electron of the sample has a scattering length of 0.282×10^{-12} cm. Σ represents the total amount of interface per unit volume of sample. The constant term $2 \cdot \pi \cdot (\rho_p - \rho_s)^2 \cdot \Sigma$ is called the Porod's limit. Figure 2b gives the variation of the specific surface area S in m²/g, derived from the Porod's limit, versus the volume fraction of particles. The

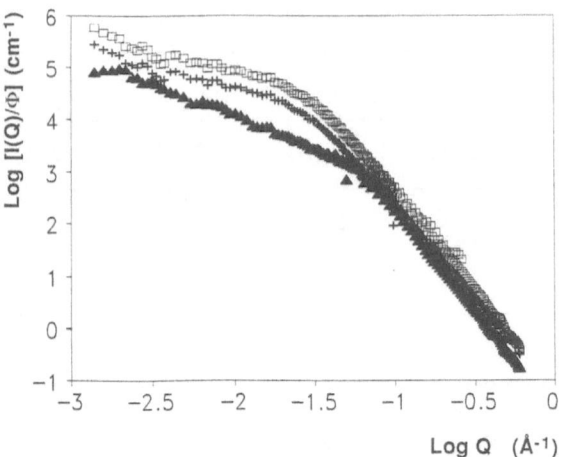

Fig. 1. Absolutely scaled scattered intensities normalized by the volume fraction of oxide particles $I(q)/\phi$ in cm — 1. The scattered intensity is measured over three decades of scattering angles using two experimental setups, for three concentrations corresponding to volume fractions of TiO_2/ZrO_2 mixed oxides particles of 0.20 (Δ), 0.06 (+) and 0.03 ($\square$). The low q part of the scattering curve is obtained using a Bonse & Hart camera. The high q part is obtained using a Guinier-Méring experimental setup with either a linear gaz detector or via image processing of a two dimensional "image plate" detector

specific surface area is not proportional to the dry volume fraction ϕ of particles, but increases as ϕ decreases. This reflects dispersions which are coarse at high concentration and tend to become finer at lower volume fractions. The observed minima on fig. 2a are due to oscillations in the form factor of the individual scattering objects. These minima are all observed at the same q values, which demonstrates that the shape of these objects does not change upon dilution. This asymptotic scattering regime is observed when $qR > 10$, where R is the typical curvature radius of the oxide / solvent interface. The value of the Porod's limit is independent of the topology of the solution: it can be a connected network as well as a dispersed set of globular particles. The only requirement for the observation of a Porod's limit is a "sharp" oxide/water interface. By "sharp" -as opposed to "diffuse" — a small angle experiment refers to an average distance between a point INSIDE bulky oxide structure and a point INSIDE pure solvent to be less than $\pi/q_{max} \approx 10$ Å. Here, the oxide water interface is determined to be smooth on a nanometric scale. Between the Porod regime and the low-q morphology regime, a cross-over region is observed.

The position of this cross-over is indicative of the typical size ξ of the dispersed material: $q_{cross-over} \approx \pi/\xi$. At $q < \pi/\xi$, the slope of the scattering gives information on the sample microstructure. Two types of microstructure can arise in concentrated solutions: a connected network or a set of interacting globular particles. If the microstructure is a connected network, the power law decay (slope in a log-log plot) observed is close to —4 for large flocculated aggregates, close to —3 for very compact lumps having few voids. Other power law decays may reflect fractal structures [14]. Two simple cases are slopes close to —1 (cylinders) or to —2 (platelets). If the network is made of particles, once volume fraction, electronic densities and specific areas are known, the shape of the particles can be determined using either model calculations on absolute scale, or shifts of the scattering maximum for a volume fraction variation at fixed specific interface Σ [15]. If the structure is a set of interacting globular particles, the low-q thermodynamic limit of the scattering mainly reflects the attractive or repulsive interactions between particles. In some cases the low-q limit can give direct information about long-range interactions between individual particles or aggregates inside the solution [16].

The standard method to distinguish between these two main types of structures is a dilution experiment: on fig. 1, the scattering curves observed for three concentrations are compared after normalization by volume fraction: $I(q)/\phi$ is compared for different volume fractions. In the case of interactions between a set of roughly identical globular particles, moving independently, the low-q part of the scattering would be completely modified when changing the volume fraction of particles. In the case of the mixed oxide sol investigated here, the spectra remain similar, indicating a more and more connected network or disordered dispersion of cylinders. The slope, which is close to —1, indicates a local microstructure similar to long monodisperse cylindrical particles. This first indication has to be checked for self-consistency as described below. From this first data set, it can be concluded that the most likely microstructure of the sol is cylindrical and there is no variation in the small angle slope which would indicate interactions between charged interfaces. In the case which is studied here, the sterical interactions are predominant. The typical size ξ of the oxide dispersion is given by $\xi = \pi/q_c$, where q_c is the value of the scattering vector at the graphically determined cross-over between the q^{-1}

and the q^{-4} regimes. It can be seen of figure 1 that ξ is slightly concentration-dependent. In the intermediate q range the cross-over region moves towards lower q values upon dilution. At low-q, all the samples behave the same way and the reduced scattered intensity decreases as q^{-1}. Figure 2a shows the results of this dilution experiment in a "Porod's plot" representation: this plot enhances the variations of the reduced scattered intensity $(I(q)/\phi)$ occurring at high q. Here ϕ is the oxide volume fraction for the three samples studied. The observed minima for the three curves, related to the average curvature of the interface, appear at the

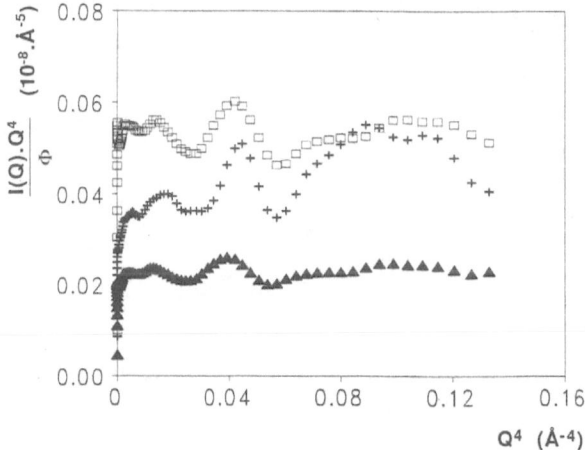

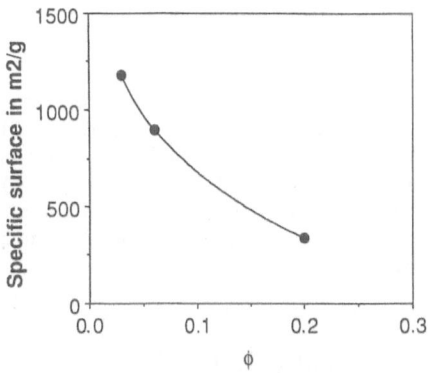

Fig. 2. Dilution experiment of a TiO_2/ZrO_2 mixed oxide sols: (a) "Porod's plot" of the data shown on fig. 1 ($\Delta\ \phi = 0.2$; $+\ \phi = 0.06$; $\square\ \phi = 0.03$). The reduction of the asymptotic limit with concentration enhances partial aggregation of the solid oxide network in the concentrated sol. (b) the asymptotic regime allows a direct and model independent measurement of the oxide/solvent area in m^2/g. A sharp decrease of this quantity is observed upon increasing solids concentration

same q values. The high q limit of the curves, also called the Porod's asymptote, increases with decreasing ϕ. Without any further assumption, the total interfacial area Σ-which has been deduced from the Porod's limit using expression {1} — is represented on figure 2b. We see here that diluting the sample with pure water increases the specific area without modifying the average curvature, because the oscillations remain at the same position. This means that dilution decreases the probability of oxide network connections. There is a reversible aggregation between the oxide interfaces, without coagulation. The values of the specific area per unit volume of the three samples are given in Table 1. On figure 3, the measured scattered intensity for the most diluted sample ($\phi = 0.03$) is compared to the calculated intensity in case of three ideal microstructures: spheres, cylinders and lamella [17]. Note that these calculations do not require any adjustable parameter. It is clear that only a dispersion of primary particles aggregating into long cylinders is the best image of the TiO_2/ZrO_2 particles dispersions. This is in agreement with the q^{-1} behaviour of the scattered intensity at small angles. Furthermore, the presence in the sol of other topologies would lead to a measured intensity different by at least one order of magnitude from the observed one. Upon increasing the solid volume fraction, particles stick on existing cylinders and increase their average radius, thus inducing a specific area reduction.

The picture of a network of disordered cylinders is therefore the only one compatible with the order of magnitude of the observed scattering. To check for self-consistency we calculated the average chord length I_c [18].

$$I_c = \frac{\pi \int_0^\infty q \cdot I(q) \cdot dq}{\int_0^\infty q^2 \cdot I(q) \cdot dq} \qquad (2)$$

For a connected or a disordered dispersion of cylinders the average chord length is the analog of the typical size of a dispersion of isolated particles. This average chord length (I_c) may be compared to a correlation length ξ. For the three volume fractions, the order of magnitude of the measured average chord length is comparable to the size

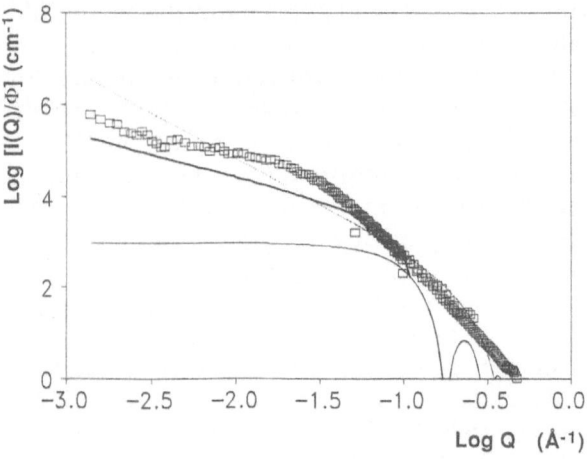

Fig. 3. The experimental data (□) measured for the most diluted mixed oxyde sample ($\phi = 0.03$) are compared to the three ideal microstructures: a dispersion of spheres (thin line), a dispersion of infinite randomly oriented and non interacting cylinders (thick line) and a perfectly dispersed infinite and flat platelets (dotted line). The characteristic size of these ideal objects are not adjustable parameters, but they are fixed for each topology, by the known volume fraction ϕ and the specific area Σ. For spheres: $R = 23$ Å, for cylinders $R = 16$ Å and for platelets total thickness $t = 16$ Å. The dispersion of cylinders is the best model for the SYNROC sol dispersion: other topologies imply a scattered intensity differing by at least one order of magnitude from the measured one

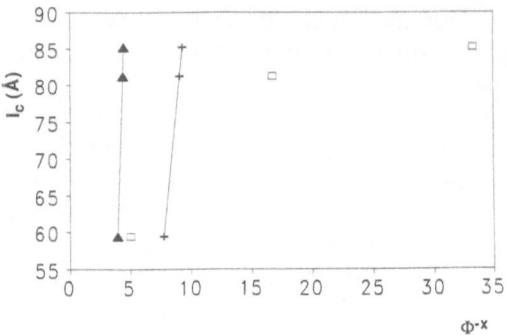

Fig. 4. Self-consistency for locally cylindrical microstructure. The average chord length I_c is plotted versus $\phi - 1$ (□), $\phi^{-1/2}$ (+), $\phi^{-1/3}$ (Δ). The non linear behaviour in ϕ^{-1} excludes a one dimensional swelling as expected for lamella

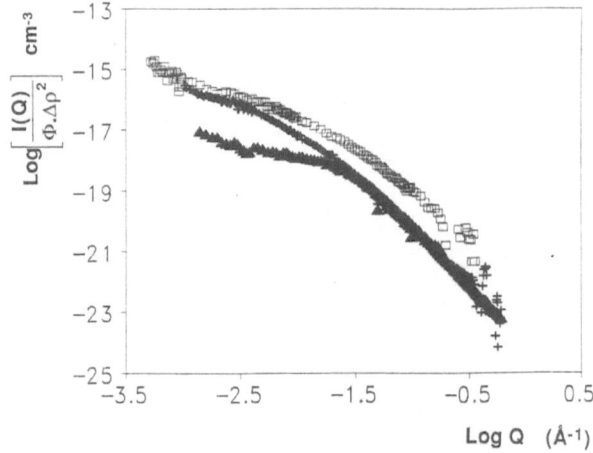

Fig. 5. Normalized scattering data comparing the most diluted TiO_2/ZrO_2 mixed oxides dispersion (Δ; $\phi = 0.03$), the "Degussa" Al_2O_3 dispersion (+) used in the precursor preparation and the real SYNROC "B" precursor (□). The effect of the small amount of alumina on the precursor is to produce branched, dendritic-like objects and a more compact structure

associated to the graphically determined cross-over (see table 1). Diluting a set of parallel charged cylinders with solvent should produce a square root dependence as predicted for polyelectrolytes [19]. Indeed figure 4 shows the difficulty to distinguish between a square root dependence ($\phi^{-1/2}$) and a dilution in the three directions of space ($\phi^{-1/3}$). This might be the case is the cylinders are not parallel to each other. But the non-linear behaviour in ϕ^{-1} completely excludes that the particles form platelets.

2) The SYNROC precursor sol

The complete SYNROC precursor, mainly from the TiO_2/ZrO_2 sols differs by the addition of alumina to the Ti-Zr oxides couple. The composition of the precursor also includes some minor quantities of other oxides (Al_2O_3, CaO and BaO), as indicated in table 1. On figure 5 the scattering by the complete SYNROC precursor sol is compared to

the scattering by a diluted TiO_2/ZrO_2 sol and the alumina sol used to prepare the precursor sol. The scattered intensity is normalized by dividing it both by the volume fraction ϕ and the contrast term $\Delta\rho^2$. With such a normalization, the correlation functions may be directly compared. The scattering by the SYNROC precursor and the alumina sol are somewhat different from that of the titanium-zirconium oxides sols. At a scale of 100 Å the structure of the SYNROC precursor is closer to the structure of the alumina sol. As the alumina sol the SYNROC precursor network does not show any typical size, no scattering peak being observed. The slope of the low-q part for both materials is close to —1.2 against —1 for the titanium-zirconium oxide sol. The high electropositivity of barium and calcium and to a lower extend of alumina seems to induce a high branching probability for the cylinders and a more compact structure. At large q there is only a smooth cross-over from the q^{-4} scattering regime to a q^{-3} behaviour indicating both sols have a more compact structure. The dense network is close to the precipitation limit [21].

3) The dried form of the precursor

We also investigated in situ the microstructure of the dried product, prior to the hot pressing step used to obtain the final SYNROC product. After drying for several hours at a temperature of 110°C, the precursor looks like a white fine powder. The scattering by a real SYNROC precursor, before and after drying, is represented in a log-log plot on figure 6. As in the case of the precursor sol described above, the scattered intensity is normalized by dividing it both by the volume fraction and the contrast term. The volume fraction of the dried precursor was taken as 0.7. With absolute units for the scattered intensity this allows for direct comparison of the specific surfaces of the liquid and dried precursor at high-q. As described above, prior to drying the scattered intensity in the low-q range decreases as $q^{-1.2}$. At high q, as expected, the scattered intensity exhibits a q^{-4} dependence. After drying two Porod's limits are observed depending on the observation scale (fig. 7). The scattered intensity in the low q range decreases as q^{-4} against $q^{-1.2}$ before drying. Scattering by the surface of large particles is observed. In addition there is an intermediate q range where the intensity behaves roughly as q^{-0}. A q^{-4} dependence is recovered at high q. In the high q range both scattering curves obtained before and after drying merge into a single one meaning that the concentrated liquid sol and the dried precursor have the same specific area Σ. This allows for determining the origin of the grains. During the drying process the objects in the precursor sol, with a measured specific surface of 3900 m²/g, aggregate to form larger particles. The coarse grains which form have a specific surface σ close to 10 m²/g the specific surface area of the precursor powder measured by BET is 540 m²/g. The specific surface

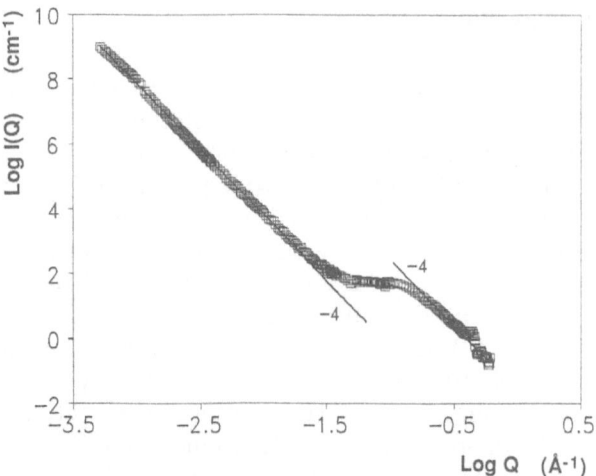

Fig. 6. Small-angle X-ray scattering by the "SYNROC" dried powder. Two Porod's limits characterized by their q-4 behaviour are observed. At low resolution, the specific area of the dry powder is about 10 m²/g. At high q a total specific interface of 3900 m²/g is measured

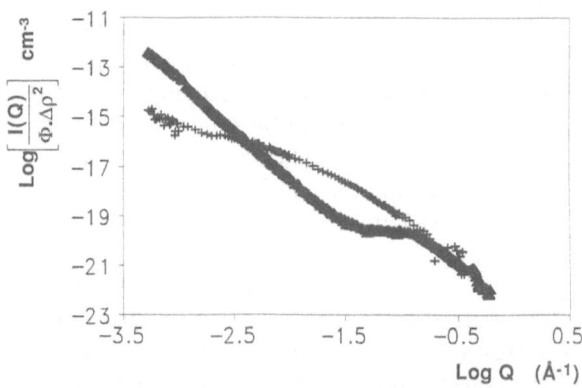

Fig. 7. Normalized scattered intensity comparing scattering before (+) and after (Δ) drying for the same sample of SYNROC precursor

area measured by x-ray scattering strongly depends on the resolution of the experiment. This represents a reduction of a factor of 400 in the specific area. All specific surfaces are derived from the absolutely scaled measurements for the scattered intensity. It is thought that the cross-over region and the q^{-0} regime observed in the intermediate-q range are related to the porosities inside the large particles.

Conclusion

We obtain three main results from this in-situ study of the microstructure of solutions of mixed oxide particles and SYNROC precursor:

(i) The TiO_2/ZrO_2 mixed oxide particles sols used as model system of the SYNROC precursor in the first part of this study show a locally cylindrical microstructure. The average cylinder diameter changes with the concentration in primary particles in the sol. With a particles volume fraction in the range of 0.03 to 0.20 the radius of the cylinders varies from 2 to 4 nm. The particles form a network of disordered and partially connected cylinders. This disordered network of cylinders is able to resist to precipitation upon reducing electrostatic repulsions due to topological reasons. Flocculation would imply breaking and reorganization of the network. The type of local microstructure, here cylindrical, is important because it determines the maximum amount of dispersed material before flocculation occurs.

(ii) The SYNROC precursor gives a slightly different SAXS spectrum from the mixed oxides particles sols. It shows a more dense and compact dendritic structure than the TiO_2/ZrO_2 mixed oxide sols. The scattered intensity in the low q range varying as $q^{-1.2}$ against $q^{-1.0}$, and a smooth cross-over from a q^{-4} regime to a q^{-3} one being observed at high q. The precursor can be seen as a network of branched disordered cylinders. The addition of other minor components as calcium and barium oxides, and especially alumina, is certainly responsible for that.

(iii) The precursor microstructure is completely changed after drying, so that two specific surfaces are observed. The network of branched cylinders collapses to form large grains. The surface of these grains scatters at low q. Their measured specific surface σ is about 10 m^2/g. At high q the total surface Σ of the grains forming objects scatters. Comparison of reduced scattered intensity by dried and non dried precursor shows that the scattering surface Σ at high q is the same than in the case of the concentrated liquid sol. So that, in the high q range the observed scattering surface after drying is the surface of the collapsed network of mixed up cylinders originally in the sol. This measured specific surface Σ of the cylinders is about 3900 m^2/g. This shows how the material starts to densify due to aggregation during drying calcination, prior to the hot pressing step. The q^{-0} behaviour observed in the intermediate q range could be due to porosities inside the grains.

Acknowledgement

We thank E. Sizgek for preparing the SYNROC samples, Dr. L. Vance A. N. S. T. O. project manager for waste conditioning for reviewing the paper, and the Uranium Trust Fund Commitee for the financial support to one of the author (R. Amal) to work in the Service de Chimie Moléculaire, C. E. A. Centre d'Etudes de Saclay, France.

References

1. Ringwood AE (1978) Safe disposal of high-level nuclear reactor wastes: a new strategy, Australian National University Press, Canberra
2. White TJ, Segall RL, Turner PS (1985) Angw Chem Int Ed Engl 24:357—365
3. Woolfrey JL, Bartlett JR, Buykz WJ (1988) Proceedings of the 2 *nd* International Conference on Ceramic Powder Processing Science, Berchtesgaden, FRG, Oct 12—14 In: Haumer, Messing, Hirano (eds) (1989) German Ceramic Soc, pp 43—50
4. Sizgek E, Bartlett JR, Woolfrey JL (1992) Ceramics, Adding the Value. Proceedings of the Interrational Ceramics Conference, Melbourne, Australia, p 1185—1191
5. Hunter RJ (1987) in "Fundations of colloid Science, volume 1", Oxford Science Publications, Clarendon Press — Oxford
6. Lambard J, Zemb Th (1991) A triple axis Bonse & Hart camera used for high resolution small-angle scattering, J. Applied Cryst, vol 24 part 5:555
7. Lesieur P, Zemb Th (1992) in "Structure and dynamics of strongly interacting colloids in solution", Chen SH, Huang JS, Tartaglia P editors, NATO ASI, serie C: mathematical and physical sciences, v. 369
8. Lambard J, Lesieur P, Zemb Th, to be published in Journal de Physique
9. Amemiya Y, Matsushita T, Nakagawa A, Satow Y, Miyahara J, Chikawa J (1988) Nuclear Instruments and Methods in Physics Research A266:645—653
10. Né F, Gazeau D, Lambard J, Lezieur P, Zemb Th, Gabriel A, accepted in J Affl Cryst
11. Zemb Th, Charpin J (1985) J Physique 46:249—256
12. Wignall GD (1991) J Appl Cryst 24:479
13. Auvray L (1991) in "Neutron, X-Ray and Light Scattering, Introduction to an Investigative Tool for Colloidal and Polymeric Systems", edited by P. Lindner and Th. Zemb, North-Holland
14. Schmidt PW (1991) J Appl Cryst 24:414—435
15. Zemb Th (1991) in: Lindner P, Zemb T (eds) Neutron, X-Ray and Light Scattering, Introduction to an Investigative Tool for Colloidal and Polymeric Systems, North-Holland
16. Belloni L (1991) in "Neutron, X-Ray and Light Scattering, Introduction to an Investigative Tool for Colloidal and Polymeric Systems", edited by P. Lindner and Th Zemb, North-Holland

Progress in Colloid & Polymer Science   Progr Colloid Polym Sci 93:247 (1993)

17. Glatter O, Kratky O, in "Small angle X-ray scattering", Academic Press, a subsidiary of Hartcourt Brace Jovanovich, Publishers.
18. Tchoubar D, Mering J (1969) J Applied cryst 2:128
19. De Gennes PG, Pincus P, Velasco RM, Brochard F (1976) J de Physique 37:1461
20. Zemb Th, Hyde ST, Derian PJ, Barnes IS, Ninham BW (1987) J Phys Chem 91:3814
21. Kallala M, Cabane B (1992) J Phys II France 2:7—25

Authors' address:

D. Gazeau
Gentre d'Etudes Nucléaires de Saclay
Service de Chimie Moléculaire
Bâtiment 125
Gif-sur-Yvette Cedex, France

Fig. 1

Direct visualization of monolayer structures with a brewster angle microscope

U. Gehlert, S. Siegel, and D. Vollhardt

Max-Planck-Institut
für Kolloid- und Grenzflächenforschung
Berlin, FRG

Key words: Monolayer — monostearin glycerol — Brewster angle microscopy

The Brewster Angle Microscope is a new device for visualizing monolayer morphologies using the zero reflectance of a water surface for p-polarized light under Brewster's angle of incidence. At the air/water interface a monolayer causes a change in reflectivity and allows the observation of monolayer structures.

Upon compression, monostearin-rac-glycerol changes its phase state on a water subphase at 35°C. With the beginning of first-order phase transition the formation of liquid phase domains becomes visible. The shape of these domains depends on growing kinetics. In the equilibrium state the domains are circular with an inner structure (Fig. 1). At higher compression rates the growth depends on the direction and two-dimensional stars are formed (Fig. 2). If the barrier is stop-

Fig. 2

ped, the system relaxes and the domains become circular again after 20 min.

Authors' address:

U. Gehlert
Max-Planck-Institut
für Kolloid- und Grenzflächenforschung
Rudower Chaussee 5
12489 Berlin, FRG

Depolarized quasielastic light scattering from reaction limited aggregates of anisotropic spherical particles

A. Vailati, D. Asnaghi, and M. Giglio

Physics Department, University of Milan, Italy

Abstract: Reaction Limited Cluster Aggregation is a well known process showing universal features: i) aggregates are fractal with a fractal dimension $d_f \simeq 2.1$, ii) the cluster mass distribution $N(m)$ decays as $m^{-\tau}$ for small m and exhibits a fast cut-off $\exp\left(\dfrac{-m}{m_c}\right)$ at large m, and iii) m_c grows exponentially in time. — In this work, we studied salt-induced aggregation of optically anisotropic spherical particles by quasielastic light scattering. By selecting the depolarized component at a low scattering angle we were able to collect only the rotational contributions to the field correlation function I_{VH}. This greatly simplifies the data interpretation, leading to a correlation function which exhibits stretched exponential behavior. By measuring I_{VH} at different aggregation time, we fully characterized the aggregation process by determining the fractal dimension, the cluster mass distribution and the aggregation kinetics in a single experiment. — Our results are in good agreement with previous experimental and theoretical works.

Key words: Aggregation — fractal — depolarized dynamic light scattering — optical anisotropy

Introduction

In order to fully characterize an aggregation process, one has to study the aggregate morphology, the clusters' mass distribution and the time evolution of the average cluster mass. None of the most commonly employed experimental techniques is completely satisfactory: T.E.M. and Coulter counting are too invasive and require sample manipulation, static and dynamic polarized light scattering give full information only when used together.

In this paper, we show how the Reaction Limited Cluster Aggregation can be fully characterized by means of a single noninvasive experimental technique, namely, depolarized dynamic light scattering from optically anisotropic aggregates.

Let us recall some of the well known features of RLCA [1, 2]. Aggregates exhibit fractal morphology with a fractal dimension $d_f \simeq 2.1$. The clusters system is polydisperse with a mass distribution

$N(m) \sim m^{-\tau} \exp\left(-\dfrac{m}{m_c}\right)$, where m is the cluster mass, m_c is a cut-off mass, and τ is a constant (masses are expressed in terms of the number of monomers composing the cluster). Moreover, the time evolution of the cut-off mass is exponential.

In this work, we present data on d_f, m_c and τ obtained by using quasielastic light scattering in conjunction with salt-induced aggregation of optically anisotropic spherical particles. The clusters too are optically anisotropic and depolarize the scattered light. Both their translational and rotational diffusion bring about fluctuations in the scattered field. Selection of the depolarized component at a low scattering angle allows to neglect the translational contribution to the field autocorrelation function $I_{VH}(t)$ since its decay is due to rotational diffusion only. As we will explain later, for large delay times I_{VH} exhibits a stretched exponential behavior, i.e.

$I_{VH} \sim \exp\left[-\left(\dfrac{t}{T_{se}}\right)^a\right]$. It will be shown that a is simply related to d_f, while the decay time T_{se} depends both on the cut off mass m_c and on d_f. Thus, from the large delay time behavior of I_{VH} measured at different aggregation times, we are able to determine the time evolution of d_f and m_c. Furthermore, the value of τ is extracted by studying the first cumulant of I_{VH} at short delay time.

Theory:
scattering from optically anisotropic aggregates

Suppose that a vertically polarized monochromatic light wave impinges on a fluid suspension of submicronic optically anisotropic particles. The response of the system to the incident field depends on the polarizability of both the particles and the solvent. If the particles are uniaxic, we can describe their response by means of two parameters, the average excess polarizability a and the optical anisotropy β [3]:

$$a = \left(\dfrac{a_{\|} + 2a_{\perp}}{3}\right) - a_s, \quad \beta = a_{\|} - a_{\perp},$$

where $a_{\|}$ is the polarizability component parallel to the particle principal axis, $a_{\perp}$ in any direction perpendicular to it, and a_s is the average polarizability of a portion of the solvent having the same volume as the particle.

The scattered field is the superposition of two components: one with the same polarization of the incident wave, the other perpendicular to it. The two corresponding autocorrelation functions are [3]:

$$I_{VV}(q,t) = N\left[a^2 \exp(-Dq^2 t) + \frac{4}{45}\beta^2 \right.$$

$$\left. \times \exp(-Dq^2 t)\exp(-6\Theta t) \right], \qquad (1)$$

$$I_{VH}(q,t) = N\frac{3}{45}\beta^2 \exp(-Dq^2 t)\exp(-6\Theta t). \quad (2)$$

In Eqs. (1) and (2) $q = \dfrac{4\pi}{\lambda}n_s \sin(\theta/2)$ is the transferred momentum, where θ is the scattering angle, n_s the solvent refractive index, λ the vacuum wavelength. D and Θ are the particle Stokes-Einstein translational and rotational diffusion coefficients, respectively, N is the number of particles in the scattering volume. Notice that Eqs. (1) and (2) are derived in the Rayleigh-Gans approximation: $\dfrac{4\pi}{\lambda}a_0(n_p - n_s) \ll 1$, where a_0 is the particle radius and n_p is the particle refractive index.

The next step is to generalize Eqs. (1) and (2) to the case of clusters, i.e., to write the suitable parameters D, Θ, a, β. We assume that the diffusion coefficients D and Θ of a fractal aggregate of gyration radius a are those of a sphere having the same radius. Keeping into account the relation between mass and radius of a fractal, $m \sim a^{d_f}$, we can write its diffusion coefficients D_m, Θ_m in terms of the cluster mass: $D_m = Dm^{-1/d_f}$, $\Theta_m = \Theta m^{-3/d_f}$.

As to the cluster mean polarizability a_m and optical anisotropy β_m, it is quite reasonable that their dependence on m is described by the following relations: $a_m \sim ma$, $\beta_m \sim \sqrt{m}\beta$. In fact, the optical axes of the particles composing the cluster are randomly oriented, consequently, the particles' contributions to the polarized scattered field add coherently, while those to the depolarized scattered field, incoherently [4].

Experimental techniques

The technique we employed requires the use of optically anisotropic spherical particles. We are very

grateful to Roberto Piazza (Dipartimento di Elettronica, Sezione di Fisica Applicata, Università di Pavia), who provided us with the suitable sample, an aqueous suspension of spheres of a polytetrafluoroethylene copolymer, characterized by an average refractive index $\bar{n} = 1.3536$ and $n_\parallel - n_\perp \simeq 10^{-2}$. The method used in these measurements is described in [5].

The particle radius is $a_0 = 45.8 \pm 3$ nm, as measured by polarized dynamic light scattering.

The sample was diluted to a volume fraction 0.4% and the aggregation was started by adding NaCl 30 mM to the sample. One correlation function was then measured every 2 h for 1 week.

The incident laser beam is vertically polarized by a polarizer placed before the cell and the scattered beam passes through an analyzer which selects the horizontal component of the field. We optically matched the solvent refractive index with the average particle refractive index, by adding glycerol to the aqueous suspension, up to a weight fraction of 17.5%. Notice that the index matching implies $a = 0$. This means that I_{VV} and I_{VH} are proportional, $I_{VV}/I_{VH} = 4/3$, since I_{VV} loses its pure translational term [5]. Thus, in this condition possible stray leakage of I_{VV} does not change the decay of the autocorrelation function we measure.

In order to neglect the decay due to the translational diffusion in (2) the condition $(q^2 D_m/6\Theta_m) \ll 1$ must be satisfied. This implies $(qa)^2 \ll 1$. Obviously, a is the largest cluster size we plan to observe. With the requirement $a \simeq 10a_0$, which corresponds to a cluster of about 100 monomers, we obtain that the scattering angle θ should be about $3°$.

Data analysis

The raw data consist of a sequence of correlation functions taken at different times during the aggregation. The measured correlation function I_{VH} is a superposition of contributions due to clusters of different size, weighted by the cluster mass distribution $N(m)$ and the cluster anisotropy β_m. If we neglect the translational diffusion terms:

$$I_{VH} \propto \frac{\displaystyle\sum_{m=1}^{\infty} m[m^{-\tau}\exp(-m/m_c)]\exp(-6\Theta m^{-3/d_f}t)}{\displaystyle\sum_{m=1}^{\infty} m[m^{-\tau}\exp(-m/m_c)]}. \qquad (3)$$

For large values of the delay time t, I_{VH} can be approximated with a stretched exponential function [6, 7]: $I_{VH} \propto \exp[-(t/T_{se})^a]$, where

$$a = d_f/(3 + d_f) , \tag{4}$$

$$T_{se} = \frac{m_c^{3/d_f}}{6\Theta} \frac{d_f}{3} \left(\frac{3 + d_f}{3}\right)^{-(3+d_f)/d_f} . \tag{5}$$

The values of d_f and m_c derived by fitting the correlation function tail with a stretched exponential are shown as a function of aggregation time in Figs. 1 and 2.

Notice that d_f exhibits a slight decrease in time. This phenomenon has been already observed [8, 9] and it is ascribed to a slow transition from the reaction limited to the diffusion limited aggregation [1]. As to the time evolution of m_c, it can be observed that the predicted exponential behavior is attained after some initial growth at variance with it. This is probably due to the fact that in the early aggregation stages the clusters' polidispersity is not sufficiently broad to be reasonably modeled by the predicted mass distribution.

Additional and important information can be extracted from the correlation function at small delay times. Consider the first cumulant of I_{VH}:

$$\Gamma_{VH} = -\frac{d}{dt} \{\ln[I_{VH}(t)]\} \big|_{t=0} . \tag{6}$$

Γ_{VH} can be easily evaluated from I_{VH} at small delay times. From Eqs. (3) and (6), we can derive the dependence of Γ_{VH} on the parameters τ, d_f, m_c and Θ:

$$\Gamma_{VH} \propto \frac{\sum_{m=1}^{\infty} m[m^{-\tau} \exp(-m/m_c)] 6 \dfrac{\Theta}{m^{3/d_f}}}{\sum_{m=1}^{\infty} m[m^{-\tau} \exp(-m/m_c)]} . \tag{7}$$

The monomer rotational diffusion coefficient Θ has been determined by low-angle depolarized dynamic light scattering from a non aggregated sample. Incidentally, from this measurement we derive the average particle radius $a_0 = 42$ nm, in good agreement with the value determined by polarized dynamic light scattering $a_0 = 45.8$ nm.

Since we measure Γ_{VH}, d_f, m_c and Θ, we can insert them into Eq. (7) and numerically solve for τ.

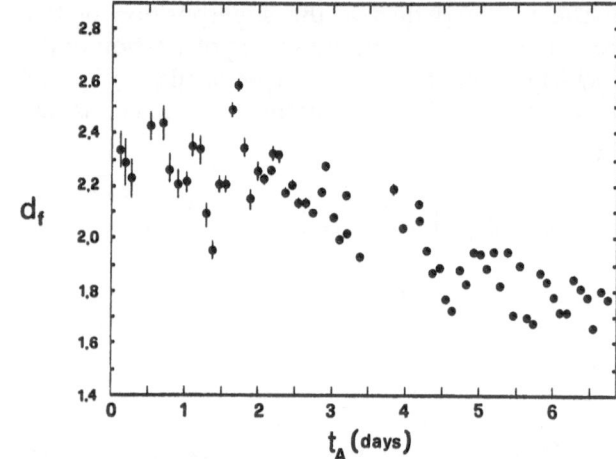

Fig. 1. The clusters' fractal dimension versus the aggregation time

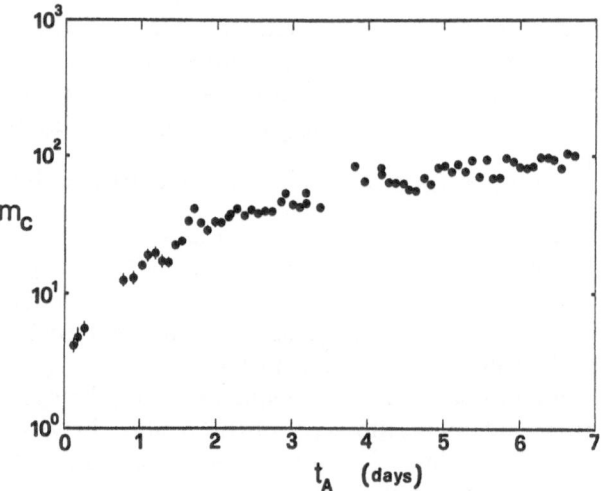

Fig. 2. Time evolution of the cut-off mass

A plot of τ versus aggregation time is shown in Fig. 3. Notice that at the very early aggregation stages the measured values of τ are affected by a large indetermination. This is probably due to the fact that the cluster mass distribution has not yet attained the proper asymptotic form. However, as the aggregation proceeds τ is eventually better defined and it approaches the value 1.5, which is in good agreement with the rather scant experimental evidence provided so far [2, 10, 11].

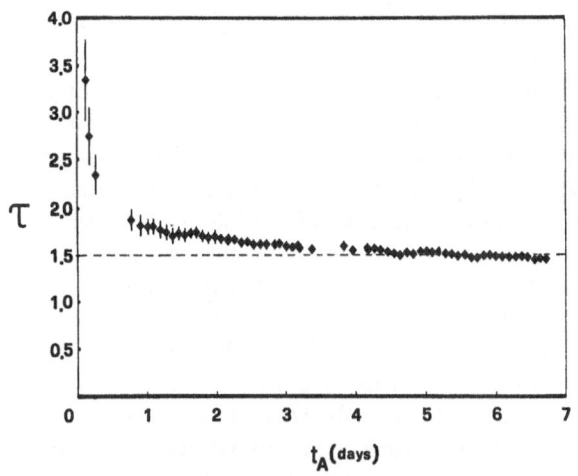

Fig. 3. τ versus the aggregation time

Authors' address:

Prof. M. Giglio
Dipartimento di Fisica
Sezione di Ottica Quantistica e Applicata
V. Celoria 16
20133 Milano, Italy

Formation of micelles and gels from block copolymers in water followed by ultra-sound velocimetry and complex densiotmetry

O. Glatter, K. Schillén, and G. Scherf

Institute of Physical Chemistry,
University of Graz, Austria

Key words: Ultra-sound velocimetry — densimetry — micelles — gel

As an additional result, we find that the value of m_c, corresponding to the time when τ begins to approach 1.5, is about 50.

Acknowledgements

Thanks are due to R. Piazza and V. Degiorgio for stimulating discussions and the loan of the sample.

This work has been supported by grants from the Ministero dell'Università e della Ricerca Scientifica e Tecnologica.

References

1. Family F, Meakin P, Vicsek T (1985) J Chem Phys 83:4144—4150
2. Lin MY, Lindsay HM, Weitz DA, Ball RC, Klein R, Meakin P (1990) Phys Rev A 41:2005—2020
3. Berne B, Pecora R (1975) Dynamic Light Scattering, Wiley, New York
4. Piazza R, Degiorgio V (1992) Opt Commun 92:45—49
5. Piazza R, Degiorgio V (1992) Physica A 182:577—592
6. Piazza R, Bellini T, Degiorgio V, Goldstein R, Leibler S, Lipowsky R (1988) Phys Rev B 38:7223—7226
7. Bellini T, Mantegazza F, Piazza R, Degiorgio V (1989) Europhys Lett 10:499—503
8. Asnaghi D, Carpineti M, Giglio M, Sozzi M (1992) Phys Rev A 45:1018—1023
9. Robinson DJ, Earnshaw JC (1992) Phys Rev A 46:2065—2071
10. von Schulthess GK, Benedeck GB, De Blois RW (1980) Macromolecules 13:939
11. Broide ML, Cohen RJ (1990) Phys Rev Lett 64:2026—2029

The PEO-PPO-PEO triblock copolymer P-85 forms complex states of aggregates in aqueous solution. Depending on temperature and concentration monomers, micelles and a body-centeredspace-cubic liquid crystalline phase can exist.

These states can be studied in detail with small-angle scattering of x-rays or neutrons, dynamic light scattering, and viscosity measurements. These experiments enable a clear description of the different states of aggregation, but are rather time consuming.

The formation of micelles from monomers can be monitored by ultra-sound velocimetry. Measurement of the time required for a short pulse to run through a well-defined sample cell is used to calculate the speed of the ultra-sonic wave. It can be shown that this speed is dependent on the number of particles dissolved in a liquid and the number is reduced by the formation of micelles. These experiments can be performed in a wide concentration range from less than 1% (w/w) up to 40%. Formation of the b.c.c. gel phase does not change the velocity of the sound waves significantly.

The highly increased viscosity in the gel phase changes the damping of the oscillations in a special densimeter. Changes in the damping forces can be used to monitor the transition from the micellar solution to the gel phase. This transition takes place within a range of a few degrees. The gel is formed at concentrations above 23% (w/w) and the width of the gel regime increases with concentration. The advantage of this technique is that only negligible forces are applied to the sample.

Authors' address:

Otto Glatter
Institute of Physical Chemistry
University of Graz
Heinrichstraße 28
8010 Graz, Austria

The study of single biological and model membranes via small-angle neutron scattering

V. I. Gordeliy*, L. V. Golubchikova[1], A. Kuklin[1], A. G. Syrykh[1], and A. Watts[2]

[1] Moscow State University, Moscow, Russia
[2] University of Oxford, UK

Abstract: In this work we return to the possible uses of small-angle neutron scattering (SANS) in the study of single membrane structure in water solutions at low membrane concentrations (i.e., in excess water). — Using the example of lipid vesicles the effectiveness of this method in the determination of membrane thickness will be shown and the limits within which this method can be used with the same aims will be set. — Most important is the fact that the use of SANS on single membranes in solution makes possible to determine the position of molecular groups with an accuracy of ~1 Å. That is, with accuracy no less than that achieved in the artificial situation on multilayer structures with the use of neutron diffraction [2, 3]. — It is important that the worked-out method is valid in the case of any single membrane, either neutral or charged, in solutions with various physico-chemical parameters.

Key words: Small-angle neutron scattering — structure — biological and lipid membranes

* Present address: Laboratoire Leon Brillouin, C.E. de Saclay, 91191 Gif-Sur-Yvette, Cedex France
Permanent address: Laboratory of Neutron Physics, Joint Institute for Nuclear Research, Dubna, Moscow District, Russia, 141980

Introduction

To determine the structure of biological objects with high resolution is a very difficult problem. Normally, it becomes possible only if one can grow a crystal of biological objects, such as proteins. The higher the constitution of a biological object, the fewer the chances to grow crystals of them. Thus, for instance, even in the case of model membranes, one can get a quasicrystal structure along the normal of membranes only at small water concentrations. But even in this case the diffraction picture is poor and contains only several diffraction reflections. In the best case, at small hydrations of neutral membranes, one manages to achieve the structural resolution of several angströms [1]. And only the use of neutron diffraction combined with deuterium labeling of molecular groups made it possible to determine the position of these groups within an accuracy of about 1 Å [2, 3]. The question still remains of whether it is correct to extrapolate the data obtained on multilayer structures (at external conditions being far from biological ones) onto biological membranes.

Some efforts to use neutron scattering for investigation of the structure of single membranes have been made. In [4, 5] small-angle neutron scattering was used to study membrane structure and to determine the thickness of lipid bilayer. In [6] the method of specular reflection of neutrons was used for the first time to study the bilayer of dimiristoyl phosphatidylcholine molecules adsorbed on the surface of quarz substrate. No doubt in this case the method produced superb results. At the same time, this work showed that the interaction between the membrane and substrate is strong enough. This means that, in more complicated cases, for instance, even in the case of protein-lipid membranes, it can cause the membrane asymmetry induced by the substrate.

In this work, we return to the possible uses of small-angle neutron scattering (SANS) in the study of single membrane structure in water solutions at low membrane concentrations (i.e., in excess water).

Using the example of lipid vesicles the effectiveness of this method in the determination of membrane thickness will be shown and the limits within which this method can be used with the same aims will be set.

Most important is the fact that the use of SANS on single membranes in solution makes it possible to determine the position of molecular groups with an accuracy of ~1 Å, i.e., with an accuracy no less than what was acheived in the artificial situation on multilayer structures with the use of neutron diffraction [2, 3].

It is important that the worked-out method is valid in the case of any single membrane, either neutral or charged, in solutions with various physico-chemical parameters.

Determination of membrane thickness from radius of gyration

The measuring of the lipid membrane thickness

In this section it will be shown that, if one knows the gyration radius of membranes along their normal, one can determine their thickness with high accuracy.

In [7] it was shown that small-angle scattering intensity on flat structures (for instance, membranes) can be, under certain conditions, approximated by the expression analogous to Guinier approximation for spheric particles:

$$I(Q) = Q^{-2} I_T(Q) , \qquad (1)$$

where Q is the neutron scattering vector, Q^{-2} is the so-called Lorentz factor of the plane, and $I(Q)$ is the so-called thickness factor. The $Q = 4\pi \sin\Theta/\lambda$, where 2Θ is the scattering angle, and λ is the wavelength of neutrons.

The thickness factor $I_T(Q)$ is determined by:

$$I_T(Q) = I(0) \exp\left[-Rc^2 Q^2\right] . \qquad (2)$$

Approximation (1) is valid at

$$2\pi/S^{1/2} < Q < 1/Rc , \qquad (3)$$

where S is the area of membrane, and Rc is the radius of gyration which is determined in the following way [8]

$$Rc^2 = \frac{\int \rho(x) x^2 dx}{\int \rho(x) dx} , \qquad (4)$$

where $\rho(x)$ is the scattering density in the direction of the normal of membrane plane.

For the membrane in solution with the scattering density ρ_S expression (4) can be rewritten as [8]

$$Rc^2 = \frac{\int_{-T/2}^{T/2} (\rho_M(x) - \rho_S) x^2 dx}{\int_{-T/2}^{T/2} (\rho_M(x) - \rho_S) dx} . \qquad (5)$$

where $\rho_M(x)$ is neutron scattering density of membrane, T is the thickness of the membrane. For a uniform distribution of scattering density from R_C we can immediately determine the thickness of the lamellar system [7]:

$$T = Rc \overline{\sqrt{12}} . \qquad (6)$$

In the case of membranes $\rho_M(x)$ is not a constant. Considering that $\rho_S = $ const, formula (5) for membranes can be rewritten as:

$$Rc^2 = \frac{Rco^2 - b_W/b_L(T^2/12)}{1 - b_W/b_L} , \qquad (7)$$

where b_L and b_W is the scattering amplitudes of the lipid molecule and of the water molecules in the volume which corresponds to the volume of one lipid molecule, and Rco is determined by formula (4) with $\rho(x) = \rho_M(x)$.

Let us estimate b_W and b_L. In the case of neutron scattering b_W depends on isotope composition of water. It is easy to see that

$$b_W = (V_L/V_{W1})/b_{W1} , \qquad (8)$$

where V_L and $V_{W1} = 29.9$ Å^3 are the volumes of one lipid and one water molecule, respectively, and b_{W1} is the scattering amplitude of water molecule. The value of V_L can be calculated from a partial volume $\bar{v}$ of dipalmitoylphosphotidylcholine (DPPC). From [10], $\bar{v} = 0.948$ cm^3 g^{-1} at the temperature 20°C. It follows then that the volume of a DPPC molecule is: $V_L = (M/Na) \bar{v} = 1156$ Å^3 (M is the molecular weight of DPPC, Na is Avogadro's number).

For DPPC molecules $b_L = 2.763 \times 10^{-12}$ cm, $b_{H_2O} = -0.168 \times 10^{-12}$ cm and $b_{D_2O} = 1.914 \times 10^{-12}$ cm. Inserting these values into (8), we get that for D_2O $b_W/b_L = 26.7$. In this case the formula (7) can be rewritten as:

$$Rc^2 = T^2/12 - Rco^2/(b_W/b_L) . \qquad (9)$$

Even in the case, when all the strongly scattering components of membrane are concentrated near its surface, $Rco < T^2/4$. Calculations of Rco from neutronographic Fourier profiles show that, in fact, $Rco^2 \sim T^2/10$. For SANS measurements in the solution of heavy water, to calculate the thickness of lipid membranes one can use the approximation $T = Rc \overline{\sqrt{12}}$.

The error in T can be evaluated from expression (9). In the case of DPPC membranes it does not exceed ~2%, at the typical bilayer thickness ~50 Å; the absolute error does not exceed 1 Å.

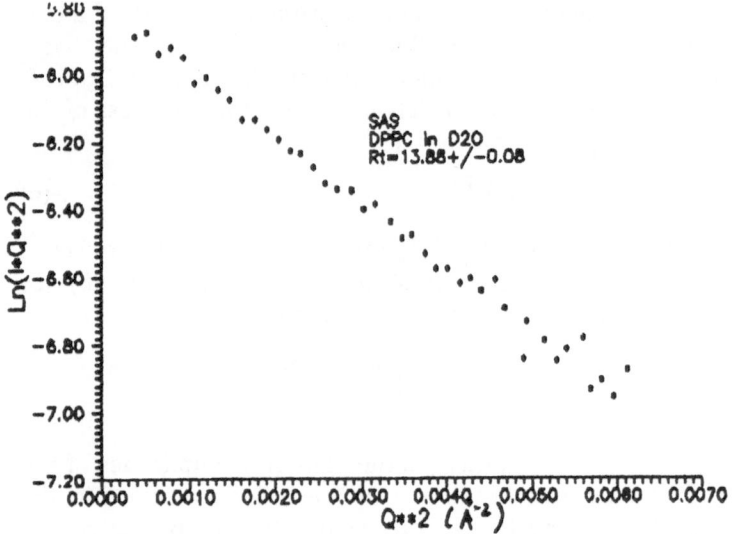

Fig. 1. The dependence of small-angle neutron-scattering intensity I for DPPC vesicles (1% (w/w) in D_2O) on the scattering vector Q

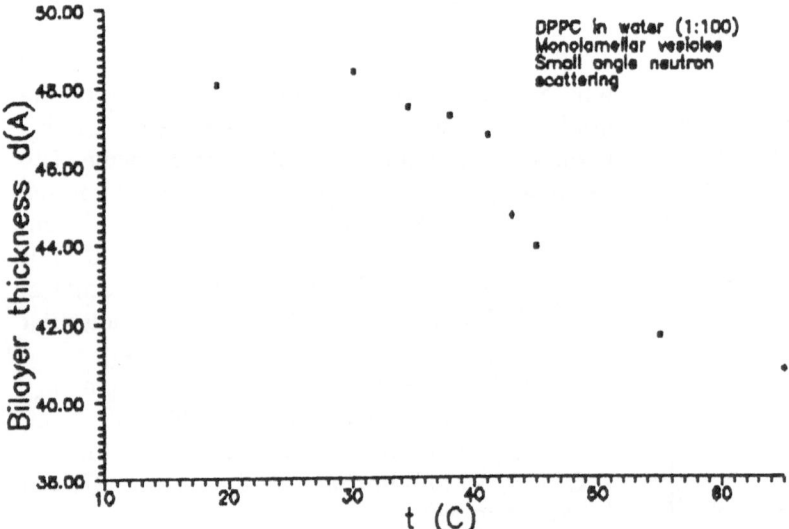

Fig. 2. The dependence of the DPPC membrane thickness d on the temperature t

Equation $T = Rc\,\overline{V12}$ coincides with the earlier received expression for the homogeneous distribution of the scattering density in the membrane. The physical essence of the coincidence is simple.

In solution, while calculating the radius of gyration, one must use the difference between the scattering densities of the membrane and solution instead of the first one (see Eq. (5)). In heavy water ρ_s is much higher than the scattering density of a lipid molecule. The reason is that the sum of positive scattering amplitudes of carbons, oxygens, nitrogen and phosphorus is practically fully compensated by the total of negative scattering amplitudes of a big number of hydrogens.

Figure 1 shows a typical dependence of $I(Q)$ for DPPC vesicles in D_2O (1% w/w) of DPPC). In these measurements the lipids of SERVA company were used. Vesicles were prepared by the method described in [4]. The measurements were carried out with the SANS instrument at IBR-2 reactor, Laboratory of Neutron Physics, Joint Institute for Nuclear Research, Dubna, Russia. The typical time of measuring one spectrum was 1 h.

Figure 2 shows the dependence measured via this method of DPPC membrane thickness $d_L(t)$ on temperature t. The statistical accuracy of the determination of membrane thickness in these experiments was ± 0.3 Å. The comparison of $d_L(t)$ with the analogous dependence attained from the measurements via X-ray diffraction on DPPC dispersions [10] shows their similarity within the limits of experimental errors. And yet, SANS has an evident advantage, because only one measurement is needed to determine the membrane thickness, and moreover, the thickness of a single membrane in solution is measured.

*The measuring of the biological
and lipid-protein membranes thickness*

We showed earlier that, in the case of the scattering density of the solution being much higher than the scattering density of the membrane, the membrane thickness is determined from radius of gyration by Eq. (6).

Yet, in the case of biological or lipid-protein membranes the ratio $\rho_S \gg \rho_M$ normally is not valid. The reason is that the scattering density of proteins is large enough for neutrons.

Further, it will be shown that, in the general case, three measurements at different contrasts (i.e., different solutions H_2O/D_2O) give an exact value of membrane thickness from the measured radii of gyration.

Let us rewrite expression (7) as:

$$Rc^2 = \frac{Rco^2 - (b_{H_2O} \cdot \chi + b_{D_2O}(1 - \chi))(T^2/12)/b_M}{1 - (b_{H_2O} \cdot \chi + b_{D_2O}(1 - \chi))/b_{D_2O})/b_M} , \tag{10}$$

where χ is the concentration of H_2O in solution. In Eq. (10) only T, Rco, and b_M are not known. Only three measurements are needed to determine these parameters. These are measurements at three different values of χ, i.e., at three different contrasts.

**Determination of molecular groups positions
in membrane in solution via small-angle
neutron scattering**

Membrane thickness is one of the important structural parameters, but not the only one in which biophysicists and biologists are interested.

Here, it will be shown, by way of measuring the radius of gyration of single membrane combined with deuterium labeling of a molecular group of the membrane component, how one can determine the membrane structure with an accuracy of about 1 Å.

Let us write the scattering density profile of membrane as $\rho(x)$. Then, for the same membrane with a localized deuterium label the scattering density can be written as:

$$\rho_D(x) = \rho(x) + n(b_D - b_M)\delta(x - x_0) , \tag{11}$$

where n is the number of substituted hydrogens, b_D and b_H are neutron scattering lengths of deuterium and hydrogen, respectively, and $\delta(x - x_0)$ is the δ-function of Dirac; x is the position of the label.

Putting $\rho_D(x)$ in (5), one gets

$$Rcd^2 = \frac{Rch^2 + n((b_D - b_H)/b)x_0^2}{1 + n(b_D - b_H)/b} , \tag{12}$$

where Rch is the radius of gyration of protonated membrane, and $b = b_L - b_W$.

If we know the lipid composition and b from independent measurements of partial lipid and water volumes, then, as follows from Eq.(12), two measurements (of Rcd and Rch) are enough to determine the position of the label.

To check the statements mentioned earlier, we measured the radii of gyration of membranes of DPPC (Fig. 1) and of DPPC with a totally deuterated choline group (DPPC-D9). The measured gyration radii of DDPC membrane is $Rch = 13.88 \pm 0.08$ Å, and for DPPC-D9 membranes it is $Rcd = 10.91 \pm 0.08$ Å.

Substituting all the known parameters in Eq. (12), we get $x_0 = 25.9 \pm 0.3$ Å.

Thermal movements spread the label. Following [2, 3], let us suppose that the label has the Gaussian distribution

$$\rho(x) \sim \exp[-(x - x_0)^2/v^2] . \tag{13}$$

Substituting (13) into (5), we get

$$Rcd^2 =$$

$$\frac{Rch^2 + n((b_D - b_H)/b))(x_0^2 + (2/\sqrt{\pi})x_0 v + 0.5\ v^2)}{1 + n(b_D - b_H)/b} . \tag{14}$$

Equation (14) is analogous to (12), but instead of x_0^2, we have

$$x_0'^2 = x_0^2 + (2/\sqrt{\pi})v x_0 + v^2 . \tag{15}$$

Progress in Colloid & Polymer Science   Progr Colloid Polym Sci 93:256 (1993)

At $v \to 0$ it follows from (14) that $x_0' = x_0$, just as it should be according to (14). Normally, for membranes $v = 1$—3 Å [2, 3].

To observe how the method of determination of the position of molecular groups works in the presence of thermal fluctuations ($v \neq 0$), we will use the data of [2]. For a DPPC membrane in the gel phase (at 20°C) v is equal to 3 ± 0.6 Å. This result was achieved via neutron diffraction.

Substituting the formerly calculated $x_0' = 25.9 \pm 0.3$ Å and $v = 3$ Å in Eq. (15), one gets $x_0 = 24.9 \pm 0.3$ Å.

In [2], via neutron diffraction on multilayer dispersions of DPPC membranes at $t = 28$°C (gel-phase) at water concentrations of 25%, it is shown that the choline group is placed at a distance of $x_0 = 24.4 \pm 0.6$ Å from the center of the membrane.

Considering that structural parameters of membranes in the gel-phase practically do not depend on temperature, and that water concentration 25% corresponds to maximum hydration, one can determine the identity of DPPC membranes in the case of [2] and this one. Comparing the value of $x_0 = 24.9 \pm 0.3$ Å to the value $x_0 = 24.4 \pm 0.3$, determined in [2], their full coincidence is evident.

Even in the case of the widest distribution of the molecular group in membrane ($\gamma = 3$ Å), the difference in label positions, determined considering $v = 0$, does not exceed 1 Å.

In summary, the proposed method of studying the structure of single (not multilayer) membranes in solution makes it possible to determine with high accuracy (~ 1 Å), not only the integral parameters of membranes (like their thickness), but the positions of membrane components as well.

We stress that the accuracy (in a sense, structural resolution) of this method is not worse than that achieved with the help of neutron diffraction on multilayer membranes [2, 3].

Also, the proposed method has an evident advantage: it makes possible to study the structure of single membranes under external conditions in which membranes exist in the cell.

Acknowledgement

The authors express their gratitude to Yu. M. Ostanevich for helpful discussions and V. G. Cherezov, M. A. Kiselev, and S. P. Yaradaykin for their help in carrying out the experiments.

References

1. Blaurock AE (1982) Biochem et Biophys Acta 650:167—207
2. Buldt G, Gally HU, Seelig J, Zaccai G (1979) J Mol Biol 134:673—691
3. Zaccai G, Buldt G, Seelig A, Seelig J (1979) J Mol Biol 134:692—706
4. Knoll W, Haas J, Stuhrmann HB, Fuldner H-H, Vogel H, Sackmann E (1981) J Appl Cryst 14:191—202
5. Sadler DM, Reiss-Husson F, Revas E (1990) Chem Phys Lipids 52:41—48
6. Johnson SJ, Bayerl TM, McDermott DC, Adam GV, Rennie AR, Thomas RK, Sackmann E (1991) Biophys J 59:289—294
7. Kratky O, Porod G (1948) Acta Phys Austriaca 2:133—137
8. Stuhrmann HB (1976) Brookhaven Symp Biol No 27
9. Pilz J, Herbst M, Kratky O, Osterhelt D, Lynen F (1970) Eur J Biochem 13:55—64
10. Inoko Y, Toshio M (1978) J Phys Soc Japan 44(6):1918—1924

Authors' address:

Dr. V. Gordeliy
Leon Brillouin Laboratory
C.E. Saclay
91191 Gif-sur-Yvette
Cedex, France

Critical micelle concentration in a ternary system AOT — water — decane as determined by small-angle neutron scattering

N. Gorski and Y. M. Ostanevich

Laboratory of Neutron Physics,
Joint Institute for Nuclear Research, Dubna, Russia

Key words: Interfaces — micelles — neutron scattering

The critical micelle concentrations (CMC) in the ternary system AOT + xH$_2$O + C$_{10}$D$_{22}$ at three water contents $x = 0$, 22.3 and 32.4 are found as CMC = 0.24 ± 0.05, 1.60 ± 0.40 and 0.76 ± 0.21 mM, respectively. For two systems with a large water content, we have observed a second critical concentration, CMC$_2 = 2.5 \pm 0.4$ and 11.6 ± 3.3 mM at $x = 22.3$ and 32.4, respectively. At the concentrations $C_{AOT} > $ CMC$_2$ the dependence of the observed characteristics of micelles on the surfactant concentration changes drastically. The first three CMCs correspond to the formation of the micelles and agree with CMC at $x = 30$, reported by other authors. The nature of the last two remains unclear for the present, though the experimental data show fast enlargement of all characteristics of the aggregates, when the solute concentration exceeds the CMC_2.

Authors' address:

Dr. N. Gorski
Joint Institute for Nuclear Research
141980 Dubna, Russia

Optimal phase behavior of water/oil blend/surfactant systems

A. Graciaa[1]), J. Lachaise[1]), G. Morel[1]),
J. L. Salager[2]), and M. Bourrel[3])

[1]) LTEMPM, Université de Pau, France
[2]) LFIRP, Universidad de los Andes, Mérida, Venezuela
[3]) Groupement de Recherches de Lacq, Artix, France

Abstract: In this paper, we study the optimal behavior of water-alcane/(alkyl)benzene blend -polyethoxylated octylphenol systems as a function of the number of ethylene oxide units of the surfactant. We find this behavior highly nonlinear and we partially interpret it by means of a differential fractionation of the surfactant. Remaining gaps from linearity suggest an interfacial oil segregation that we find favorable to polar and small (alkyl)benzene molecules.

Key words: Optimal phase behavior — optimal microemulsion — oil blend — surfactant partitioning — oil segregation — alcane — (alkyl)benzene

Introduction

In most industrial applications the organic phases of microemulsions are composed of oil blends which must be modelized to obtain a better understanding of the phase behavior of the systems. Thus, for instance, in enhanced petroleum recovery formulation of efficient microemulsions must consider the very complex composition of petroleum. Thus, scientists have tried to establish relations of equivalence between petroleum and model oil

blends towards the optimum behavior of microemulsions [1]. These blends are generally composed of two, sometimes three alcanes, alkylbenzenes or alkylcyclohexanes [2—6].

The optimal behavior of homologous oils the molecular volumes of which are similar, has been found to be linear in function of the parameter which provides the optimum. But nonhomologous oils, or oils which have very different molecular volumes, present nonlinear optimum behaviors which are not well understood [1, 7].

In this paper, we study the optimal behavior of water-alcane/(alkyl)benzene blend-polyethoxylated octylphenol systems in function of the number of ethylene oxide units of the surfactant. We examine this behavior in the light of the differential fractionation of the surfactant components.

Experimental

Water was deionized and redistilled. Alcanes (tetradecane, octane), and (alkyl)benzenes (benzene, butylbenzene, octylbenzene) were manufactured by Fluka.

The ethoxylated octylphenol nonionic surfactants were manufactured by Seppic and are marketed under the trade name Montanox. They are composed of oligomers which differ in their number of ethylene oxide units; these oligomers are distributed according to a Poisson distribution [8] that we have verified by HPLC.

Aqueous (or organic) solutions of the surfactants were prepared and mixed with an oil blend (or an aqueous phase) to achieve a water/oil volume ratio equal to 1. The surfactant weight percentage in the systems was 5%.

The systems were placed into sealed tubes which were maintained at $(25.0 \pm 0.1)°C$. The tubes were agitated twice a day during 1 week, then left quiet during 2 months to ensure that equilibrium was attained [9, 10].

At equilibrium, the systems appear under one of the three Winsor types [11, 12]. Optimization is found by varying the average number of ethylene oxide units of the surfactant. It is obtained when the excess organic and aqueous phases have simultaneously identical volume, for a value of the average number of ethylene oxides that we shall call bulk EON*.

The surfactant fractionation between the different phases was measured by gas chromatography with an ionization detector [13]. The compositions of the organic phases in the excess phase or in the microemulsion phase were determined with the same technique.

Results and discussion

Tetradecane was successively mixed with hexane, benzene, butylbenzene, and octylbenzene to form binary mixtures the tetradecane molar fractions of which were varied from 0 to 1. The determination of the bulk EON* of all the systems was performed under the experimental conditions fixed above. The variations of the bulk EON* versus the tetradecane molar fractions in the binary mixtures are reported in Fig. 1. These variations are linear for the tetradecane/hexane blend; they are nonlinear for tetradecane/benzene, tetradecane/butylbenzene and

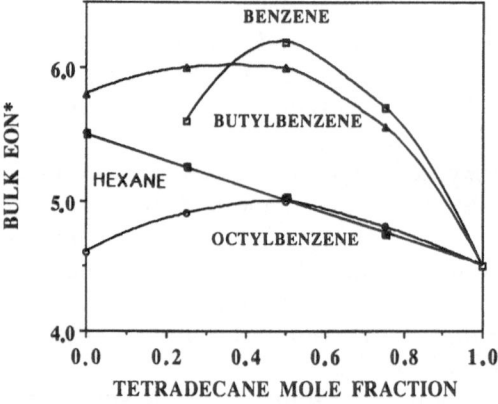

Fig. 1. Variation of the bulk EON* as a function of the tetradecane mole fractions in benzene/tetradecane, butylbenzene/tetradecane, and octylbenzene/tetradecane blends

tetradecane/octylbenzene blends. Each of these three curves presents one maximum. The difference from linearity is the higher as the number of carbons of the (alkyl)benzene is small. Furthermore, it must be noticed that for tetradecane/benzene blends the optimization is impossible for tetradecane molar fractions lower than about 0.25; then the systems are of the Winsor I type.

Surfactant fractionation

We have already reported on several occasions that the organic phases of water/oil/polyethoxylated

octylphenols can contain significant quantities of oligomers [13, 14]. This fractionation depends on the ethylene oxide number of the oligomer, on the water/oil couple, on temperature. It is ruled by partitioning coefficients [8, 10, 15] and critical micelle concentrations (CMCs) of the different oligomers of the surfactant. Critical micelle concentrations are known [16], but partitioning coefficients are unknown for the water/oil couples considered here.

We have directly measured the fractionation of the oligomers, the ethylene oxide numbers of which were included between 2 and 7. We have calculated from these results the partitioning coefficients of these oligomers between the studied water and the different oil blends according to a procedure described elsewhere [17]. It is found that these partitioning coefficients vary linearly as a function of the ethylene oxide number.

For fractionations of the oligomers the ethylene oxide numbers higher than 7 are too small to be measured with a sufficient precision. We have been obliged to calculate them, on the one hand, from linear extrapolations of the determined partitioning coefficients, and on the other hand, from the known CMCs. This calculus is the inverse of the precedent.

Finally, exact knowledge of the partition of all the oligomers allows to determine the effective composition of the surfactant at the interfaces. We have found that this composition differs from the initial surfactant composition, so that we can define an interfacial EON*.

When, for the studied systems, we report interfacial EON* versus tetradecane molar fractions in the blends, the maxima vanish and the variations get closer to linearity (Fig. 2). Thus, surfactant fractionation contributes to the nonlinear variations of the bulk EON* as a function of the tetradecane molar fraction.

Oil segregation

We observe in Fig. 2 that for tetradecane molar fractions lower than about 0.5, interfacial EON* is practically constant and equal to the bulk EON* of the pure (alkyl)benzene. This result suggests that for these tetradecane molar fractions the interfacial surfactant would be principally in contact with (alkyl)benzene. Close to the interfacial zone the oil blend could be poorer in tetradecane than in the

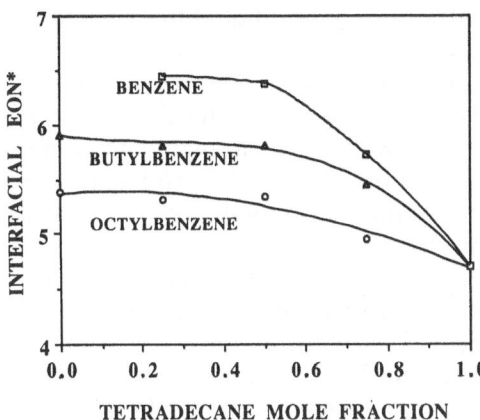

Fig. 2. Variation of the interfacial EON* as a function of the tetradecane mole fractions in benzene/tetradecane, butylbenzene/tetradecane, and octylbenzene/tetradecane blends

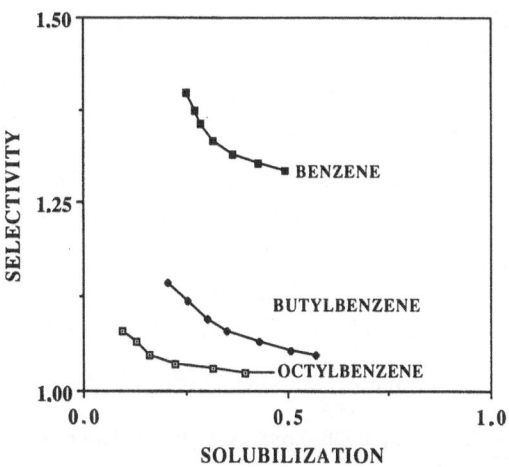

Fig. 3. Interfacial selectivity as a function of solubilization for benzene, octylbenzene and butylbenzene

bulk of the blend, which could enhance the linearization of the interfacial EON* variations.

It is impossible to determine with precision the composition of the oil blend in the interfacial region because the optimal solubilization (ratio of the solubilized oil blend volume by the interfacial surfactant volume in the optimal microemulsion [18]) is high (about 3). Then, the interfacial oil volume is small compared with the bulk oil volume, and the difference between interfacial and bulk compositions cannot be detected easily. On the other hand, this difference could be detected in systems having low solubilization parameters, as have Winsor I systems located far from the optimum. We have obtained such systems by varying the tetradecane/(alkyl)benzene weight proportion of systems formed with a clearly hydrophile surfactant such as the 9 times ethoxylated octylphenol. We have measured the solubilizations of these systems and their selectivities (defined as the ratios of the (alkyl) molar fractions in the dispersed oil phase of the microemulsion and in the excess organic phase). The results are reported in Fig. 3. We observe that the (alkyl)benzene selectivities are higher than 1, which means that (alkyl)benzene molecules are preferentially located in the interfacial zone. For a given solubilization the selectivity is the higher as (alkyl)benzene molecules are polar and small.

Iterative interfacial oil segregation could be used to separe or purify organic phases, as we have recently proposed for surfactants [19]. Indeed, it would be sufficient to balance again with a suitable surfactant the oil blend extracted from the optimal microemulsion phase.

Conclusion

The highly nonlinear optimal phase behavior of systems composed of water, tetradecane/(alkyl)-benzene blend, polyethoxylated octylphenol systems can be partially interpreted by the differential fractionation of the surfactant oligomers. Remaining gaps in linearity suggest an interfacial oil segregation that we measured on systems presenting small solubilizations. We have found that this segregation is favorable to polar and small (alkyl)benzene molecules; it could serve to separate or purify organic blends.

References

1. Maura C, Puerto, Reed RL (1983) Soc Pet Eng, Paper SPE 10678
2. Cayias JL, Schechter RS, Wade WH (1976) Soc Pet Eng J 351—357
3. Cayias JL, Schechter RS, Wade WH (1977) J Colloid Interf Sci 59:31—38
4. Cash L, Cayias JL, Fournier G, McAllister D, Schartz T, Schechter RS, Wade WH (1977) J Colloid Interf Sci 59:39—44
5. Salager JL, Morgan JC, Schechter RS, Wade WH (1979) Soc Pet Eng J 107—115
6. Bourrel M, Verzaro F, Chambu C (1987) SPE Reservoir Engineering 2(1):41—53
7. Bourrel M, Schechter RS (1988) Microemulsions and Related Systems, M. Dekker, New York

8. Schick MJ (1966) Nonionic Surfactants p 45, M. Dekker, New York
9. Bourrel M, Graciaa A, Schechter RS, Wade WH (1979) J Colloid Interf Sci 72(1):161—163
10. Graciaa A, Barakat Y, Schechter RS, Wade WH, Yiv S (1982) J Colloid Interf Sci 89:217
11. Winsor PA (1948) Trans Faraday Soc 44:367—398
12. Winsor PA (1954) Solvent Properties of Amphiphilic Compounds, Butterworths Scientific Publications, London
13. Graciaa A, Lachaise J, Sayous JG, Grenier P, Yiv S, Schechter RS, Wade WH (1983) J Colloid Interf Sci 93:474—486
14. Graciaa A, Lachaise J, Marion G, Schechter RS (1989) Langmuir 5:1315—1318
15. Crook EH, Fordyce DB, Trebbi GF (1965) J Colloid Sci 20:191—204
16. Crook EH (1963) J Phys Chem 67:1987
17. Sayous JG (1983) Thesis, University of Pau, France
18. Reed RL, Healy RN (1977) In: Shah DO, Schechter RS (eds) Improved Oil Recovery by Surfactant and Polymer Flooding p 383—437. Academic Press, New York
19. Graciaa A, Lachaise J, Marion G, Bourrel M, Rico I, Lattes A (1989) Tenside Surfactants and Detergents 26(6):384—386

Authors' address:

Professor J. Lachaise
L.T.E.M.P.M.
Centre Universitaire de Recherche Scientifique
Avenue de l'Université
64000 Pau, France

Studies of PS latex demonstrated that the surfactant molecules adsorbed on the particle surface significantly contribute to the measured scattering intensities [1]. For PMMA latex and PS latex structure factors were obtained up to high volume fractions (20%). The fits of the respective data were made using a repulsive coulomb potential (RMSA approximation).

PS/PMMA latexes were prepared by seeded emulsion polymerization (seed: PS particles) [1, 2]. The existence of a core-shell structure could be established for the latexes under investigation. It seemed that polymerization of MMA by means of the absorption method led to a relatively diffuse interface between PS and PMMA — in contrast to polymerization of MMA under starved conditions.

References

1. Grunder R, Urban G, Ballauff M (1993) Colloid Polym Sci 271:563
2. Grunder R, Kim YS, Kranz D, Müller HG, Ballauff M (1991) Angew Chem 103:1715; (1991) Angew Chem Intern Ed 30:1650

Authors' address:

M. Ballauff
Polymer-Institut
Universität (TH) Karlsruhe
Kaiserstraße 12
76128 Karlsruhe, FRG

Small-angle x-ray analysis of latex particles

R. Grunder, G. Urban, and M. Ballauff

Polymer-Institut, Universität Karlsruhe, FRG

Key words: Latex particles — small-angle x-ray scattering

A small-angle x-ray scattering (SAXS) study of polystyrene (PS), polymethylmethacrylate (PMMA), and polystyrene (PS)/polymethylmethacrylate (PMMA) latex particles is reported. The scattering experiments were performed in the q-range from 0.037 nm^{-1} to 1.5 nm^{-1} ($q = (4\pi/\lambda) \sin(\theta/2)$; θ: scattering angle) using a Kratky camera. In the course of the interpretation of the scattering data additional information from electron microscopy, ultracentrifugation, and soap titration was taken into account.

Molecular approach to the curvature elasticity of lipid bilayers and biological membranes

T. D. Gurkov, P. A. Kralchevsky, and I. B. Ivanov

Laboratory of Thermodynamics and Physico-Chemical Hydrodynamics, University of Sofia, Faculty of Chemistry, Sofia, Bulgaria

Key words: Biomembranes — lipid bilayer

In this work, we consider the bending elasticity of biomembranes taking into account the interactions between the constituent molecules. The lipid bilayer is represented as composed of two layers of polar heads and a thin curved hydrocarbon film consisting of the lipid tails. The polar heads are regarded as hard disks carrying vertically oriented effective dipole moments which repel each other. The hard disk contribution in the bending elastic modulus is evaluated from the relative stretching of the two monolayers upon bending, with the stretching modulus taken from II/A experimental data of other authors. The magnitude of the dipole-dipole repulsion is assessed numerically. The II/A curves provide values for the interaction parameter in the equation of state of the surface layer. Those experimental data also give evidence for the existence of dimers probably due to the horizontally oriented component of the dipole moment of the lipids. The effective vertical dipole moment of the dimers resulting from II/A curves coincides with the outcome of surface potential measurements. This fact points out to the irrelevance of the interaction between horizontal dipoles as far as the long-range interaction constant in the surface equation of state is concerned.

Interactions between the hydrocarbon film and the surrounding water phases provide separate contribution in the bending constant. This is evaluated on the basis of the thin film thermodynamics, accounting for i) the curvature-dependent interfacial tension and the surface bending moment coming from the contact between the water and the hydrocarbon; ii) the disjoining pressure due to the long-range interaction excess. The latter contribution turns out to be negligible.

The resulting total bending elastic modulus is close to the experimentally determined one.

Authors' address:

Ivan B Ivanov
Laboratory of Thermodynamics
Faculty of Chemistry
University of Sofia
J. Boucher Avenue 1
BG-1126 Sofia, Bulgaria

Amphiphilic alkyl piperazine derivatives — synthesis, physical, and colloidal properties of new ampholytic surfactants and bactericides

K. Haage[1]), H. Fiedler[1]), R. Wüstneck[2]), and B. Weiland[1])

[1]) Max-Planck-Institute for Colloid and Interface Science, Berlin-Adlershof, FRG
[2]) KAI e. V., Berlin-Adlershof, FRG

Key words: Ampholytic surfactant — maleic monoester, bacteriostatics — adsorption parameter — surface tension — critical micelle concentration (CMC)

Maleic acid mono [(4-alkyl-piperazine-2-yl)ethyl] esters and mono-4-alkyl-piperazinides with longer alkyl chains show distinct surface activity. The former (II) are interesting model surfactants for the complex behavior of an ampholytic structure, which contains more than one center of basicity in the hydrophilic part. As the equilibrium between the different ionic species (I-IV, scheme 1) depends on the pH value, the surface properties of such an ampholytic system are a result of superimposing effects. This complexity was investigated by testing the series of homologues of the structure II (R varies from C_8H_{17} to $C_{13}H_{27}$, i.e., even and odd carbon number).

The precursor for II, (4-alkyl-piperazine-2-yl)-ethanol, was synthezised via alkylation of 2-piperazinyl-ethanol with alkylbromides. The esterification was carried out with maleic anhydride in molar ratio in ether solution.

The described maleic acid derivatives are potential bacteriostatics. For II the maximum of activity against *Escherichia coli* and *Staphylococcus aureus* was found at the alkyl chain length of $R = C_9H_{19}/C_{10}H_{21}$.

The pKa values of the surfactants were determined by potentiometric titration in water and in aqueous methanol, so that the content of the different ionic forms I-IV in the solution at definite pH values was estimated within the pH range from 0.5 to 10. Depending on the pH value of the solution the adsorption behavior differs extremly at the air/solution interface. The betaine-like structure II at

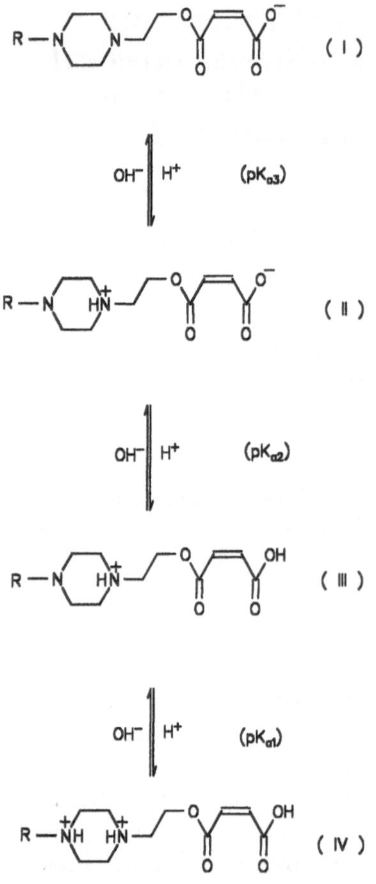

Scheme 1. Protonation equilibrium

Spinodal decomposition of polymer blends in different length scales

Th. Hack, M. Stamm, and V. Abetz

Max-Planck-Institut für Polymerforschung, Mainz, Germany

Different blends of polystyrene and poly(cyclohexylacrylate-stat-butylmethacrylate) have been investigated by small-angle neutron scattering (SANS), double crystal diffractometry (DCD), small-angle light scattering (SALS) and light microscopy (LM) in the one- and two-phase region. The Flory-Huggins interaction parameter χ was determined by SANS in the one-phase region for different temperatures and compositions using the RPA-formula of de Gennes [1]. From those experiments, the spinodal line could be determined.

The time evolution of the structure factor after temperature jumps into the two-phase region was studied by SANS, DCD, SALS and LM. A time behavior, as predicted by Cahn and Hilliard for the early stages [2], could not be measured. The observed coarsening of the concentration fluctuations over a range of length scales between 60 nm and 30 µm, as well as the scattered intensities could be described with scaling laws in accordance with theory [2—4].

Reverse jumps from the two-into the one-phase region have been investigated by SALS. During the dissolution of the inhomogeneities a further coarsening is observed, while the amplitudes decrease. This is in qualitative agreement with predictions by Akcasu [5]. Details are published elsewhere [6].

the isoelectric point (pH 6.2) has the highest surface activity. Because of that and of the low concentration of other ionic species in the bulk phase this pH value was used to determine adsorption parameters of the whole homologous series.

A special method of determining equilibrium values of the surface tension was used. The Frumkin isotherm was selected to determine the adsorption parameter using the σ-log c curves. The CMC, the minimum area demanded by one molecule adsorbed (Ω_F), and the standard free energy of adsorption ΔG^0 show an even/odd alternation like the melting points of these surfactants.

Authors' address:

Dr. habil. Klaus Haage
Max-Planck-Institut für Kolloid- und
Grenzflächenforschung
Rudower Chaussee 5
D 12489 Berlin-Adlershof, FRG

References

1. de Gennes PG (1979) "Scaling concepts in Polymer Physics", Cornell University Press, Ithaca
2. Binder K (1991) Spinodal Decomposition. In: Cahn RW, Haasen P, Kramer EJ (eds) Materials Science and Technology, Vol 5, Phase Transformations in Materials, Weinheim
3. Furukawa H (1985) Adv Phys 34:703—750
4. Hashimoto T, Takenaka M, Jinnai H (1991) J Appl Cryst 24:457—466
5. Akcasu AZ, Erman B, Bahar I: submitted to Makromol Chem

6. Hack T, Stamm M, Mortensen K, Siol W, Abetz V, Acta Polymerica, to be published

Authors' address:

Th. Hack
Max-Planck-Institut für
Polymerforschung
Postfach 3148
55021 Mainz, FRG

SANS studies of hydrating cement pastes

F. Haeussler[1])*, F. Eichhorn[2]), and H. Baumbach[3])

[1]) Department of Natural Sciences,
  Leipzig University of Technology, Germany
[2]) Research Centre Rossendorf Inc., Dresden, Germany
[3]) Institute for Nondestructive Testing Saarbrücken
  (FhG), Saarbrücken, Germany

Key words: Cement paste — hydration — interfaces — neutron scattering — structure

On the spectrometer MURN at the pulsed reactor IBR-2 a hydrating Portland cement paste was studied by small-angle neutron scattering (SANS). By using the TOF-method, a momentum transfer from 0.07 nm^{-1} to 7 nm^{-1} is detectable [1]. The SANS effect is influenced by all subunits existing in hydrating cement pastes. In the measured Q-region the hardening cement paste does not show a Porod-like behavior of SANS-curves. In contrast, the Porod's potential law holds for dry powder samples of clinker minerals and silica fume [2]. By measuring the characteristics of the scattering curves (potential behavior, the radius of gyration, and the macroscopic scattering cross-section at $Q = 0$ nm^{-1}) at different times after the onset of hydration, some evolution of the inner structure of the hardened cement paste can be noted. In [3], on the study of the potential behavior of the scattering curves, Kriechbaum et al. pointed to the fractal nature of the hydration products.

*) present address: Joint Institute for Nuclear Research,
  Frank Laboratory of Neutron Physics, 141980 Dubna,
  Moscow Region, Russia

Acknowledgement

The work reported has been performed with partial support of the Bundesminister für Forschung und Technologie through grant no. 03-BA3LEI. The authors are fully responsible for the content of this publication.

References

1. Ostanevich YuM (1988) Macromol Chem Maromol Symp 15:91—103
2. Häußler F, Eichhorn F, Baumbach H (1991) In: Annual Report of the LNP (1991), Dubna (Russia), pp 165—168
3. Kriechbaum M et al. (1989) Progress Colloid Poly Sci 79:101—105

Authors' address:

Dr. Frank Eichhorn
Research Centre Rossendorf Inc.,
POB 510119
01314 Dresden, Germany

Dr. Frank Häußler
Joint Institute for Nuclear Research
Frank Laboratory of Neutron Physics
141980 Dubna, Moscow Region, Russia

Study of the structure of water-soluble amorphous silicates

M. Hasznos-Nezdei

Institute of Chemical Engineering of the Hungarian Academy of Sciences

Abstract: Amorphous silicates produced from water-glass solutions are applied in many areas of industry. Sodium silicates are used in detergent formulations and potassium silicate powders in paints. These spray-dried powders are water-soluble. The difference in the inner structures of the amorphous water-glass powders can be checked by the thermoanalytical method.

Key words: Spray drying — silicate structure — water-glass powders — dehydration — thermoanalysis

Experimental

The water-soluble amorphous silicates are synthetic products; they can be produced by spray-drying from water-glass solutions [1, 2]. Their compound can be characterized with the following general formula:

$$x M_2O \cdot y SiO_2 \cdot z H_2O \, ,$$

where M is alkali metal, in general, sodium or potassium. The ratio of y/x is between the value of 2 and 4.

The drops of water-glass with sol-state are partly dehydrated during spray-drying and they become amorphous particles. The behaviors of the water-glass powders can be changed by the drying parameters between wide limits. Their water content is determined by outlet drying temperature, and their bulk-density is determined by inlet temperature. The dehydration of the water-glass drops take place during a short time period (10—50 s). First, the surface of the drops is dehydrated. The water leaves the drops as liquid or vapor phase, depending on outlet drying temperature. The remaining water of maximum 20% in the powder can be found in free and bonded forms. Their ratio can be followed by thermoanalytical studies.

Figure 1 shows the derivatogram of the amorphous powder produced from water-glass solutions with modulus of 3,3—3,5 (SiO_2/Na_2O ratio), with concentration of 38,0—40,0 Be⁰, and with technical quality. The powder is produced in a pneumatic pulverizer [3].

Results

The free and bonded water contents of the sample are separate. At low temperature (under 150°C) the free water leaves the system [4]. Its value is about 6% in the sample. Lower outlet drying temperature (with 15°C) was applied in the case of another sample and this was indicated by the lower endothermic peak (79°C). The free water content quarantees the solubility of amorphous silicate powders in water. The solubility in water is higher — in the case of same water content — if the ratio of SiO_2/Na_2O is lower.

The water-glass powders are dehydrated gradually above 130—150°C. In this case the silanol groups are transformed to the siloxane groups. The dehydration process is finished above 500°C. This period contains two steps, indicating that the water bonded to the silanol group exists both as adsorbed and as structural water (like the precipitate silicas).

References

1. Christophliemk P (1985) Glastechn Ber 58:308—314
2. Masters K (1984) Spray Drying Handbook Wiley, New York, pp 563—564
3. Hung Pat 179—554
4. Dent Glasser LS, Lee CK (1971) J Appl Chem Biotechnol 21:127—133

Author's address:

Magdolna dr Hasznos-Nezdei
Institute of Chemical Engineering
of the Hungarian Academy of Science
P.O. BOX 125
8201 Veszprém, Hungary

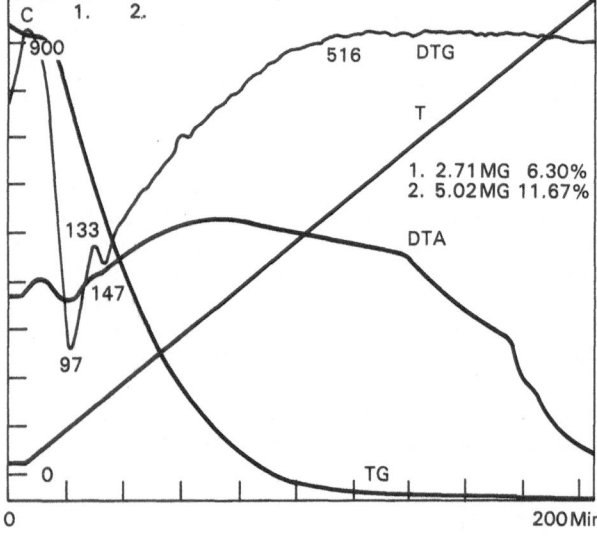

Fig. 1. Thermoanalytical curve of spray-dried sodium silicate (wt. ratio SiO_2/Na_2O = 3,3—3,5)

Configuration transition in thin gelatin layers

H. Hermel and A. Seeboth

Max-Planck-Institute for Colloid and Interface Research, Berlin, FRG

Abstract: By cold-drying thin gelatin layers form a matrix with triplehelical fibrils of preferred orientation. This orientation is variable and depends on the thickness of the layer. Indeed, this is no typical effect of the layer thickness, but it is related to the residual water content (moisture) in the nominally dry layers. These correlations are be discussed. The results are of interest for the orientation of the mesogenic phase of LC molecules dispersed as droplets in the gelatin layers.

Key words: Thin layer — matrix — gelatin layers, preferred orientation — dispersed liquid crystals

Gelatin forms optically transparent, mechanically stable and storable layers. If the layers are produced from aqueous solution (sol) by cold drying, triple helices and collageneous fibrils are reconstituted. They are of preferred orientation (matrix) in the dry layer. We embedded liquid crystals (LC) as a discrete phase (droplets) in the gelatin layer and could show that the gelatin matrix imposes its preferred orientation on the LC molecules in the droplets, initiated by the interaction between the triple helices of gelatin and the LC at the interface protein/LC [1] (Fig. 1).

The preferred orientation of the gelatin matrix is variable. Variation can be achieved by neutral salts [2]. However, different matrix orientations are developed also during layer formation in the cold-drying process.

The cold-drying proceeds via the gel phase. Our conoscopic measurements [3] show that the gel, abundant in water, is optically isotropic. Hence, it follows that the triple helical fibrils have random orientation in the gel network. With increasing dehydration the gel is turning anisotropic. As the measurements show, there is a preferred orientation of the fibrils parallel to the plane of the gel. The same is reported in the literature also for the dry layer [4].

Therefore, it was surprising that the conoscopic measurements of the dry layers not only revealed parallel, but also other preferred orientations of the fibrils. They can be related to different ranges of layer thickness, (Fig. 2 and Table 1).

This variable preferred orientation of the fibrils is not a typical effect of the layer thickness, but it correlates with the residual water (moisture) in the nominally dry layer. We found that the moisture varies with the layer thickness. That can be observed in a small range (12.0 to 17.5% moisture content) and, moreover, only with cold-dried layers, but not

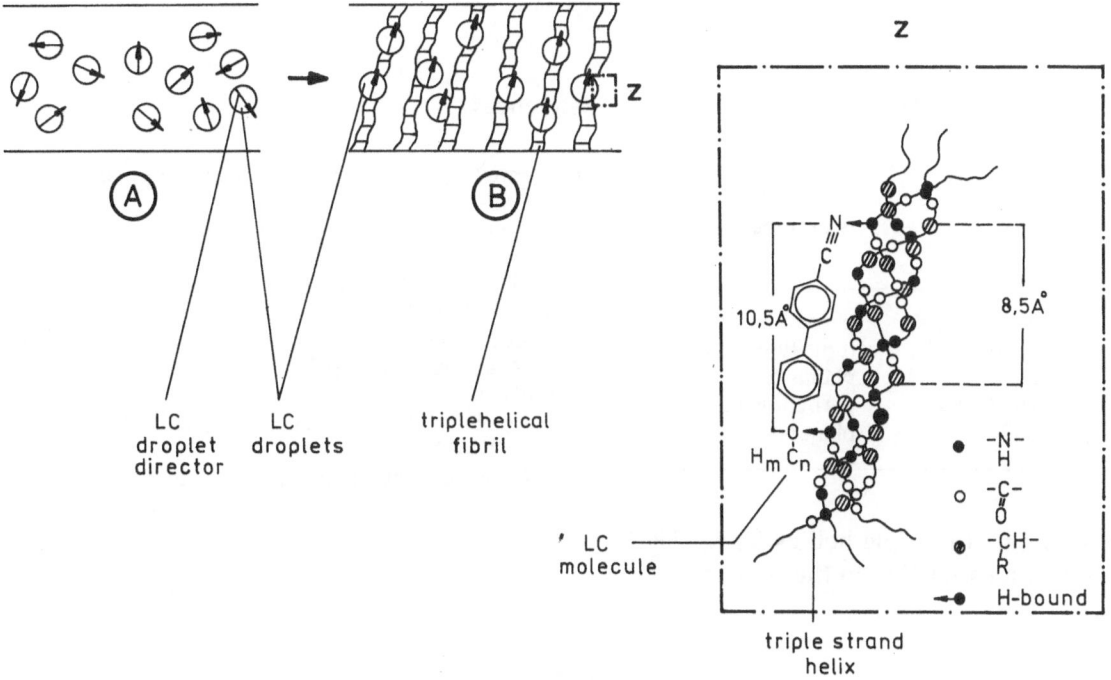

Fig. 1. (Scheme) LC dispersed as discrete phase (droplets). A) Random droplet director orientation; B) In the gelatin matrix preferred droplet director orientation. The gelatin matrix forces its preferred orientation of the LC molecules, initiated by the interaction at the interface. Z) The detail drawing shows the interaction at the interface with cyanobiphenyl, for example

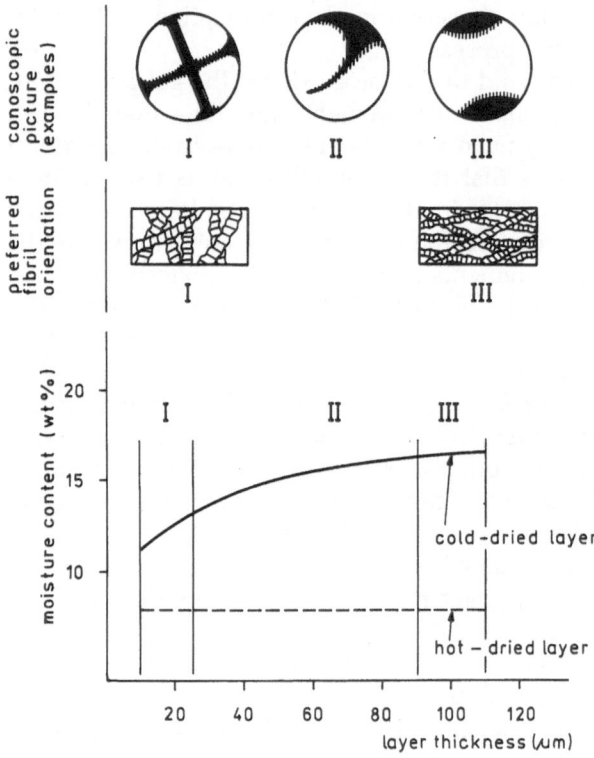

Fig. 2. (Scheme) Moisture in dependence of the layer thickness and the corresponding conoscopic pictures and triplehelical fibril orientations of the cold-dried layer matrix

Table 1. Correlation between layer thickness and preferred matrix orientation

Range of layer thickness [µm]	Tilt angle [degree]	Preferred orientation is dominant
10—25	76, 84, 85, 86,	perpendicular
25—90	6, 10, 12, 13, 20, 27, 36, 44, 48, 72, 83,	transition from planar to perpendicular
90—110	< 7	planar

with hot-dried layers (no triple helices form) (Fig. 2). This points to a correlation to the triple helical fibrils.

What is the reason?

Above the critical gel concentration (between 0.97 and 0.87 wt-% at 4°C) the fibril thickness decreases with increasing sol concentration [5]. Therefore, the

fibrils became thinner with increasing layer thickness. This condition for the production of our differently thick layers was fulfilled [6]. A network of fragile fibrils is highly fine-meshed compared with a network of coarse fibrils. The cavities in the gel network shrink when the gel is dehydrated to the dry layer, but they are later filled with non-jelly-ing aqueous sol phase of the gelatin. According to present findings, the fine-meshed fibril network obviously retains more moisture than the coarse-meshed. This leads to an increase of the internal osmotic pressure, corresponding with findings of Cogrossi [7].

Therefore, the fibril thickness, the water enclosed in the cavities, and the internal layer pressure we regard to be the reason for the fibrils orientation in the layer:

— planar: fragile fibrils, more moisture, high internal layer pressure;
— perpendicular: coarse fibrils, less moisture, low internal layer pressure, while the transition between the two extreme values is smooth.

We are investigating how these results can be used for the orientation of LC dispersed in gelatin layers.

References

1. Hermel H, Seeboth A (1992) Journ Appl Polymer Sci 46:143—146
2. Hermel H, Seeboth A, Kersten B, Legutke H (1991) Journ Imag Sci 35:87—89
3. Seeboth A, Hermel H (1989) Thin Solid Films 173:L119—L121
4. Kogure M et al. in Cox RJ (1976) Photographic Gelatin II, Academic Press, p 131
5. see 4, p 37
6. Hermel H, Seeboth A (1993) Thin Solid Films 223:371—374
7. Cogrossi C (1985) Journ Imag Sci 29:23—26

Authors' address:

Dr. sc. nat. Horst Hermel
MPI for Colloid and Interface Research
Rudower Chaussee 5
D-12489 Berlin, FRG

Progress in Colloid & Polymer Science   Progr Colloid Polym Sci 93:267 (1993)

Characterization of worm-like micelles with solubilized dye molecules using static and dynamic light scattering

B. Herzog[1], K. Huber[1], and A. R. Rennie[2]

[1] Ciba-Geigy AG, Chemicals Division, K-420.504, Basel, Switzerland
[2] Polymers and Colloids Group, Cavendish Laboratory, University of Cambridge, United Kingdom

Abstract: A water-insoluble dye has been solubilized using the cationic surfactant tetradecyltrimethylammonium bromide. At certain concentrations of added potassium bromide giant micelles are present containing large amounts of the dye. The micelles grow when surfactant/dye-concentration is increased. Analysis of static and dynamic light-scattering results suggests a worm-like structure of the micelles.

Key words: Solubilization — surfactant — worm-like micelles — static light scattering — dynamic light scattering

Introduction

It has been shown previously that micelles of tetradecyltrimethylammonium bromide (TTAB) containing water-insoluble dye molecules exhibit considerable one-dimensional growth in the presence of potassium bromide when surfactant/dye-concentration is increased. Flexibility and polydispersity enlarge with the amount of added salt [1]. The present contribution presents static and dynamic light scattering results on the same system.

Experimental

Preparations of TTAB/dye-solutions have been performed as described in [1]. The dye is 4-[[4-[[- (2-hydroxyethoxy)phenyl]azo]-5-methoxy-2-methyl-phenyl]azo]-phenol. Static light-scattering measurements have been carried out with an ALV-1800 apparatus at 19 scattering angles (26°C—143°C) and with a krypton ion laser as light source (647 nm). Radii of gyration have been obtained by taking the initial slope of a plot according to

$$K \cdot (c - CMC)/R_\Theta$$

$$= 1/M_w \cdot (1 + <R_g^2>_z \cdot q^2/3) , \qquad (1)$$

where $K = (2\pi n_0)^2 \, (dn/dc)^2/(N_A \lambda_0^4)$ with n_0 the refractive index of the medium, dn/dc the refractive increment of the solution, N_A Avogadro's constant, and λ_0 the wavelength of incident light; c is the TTAB/dye-concentration and CMC the critical micelle concentration of TTAB in g/cm³, R_Θ is the Rayleigh ratio at angle Θ, M_w the weight average of the molecular weight, $<R_g^2>_z$ the z-average of the radius of gyration, and the scattering vector is defined by $q = 4\pi n_0 \sin(\Theta/2)/\lambda_0$.

Dynamic light-scattering experiments have been performed at five angles (30°C—150°C) using a Malvern PCS 160 SM photon correlation spectrometer with a Malvern K7032 correlator and a helium-neon laser (633 nm) as light source. Apparent diffusion coefficients D_q have been evaluated by cumulants analysis. Translational diffusion coefficients D_0 have been obtained by extrapolation of D_q to $q = 0$ according to:

$$D_q = D_0(1 + C < R_g^2 >_z q^2) . \qquad (2)$$

The dimensionless quantity C, which is sensitive to chain stiffness [2], can be obtained from the slope of a plot of D_q against q^2 using the corresponding value of R_g obtained from static light scattering. From D_0 the hydrodynamic radius can be calculated via the Stokes-Einstein equation:

$$R_h = kT/ (6\pi\eta D_0) , \qquad (3)$$

where η is the solvent viscosity and kT has the usual meaning.

Results and discussion

Tables 1 and 2 show the results of static and dynamic light-scattering measurements at three TTAB/dye-concentration at 0.25 M and 0.5 M KBr.

Radii of gyration and hydrodynamic radii increase with TTAB/dye-concentration, indicating micellar growth. Micelles are larger at the higher KBr-concentration. The ratio $\rho = R_g/R_h$ is a structure sensitive parameter. For monodisperse spheres its theoretical value is 0.778 and for flexible chains 1.5. An increase in chain stiffness or polydispersity leads to values higher than 1.5 [2]. The results show that TTAB-micelles in the presence of salt behave as worm-like chains. The quantity C is sensitive to chain flexibility. It has been calculated for polymer

Table 1. Results of DLS- and SLS-experiments at 0.25 M KBr

c (g/cm^3)	D_0 (cm^2/s)	R_h (nm)	R_g (nm)	ρ	C
$4 \cdot 10^{-4}$	$1.28 \cdot 10^{-7}$	19.1	38.0	1.99	—
$8 \cdot 10^{-4}$	$7.28 \cdot 10^{-8}$	33.7	58.3	1.73	0.303
$1.5 \cdot 10^{-3}$	$4.09 \cdot 10^{-8}$	59.8	91.0	1.52	0.226

Table 2. Results of DLS- and SLS-experiments at 0.5 M KBr

c (g/cm^3)	D_0 (cm^2/s)	R_h (nm)	R_g (nm)	ρ	C
$4 \cdot 10^{-4}$	$1.11 \cdot 10^{-7}$	22.1	46.2	2.09	—
$8 \cdot 10^{-4}$	$6.77 \cdot 10^{-8}$	36.2	72.5	2.00	0.155
$1.5 \cdot 10^{-3}$	$3.47 \cdot 10^{-8}$	73.3	116.6	1.59	0.145

Table 3. Parameters C and reduced lengths $L/(2\,a)$ for TTAB/dye-micelles

$c(KBr)$ (M)	a (nm)	$c = 8 \cdot 10^{-4}$ g/cm^3			$c = 1.5 \cdot 10^{-3}$ g/cm^3		
		L (nm)	$L/(2\,a)$	C	L (nm)	$L/(2a)$	C
0.25	391	219	0.28	0.303	274	0.35	0.226
0.5	71	271	1.91	0.155	525	3.70	0.145

chains in reference [2] as function of the reduced length $L/(2\,a)$, where L is the contour length and a the persistence length. Persistence lengths of TTAB/dye-micelles have been evaluated previously from overall dimensions and particle form factors [1]. The experimental values of L, a, $L/(2\,a)$ and C are listed in Table 3. Figure 1 shows calculations of reference [2] and experimental results of this work.

C reflects rotations and internal motions of the chains. Experimental values of C are of reasonable magnitude, but when plotted against $L/(2\,a)$ they disagree with theoretical curves. Unlike polymers, micelles are in dynamic equilibrium with their monomeric units and their internal dynamics will probably affect C.

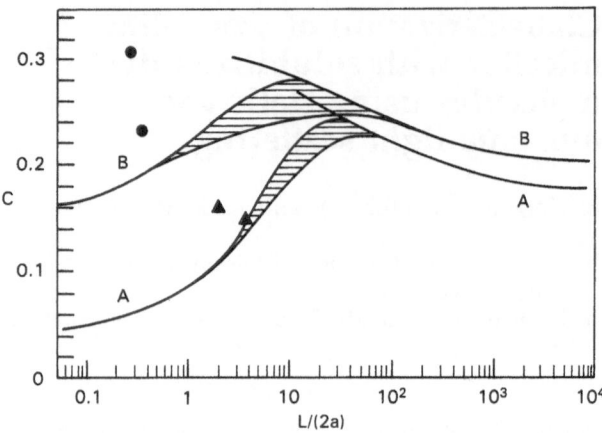

Fig. 1. Calculated dimensionless quantity C plotted against $L/(2\,a)$; curve A for monodisperse ($M_W/M_N = 1$) and curve B for polydisperse distribution ($M_W/M_N = 2$); experimental values have been taken from table 3, circles: $c(KBr) = 0.25$ M, triangles: $c(KBr) = 0.5$ M

Conclusions

Values of ρ obtained with TTAB/dye-micelles at 0.25 M and 0.5 M KBr indicate a worm-like structure. Experimentally obtained quantities C are of reasonable magnitude, but deviate from theoretically expected dependence on reduced length. This may be due to the particular nature of internal dynamics of micelles.

References

1. Herzog B, Huber K (1992) Progr Colloid Polym Sci 89:87—88
2. Schmidt M (1984) Macromolecules 17:553—560

Authors' address:

Dr. Bernd Herzog
Ciba-Geigy AG
Chemicals Division
K-420.504
CH-4002 Basel, Switzerland

Coagulation of polymer colloids by immuno gamma globulin molecules

A. Fernández-Barbero, M. Cabrerizo, R. Martínez, and R. Hidalgo-Alvarez

Biocolloids and Fluid Physics Group, Department of Applied Physics University of Granada, Spain

Abstract: This work deals with certain kinetic aspects of the coagulation of polymer colloids by immuno gamma globulin molecules. Negatively and positively charged polystyrene microspheres were used as a carrier of antibody molecules and the coagulation was studied experimentally by measuring the change in turbidity with time after addition of various amounts of rabbit immuno gamma globulin (IgG). Coagulation occured when those pretreated latexes were mixed with a solution containing IgG molecules. The initial change in turbidity with time has been measured upon addition of various amounts of rabbit IgG to a constant amount of positively and negatively charged polystyrene latex. The general form of these curves does not deviate too far from a parabola. In all cases the coagulation rate is maximum with a surface coverage of one-half, and minimum at full coverage. To explain the experimental data, we have used a theory developed by Singer at al. which considers the adsorption of IgG on the polymer colloids and the coagulation of colloid particles as two consecutive processes. Two different mechanisms have been proposed for the action of protein coagulants, i.e., charge neutralization and bridge formation. — Particle concentration plays an important role in the coagulation kinetic; highest sensibility is found at 10^{10} particles/cm^3. Also, pH exerts a decisive influence in the coagulation kinetic of polymer colloids by IgG molecules. The rate of coagulation at pH 5.0 is much larger than that at pH 9.0. Also, from the straight lines expressing the relation between the time variation of turbidity and the latex concentration the number of IgG molecules adsorbed onto each latex particle can be derived. At pH 9.0 that number is 1050 on negatively charged polystyrene latex particles. Finally, the reaction rate constant of two primary particles forming a doublet was determined by Lichtenbelt et al. theory.

Key words: Polymer colloids — IgG molecules

List of symbols

a	= particle radius
C_1	= extinction cross section of singlets
C_2	= extinction cross section of doublets
K_{11}	= kinetic constant of doublet formation
N	= adsorption sites on PS bead
N_{10}	= initial number of *PS* beads
P_0	= initial protein (IgG) concentration
PS	= polystyrene
τ	= turbidity
$\left(\dfrac{d\tau}{dt}\right)_0$	= initial variation of turbidity with time
λ	= wavelength

Introduction

Protein binding to polymer surfaces is the necessary precursor to a large number of commercially, industrially, and biologically important adhesion phenomena, including biofilm formation, thrombogenesis on medical implant materials and adsorption of antibodies or antigens onto polymer colloids. In the last case, monodisperse polymer microspheres are used as a carrier of diagnostic reagents for clinical examination of a great variety of analities [1—4]. This work deals with certain kinetic aspects of the coagulation of polymer colloids by immuno gamma globulin (IgG) molecules. This study is an effort to establish the agglutination mechanism of organic monodisperse colloids like the polystyrene latices (PS) by proteins. With this aim negatively and positively charged polystyrene beads have been used as carriers of antibody molecules.

Of the various experimental methods available for studying the colloidal stability of polymer beads, we have used turbidity. The effects of particle concentration and pH on coagulation of polystyrene beads have been studied and an attempt has been made to explain the data in terms of the number of IgG molecules adsorbed onto each polystyrene bead.

Materials and methods

Polystyrene beads

Negatively charged polystyrene beads were from Rhône Poulenc (Estapor K030), with an average particle diameter of 297 ± 7 nm and a solid content of 5.05% (w/w). The surface charge density ($\sigma_0 = 6.9$ μC/cm^2) was determined by conductometric and potentiometric titrations. Cleaning of polystyrene dispersions was carried out by ion-exchange.

Positively charged polystyrene beads were prepared essentially according to Goodwin et al. [5] with some slight modifications [6]. The average particle diameter, as obtained by electron microscopy was determined to be 265 ± 15 nm, a solid content of 5.84% (w/w) and $\sigma_0 = 20.0 \pm 0.5$ µC/cm^2.

Antibodies

The antibodies were kindly donated by Biokit S. A. (Spain). Rabbit polyclonal IgG was purified from normal rabbit serum by ammonium sulphate fractionation, followed by anion exchange chromatography. The purified rabbit IgG was stored at −20°C.

Methods

The isoelectric points (IEP) of IgG preparation were determined by isoelectric focusing (IEF) performed on the Phast-System equipment (Pharmacia) using a Phast Gel-1 IEF polyacrylamide media (Pharmacia) that covers the pH range 3—9. The IEP values obtained for the rabbit IgG are in the range 6.1—8.7.

The rate of coagulation is studied by measuring the turbidity as a function of time after addition of various amounts of IgG, using a Spectronic 601 spectrophotometer (1).

Results and discussion

The initial change in turbidity with time has been measured upon addition of various amounts of rabbit IgG to three different amount of anionic polystyrene beads. The results at pH 9 are shown in Fig. 1. The general shape of these curves does not deviate too far from a parabola, although the right-hand size of the curves seems to be somewhat too high, i.e., at surface coverages between half and unity, coagulation seems to be somewhat faster. These results can be reasonably well explained by the theory developed by Singer et al. [7]. According to these authors,

$$\left(\frac{d\tau}{dt}\right)_0 = KN_{10}^2 \frac{P_0}{NN_{10}} \left[1 - \frac{P_0}{NN_{10}}\right] \qquad (1)$$

where $K = (1/2\ C_2 - C_1)\ K_{11}'$ is a constant independent of latex and protein concentrations (2).

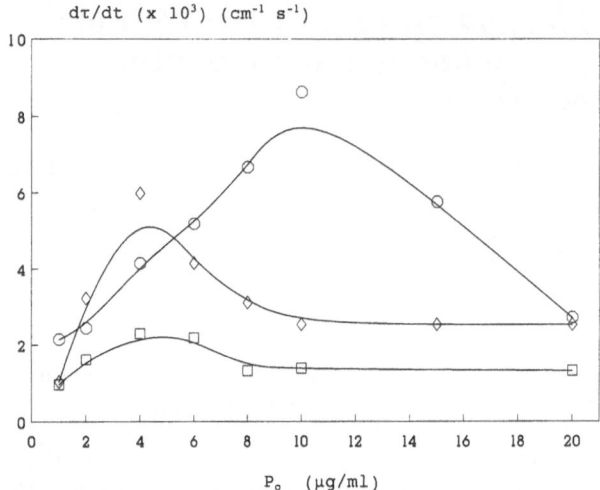

Fig. 1. The initial slope of the turbidity with time curve as a function of the initial protein concentration for three different particle concentration of anionic latex. ○ 1.27 × 10^{10} particles cm^{-3}, ◇ 1.00 × 10^{10}, □ 0.75 × 10^{10}

1) Variation of turbidity was measured at 20°C using incident light with a wavelength of 620 nm for the R-P beads, and 500 nm for the cationic polystyrene beads. These values of λ were selected for two reasons: first, to get an adequate sensibility in the turbidity variations which increases at low wavelengths, and second, the α-parameter value (= $2\pi a/\lambda$) to use the RGD theory.

2) From Eq. (1) we can distinguish three limit cases:
   a) If N_{10} and N remained constant, then $(d\tau/dt)_0$ should depend on P_0 according to a parabolic curve.
   b) If the surface coverage ($\Theta = P_0/NN_{10}$) remained constant, then $(d\tau/dt)_0$ should depend on N_{10}^2 linearly,

$$\left(\frac{d\tau}{dt}\right)_0 = [K\Theta\ (1 - \Theta)]\ N_{10}^2 . \qquad (2)$$

   c) If the initial protein concentration (P_0) remained constant, then $(d\tau/dt)_0$ should depend on N_{10} linearly,

$$\left(\frac{d\tau}{dt}\right)_0 = -\left(\frac{KP_0^2}{N^2}\right) + \left(\frac{KP_0}{N}\right) N_{10} . \qquad (3)$$

It must be noted that the highest sensibility is achieved at a particle concentration of 1.0×10^{10} particles cm^{-3}, and that the maximum of each curves corresponds to a surface coverage of around 50%.

Further experiments to test Eq. (1) have been performed with positively charged polystyrene beads at different pH values. First, case (a), the results at constant initial particle concentration ($N_{10} = 1.91 \times 10^{10}$ part cm^{-3}) are shown in Fig. 2. Also, in this case, the general shape of these curves does not deviate too far from the parabolic shape they should have, according to Eq. (1). Hence, equation derived by Singer et al. [7] is able to explain the coagulation of polystyrene beads independently of the electric charge of the IgG molecules and polymer carriers.

Second, case (b), according to Eq. (1), plots of $(d\tau/dt)_0$ versus $(N_{10})^2$, measured at constant surface coverage (P_0/N_{10}), should give straight lines through the origin and, indeed, the experimental

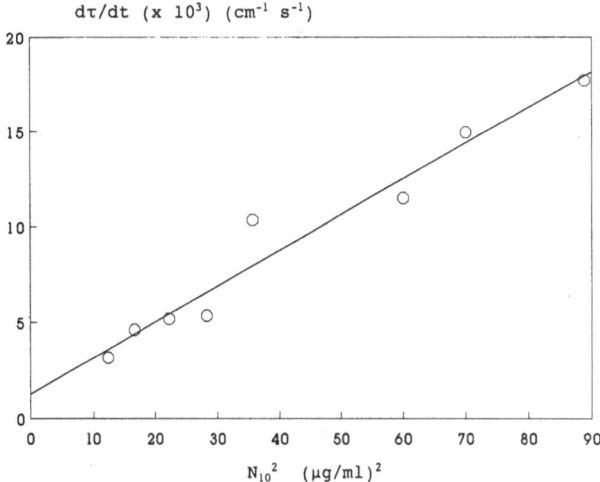

Fig. 3. The initial rate of coagulation as a function of square of anionic latex concentration

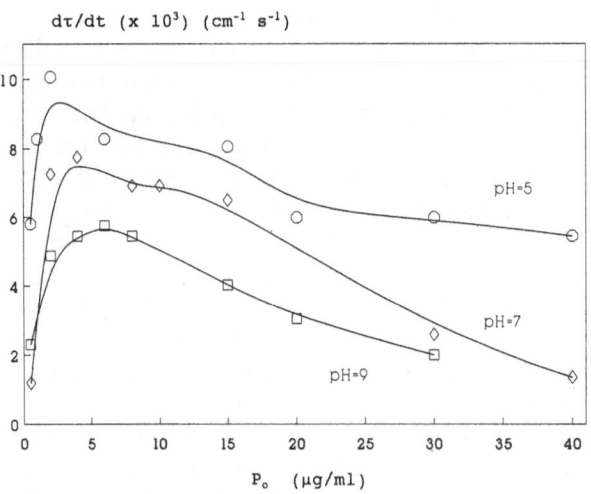

Fig. 2. The initial slope of the turbidity with time curve as a function of protein concentration for three different pH and cationic latex

points in Fig. 3 indicate these straight lines. Identical dependence between $(d\tau/dt)_0$ and $(N_{10})^2$ is found with cationic PS beads, as can be seen in Fig. 4.

Third, case (c), the rate of coagulation is measured at constant protein P_0 concentration, varying the PS bead concentration, and according to Eq. (1) plots of $(d\tau/dt)_0$ versus N_{10} should again give

straight lines. In Fig. 5 the experimental results are plotted for an initial protein concentration of 1 μg/ml, and again satisfactory a straight line can be drawn through these points.

In the plot of $d\tau/dt$ versus N_{10} (Fig. 5), according to Eq. (3), the maximum number N of IgG molecules adsorbed per anionic PS bead and the K-constant can be obtained. These parameters are 1050 IgG molecules per PS bead and 6.66×10^{-23} s^{-1} cm^5 particle^{-2}. From the K-value the kinetic constant K'_{11} of doublet formation is derived. The coefficients C_2 and C_1 are obtained using the RGD

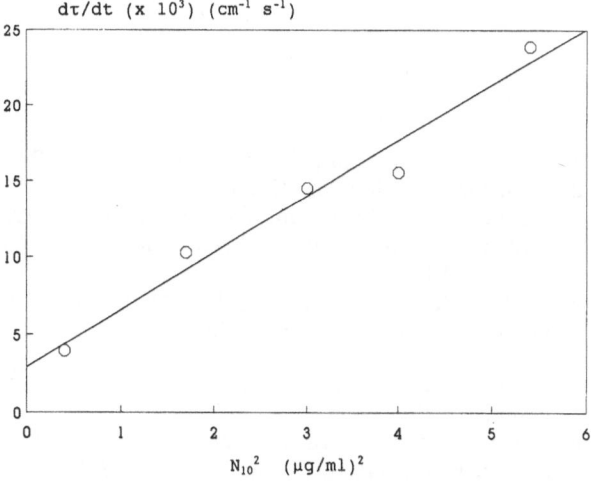

Fig. 4. The initial rate of coagulation as a function of square of cationic latex concentration

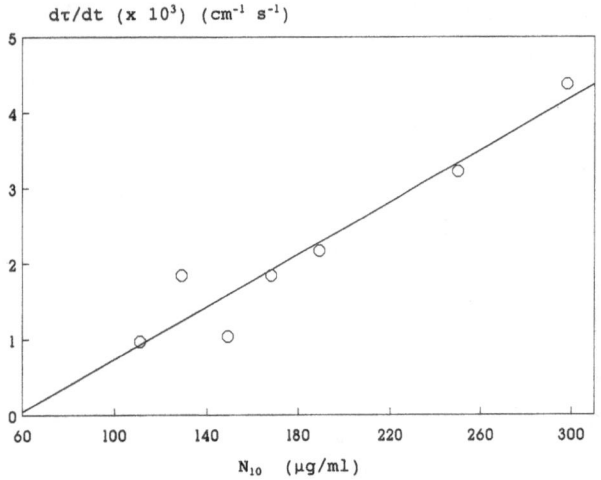

dτ/dt (x 10³) (cm⁻¹ s⁻¹)

Fig. 5. The initial rate of coagulation as a function of the anionic latex concentration at constant initial protein concentration

theory of light dispersion [8]. Therefore, K_{11}' is equal to $(2.2 \pm 0.8) \times 10^{-12}$ s⁻¹ cm³ particle⁻¹. K_{11}' contains the usual factors of diffusion-controlled coagulation such as the diffusion constant of the PS particles, double-layer repulsion, and van der Waals attraction [7]. In the case of the positively charged PS beads the kinetic constant is $(4.5 \pm 1.1) \times 10^{-12}$ s⁻¹ cm³ particle⁻¹ and N of IgG molecules per PS bead is 1630. Both values of K_{11}' are different because of the number of IgG molecules adsorbed on each type of PS beads.

Acknowledgement

We are grateful to Biokit S. A. (Barcelona, Spain) for research support through grant number BK-147.

References

1. Singer JM, Plotz CM (1956) Am J Med 21:888—892
2. Masuzawa S, Itoh Y, Kimura H, Kobayashi R, Miyauchi Ch (1983) J Immunol Methods 60:189—196
3. Millán JL, Nustad K, Nørgaad-Pedersen B (1985) Clin Chem 31:54—59
4. Serra J, Puig J, Martín A, Galisteo F, Gálvez MªJ, Hidalgo-Alvarez R (1992) Colloid Polym Sci 270:574—583
5. Goodwin JW, Ottewill RH, Pelton R (1979) Colloid Polym Sci 257:61—69
6. Galisteo F, de las Nieves FJ, Cabrerizo MA, Hidalgo-Alvarez R (1990) Progr Colloid Polym Sci 82:313—320
7. Singer JM, Vekemans FCA, Lichtenbelt JWTh, Hesselink FTH, Wiersema PH (1973) J Colloid Interface Sci 45:608—614
8. Lichtenbelt JWTh, Ras HJMC, Wiersema PH (1974) J Colloid Interface Sci 46:522—526

Authors' address:

A. Fernández-Barbero
Biocolloids and Fluid Physics Group
Department of Applied Physics
University of Granada
E-18071 Granada, Spain

Characterization of aqueous dissolved biopolymers

A. Huber[1]), W. Praznik[2]), and G. Cvirn[1])

[1]) Institut für Physikalische Chemie der KF-Universität, Graz/Austria
[2]) Institut für Chemie der Universität für Bodenkultur, Vienna/Austria

Key words: Polysaccharides — chromatography-multi-detection — polymer-coil-size — polymer-coil-shape

As is well known, molecular characteristics like structure, particle size/shape and interactive properties of aqueous dissolved biopolymers like polysaccharides are closely correlated with macroscopic polymer qualities. Thus, the investigation of molecular characteristics is an initial step in any comprehensive characterization of these materials. Therefore, the basic polymer characteristics have to be correlated with appropriate biochemical and physicochemical parameters and, finally, adequate experimental techniques have to be applied to yield the required information in terms of these parametervalues. Generally, a combined application of isolation/purification/separation-techniques, biochemical processing and analysis, and physicochemical analysis has to be applied to reach this goal. From the physicochemical point of view especially the characterization of already purified and pretreated aqueous dissolved polysaccharide-fractions by means of size-exclusion chromatography (SEC) and simultaneous multidetection (e.g., universal mass by DRI, specific mass by UV/VIS, ultrasensitive universal mass by evaporization/light scattering, optical contrast by an interferometric refractometer at fixed wavelength, molecular weight by LALLS, particle size by MALLS, particle shape and branching characteristics by viscosity) enables successful access to a

Table 1. Comparison of molecular characteristics of aqueous dissolved polysaccharides Inulin, Sinistrin and Dextran

Sample	Mw ±3% [g/mol]	Mn ±5% [g/mol]	Mw/Mn ±5%	$[\eta]$ [ml/g]	SMH a ±10%	SMH K ±20% [ml/g]	$R_e w$ ±10% [nm]	$\rho_e w$ ±10% [g/ml]
Inulin	6100	5100	1.20	5.6	0.23	0.804	1.74	0.45
Sinistrin	6700	5900	1.14	4.0	0.43	0.090	1.62	0.64
Dextran	6400	5300	1.21	7.9	0.50	0.098	1.96	0.35

Inulin:	Fructan, main chain: $\beta2 \rightarrow 1$, linear
Sinistrin:	Fructan, main chain: $\beta2 \rightarrow 1$, long-chain-/short-chain branched: $\beta2 \rightarrow 6$
Dextran:	Glucan, main chain: $\alpha1 \rightarrow 6$, short-chain branched: $\alpha1 \rightarrow 2$, $\alpha1 \rightarrow 3$, $\alpha1 \rightarrow 4$

	SEC-DRI/LALLS + Univ. Cal. $R_e w$ ±10% [nm]	Quasielastic Light Scattering R_H ±5% [nm]	Small Angle X-ray Scattering R_g ±5% [nm]
Inulin	1.74	1.92	1.76
Sinistrin	1.62	1.89	1.54

wide range of such characteristic polymer-parameters [1—6].

For three different polysaccharides (Inulin, Sinistrin and Dextran) of about identical weight average molecular weight (Mw) and number average molecular weight (Mn) an extended characterization was performed combining results from SEC-DRI/LALLS-investigations and data from universal calibration. The results are summarized in Table 1. For aqueous dissolved polysaccharide samples as well, different kinds of main chain linkages as different branching characteristics for identical main chain linkages could be distinguished, yielding quite different values for Staudinger/

Mark/Houwink (SMH) constants K and a, polymer coil dimensions ($R_e w$: sphere-equivalent weight average particle radius), and coil-density ($\rho_e w$: weight average mass density of the equivalent sphere). Additionally, it was demonstrated that the obtained average particle radii ($R_e w$) were reasonable compared to results from quasielastic light scattering (R_H: hydrodynamic sphere equivalent radius calculated from D_T via Stokes/Einstein) and small-angle x-ray scattering (R_g: radius of gyration).

The linear fructan Inulin shows up with a high value for Staudinger/Mark/Houwink (SMH) K and

small value for a, indicating a stiff cancel molecule. A close-to-rodlike shape of inulin is even supported by the relatively high values for $R_e w$ and the low mass density $\rho_e w$ of the particles. The branched fructan Sinistrin appears as a dense coil (high a, small K, small $R_e w$ and high $\rho_e w$). The branched glucan Dextran shows, similar to Sinistrin, high values for a and small values for K, but differs significantly from Sinistrin in the values for $R_e w$ and $\rho_e w$ caused by different monomers (fructose, glucose) and different kind of glycosidic linkages ($\beta2 \rightarrow 1$, $\alpha1 \rightarrow 6$) forming these polymers (fructans, glucans).

References

1. Praznik W, Beck RHF (1985) J Chromatogr 348:187—197
2. Praznik W, Beck RHF, Spies T, Huber A (1988) Chromatographia 26:359—604
3. Eigner WD, Abuja P, Beck RHF, Praznik W (1988) Carbohydr Res 180:87—95
4. Heinze B, Praznik W (1991) J Appl Polym Sci: Appl Polym Symp 48:207—225
5. Huber A (1991) Biochem Soc Trans 19:505—506
6. Huber A (1992) In: Kulicke WM (ed) Analysis of Polymers/Molarmass and molar-mass distribution of

polymers, polyelectrolytes and latices. Hüthig & Wepf Verlag, Basel, Heidelberg, New York 61, pp 248—270

Authors' address:

Dr. Anton Huber
Institut für Physikalische Chemie
Heinrichstrasse 28
A-8010 Graz, Austria

References

1. Graf C (1991) J Chem Phys 6284:95(9)
2. Provencher SW (1982) Comput Phys Commun 27:213

Authors' address:

C. Johner
Fakultät für Physik
Universität Konstanz
Postfach 5560
78434 Konstanz, FRG

Long-time behavior of the field correlation function of aqueous suspensions of rodlike fd-virus particles

C. Johner, M. Deggelmann, C. Graf,
M. Hagenbüchle, U. Hoß, C. Martin,
and R. Weber

Fakultät für Physik, Universität Konstanz, Germany

Key words: Rodlike fd-virus particles — dynamic light scattering

Dynamic light scattering (DLS) measurements on aqueous suspensions of rodlike fd-virus particles (length = 880 nm, diameter = 9 nm) in a concentration regime between 1 c^* and 8 c^* (1 c^* = 1 particle/length^3) are reported. The short-time behavior of the field correlation function $g_E(q, t)$ of aqueous fd-suspensions has been examined recently [1]. The present work gives an account of the long-time behavior of $g_E(q, t)$ at different ionic strengths. The correlation function is analyzed by the program CONTIN [2] which fits the experimental data with a spectrum of exponentially decaying functions,

$$g_E(q, t) = \int_0^\infty e^{-\lambda t} s(\lambda) d\lambda.$$ Via $\lambda = D_{app} q^2$ $D_{app}(q)$ can

be identified with an apparent diffusion coefficient. We found no significant concentration dependence of the diffusion coefficient at samples with totally screened Coulomb interaction. D_{app} lies somewhat below the values obtained from the measurements of the short time part of $g_E(q, t)$. Further the decrease of the apparent diffusion coefficient with decreasing salt concentration is exposed. The data are compared with present theories.

New Resonance in the electric response function of mesogels

D. Vollmer and H.-F. Eicke

Inst. for Physical Chemistry, University of Basel, Switzerland

Abstract: The complex permittivity of a water-in-oil microemulsion containing block-copoly(oxyethylene/isoprene/oxyethylene) is measured as a function of frequency and temperature. A new resonance in the imaginary part of the dielectric permittivity is observed for temperatures well below the percolation threshold.

Key words: Microemulsions — mesogels — dielectric-spectroscopy

Water/AOT(sodium di-2-ethylhexylsulfosuccinate)/iso-octane systems form water-in-oil microemulsions in a certain range of temperature and composition. They form selective solvents for both components of block-copoly(oxyethylene/isoprene/oxyethylene), i.e., a triblock copolymer build up by a hydrophilic(POE)-hydrophobic(PI)-hydrophilic (POE) sequence. The copolymer connects microemulsion-droplets so that viscoelastic networks (mesogels) are formed at sufficiently high copolymer concentrations. Components of the system are described elsewhere [1].

Percolation of droplets is observed when water volume fraction or temperature of the system is increased. As a consequence, one observes a peak in the dielectric loss factor ε'' [2,3]. Here, we report on an additional resonance in the dielectric loss factor below the percolation threshold for mesogels.

The complex permittivity $\varepsilon' + i\varepsilon''$ is measured as function of frequency with a Hewlett-Packard HP 4192A impedance analyzer operating in the frequency range 10 Hz to 1.3 10^7 Hz. The measuring

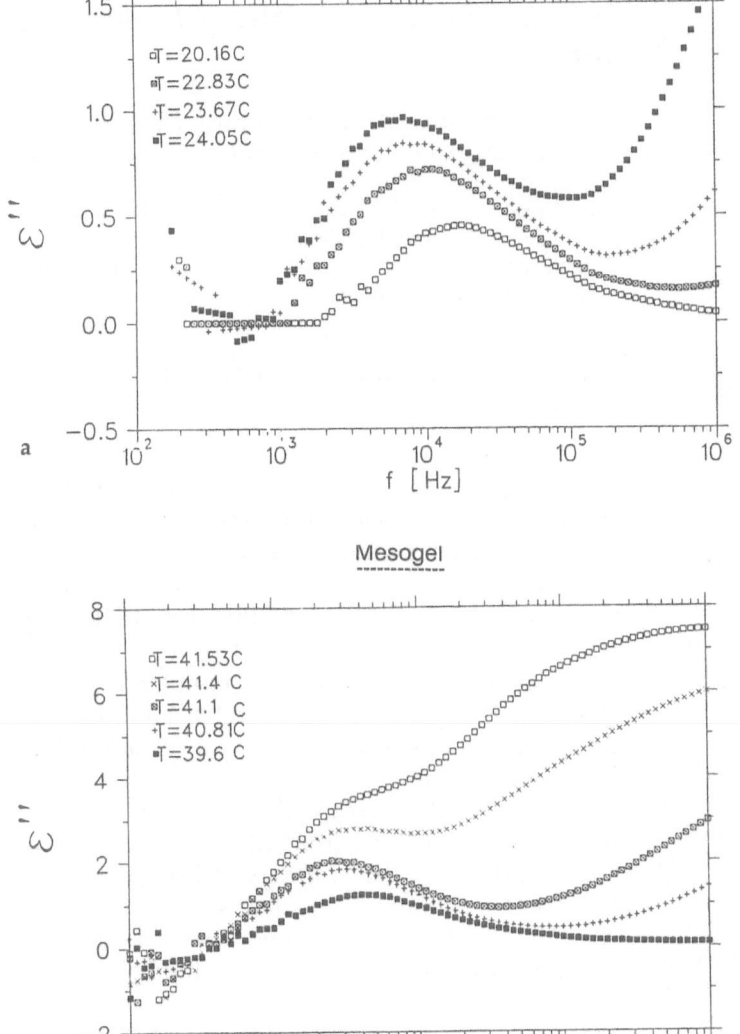

Fig. 1. Frequency and temperature dependence of the imaginary part of the dielectric permittivity for the microemulsion (a) and the mesogel (b). Sample composition: 0.9 mol dm^{-3} AOT H$_2$O/AOT(mol/mol) = 61.8. The mesogel contains, on the average, nine copolymers per droplet. The triblock copolymer consists of two POE blocks with number average molecular weight M_n = 10390 g mol^{-1} each, and a PI block with M_n = 49200 g mol^{-1}

cell consists of a glass bulb containing two parallelly arranged platinized electrodes and is thermostated with temperature stable within 0.02 K.

Figure 1 shows the imaginary ε'' part of the permittivity for a microemulsion and a mesogel as function of frequency and temperature. Between 10^3 and 10^4 Hz ε'' shows a maximum, respectively, a shoulder before it rises to the previously observed maximum that is connected with percolation [2, 3]. For the mesogel the resonance becomes more pronounced and shifts towards lower frequencies (Fig. 1b).

Since the low-frequency conductivity has been subtracted, the dielectric loss contribution to ε'' is observed. We assume that the resonance is connected with double-layer relaxation and its variation is due to influence of highly soluble POE-blocks.

Acknowledgement

This research has been supported by the Swiss National Science Foundation

Progress in Colloid & Polymer Science   Progr Colloid Polym Sci 93:276 (1993)

References

1. Eicke H-F, Quellet Ch, Xu G (1989) Colloids Surf 36:97—107
2. Eicke H-F, Geiger S, Sauer FA, Thomas H (1986) Ber Bunsenges Phys Chem 90:872—876
3. van Dijk MA, Casteleijn G, Joosten JGH, Levine YK (1986) J Chem Phys 85:626—631

Authors' address:

Doris Vollmer
Institute of Physical Chemistry
Klingelbergstr. 80
4056 Basel, Switzerland

Langmuir-Blodgett films of immunoglobulins — perspective material for biosensors

L. A. Feigin, R. L. Kayushina, T. B. Dubrovsky[1], and V. V. Erokhin

Institute of Crystallography, Russian Academy of Sciences, Moscow, Russia
[1] Bach Institute of Biochemistry, Russian Academy of Sciences, Moscow, Russia

Key words: LB films — immunoglobulins

Monomolecular Langmuir-Blodgett antibody films were prepared by Langmuir Trough (Joyce-Loebl) under a surface pressure of 20 mN/m at room temperature. Glutaraldehyde coupling LB films were then deposited onto a piezoelectric crystal modified with a thin layer of siloxane polymer carrying a high reactivity amino groups for the covalent attachment of a monolayer of antibody and protection of the crystal surface from solvent penetration. The event of antibody-antigen binding was studied by the microgravimetric method, using a mass-sensitive quartz resonator to detect and measure the amount of antigen in a test solution. The detection limit of the developed methods and the range of application of the piezoelectric resonators as sensing elements in the immunoassay are discussed. The antibody-antigen binding was estimated in solution with the antigen concentration in the range from 0.01 mg/ml to 0.0005 mg/ml.

Authors' address:

L. A. Feigin
Institute of Crystallography
Russian Academy of Sciences
Leninsky pr. 59
117333 Moscow, Russia

The time-dependent self-organization of immunoglobulins IgG and IgM monolayers at the air-water interface

G. B. Sukhorukov[1]), T. B. Dubrovsky[1]) R. L. Kayushina[1]), V. A. Lapuk[2]), and Yu. I. Khurgin[2])

[1]) Institute of Crystallography, Russian Academy of Sciences, Moscow, Russia
[2]) Institute of Organic Chemistry Russian Academy of Sciences, Moscow, Russia

Key words: Immunoglobulins — self-organization — interface — protein films

The process of immunoglobulin (Ig) monolayer formation at the air-water interface was studied. We had measured simultaneously surface pressure and the protein contents at interface determined by a new technique based on the registration of fluorescence of Eu-labeled IgG. The kinetics of self-organization of monolayer was monitored during 10—15 h. The concentration of initial solutions was 0.8 and 2.2 µg/ml. There are two stages of the film formation. At the first stage (1—2 h) the surface protein concentration reaches its maximal value 0.2—0.3 pmol/cm^2 and the surface pressure is not changed significantly. At the second stage (2—15 h) the pressure increases and levels off at the value 10—12 mN/m (2—5 h) and is not changed after that (5—15 h). During the second stage the IgG surface density is almost constant (0.2—0.3 pmol/cm^2). This corresponds to the planar layout of IgG molecules. A similar shape of kinetic curves was obtained for IgM.

We propose that IgG molecules in self-organized monolayers have a better ordering than monolayers compressed by Langmuir technique.

Authors' address:

G. B. Sukhorukov
Institute of Crystallography
Russian Academy of Sciences
Leninsky pr. 59
117333 Moscow, Russia

Progress in Colloid & Polymer Science   Progr Colloid Polym Sci 93:277 (1993)

Synthesis and characterization of multiphase latexes prepared by two-step emulsion polymerization

Y. S. Kim, and M. Ballauff

Polymer-Institut, Universität (TH) Karlsruhe, FRG

In this work core/shell (polystyrene(PS)/polymethylmethacrylate (PMMA)) latexes and inverse latex systems (PMMA/PS) were prepared by two-step emulsion polymerization. In order to avoid complications when characterizing the structured latexes, monodisperse PS and PMMA latexes of various diameters were prepared by seeded emulsion polymerization and used as core systems. In the second-step polymerization the formation of a new crop of latexes is avoided by the controlled level level of surfactant and initiator. For the characterization of the core, core/shell, and inverse system various experimental methods were used, e.g., electronmicroscopy (TEM, thin-sectioning, shadowcasting, SEM) [1], small-angle x-ray scattering (SAXS) [2,3], ultracentrifugation (UC), soap titration [1] and measurement of densities and refractive indexes. From the shadow-casting a very characteristic behavior of PMMA phase could be observed, and this was used for the determination of surface polymer phase of structured latex particles [1].

All these experimental results showed that the given system PS/PMMA exhibits the assumed core/shell structure. [1, 2] However, in the inverse system PMMA/PS more complicated, heterogeneous structures were observed [1], e.g., depending upon the addition method of the second monomer and particle size of the core.

References

1. Kim YS, Ballaufff M, in preparation
2. Grunder R, Kim YS, Kranz D, Müller HG, Ballauff M (1991) Angew Chem 103:1715, Angew Chem Intl Ed 30:1650
3. Grunder R, Urban G, Ballauff M (1993) Colloid Polym Sci 271:563

Authors' address:

M. Ballauff
Polymer-Institut
Universität (TH) Karlsruhe
Kaiserstraße 12
76128 Karlsruhe, FRG

Formation of colloidal gold by using acetylenic glycol nonionic surfactants

S. Sato, H. Sezaki and H. Kishimoto

Faculty of Pharmaceutical Sciences, Nagoya City University, Japan

Key words: Colloidal gold — nonionic surfactant — surfynol — transmission electron micrograph (TEM) — UV-VIS spectrum

Recently, colloidal gold has been recognized anew as an important material for chemical and medical technologies [1, 2], and an improvement of its preparation method is desivable [1, 3]. We chose acetylenic glycol nonionic surfactants, Surfynol 465 and 485 (Air Product and Chemicals, USA) [4, 5] as moderate promoter and stabilizer for colloidal gold and obtained reproducible results for its formation.

Equal volumes of $HAuCl_4$ aq. (0.1 − 5 mM) and Surfynol 465 or 485 aq. (3 − 100 mM) were mixed at 25 °C and other temperatures. After ca. 10 min or more of lag time, which depended on the concentrations, the characteristic absorption peak at 530 nm appeared, changing the tint of mixture to pink, wine red, and dark red, and showing the formation of colloidal gold, which was confirmed by transmission electron microscopy (TEM). The colloidal gold formation took a whole day or more to complete, making possible its kinetic study by spectroscopy and TEM. The formation of colloidal gold proceeded linearly to $HAuCl_4$ concentration. With regard to the quality of colloidal gold, the optimal molar ratios of Surfynol 465 and 485 to $HAuCl_4$ were ca. 30 and 10, respectively. Surfynol 485, having longer polyoxyethylene chains, was superior to Surfynol 465, which has shorter ones, in the role of promoting the reaction. Colloidal gold, thus obtained, was stable, at least for several months, at room temperature with or without dialysis against distilled and deionized water. Its particle size was nearly homogeneous and became smaller (10 nm to 5 nm by TEM) by increasing the reaction temperature, although the particle had the tendency to coagulate at the higher temperature, e.g., 50 °C. It was found also that colloidal gold is similarly formed with light shielding.

References

1. Yonezawa Y, Sato T, Ohno M, Hada H (1987) J Chem Soc Faraday Trans I 83:1559

2. a) Hodges GM, Southgate J, Toulson EC (1988) In: Albrecht RM, Hodges GM (comps) Biotechnology and Bioapplications of colloidal gold. Scanning Microscopy International, Chicago, pp 1—18. b) Horisberger M, ibid, pp 19—40
3. Kurihara K, Kizling J, Stenius P, Fendler JH (1983) J Am Chem Soc 105:2574
4. Sato S, Kishimoto H (1988) J Colloid Interface Sci 126:108
5. Sato S (1989) J Phys Chem 93:4829

Authors' address:

Dr. Shizuko Sato
Faculty of Pharmaceutical Sciences, Nagoya City University, Tanabe-dori 3—1, Mizuho-ku
Nagoya, 467, Japan

Nonequilibrium thermodynamics of interfaces and kinetics of phase transitions

N. N. Kochurova, and A. I. Rusanov

Chemistry Faculty, St. Petersburg University, Russia

Key words: Nonequilibrium thermodynamics — interface — phase transition

Introduction

Nonequilibrium surface phenomena are very important in biological membrane diffusion, in charge transition, etc. The results of an experimental and theoretical investigation of nonequilibrium processes on fresh surface are presented.

Gibb's equation for nonequilibrium interface

We investigated the most general and the practically important case of the adsorption, the chemical reactions, and the polarization of the liquid surface (as the result of the polar molecules' orientation and the formation of the ionic double layer) [1]. The influence of the formation of the ionic double layer on the surface polarization is not taken into consideration in some papers, for example, in [2].

With using methods of irreversible thermodynamics these processes are derived. It is shown that the correlation between the changes of dynamic surface tension γ and electric surface potential χ exists. The linear term in this relation bears evidence of the influence of the ionic layer on the nonequilibrium surface potential [1].

These results have large practical interest, because they help to find out the mechanism of the electric surface potential formation from experiment. If the experimental correlation $\Delta\gamma - \Delta\chi$ has no $\Delta\chi$ in the first power, then the surface potential is formed by means of the orientation of the polar molecules. If the experimental dependence $\Delta\gamma - \Delta\chi$ has $\Delta\chi$ in the first power, then the surface potential is formed by means of the diffusion.

Dependence of the nonequilibrium surface tension on concentration

Surface electrization takes place in many processes of flow, which causes the appearance of the additional term $-q\,d\chi$ (q is the surface density of the charge) in the generalized adsorption equation for dynamic surface [3]. The dependence of the surface tension from concentration c may be determined from the equation

$$\frac{d\gamma(t,\,c)}{dc} = -\frac{RT}{c}\,\Gamma\,(t,\,c) - q\,\frac{d\chi(t,\,c)}{dc}\,,$$

where T is the temperature, t is the time (the surface age). The experimental result is shown in Fig. 1. The curve has a minimum, which is a well known Jones-Ray effect on electrolyte aqueous solutions, and may be explained in terms of surface electrization [3]. It is known that if the adsorption of NaCl is negative ($\Gamma < 0$), then $\Delta\gamma$ will increase with concentration. But the second term in the right part of equation reduces $\Delta\gamma$, because $q < 0$ and $\dfrac{d\chi}{dc} < 0$ at $c < 10^{-3}$ mol/l [4]. Thus, the minimum arises.

Connection between the surface structure relaxation and the evaporation-condensation kinetics

Our experimental data show that the evaporation and condensation rate i dependence from the surface properties of liquids is a result of the so-called condensation coefficient f [5]: $i \sim f$.

The condensation coefficient f may be evaluated as [6]:

Progress in Colloid & Polymer Science   Progr Colloid Polym Sci 93:279 (1993)

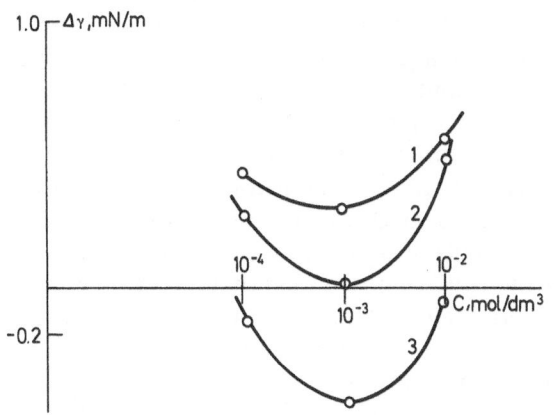

Fig. 1. Dependence $\Delta\gamma$ from the concentration C and the surface age t for aqueous solutions of NaCl: 1 — t = 250 ÷ 500 s; 2 — t = 17 s; 3 — t = 12 s

$$f = \exp\left(-\frac{\gamma_0 - \gamma_\infty}{\kappa T} \cdot a\right)$$

$$\cdot \frac{1}{1 + x\left[\exp\left(-\frac{\gamma_0 - \gamma_\infty}{kT} a\right) - 1\right]} .$$

Here, γ_0 is the dynamic surface tension of the fresh (at $t = 0$) liquid surface on which the condensation of vapor may be carried out, γ_∞ is the equilibrium surface tension, a is the area of the molecula on the liquid surface, $x = 0.4 - 0.5$, and k is Bolzman's constant.

Thus, the surface tension as the main property of the surface and its structure determines the rate of phase transition. Our equation shows that the evaporation (condensation) rate of fresh non-equilibrium surface of water (γ_∞ is replaced by γ_t ≈ 100 dyn/cm) is 10 times higher than the rate in the case of the equilibrium static surface of water. This result conforms with the experiments [7].

References

1. Kochurova NN, Rusanov AI (1981) Colloid Int Sci 81:297—303
2. Sanfeld A (1968) Introduction to thermodynamics of charged and polar layers, Willey, London, pp 170—203
3. Kochurova NN, Rusanov AI, Myrsakhmetova NO (1991) Dokl AN SSSR 316:1425—1428
4. Lopatenko CB, Kontush CM (1984) Izvestia AN SSSR, Seria Energia and transport, NI, pp 151—155
5. Stuke B (1959) Zeit f Electrochem 63:140—145
6. Kochurova NN, Kalashnikova TN, Rusanov AI (1977) J Fis Chim 51:1790—1793
7. Hiekman KCD (1954) Industand Eng Chem 46:1442—1447

Authors' address:

Prof. Natalia Kochurova
Chemistry Faculty
St. Petersburg University
199034 St. Petersburg, Russia

Hydration repulsion between lipid bilayers modified by non-ionic alkyl oligo (ethylene oxide) surfactants

B. König and G. Klose

FB Physik, Universität Leipzig, FRG

Key words: Lipid — alkyl oligo (ethylene oxide) — membranes — hydration forces

Model membrane systems of 1-palmitoyl 2-oleoyl-sn-glycero-phosphatidyl-choline (POPC) containing oligo(oxyethylene) mono-alkyl ethers of the general type $C_{12}H_{23}O(CH_2CH_2O)_nH$ ($C_{12}EO_n$) with n = 2, 4, 8 and a molar detergent/lipid ratio $R_{A/L}$ = 0.5 were prepared. The appropriate aqueous dispersions were in the liquid crystalline L_a phase at T = 25 °C (molar ratio water/lipid $R_{W/L}$ from 6 to 150). The repeat distance d of the samples was determined by x-ray diffraction [1]. By using the Luzzati approach [2] the thickness of the membrane bilayer d_L and of the intermembrane water layer d_W were derived from d. The empirical hydration pressure was measured by the osmotic stress technique [3] and fitted according to:

$$P = P_0 \exp\left\{-\frac{d_W}{\lambda}\right\} . \tag{1}$$

The results obtained suggest a location of the $C_{12}EO_n$ molecules with their alkyl chain embedded in the membrane interior and with the EO-segments extending to the polar interface and water region. The incorporation of $C_{12}EO_n$ causes a thinning and a lateral expansion of the lipid membrane as well as an increase of the water layer thickness. Compared to pure POPC, the hydration forces of the mixed systems are characterized by smaller P_0 but larger λ. The modifications of the hydration force are presumably due to changes in the interfacial water structure by laterally spacing the lipid molecules as well as by direct interaction of the polar EO segments with lipid head groups and water via hydrogen bonding. The comparably long EO-chain of $C_{12}EO_8$ causes an additional steric repulsion and increases the membrane flexibility. At high water content the measured force is dominated by an undulation force.

References

1. Klose G, König B, Schulze G, Degovics G (1988) Chem Phys Lipids 47:225—234
2. Luzzati VA (1968) In: Chapman D (ed) Biological Membranes, Academic Press, NY, pp 71—123
3. Rand P, Parsegian VA (1989) Biochim Biophys Acta 988, 351—376

Authors' address:

Prof. Dr. G. Klose
Universität Leipzig
FB Physik
Linnéstraße 5
04103 Leipzig, FRG

Some results of investigating physico-chemical properties of aqueous solutions of both the synthetic surfactant sodium dodecyl sulfate (SDS) and each of two surfactants obtained from assimilating processes, sophorose lipid or rhamnose lipid, in a concentration range that involves micelle formation, are presented and discussed in this paper. Measurements of solubilities, surface tensions, conductivities, pH values, and enthalpies of dilution at 25°C were evaluated for this purpose. An improved method using a concentration gradient is proposed in order to perform very precise conductivity and calorimetric measurements in flow versions. The investigations point out that the low solubility of glycolipids in water may be improved considerably by SDS. Due to the observed concentration dependencies of all regarded properties, a formation of mixed micelles consisting of SDS and glycolipid monomers can be supposed. This aggregation takes place at about one order of magnitude in concentration lower than that at which the critical micelle concentration of SDS is located. The surfactant mixtures in solution exhibit a nonideal behavior.

Authors' address:

Dr. Simone König, Prof. Dr. K. Quitzsch
Universität Leipzig
Fachbereich Chemie
Technikum Analytikum
Linnéstr. 3
D-04103 Leipzig, FRG

Physico-chemical investigations in systems of composition biosurfactant/sodium dodecyl sulfate/water around the critical micelle concentration

S. König[1]), K. Quitzsch[1]), R. Hommel[2]), D. Haferburg[2]), and H.-P. Kleber[2])

[1]) Department of Chemistry, University of Leipzig
[2]) Department of Biological Sciences, University of Leipzig, FRG

Key words: Biosurfactant — critical micelle concentration — mixed micelles — physico-chemical methods

Analysis of electrophoretic mobility data for biological cells with a new membrane model

T. Kondo and H. Ohshima

Faculty of Pharmaceutical Sciences and Institute of Colloid and Interface Science, Science University of Tokyo, Japan

Abstract: A new membrane model was proposed to analyze the electrophoretic mobility data of biological cells. The model assumes that fixed charges are uniformly distributed in a planar surface layer on the membrane lipid core and that electrolyte ions can penetrate into the layer. Based on this membrane model, an approximate analytical expression that directly relates the electrophoretic mobility of biological cells to the charge density, N, in the surface region was derived. The expression involves the Donnan potential and the surface potential in the region, both of which are a function of N and electrolyte concentration, and a parameter relating to the depth of fluid drag in the region, λ. — The best curve fitting for the experimentally obtained electrophoretic mobility values for human erythrocytes and guinea-pig polymorphonuclear leucocytes as a function of electrolyte concentration using appropriate values of N and λ gave the charge density in the surface region of both types of cells to show that the distribution of the negative charges arising from the acidic groups is fairly uniform, while the positive charges arising from the basic groups increase in the density with increasing inward distance from the boundary of the surface region and the medium.

Key words: Electrophoresis — biological cells — membrane model — charge distribution

Introduction

Cell electrophoresis is a convenient method to obtain information about the surface structure of the cellular organization. The electrophoretic mobility of a biological cell is due to the presence of fixed ionogenic groups and adsorbed ions in the surface region. Electrophoretic mobility data for biological cells can provide us with useful information on the charge distribution in the cell surface region if a proper membrane model is used to analyze the data.

In this article, we propose a new membrane model[1—3] which assumes a uniform distribution of the membrane-fixed charges through a cell surface layer of finite thickness and the penetrability of electrolyte ions into the layer, and describe the results of analysis of the electrophoretic mobility data for human erythrocytes and guinea-pig polymorphonuclear leucocytes with the use of a formula [4] that directly relates the electrophoretic mobility to the charge density in the cell surface region which was derived on the basis of the new membrane model.

Experimental

Blood cells

Fresh human blood samples of groups A, B, AB, and O were supplied by the Japanese Red Cross Blood Center. The fresh blood was diluted with 0.9% (w/v) NaCl solution and centrifuged to collect erythrocytes. The collected cells were washed four times with the NaCl solution in the centrifuge and the washed cells were dispersed in the NaCl solution at a volume concentration of 30%. A small portion of the cell suspension was added to each of the buffer solutions with different pH values and ionic strengths to give 0.1% cell suspension. The buffer solutions used were phthalate(pH 3.0) and veronal (pH 5.5, 7.5, 9.5) buffer solutions, the ionic strength of which was varied from 0.005 to 0.154 with the addition of NaCl. All buffer solutions were made isotonic by adding appropriate amounts of sucrose.

Suspensions of guinea-pig polymorphonuclear leucocytes were prepared according to the method of Kakinuma [5]. The peritoneal exudates of male guinea-pigs weighing 300—400 g were collected and washed. After erythrocytes were removed from the resultant suspension by a hypotonic treatment, guinea-pig polymorphonuclear leucocytes were isolated by a density-gradient centrifugation method. The isolated leucocytes were dispersed in buffer solutions at a suitable concentration. The buffer solutions employed were acetate(pH 2.5, 3.0, 4.0, 4.5, 5.5), phosphate(pH 7.4) and carbonate(pH 9.0), the ionic strength of which was varied from 0.0015 to 0.154 by adding NaCl. All buffer solutions were made isotonic with the addition of appropriate amounts of sucrose.

Electrophoretic mobility measurement

Electrophoretic mobility measurements were made on a Pen Kem System 3000(USA) at 310 K for human erythrocytes and at 298 K for guinea-pig polymorphonuclear leucocytes. Mobility readings were repeated at least 64 times (32 times in both directions) for each sample and the readings were averaged.

Results and discussion

The charge density in the surface region of both types of cells was estimated from electrophoretic mobility data by means of the following formula [4].

$$\mu = \frac{\varepsilon_r \varepsilon_0}{\eta} \frac{\Psi_0/\kappa_m + \Psi_{DON}/\lambda}{1/\kappa_m + 1/\lambda} + \frac{zeN}{\eta\lambda^2} , \qquad (1)$$

where Ψ_0 is the potential at the boundary of the surface region and the medium, Ψ_{DON} is the Donnan potential, κ_m is the Debye-Huckel parameter of the surface region, λ is a parameter whose reciprocal has the dimension of length, z and N are the valence and the density of charged groups expressed as the algebraic sum of positive and negative charges in the surface region, η is the viscosity and ε_r and ε_0 are the relative permittivity of the medium and the permittivitiy of a vacuum, respectively. The Donnan potential, the surface potential and the Debye-Huckel parameter of the surface region are given as

$$\Psi_{DON} = kT/e \ 1n(zN/2n + [(zN/2n)^2 + 1]^{1/2}) \qquad (2)$$

$$\Psi_0 = kT/e(1n\{zN/2n + [(zN/2n)^2 + 1]^{1/2}\}$$

$$+ 2n/zN\{1-[(zN/2n)^2 + 1]^{1/2}\}) \qquad (3)$$

$$\kappa_m = \kappa\{(zN/2n)^2 + 1\}^{1/4} , \qquad (4)$$

where κ is the Debye-Huckel parameter of the medium.

In order to determine the orders of magnitude of N and $1/\lambda$, curve fitting was performed for the mobility data at different pH values and ionic strengths of the medium using various values of N and λ. As a result, the best fitting was obtained when N ranged from 0.011 to 0.016 M and $1/\lambda$ from 2.5 to 3.0 nm for human erythrocytes of all blood groups at pH values above 5.0. The agreement between the calculated and observed mobilities is good, as is typically shown in Fig. 1 for the data at pH 7.5. The result implies that the distribution of the acidic groups is fairly uniform, because each curve fitting was done using a constant N, which corresponds to a uniform distribution of the acidic groups in the surface region [6].

Similarly, the values of N ranging from 0.015 to 0.020 M and of 2.0 nm for $1/\lambda$ gave the best fitting for the observed mobilities of guinea-pig polymorphonuclear leucocytes at pH values above 5.0, again implying a fairly uniform distribution of the acidic groups in the surface region of the cells. A typical example is shown in Fig. 2 for the mobility data at pH 7.4 [7].

On the contrary, no pair of N and $1/\lambda$ that gives a good agreement between the calculated and observed mobilities was found at pH values below 4.5. This strongly suggests that the density of the basic groups is not uniform in the surface region because the contribution from these groups to the charge density is superior to that from the acidic

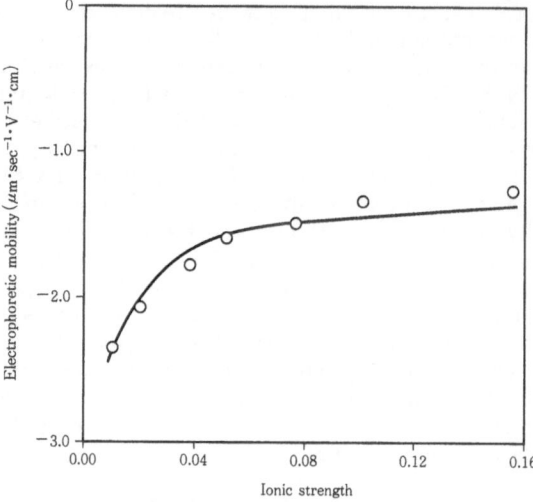

Fig. 1. Electrophoretic mobility-ionic strength curve for human erythrocytes (blood group A) at pH 7.5. Open circles represent the experimentally observed values

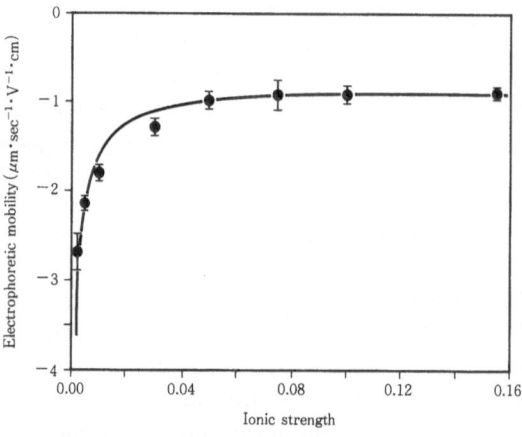

Fig. 2. Electrophoretic mobility-ionic strength curve for guinea-pig polymorphonuclear leucocytes at pH 7.4. Closed circles represent the experimentally observed values

groups in the acidic media where dissociation of the acidic groups is suppressed to a great extent. The charge density at an acidic pH calculated fixing the value of $1/\lambda$ at a constant value was found to increase as the ionic strength of medium decreases or the Debye length increases, suggesting that the basic groups are concentrated more in the inner part than in the outer part of the surface region for both types of cells.

References

1. Ohshima H, Ohki S (1985) Biophys J 47:673—678
2. Ohshima H, Kondo T (1986) Colloid Polym Sci 264:1080—1084
3. Ohshima H, Kondo T (1987) J Colloid Interface Sci 116: 305—311
4. Ohshima H, Kondo T (1989) J Colloid Interface Sci 130:281—282
5. Kakinuma K (1968) Jap Exp Med 38:165—167
6. Kawahata S, Ohshima H, Muramatsu N, Kondo T (1990) J Colloid Interface Sci 138:182—186
7. Nagahama T, Muramatsu N, Ohshima H, Kondo T (1992) Colloids and Surfaces 67:61—65

Authors' address:

Professor T. Kondo
Faculty of Pharmaceutical Sciences
Science University of Tokyo
Shinjuku-ku
Tokyo 162, Japan

by a stepper motor via the ring onto the interface. The movement of the ring is registered by a local-sensitive photodiode. Due to the sensitivity of the photodiode and the analog/digital transverter the circular movement of the ring can be measured with an accuracy of ± 0.01 degrees, depending on the chosen amplitude of the deformation.

The apparatus can be used for simple ramp-type deformations as well as studies with a constant shear tension. Measurements with harmonic deformations are also possible.

Different applications of the set-up are discussed. The experimental data show the high reproducibility and accuracy of the apparatus.

Authors' address:

Dr. Jürgen Krägel
Max-Planck-Institut für Kolloid-
und Grenzflächenforschung
Rudower Chaussee 5
D-12489 Berlin, FRG

Automated surface shear rheometer

J. Krägel[1]), S. Siegel[1]), R. Miller[2]), M. Born[2]), B. Ehmke[3]), and K.-H. Schano[4])

[1]) KAI e.V./WIP, Berlin, FRG
[2]) MPI of Colloid and Surface Science, Berlin, FRG
[3]) Kleinwerkzeug-, Maschinen- u. Apparatebau, Berlin, FRG
[4]) Automation Service and Consulting, Berlin, FRG

Key words: Surface rheology — adsorption layers — polymer surfactant interaction

Rheological investigations at fluid interfaces provide information about interactions in and structure of adsorption layers formed by polymers or biopolymers and their mixtures with surfactants. These studies require equipment which does not disturb the structure during the measuring procedure. Thus, the necessary accuracy of such devices has to be very high as only small interfacial deformations are allowed during the rheological measurements.

A new surface shear rheometer was designed based on the principle of a ring hanging at a torsion wire. The wire transfers the deformation, produced

Static and dynamic adsorption behavior of n-alkyl dimethylammonio acetic acid bromides at liquid interfaces

J. Kriwanek[1]), R. Wüstneck[2]), and R. Miller[1])

[1]) Max-Planck-Institute of Colloid and Surface Science, Berlin, FRG
[2]) KAI e.V., Berlin, FRG

Key words: Adsorption kinetics — frumkin isotherm — betains at liquid interfaces

In an aqueous solution of pH = 7 the homologous n-alkyl dimethylammonio acetic acid bromides behave like betains. Their adsorption behaviour at the aqueous solution — air interface was studied by measurements of equilibrium surface tensions using the ring tensiometry. The interpretation of the data based on the Frumkin isotherm yields no specific interaction in the adsorption layers for chain lengths up to 12 alkyl groups. For longer chain lengths (13 to 16 alkyl groups) a linear increase of the Frumkin interaction parameter is observed, while the absolute values are comparatively small. The molecular area of all members of the series studied ($C_8 - C_{16}$) is constant and amounts to an average value of 44.6 Å²/molecule.

Dynamic surface tension measurements using an automated drop volume tensiometer are performed for the homologs with medium alkyl chain lengths. For these samples the adsorption process can be observed by this method in an excellent way. The studies are made at different concentrations at the solution — air and solution — decane interfaces. The obtained data are interpreted by quantitative models and the adsorption mechanisms are discussed in terms of diffusion controlled adsorption.

Authors' address:

Dr. rer. nat. habil. R. Miller
Max-Planck-Institute
of Colloid and Surface Science
Rudower Chaussee 5
Berlin-Adlershof
D-12489 Berlin, FRG

Table 1. Solid-to-liquid transitions. Relative volume (V) and compressibility (K) effects

	$\Delta V/V$	$\Delta K/K$	$\dfrac{\Delta K/K}{\Delta V/V}$	Refs.
Globular proteins	0.01	0.30	30	1, 2
DPPC vesicles	0.04	0.15	4	3—5
Macroscopic organics	0.05—0.1	0.5—1.0	10	Reviews and Handbooks

Theoretical expectation:

three-dimensional system	$\dfrac{\Delta K/K}{\Delta V/V} > 7$
two-dimensional system	$\dfrac{\Delta K/K}{\Delta V/V} > 10.5$

Volume and compressibility effects of phase transitions in lipid bilayers compared to that in proteins in aqueous solutions

D. P. Kharakoz

Institute of Theoretical and Experimental Biophysics
Russian Academy of Science, Pushchino, Russia

Key words: Phase transition — lipid membranes — proteins

Gel-to-(liquid crystalline) transition in lipid membranes is a first-order phase transition. It is of interest to compare volume and compressibility effects of this transition with that of melting of proteins and macroscopic organics. Experimental data on the relative changes of volume and compressibility effects are presented in Table 1. The data are normalized per the intrinsic volume and compressibility of lipid and proteins. The data on the proteins are related to the native-to-(molten globule) transition, which is known to be a first-order phase transition.

The theoretical expectation for the compressibility/volume ratio is obtained from a consideration of known intermolecular interaction potentials. It has been taken into account that, in addition to the change of average intermolecular distances upon melting which lead to a compressibility increase, there is a contribution to the compressibility from relaxation processes occurring in the liquid state.

Relative volume and compressibility effects of the melting of proteins are by an order of magnitude smaller than that of typical "solid-to-liquid" phase transitions in macroscopic molecular systems. This fact is due to a large amount of water (about 50 wt. %) penetrating the molten globule [2] and, thus, compensating the effects of melting.

Gel-to-(liquid crystalline) transition in DPPC lipid bilayers is accompanied by a large relative volume effect (4%) similar to that for macroscopic organics. At the same time, intrinsic compressibility of the bilayer increases by only about 15%, less then that of proteins (Table 1). The ratio $(\Delta K/K)/(\Delta V/V) = 4$ is unexpectedly low.

References

1. Kharakoz DP, Sarvazyan AP (1980) Studia Biophysica 79:179—180
2. Kharakoz DP, Bychkova VE, unpublished

3. Mitaku S (1978) Ikegami A & Sakanishi, A Biophys Chem 8:295—304
4. Laggner P, Lohner K, Degovics G, Muller K, Shuster A (1987) Chem Phys Lipids 44:31—60
5. Kharakoz DP, Colotto A, Lohner K, Laggner P, J Phys Chem (accepted)

Author's address:
D. P. Kharakoz
Institute of Theoretical and
Experimental Biophysics
Russian Academy of Science
142292 Pushchino, Russia

Interactions between water in oil droplets filled by native or chemical modified cytochrome c

G. Cassin, K. M. Larsson, S. Illy, and M. P. Pileni*

Université P et M Curie, Laboratoire SRSI, Paris, France CEN Saclay, DRECAM, SCM, Gif sur Yvette, France

Abstract: In this paper it is shown that the solubilization of native and chemically modified cytochrome c differing by their number of elementary charge in reverse micellar solution induces changes in the percolation onset and the intermicellar interaction potential.

Introduction

A liquid interface can be formed by mixing surfactant, water and oil, where the surfactant forms a monolayer dividing the water and oil microdomains. One intensively studied surfactant is the sodium bis(2-ethylhexyl) sulphosuccinate (AOT). At low water contents this surfactant usually forms spheroidal aggregates called reversed micelles. Reverse micelles have been shown to serve as hosts for proteins and enzymes [1—4]. Two important parameters to describe the reverse micelles are the water to surfactant ratio; $W = [H_2O]/[AOT]$, and the water volume fraction, ϕ_w, defined as: $\phi_w = V_{aq}/V_T$ where V_{aq} and V_T are the aqueous and total volume respectively.

Native cytochrome c from horse heart is a protein highly positively charged. Most of the positive charges are situated in the electron-transfer interacting domain, whereas the negatively charged residues are placed on the opposite side of the protein molecule, which makes the protein a very strong dipole. The positive charges near the hemic cleft are physiologically very important since the reaction partners of cytochrome c, as well as the inner mitochondrial membrane, have a negative charge.

It has been previously shown that cytochrome c interacts with the AOT surfactant layer [5—6] and induces percolation process at lower temperature and water volume fraction as compared with unfilled droplets [7]. These phenomena have been attributed to the location of the protein at the interface. In order to investigate the nature of these interactions, we have made two derivates of cytochrome c which have a negative net charge, as compared to the native protein. This makes it possible to vary the net charge of the protein while keeping the pH constant.

Cytochrome c modification and characterization

Ferricytochrome c is acetylated and succinylated by the method of Finkelstein et al. [8]. The theoretical charge of the native cytochrome c was calculated from the chemical composition [9], assuming that all polar amino acids are situated on the surface. The heme propionate groups give no contribution to the charge since they are deeply buried within the protein matrix [10] and the tyrosine groups are assumed nondissociated. The degree of modification was determined by titration of the numbers of free $-NH_2$ groups using an OPA reagent [11]. At pH 8.4 the acetylated protein was found to have a net charge of approximately -3 and for the succinylated form -12, whereas the native cytochrome had a positive net charge of $+8$.

Structural studies of reverse micelles containing cytchrome c derivatives

At fixed temperature, the increase of the water volume fraction induces an increase in the conductivity by several orders of magnitude. This is the percolation phenomenon corresponding to the formation of an infinite cluster [12]. The conductivity measurements of reverse micelles filled by various cytochrome c derivatives (Fig. 1) show changes percolation onset compared to empty micelles: the native cytochrome c induces a percolation onset at low water volume fraction ($\Phi_p = 10\%$), whereas acetylated cytochrome c does not change the percolation onset compared to empty micelles ($\Phi_p = 15\%$) and the succinylated derivatives induces a shift in the percolation at higher polar volume fraction($\Phi_p = 25\%$). This behavior of the percolation onset of reverse micelles in the presence of various cytochrome c derivatives indicates a change in the interaction potentials between filled water droplets compared to empty reverse micelles. To estimate such changes in the interaction potentials, SAXS experiments have been performed at a given water

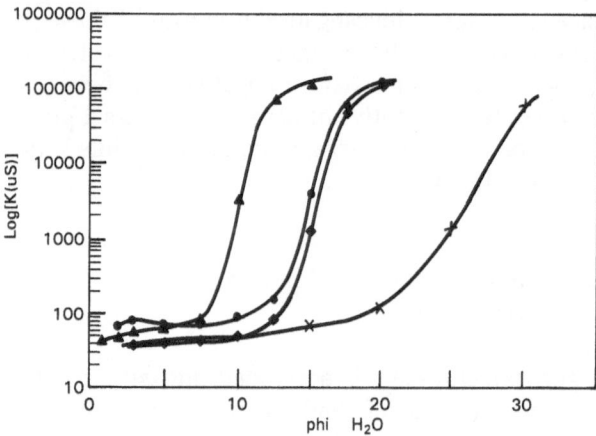

Fig. 1. Variation of the conductivity with the water volume fraction of AOT reverse micelles in nonane solvent, in the abscence (◇) and in the presence of native (△), acetylated (●), succinylated (+) cytochrome *c* derivatives

volume fraction at 22°C in nonconducting regime (7.5%). In order to determine the intermicellar potential, the spheres adhesives potential is used which is the limit of a squarewell potential [13]. Using the Percus — Yevick (PY) approximation, the explicit expression of the intermicellar structure factor is obtained by assuming that microemulsions can

be considered as a one component macro-fluid of interacting particles with potential in the continuous solvent. Using Baxter's formalism the analytical expression of the structure factor and the sticky parameter, τ^{-1}, can be deduced [14]. The analytical expression of $S(q)$ is in good agreement with Monte-Carlo simulation [15—16]. The SAXS data are reasonably well fitted for a given sticky parameter (Fig. 2). Table 1 gives the sticky parameter value, τ^{-1}, that is to say, the effective intermicellar attraction for micelles containing the various cytochrome *c* derivatives. A strong increase in the attractive interactions by solubilization of native cytochrome *c* can be observed. With acetylated derivative, the attractive interactions are similar to those obtained with empty droplets. With succinylated derivative the attractive interactions strongly decrease. These data are in good agreement with the change in the percolation onset described above and could be explained in terms of average location of cytochrome *c* derivatives in the droplet: the native cytochrome *c*, highly positively charged is situated at the AOT interface, whereas the negatively charged derivatives are partially dissolved in the aqueous core of the reverse micelle. The differences in the behavior between the two negative charge cytochrome *c* derivatives could be

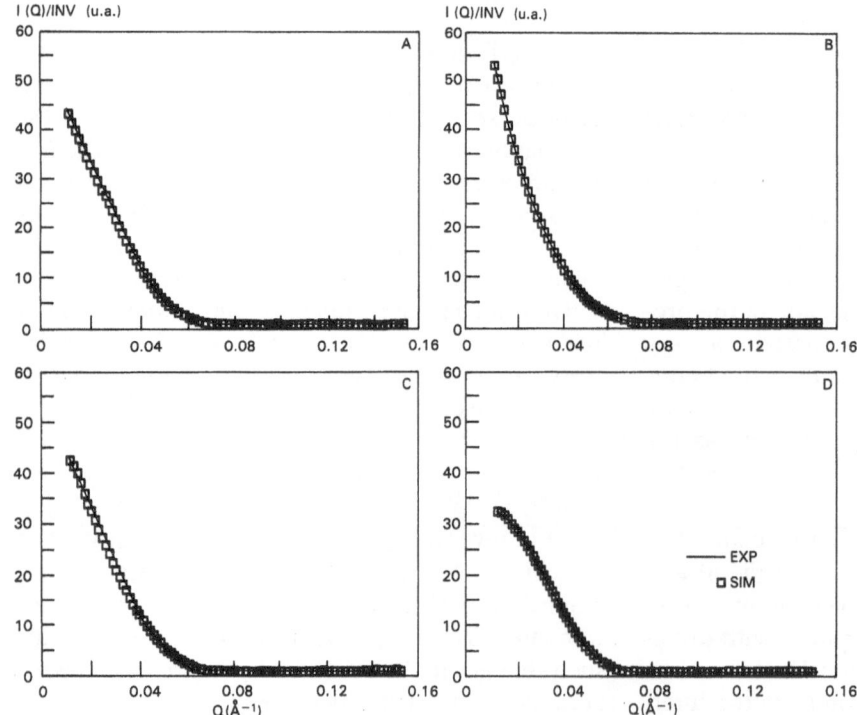

Fig. 2. Simulated and experimental SAXS spectra obtained at a water volume fraction equal to 7.5%, w = 40 of AOT reverse micelles in nonane bulk solvent for empty micelle (*A*), micelle filled by native (*B*) acetylated (*C*), and succinylated (*D*) cytochrome *c* derivatives

Table 1. Percolation onset and sticky parameter for native and modified cytochrome c.

Micelles containing	Native	Acetyl	Succinyl	Empty
Net charge	+8.0	−3.2	−12	0
Percolation onset	10%	14%	25%	15%
Sticky parameter (τ^{-1})	5.88	4.44	2.5	4

explained by the presence of very high number negative charges on sucinylated derivative which could behave as an increase in the ionic strength in the water pool. Actually, the addition of a salt such as sodium chloride in empty micelles induces an increase in the percolation onset compared to that obtained in the absence of salt.

Conclusions

We have shown that solubilization of various cytochrome c derivatives in reverse micelles induces structural changes due to a change in the attractive interaction potentials between droplets.

References

1. Shield JW, Ferguson HD, Bommarious AS, Hatton TA (1986) Ind Eng Chem Fund 25:603
2. Luisi PL, Giomini M, Pileni MP, Robinson BH (1988) Biochem Biophys Acta 947:209
3. Martinek K, Klyachko NL, Kabanov AV, Khmelnitsky YL, Levashov AV (1989) Biochim Biophys Acta 981:161
4. Pileni MP (1989) Structure and reactivity in reverse micelles. Elsiver, Amsterdam
5. Pileni MP, Zemb T, Petit C (1985) Chem Phys Lett 178:414—420
6. Vos K, Laane C, Weijers SR, Van Hoek A, Veeger C, Visser JWG (1987) Eur J Biochem 169:259
7. Huruguen JP, Authier M, Greffe JL, Pileni MP (1991) Langmuir 7:243
8. Finkelstein E, Rosen GM, Patton SE, Cohen MS, Rauckman EJ (1981) Biochem Biophys Res Comm 102:1008
9. Dickerson RE, Timkovich R (1975) Boyer PD (ed) Academic Press, pp 397
10. Buchnell GW, Louie GV, Brayer GD (1990) J Mol Bio 21:585
11. Church FC, Porter DH, Catignani GL, Swaisgood HE (1985) Analyt Biochem 146:343
12. Huang JS, Safran SA, Kim MW, Grest G, Kotlarchyk M, Quirke N (1984) Phys Rev Lett 53:592
13. Baxter RJ (1968) J Chem Phys 49:2770
14. Regnaut C, Ravey JC (1989) J Chem Phys 91:1211
15. Seaton NA, Glandt ED (1985) J Chem Phys 87:1785
16. Kranendonk WGT, Frenkel D (1988) Mol Phys 64:403

Authors' address:

M. P. Pileni
Université P et M Curie
Laboratoire SRS II
Bâtiment F (74)
4 place Jussieu
F 75005 Paris, France

A superconducting layered polyelectrolyte with strong intracrystalline swelling

A. Lerf and W. Biberacher

Walther-Meißner-Institut, Garching , FRG

Abstract: We report on the first experimental evidence for osmotic swelling in a layered disulfide interaction compound

Key words: Layered dichalcogenides — intercalation compounds — polyelectrolytes — osmotic swelling

The electrically conducting layered disulfides intercalate hydrated (solvated) metal ions. These compounds show polyelectrolyte behavior [1] like the clay minerals, and some of them form colloids [1—3]. But, in contrast to the colloid-forming sodium montmorillonite, a strong intracrystalline swelling as a precursor phenomenon for the colloid formation was not observed in the case of the sulfides.

By choosing tetramethylammonium as the charge compensating interlayer cation, a system is found where the interlayer distance increases from about 3000 pm to about 8000 pm (the highest layer distance observed for intercalation compounds of the layered dichalcogenides) by reducing the electrolyte concentration following a $c^{-1/2}$ law like in the case of the montmorillonite [4]. The intercalation is carried out by treating $2H - TaS_2$ (particle diameter 100—200 µm) with 0.06 M to 0.008 M aqueous solutions of $[N(CH_3)_4]OH$. At concentrations above 0.06 M a layer distance of about 1700 pm is observed, significantly higher than the value (about 1130 pm) measured for intercalation in 0.001 M solutions or after washing the samples with deionized water and drying in air. The difference indicates an additional uptake of $[N(CH_3)_4]OH$ in the first case, which is confirmed by chemical analysis. The strong expansion of the interlayer space shrinks to 1700 pm reversibly (including the reappearance of the 101 reflections) if dilute solution are replaced by a 0.25 M solution of $[N(CH_3)_4]OH$.

Washing the 1700 pm-phase with deionized water once or twice leads to a dramatic swelling of the solid. The solid intercalated with a 0.001 M $[N(CH_3)_4]OH$ disperses in part. The remaining solid is a transparent, porous material of red color (in contrast to the metallic blue starting material), indicating a strong exfoliation of the crystals. X-ray diagrams show only the Bragg reflections of the lattice constants in the layer planes.

Preliminary measurements of the transition temperature to the superconducting state show T_c-values in the order of 3—4 K, even for the solids prepared in the dilute solutions.

References

1. Lerf A, Schöllhorn R (1977) Inorg Chem 16:2950
2. Murphy DW, Hull GW (1975) J Chem Phys 62:973
3. Joensen P, Frindt RF, Morrison SR (1986) Mat Res Bull 21:457
4. Norrish K (1954) Disc Faraday Soc 18:120

Authors' address:

A. Lerf
Walther-Meißner-Institut
D-8046 Garching, FRG

Microemulsion gels obtained by the sol-gel method using metal alkoxides

P. Lianos and D. Papoutsi

School of Engineering, Physics Section
University of Patras, Greece

Abstract: Microemulsion gels have been produced by the sol-gel method using silicium tetramethoxide and titanium isopropoxide and water-in-oil microemulsions made of 1-pentanol (or cyclohexane and 1-pentanol) and sodium dodecylsulfate. Some preliminary results are presented concerning methods of preparation and photophysical characterization.

Key words: Sol-gel method — microemulsion gels

Introduction

The hydrolysis of Si, Ti, and other metal alkoxides leads to a release of alcohol and formation of an oxide through the following general scheme:

$$M(OR)_4 + 4\,H_2O \rightarrow M(OH)_4 + 4\,ROH \quad \text{hydrolysis}$$

$$M(OH)_4 \rightarrow MO_2 + 2\,H_2O \quad \text{condensation},$$

where M is the metal, for example, Si or Ti. If $M(OH)_4$ is kept in suspension, a rather slow condensation process takes place leading to a gel and, after evaporation of the entrapped liquid species, to a xerogel. This is the basis of the sol-gel method [1] for the formation of inorganic polymers (glasses and ceramics). Materials produced by this method have the advantage of high purity, low temperature processing, ability to be easily doped by a great variety of doppants and ability to be easily handled and shaped in any desired form (blocks, spheres, disks, coatings, etc.).

Two major problems must be solved to control a sol-gel process:

1) A choice must be made of the composition of the original sol, taking into account that water and $M(OR)_4$ do not mix. However, a clear gel can only derive from a clear sol. Clear sols are thus obtained by addition of an alcohol or by forming a microemulsion [2, 3] (in our case a water-in-oil, w/o, microemulsion). Then doppants of different degree of hydrophilicity (or hydrophobicity) can be simultaneously introduced in the original sol or the microemulsion (which in the sol-gel process is still called a sol).

2) $M(OH)_4$ must not precipitate, since condensation then is negligible, if not impossible. This question is particularly important with transition metals such as Ti which hydrolyze rapidly and precipitate. Controlled hydrolysis of $Ti(OR)_4$ becomes very successful with the dispersion of water in a w/o microemulsion. Then, the alkoxide dissolves in the continuous phase.

Here, we present some results on microemulsion gels formed with $Si(OCH_3)_4$ and $Ti(iso\text{-}OC_3H_7)_4$

and simple w/o microemulsions containing 1-pentanol (or cyclohexane and 1-pentanol), and sodium dedecylsulfate (SDS).

Cyclohexane-pentanol + tetramethoxysilane

W/o microemulsions were made using a constant water/surfactant ratio equal to 2.5. All compositions were chosen close to the demixion line in the pseudo-ternary phase diagram. Different samples are distinguished by the wate volume fraction Φ_w. The volume fraction of $Si(OCH_3)_4$ was always 0.2. The temperature was 20°C. We have recorded the following observations:

— Clear gels are formed for $\Phi_w > 0.1$
— Gelling occurs within a few hours after mixing. Gelling can be visually observed and it can be monitored by various methods. One way, for example, is measuring the luminescence decay time of $Ru(bpy)_3^{2+}$ co-dissolved in the original sol (aerated). As gelation proceeds and the luminophore is immobilized, the decay probability decreases, increasing the decay time τ_0. Thus, for $\Phi_w = 0.25$, τ_0 changed from only 261 ns to 1090 ns within 3 h and remained unchanged at later times. Gelation can also be monitored by observing pyrene excimer formation, even though higher probe concentrations are then necessary.
— Molecular interactions can be studied in the transparent matrix of a microemulsion gel, in relation with their possible optical applications. We have investigated electron-transfer quenching of the $Ru(bpy)_3^{2+}$ luminescence by MV^{2+}. The distribution of the acceptor molecules around the excited luminophore is not homogeneous. One way to model it is to consider it fractal. In that case the time-correlated luminescen intensity is given by [4]:

$$I(t) = I_0 \exp(-k_0 t) \exp(-A\ln^d(at)), \qquad (1)$$

where the exponent d of the logarithm is the fractal dimension of the distribution, while k_0, A, and a are constants. We have found that high concentrations of the MV^{2+} (as measured in the original sol) ranging around and above 10^{-2} M are necessary for efficient electron transfer. Analysis with Eq. (1) gave very inhomogeneous distributions with fractal dimension values ranging below 1. The problem is being studied further in our laboratory.

Pentanol-SDS w/o microemulsions + Titanium isopropoxide

In these samples, 1-pentanol is the continuous phase (cf. ref. [3]). We have prepared w/o microemulsions using different water and SDS concentrations. The quantity of the added $Ti(iso\text{-}OC_3H_7)_4$ varied accordingly. We have recorded the following observations:

— When the amount of water was relatively large, $Ti(OH)_4$ immediately formed and precipitated.
— No precipitation was observed below 3.7 wt% SDS and 6.8 wt% water.
— A non-transparent white monolith can be formed with an alkoxide-to-water molecular ratio up to about 0.15. When heated to 500°C, all organic content is burned out and a white powder is obtained with all spectroscopic (light absorption and emission) characteristics of rutile.
— A transparent yellow monolith is formed at higher alkoxide/water ratios. Its light absorption increases abruptly below 400 nm. However, it gives no fluorescence. The gel, which can be shaped in any desired form, can be used as a cutoff light filter. The nature of the coloring material is under further investigation.
— A comparison with samples prepared in pure alcohol (e.g., methanol or ethanol, cf. ref. [5]) without SDS shows that, in w/o microemulsions, higher water/alkoxide ratios can be used to obtain a gel without precipitation.
— Gels are obtained within hours.
— In pure alcohol presence of acid is necessary in order to charge the colloidal particles formed and keep them in suspension. No acid addition is necessary in w/o microemulsions.

References

1. Hench LL, West JK (1990) Chem Rev 90:33—72
2. Guizard C, Larbot A, Cot L, Perez S, Rouviere J (1990) 87:1901—1922
3. Friberg SE, Yang CC, Sjoblom J (1992) 8:372—376
4. Blumen A, Klafter J, Zumofen G (1986) In: Zschokke I (ed) Optical Spectroscopy of Glasses, D. Reidel Pubershing Co, p 199—265
5. Anderson MA, Gieselmann MJ, Qunyin X (1988) J Membr Sci 39:243—285

Authors' address:

Panagiotis Lianos
School of Engineering
Physics Section
University of Patras
GR-26500 Patras, Greece

Progress in Colloid & Polymer Science

Progr Colloid Polym Sci 93:294 (1993)

Phase stability and structure of non-toxic microemulsions

M. A. López Quintela, L. Liz, and J. Blanco*

Depts. of Physical Chemistry and
*Pharmaceutical Technology
University of Santiago de Compostela, Spain

Key words: Microemulsions — phase behavior — structure — SANS

For many pharmaceutical and medical applications, there is a need of nontoxic microemulsions. Most of the research work until now has been carried out using surfactants like AOT, C_iE_j, etc., which are not suitable to be applied directly for such aims.

In this paper, we report on our studies of the phase behavior and the structure, studied by electrical conductivity and small-angle neutron scattering (SANS) measurements of microemulsions composed of: an isotonic aqueous solution of NaCl (9 gr/l), ethyl oleate as an oil, a commercial surfactant containing a mixture of glycerides and polyglycides of caprylic and capric acids, and a long-tail cosurfactant, polyglycerol isostearate.

In all the samples, the volume ratio between surfactant and cosurfactant was kept constant at the value 3:1, so that for the ternary phase diagrams we considered the mixture as a single component.

We have studied the different phases existing in the whole pseudo-ternary phase diagram in the temperature range between 16° and 60°C. Such studies showed the presence of a relatively large isotropic one-phase domain below 42°C, which gets smaller over such a temperature and disappears at 50°C. Apart from this one-phase domain, two different two-phase domains are found in the diagram.

There is a first two-phase domain for low surfactant + cosurfactant concentrations. In such a domain, one can observe the coexistence of two microemulsions with the same structure and different droplet concentration, as could be expected for systems with interactions in the dispersed phase [1]. These microemulsions are of the w/o type in the vicinity of the oil corner, and of the o/w type in the vicinity of the water corner.

The second two-phase domain is observed near the surfactant corner, and it is formed by a lower aqueous phase and an upper lamellar phase.

Within the one-phase domain, by performing electrical conductivity measurements, we detected the presence of w/o, o/w and bicontinuous microdomains, according to the relative amounts of water and oil in their composition.

In order to obtain further knowledge on the structure of the system, we performed SANS measurements on several samples within the one-phase domain and at different temperatures. Data analysis via models for spheres, ellipsoids, and cylinders, with and without interactions [2], failed. A qualitative analysis of the scattering curves shows the presence of attractive interactions between microdomains that increase with temperature. Fractal analysis [3] of the data corresponding to the more diluted samples showed the presence of fractal microdomains with surface fractal dimensions around the value 2.3. The average size of such objects is about 16 nm diameter, although the presence of aggregates of these objects is possible as well.

References

1. Cazabat AM (1985) In: Physics of Amphiphiles: Micelles, Vesicles and microemulsions. XC Corso, p 723
2. Chen SH (1986) Ann Rev Phys Chem 37:351
3. López Quintela MA, Buján Núñez MC (1991) J Phys I France 1:1251

Authors' address:

M. A. López Quintela
Dept. of Physical Chemistry
University of Santiago de Compostela
E-15706 Santiago de Compostela, Spain

Stabilization of non-lamellar structures in phospholipid membranes by squalene

K. Lohner, G. Degovics, E. Gnamusch*,
F. Paltauf*, and P. Laggner

Institute of Biophysics and X-Ray Structure Research,
Austrian Academy of Sciences, Graz, Austria
* Institute of Biochemistry and Food Chemistry,
  Technical University of Graz, Graz, Austria

Key words: Phospholipid polymorphism — hydrophopic molecules — differential scanning calorimetry — x-ray diffraction

Physico-chemical studies on the interaction between hydrophobic molecules and phospholipid membranes have been performed for a variety of reasons, e.g., health aspects — lipid soluble drugs, anesthetically active compounds — or to get insight into fundamental questions of membrane properties — stabilization of phases, mechanisms of phase transitions [1]. Here, we present a study on the lipid polymorphism of aqueous dispersions of 1-stearoyl-2-oleoyl-phosphatidylethanolamine (SOPE) and 1-palmitoyl-2-oleoyl-phosphatidylcholine (POPC) (95/5 mol/mol) in the presence of squalene, an intermediate in the biosynthesis of sterols.

SOPE/POPC model membranes undergo a lamellar gel (L_β) to liquid-crystalline (L_a) and a lamellar to inverse hexagonal (H_{II}) phase transition at 30° and 64°C, respectively. With increasing concentration of squalene the main transition temperature was decreased from 30°C to 28.1°C for samples containing 5 mol% squalene, indicating a higher affinity for the fluid phase. The phase transition enthalpy was not affected as detected by high precision differential scanning calorimetry. The structure of the phospholipid aggregates was determined by small-angle x-ray diffraction experiments showing only a minor increase of the lamellar repeat distance of the liquid crystalline phase for the squalene containing samples. Upon incorporation of 5 mol% squalene the lamellar to non-lamellar transition is shifted to the physiological temperature range (about 39°C). Still further increase of the squalene concentration (10 mol%) leads to an overlapping of both the L_β to L_a and L_a to H_{II} phase transition. Analysis of the x-ray experiments showed that the size of the tubes of the inverse hexagonal phase is linearly increasing up to 5 mol% squalene. At higher squalene concentration only a relatively small growth to the H_{II} tubes is observed, probably due to limiting partition of squalene into the hydrophobic core of the membrane.

The stabilization of the non-lamellar phase by squalene can be explained on the basis of the model established by Gruner [2]. The author argued that owing to the geometry of the hexagonally packed lipid tubes in the H_{II} phase the phospholipids have to adopt different lengths in order to fill the hydrophobic volume. This results in an anisotropic packing environment, an unfavorable energetical situation. However, the constraints of the hydrocarbon chains can be reduced by filling the interstitial space between the hexagonally packed phospholi-

pid tubes with hydrophobic molecules like squalene. Recent H^2 NMR [3] and neutron diffraction [4] experiments showed that alkanes indeed dissolve in the most disordered region of the hydrophobic core, which are the perimeters and corner regions of the hexagons.

References

1. Lohner K (1991) Chem Phys Lipids 57:341—362
2. Gruner SM (1981) Proc Natl Acad Sci USA 82:3665—3669
3. Siegel DP, Banschbach J, Yeagle PL (1989) Biochemistry 28:5010—5019
4. Turner DC, Gruner SM, Huang JS (1992) Biochemistry 31:1356—1363

Authors' address:

Karl Lohner
Institute of Biophysics and
X-Ray Structure Research
Austrian Academy of Sciences
Steyrergasse 17/VI
A-8010 Graz, Austria

Structure analysis of Langmuir-Blodgett films of phospholipid bilayers

K. Lohner[1]), O. Konovalov[2]), V. V. Erokhin[2])
I. V. Myagkov[3]), V. I. Troitzky[3]), T. I. Berzina[3],
L. A. Feigin[2]), and P. Laggner[1])

[1]) Institute of Biophysics and X-Ray Structure Research, Austrian Academy of Sciences, Graz, Austria
[2]) Institute of Crystallography, Russian Academy of Sciences, Moscow, Russia
[3]) Institute of Physical Problems, Zelenograd, Russia

Key words: Phospholipids — oriented layers — Langmuir-Blodgett films — x-ray diffraction — x-ray reflectivity — electron diffraction

Phospholipid bilayers are the matrix of biological membranes. It is well known that, owing to their amphiphilic character, phospholipids are able to form condensed monolayers at the air/water interface [1]. Therefore, they were already used as substrates for the formation of Langmuir-Blodgett (LB) films [2]. Such films are of growing interest for molecular devices or simply for being used as model systems to study biological processes. In this study 1,2-dipalmitoyl-phosphatidic acid (DPPA) and -phosphatidylcholine (DPPC), two phospholipids with identical hydrophobic side-chains, but differing in their headgroup composition, as well as 1,2-dihexadecyl-phosphatidylcholine (DHPC), the ether analog of DPPC, were tested as substrates for the preparation of well-defined LB films.

Monolayer studies were performed on various subphase conditions, to obtain surface pressure as well as surface potential -area isotherms of the system and to determine the regime of deposition of molecular layers. Both, Langmuir-Blodgett and Langmuir-Schäfer techniques were applied for the deposition on various solid substrates, which had been treated with silanizing agents. DPPA, which is negatively charged at physiological conditions, was more easily transferred onto a solid substrate, especially in the presence of divalent cations, as compared to the zwitterionic phospholipids DPPC and DHPC. The spacing of the LB superlattice of DPPA was about 52 Å as calculated from x-ray diffraction patterns consisting of several peaks. This value is also in good agreement with previously published data obtained by IR spectroscopy [3]. Data that we gained from x-ray reflectivity measurements gave the same results. Such a lamellar repeat unit can be explained by the formation of a Y-type film. The total thickness of the LB film was determined from the Kiessig fringes, indicating a transfer rate of about 1. Information on the hydrocarbon chain packing in the layers of the LB film was gained from electron diffraction experiments, which showed typical patterns for a hexagonal packing of the phospholipid side-chains.

For DPPA, we were able to prepare reproducible films up to 100 layers, i.e., 50 bilayers, which could not be obtained from either DPPC nor DHPC. On the contrary, quite frequently the layers came off on the downstroke. In such cases, where we succeeded in the transfer of these lipids onto the siliconized substrate, the x-ray diffraction patterns were of poor quality. Only one broad peak could be detected, which is indicative for a less ordered layer structure as compared to the DPPA LB films.

References

1. Gaines JrGL (ed) (1966) Insoluble monolayers at liquid-gas interfaces. Wiley-Interscience, New York
2. Ulman A (ed) (1991) An introduction to ultrathin organic films. From Langmuir-Blodgett to self-assembly. Academic Press Inc, San Diego
3. Cui DF, Howarth VA, Petty MC, Ancelin H, Yarwood J (1990) Thin Solid Films 192:391—396

Authors' address:

Karl Lohner
Institute of Biophysics and
X-Ray Structure Research
Austrian Academy of Sciences
Steyrergasse 17/VI
A-8010 Graz, Austria

Temperature-dependence of spreading monolayers of aliphatic alcohols at the water/air interface

P. Lo Nostro, G. Caminati, and G. Gabrielli

Department of Chemistry, University of Florence, Italy

Abstract: The anomalous temperature-dependence of spreading isotherms of tetradecanol ($C_{14}OH$) and tetradecanol/oleic acid mixtures in monolayer at the water/air interface has already been reported in previous papers. In fact, an anomalous lowering of the molecular area of the film was observed between 30° and 35°C. This behavior was ascribed to a hydration-dehydration process of the interfacial polar head groups. — In the present work a systematic study of spreading monolayers of 1-hexadecanol, 1-octadecanol and 1-eicosanol ($C_{16}OH$, $C_{18}OH$ and $C_{20}OH$ respectively) at four different temperatures (20°, 25°, 30° and 35°C) at the water/air and urea 0.5 M/air interfaces is reported. The use of urea 0.5 M as a subphase is justified by its properties as a waterstructure breaker. — The experimental results show that the anomalous behavior of the monolayers as a function of the temperature depends both on the nature of the subphase and on the hydrophobic chain length of the alcohol. Discussion of the results confirms that a hydration-dehydration process of the polar head groups is involved in the observed phenomenon.

Key words: Spreading monolayers — aliphatic alcohols — urea — hydration effect — hydrophilic-hydrophobic balance

Introduction

The regular behavior of spreading monolayers at the water/air interface consists in the shift of the π/A

curve toward higher area values as the temperature rises [1]. However, the isotherms of 1-tetradecanol ($C_{14}OH$) films show an "anomalous" temperature-dependence, consisting of a remarkable shift of the π/A curves toward lower areas and pressures [2, 3]. This phenomenon has been ascribed [4, 5] to the hydration-dehydration effect of the polar head groups, which depends on the temperature, but an extensive explanation of all the involved factors is still necessary.

The purpose of the present work is to study the effect of the hydrophilic-hydrophobic balance of the amphiphile, and of the interaction betweeen the polar head groups and the water molecules of the hydration shell. For this reason, we will examine the surface behavior of three different aliphatic alcohols, namely, 1-hexadecanol (cetyl alcohol, $C_{16}OH$), 1-octadecanol (stearilic alcohol, $C_{18}OH$), and 1-eicosanol (arachidyl alcohol, $C_{20}OH$) as a function of the temperature and of the subphase, by comparing the isotherms of $C_{16}OH$, $C_{18}OH$ and $C_{20}OH$ spread onto water or urea 0.5 M, at 20°, 25°, 30° and 35°C. The choice of urea was made because of its well-known property as a water-structure breaker. The effect of an aqueous solution of sucrose as a subphase is currently being studied.

Materials and methods

1-hexadecanol, 1-octadecanol, 1-eicosanol and urea were purchased from Carlo Erba-Farmitalia

(Milano, Italy) and used without any further purification. Surface pressure measurements were performed according to the method already described in previous papers [6].

Results and discussion

We recorded the π/A isotherms of $C_{16}OH$, $C_{18}OH$ and $C_{20}OH$ at 20°, 25°, 30° and 35°C at the water/air and urea 0.5 M/air interfaces; in the case of $C_{16}OH$/water monolayers (see Fig. 1), the curves shift toward lower areas and pressures as the temperature rises, with a dramatic lowering of the collapse pressure (π_c) and of the limit area (A_{lim}) as the temperature increases from 30° to 35°C. On the other hand, in the case of $C_{18}OH$/water and $C_{20}OH$/water films this effect disappears, and the isotherms rise again as a function of the temperature, following the well-known behavior of monolayers.

Figure 1 also reports the isotherms of $C_{16}OH$ films at the urea 0.5 M/air interface at different temperatures. The anomalous temperature-dependence of the π/A isotherm is reduced but still present, showing also an increase of the expanded phase of the film at high temperatures.

The isotherms of $C_{18}OH$ and $C_{20}OH$ films spread onto an aqueous solution of urea show the regular temperature-dependence. Since both 1-octadecanol and 1-eicosanol monolayers do not show any

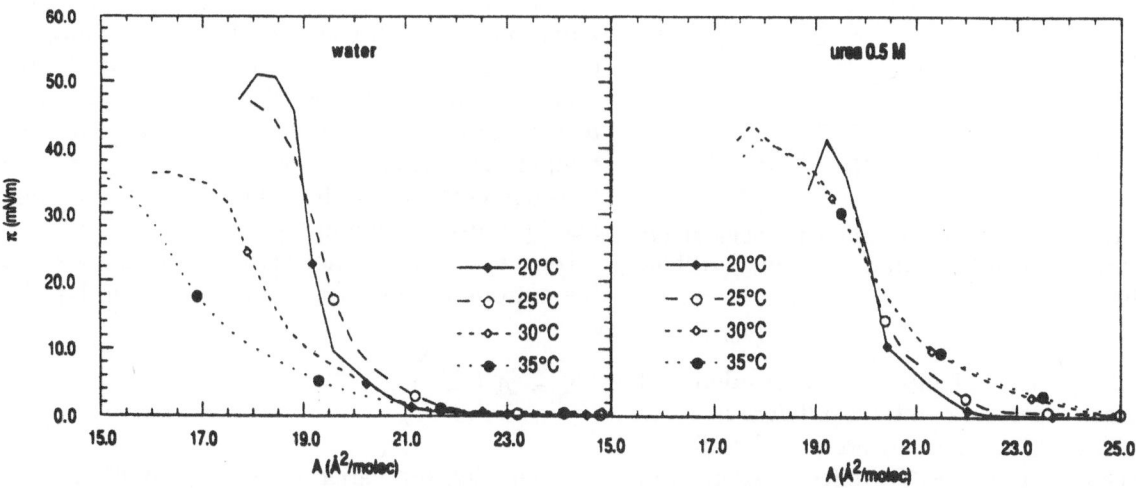

Fig. 1. π/A isotherms of 1-hexadecanol films at the water/air and urea 0.5 M/air interfaces at different temperatures

relevant anomalous temperature-dependence, we will focus our attention only on the $C_{16}OH$ case.

Figure 2 reports the A_{lim} values of $C_{16}OH$ monolayers as a function of the subphase and of the temperature. The plot clearly shows that the effect of T on the properties of the monolayer is more evident in the case of water than urea 0.5 M.

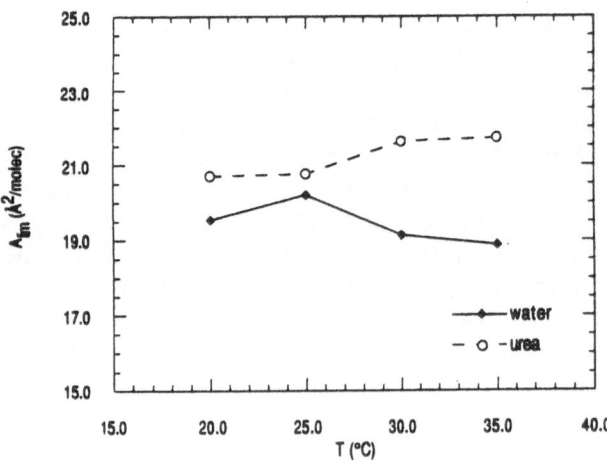

Fig. 2. Limit area of 1-hexadecanol films at the water/air and urea 0.5 M/air interfaces as a function of the temperature

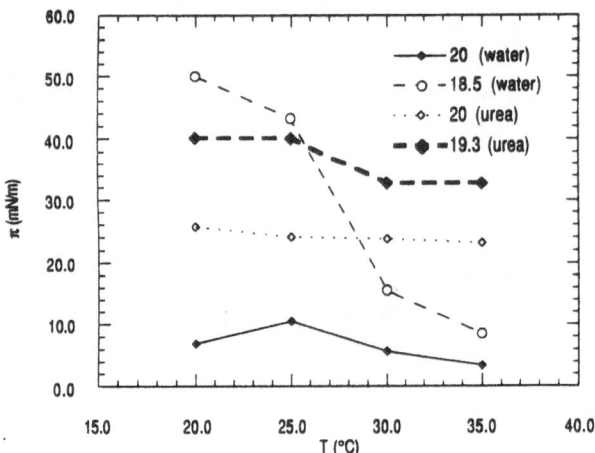

Fig. 3. Surface pressure of 1-hexadecanol films at the water/air and urea 0.5 M/air interfaces as a function of the temperature

Figure 3 reports the temperature-dependence of the surface pressure of $C_{16}OH$/water and $C_{16}OH$/urea films at constant areas and indicates that π decreases as T rises. The effect is stronger in the case of water and, especially, when the film is about to collapse, i.e., in the condensed phase.

However this effect cannot be ascribed to the collapse because π is lower than π_c, and because there is no variation of π as a function of time [7]. The calculation of the compressibility modulus defined according to

$$K = -A \left(\frac{\partial \pi}{\partial A} \right)_{P,T} \tag{1}$$

of all π/A curves indicates that the anomalous temperature-dependence is not due to a change in K, that is of the bidimensional phase of the film. Therefore, the behavior of A_{lim} and π of $C_{16}OH$ monolayers at the water/air interface can be interpreted as the consequence of the dehydration of the alcoholic polar head group, that is the transfer of some water molecules from the coordination shell to the bulk phase, accompanied by the rearrangement of the packing of the amphiphilic molecules at the interface and the contraction of the film of about $2 \div 3 \text{ Å}^2$/molec, as the temperature changes from 25° to 35°C. Considering that the average area of a water molecule is about 7.3 Å^2 [4], the film contraction corresponds to the loss of one water molecule per three molecules of amphiphile at the interface.

When the subphase is constituted by an aqueous solution of urea 0.5 M, the effect is largely reduced, but still present. The property of urea as a water-structure breaker is well-known in the literature; the fact that urea reduces the structure of the water surface can explain why the anomalous effect recorded for $C_{16}OH$/water monolayers is weaker in the case of $C_{16}OH$/urea films. We must recall however that the effect of urea on the interaction between water molecules and polar head groups of amphiphiles is still controversial [8]. Anyhow, we can suppose a direct interaction of urea that replaces some water molecules at the interface. This result confirms the hypothesis that the unusual temperature-dependence of the isotherms is related to the interactions of the subphase with the polar head groups of the amphiphile.

For a spreading monolayer at a constant area A, the entropy of the surface is defined as in Eq. (2):

$$S_s = A \left(\frac{\partial \pi}{\partial T} \right)_A . \tag{2}$$

Calculating the variation ΔS_s (in J mol^{-1} K^{-1}) of the transition below and above 30°C for two different area values (one in the expanded phase

Table 1.

	$\Delta S_s^{\text{water}}$ ($A = 20$)	$\Delta S_s^{\text{water}}$ ($A = 18.5$)	ΔS_s^{urea} ($A = 20$)	ΔS_s^{urea} ($A = 19.3$)
below 30°C	-5.56 J mol^{-1} K^{-1}	-12.86	-7.34	-4.71
above 30°C	$+1.64$ J mol^{-1} K^{-1}	$+12.64$	$+7.32$	$+4.69$

region and the other in the liquid intermediate phase region), we found the data reported in Table 1. Assuming that the anomalous behavior of 1-hexadecanol films is due to a variation of the number of water molecules in the hydration shell of the polar head groups, ΔS_s must be associated to a different packing of the alcohol molecules at the water/air interface and to different interactions between the head groups of the amphiphile and the coordinating molecules of the subphase. In fact for $T < 30°C$, $\Delta S_s < 0$, which reveals the ordering effect of the amphiphilic molecules on the water surface. This is in agreement with the hypothesis of the transfer of some water molecules from the hydration shell of the monolayer to the bulk phase and the consequent closer packing of the amphiphile chains at the water/air interface. When $T > 30°C$, thermal agitation produces a disordering effect which results in $\Delta S_s > 0$. Comparing ΔS_s calculated for the same subphase but in different phase regions, we observe that — in the case of pure water — the absolute value is higher for the expanded phase region and lower for the liquid intermediate phase. On the other hand, in the case of urea 0.5 M, we find the opposite trend, which is probably due to the fact that urea replaces some water molecules at the interface.

Finally, the fact that the anomalous temperature-dependence of long-chain aliphatic alcohols $C_n OH$ isotherms is quite negligible for $n > 16$ can be ascribed to the different hydrophilic-hydrophobic balance. In fact, in the case of $C_{18}OH$ and $C_{20}OH$ the interactions between aliphatic chains of different molecules are greater than in the case of $C_{14}OH$ and $C_{16}OH$, where the interactions of the —OH polar groups with the water molecules of the surface are overwhelming.

Conclusions

On the basis of the experimental results the following conclusions can be drawn:

1) the anomalous temperature-dependence of 1-hexadecanol films at the water/air and urea/air interface is shown by the π/A isotherms, due to a hydration-dehydration effect that occurs between 30° and 35°C. The compressibility modulus indicates that this phenomenon is not due to a bidimensional phase transition. 1-tetradecanol presents the same phenomenon, whereas 1-octadecanol and 1-eicosanol do not show this behavior. Therefore, the main factor involved must be the hydrophilic-hydrophobic balance between the alcoholic head group (—OH) and the $(-CH_2-)_n$ alkyl chain of the amphiphile.

2) this unusual effect disappears if the subphase is replaced by an aqueous solution of urea. In this case, either a change of the water structure or a simple replacement of some water molecules by urea molecules produce different interactions of the surface with the polar head groups. This result is consistent with the hypothesis that the behavior of 1-hexadecanol films at the water/air interface is due to the interaction of the alcoholic head group with the water molecules.

3) A_{lim}/T and π/T plots and the calculation of the surface entropy S_s confirm the hypothesis of a process lead by the interactions at the interface of the monolayer.

References

1. Adamson A (1976) In: Wiley John, Physical Chemistry of Surfaces. New York, pp 134—145
2. Gabrielli G, Senatra D, Caminati G, Guarini GGT (1988) Colloid Polym Sci 266:823—831
3. Caminati G, Senatra D, Gabrielli G (1991) Langmuir 7:1969—1974
4. Steinbach H, Sucker C (1980) Adv Coll Interface Sci 14:43—65
5. Caminati G, Senatra D, Gabrielli G (1991) Langmuir 7, pp 604—607
6. Lo Nostro P, Niccolai A, Gabrielli G (1989) Progr Colloid Polym Sci 79:43—48

7. Gaines G (1966) In: Prigogine I, Insoluble Monolayers at Liquid-Gas Interfaces. New York, pp 144—151

8. Baglioni P, Ferroni E, Kevan L (1990) J Phys Chem 94:4296—4298, Baglioni P, Rivara-Minten E, Dei L, Ferroni E (1990) J Phys Chem 94:8218—8223, Baglioni P, Dei L, Lo Nostro P, Kevan L, Langmuir (submitted)

Authors' address:

Dr. P. Lo Nostro
c/o Dip. Chimica
via Gino Capponi, 9
I-50121 Firenze, Italy

Multiequilibria of 1 H-benzimidazole 2-(2′-furanyl) neutral and protonated forms in the presence of amphiphylic aggregates

A. Lopes[1]), A. L. Maçanita[1]), F. S. Pina[2]), and E. C. C. Melo[1])

[1]) Centro de Tecnologia Química e Biológica — IST OEIRAS, Portugal
[2]) Faculdade de Ciências e Tecnologia da U. N. L. Monte da Caparica, Portugal

Key words: Acid-base multiequilibria — partition coeficients — fluorescence anisotropy — amphiphylic aggregates

Fuberidazole (1-H-benzimidazol 2-(2′-furanyl)) is a fungicide in widespread use. The photodegradation pattern of fuberidazole in water and organic media is quite different. This observation induced us to investigate how this pesticide partitions between aqueous and amphiphylic aggregates or other hydrophobic structures which may be present in natural waters.

In the pH range of natural aqueous media, both protonated and neutral forms of fuberidazole exist in solution (pK_a = 5.0). This acid-base equilibrium becomes more intricate when lipophilic aggregates, like phospholipid bilayers or surfactant micelles are present, because we need to account for the solubility of Fub and FubH$^+$ in the non aqueous phase.

In the present work, we measure the solubility of fuberidazole in natural water: surfactant systems as a function of the pH. In Table 1, we present the values of K_p obtained using the fluorescence

Table 1. Equilibrium constant for Fub and FubH$^+$ in C$_{12}$E$_{10}$: water, SDS: water and HTAC: water systems at 25°C

Surfactant system	K_p	K_p^+
C$_{12}$E$_{10}$: water	190 ± 20	3.8 ± 0.2
SDS: water	308 ± 25	11600 ± 1000
HTAC: water	850 ± 80	—

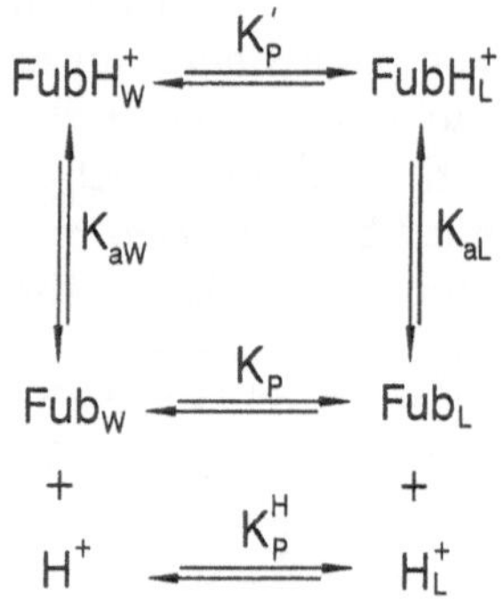

Fig. 1

anisotropy method in buffered soultions of sodium dodecylsulfate SDS, polyoxyethylene-10 lauryl ether, C$_{12}$E$_{10}$, and hexadecyltrimethylamonium chlorid, HTAC.

The partition of Fub and FubH$^+$ between micelle and water as a function of the volume fraction of lipidic media is finally discussed as a function of pH.

Authors' address:

Antonio Lopes
Centro de Tecnologia Quimica e Biologica
Rua de Quinta Grande 6, Ap 127
P-2780 Oeiras, Portugal

The colloid dispersion of natural and synthetic resins in water

N. Lysogorskaia

St. Petersburg, Russia

The possibility of the spontaneous colloid dispersion of the given projects is obtained by independent optical methods, activity by temperature, addition of various surfactants, acid, and liquor medium.

The regularity of this phenomenon was established, not only in different resins, but in some other solid substances, considered, in practice, to be previously insoluble in water. The interphase tension estimation, the specific surface phase separation, the energy of the process activation carried out, for example, in space dispersion showed that the homogeneous surfactant solution and mycelerned to concentration, five times higher KKM, promote the resins' dispersion, nonstable lyophobic systems formation, and the free surface system energy prevails over entropy solution factor.

But at concentrations surfactant in order and higher KKM, in prevailing entropy solution factor the stable lyophiling micro-emission with over low meaning of the interphase tension is formed.

The latter has a great applied meaning, for example, for the reducing cellulose resin ousity.

The electronno-microscopical investigations of the surfactant solutions in mediums with various pH (from 2 to 12) was carried out, which was subjected by high temperature treatment (up to 160°C).

Author's address:

N. Lysogorskaia
Institute for Pulp and Paper Industry
ul. Ivana Chernykh
198 092 St. Petersburg, Russia

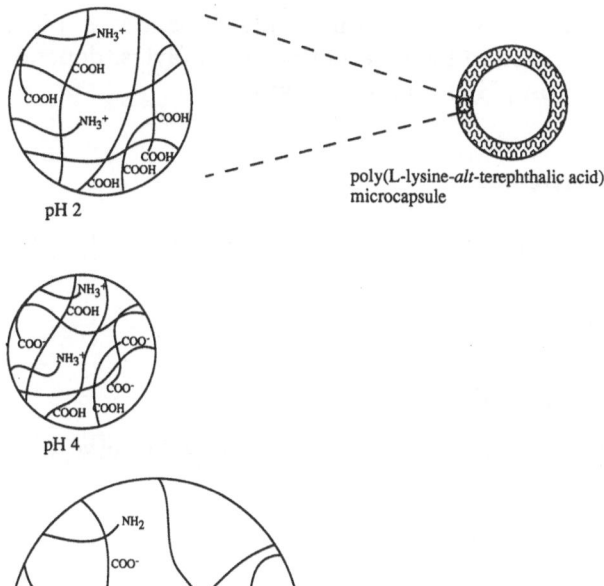

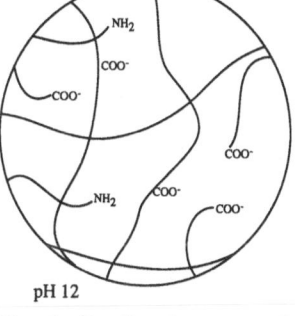

Fig. 1. Predicted structural change of poly(L-lysine-*alt*-terephthalic acid) microcapsule membrane at various pH

Dependence on pH of permeability towards electrolyte ions of poly(L-lysine-*alt*-therephthalic acid) microcapsule membranes

K. Makino, E. Miyauchi, Y. Togawa, H. Ohshima, and T. Kondo

Faculty of Pharmaceutical Sciences,
Science University of Tokyo

Key words: pH-dependence — permeability — microcapsule membrane — electrolyte ions

Permeation of electrolyte ions through poly-(L-lysine-*alt*-terephthalic acid) microcapsule membranes was studied as a function of the pH of the medium at different ionic strengths. When the pH of the medium was varied, the permeation rate for 5-sulfosalicylate anions as well as phenyltrimethyl-ammonium cations was slow at pH values lower than 4, showed a sudden and large increase in the pH range between 4 and 6, and thereafter remained unchanged. This remarkable change in the permeation rate was found to be produced by an abrupt increase in the microcapsule size observed in the same pH range. Increase in the ionic strength of the medium at pH values higher than 6 increased the rate of anion permeation, but decreased the rate of cation permeation due to increase in the screening effect of salt ions on the negative charges in the microcapsule membranes. These phenomena suggest the structure of the microcapsule membranes is varied depending on pH and ionic strength, which is caused by a change of charge density in the microcapsule membranes, as schematically shown in Fig. 1. In poly(L-lysine-*alt*-terephthalic acid) microcapsule membranes, a polymer chain has a carboxyl group and an amino group at its terminals.

Such a phase transition phenomenon is often observed in hydrogels of polyelectrolytes, depending on pH and ionic strength.

References

1. Miyauchi I, Togawa Y, Makino K, Ohshima H, Kondo T (1992) J Microencapsulation 9:329—333

Authors' address:

Dr. Kimiko Makino
Faculty of Pharmaceutical Sciences
Science University of Tokyo
12 Ichigaya Funagawara-machi Shinjuku-ku Tokyo 162
Japan

Self-organization in water of bipolar phosphate amphiphiles

A. Kokkinia[1]), C. M. Paleos[1]), A. Malliaris[1]), and A. Xenakis[2])

[1]) NRC "Demokritos", Agia Paraskevi, Athens, Greece
[2]) The National Hellenic Research Foundation, Athens, Greece

Abstract: The self association in water of dodecane-1,12- and hexadecane-1, 16-bis(phosphate disodium) salts, has been investigated by means of physicochemical methods including electrical conductivity, video-enhanced microscopy, absorption spectroscopy and fluorescence probing. No evidence of true micellar formation was obtained. Instead, the results have indicated that aggregates of widely varying sizes are formed in aqueous media.

Key words: Phosphate bolaamphiphiles — molecular aggregates — video enhanced microscopy — fluorescence probing

Introduction

Bipolar amphiphiles, or bolaamphiphiles, have recently received increasing attention in view of their application in modifying the stability of vesicles when incorporated in their membrane [1]. Thus, it has been found that certain bolaamphiphiles stabilize vesicles by spanning their membranes [2], in an analogous manner by which thermophilic and acidophilic bacteria are stabilized, i.e., by the incorporation of bolaamphiphiles [1]. On the contrary, it was found that other non-ionic bolaamphiphiles, both monomeric and polymeric, have a disrupting effect on vesicles [3, 4], evidently by penetrating the vesicular membrane in a way that affects its segregation.

The literature on the aggregational behavior in water of simple ionic and non-ionic bolaamphiphile systems is rather limited. In studies initiated as early as mid 70's, the formation of micelles from these bolaamphiphiles has been shown to depend on the length of their aliphatic chain. Thus, the diquaternary ammonium salts ($X^- Me_3N^+(CH_2)_n N^+ Me_3 X^-$) with $n = 22$ appear to form normal micelles in water [5] while for $n = 12$ or 16, although surface activity was demonstrated, the evidence of normal micellization was ambiguous [6—8].

In the present study, we have investigated the association in water of simple bolaamphiphiles, i.e., of dodecane 1,12- and hexadecane 1,16-bis-(phosphate disodium) salts, ($C_{12}P$ and $C_{16}P$, respectively, shown in Fig. 1) by several physicochemical techniques including electrical conductivity, video-enhanced microscopy and absorption, and fluorescence probing.

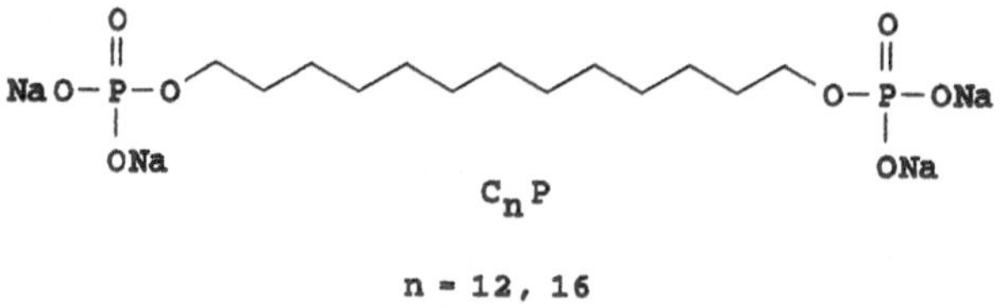

Fig. 1. Chemical structure of bolaamphiphiles $C_{12}P$ and $C_{16}P$

Experimental

The synthesis of the bolaamphiphiles of this study is obtained by interacting phosphorous oxychloride with the corresponding 1,12-dodecane and 1,16-hexadecane diols [9]. The sodium salts were formed by neutralization of the aqueous dispersions of the acids with NaOH. $C_{12}P$ was recrystallized from a mixture of ethylacetate:ethanol (10:1) and $C_{16}P$ from absolute ethanol. The purity of both bipolar surfactants was confirmed by C, H, and P elemental analysis. Electrical conductivity was measured with a Metrhom Herisau E512 conductometer in conjunction with a thermostated cell capable of regulating temperatures at .5°C. Triply distilled water having electrical conductivity less than 2—3 µS/cm was used throughout this work.

The opacity of the solutions was detected using the Varian-Cary 210 spectrophotometer. Zone refined pyrene (Py, Aldrich 99%) was used as fluorescence probe, while fluorescence spectra were recorded with a JASCO FP-777 fluorometer. The experimental setup and the method of the video-enhanced microscopy have been described in previous articles [10, 11].

Results and discussion

Electrical conductivity measurements of aqueous solutions of surfactants $C_{12}P$ and $C_{16}P$ did not demonstrate a regular clear-cut breaking point in the conductivity vs concentration plots in both cases (Fig. 2). Instead, the continuous small curvature, in both solutions, from salt concentration ca. $2 - 3.5 \times 10^{-2}$ M down to zero, suggests the possibility of formation of a wide range of aggregates from true oligomers to very large conglomerates.

Static fluorescence spectra of pyrene solubilized in aqueous solutions of the two surfactants [12, 13] have provided strong evidence for the formation of

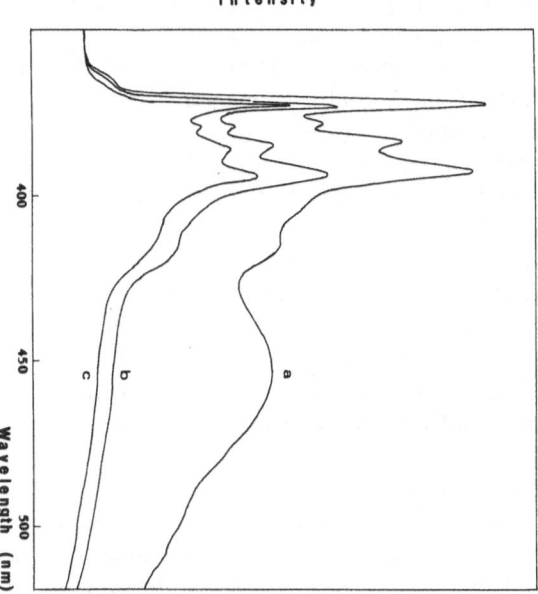

Fig. 3. Fluorescence spectra of 0.5×10^{-5} M Py in 3×10^{-3} M solution of $C_{12}P$, (a) before filtration, (b) and (c) after filtration through a 8000 A and a 3000 A Millipore filter, respectively

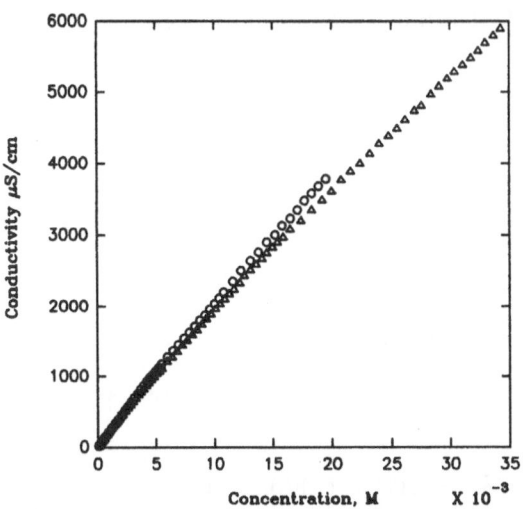

Fig. 2. Electrical conductivity vs concentration plots of $C_{12}P$ ($\triangle\triangle$) and $C_{16}P$ (∞) dissolved in water

large aggregates. Fig. 3 shows the emission spectrum of $[Py] = 0.5 \times 10^{-5}$ M in a $[C_{12}P] = 3 \times 10^{-3}$ M solution. Excimer pyrene fluorescence is predominant in these spectra despite the very low concentration of the fluorophore. This indicates that extremely large aggregates are present in the solution, so that the Poisson distribution of Py in the aggregates allows occupation of a large number of aggregates by more than two pyrene molecules, which produce the excimers. To further confirm this interpretation of the excimer fluorescence we have passed the solution through millipore filters of 8000, 3000, 2200, and 1000 A and recorded the fluorescence after each filtration. It was found that the fluorescence intensity ratio of the monomer to the excimer emission (I_M/I_E) was continuously increasing (Fig. 3). This means that each filtration removed a number of aggregates with sizes corresponding to the pores of the particular filter. Evidently, with these aggregates the solubilized pyrene molecules responsible for excimer fluorescence were removed as well, and, therefore, the ratio I_M/I_E was enhanced. Similar behavior of the fluorescence was observed when the large aggregates were removed by centrifugation (g = 17000).

When to the original surfactant solution ($[C_{12}P = 3 \times 10^{-2}$ M), which included pyrene ($[Py] = 10^{-5}$ M), surfactant solution of the same concentration but without pyrene in it, was added, reducing $[Py]$ to ca. 1/3 of its initial concentration, the excimer fluorescence diminished.

Finally, the fluorescence spectrum of pyrene in 3×10^{-2} M aqueous solution of $C_{12}P$ did not show any abrupt change when the solution was diluted with water down to ca. 7.5×10^{-3} M. This corroborates the findings of the study of the electrical conductivity, i.e., that a true critical micellar concentration does not exist.

To further confirm the formation of large aggregates in these systems we have examined solutions using the technique of video enhanced microscopy. By this method, we were indeed able to detect large particles moving in the field of view. The presence of large particles in the solution was also confirmed by absorption spectroscopy. Thus, a broad absorption band was observed from these solutions in the region of 4000 A, attributed to light scattering.

Both bolaamphiphiles $C_{12}P$ and $C_{16}P$ appear to behave in similar manner in all respects. Note that the corresponding acids were not adequately soluble in water to produce solutions appropriate for study. In conclusion, our results have indicated that these two bis (phosphate) sodium salts $C_{12}P$ and $C_{16}P$ form in water aggregates of various sizes from very small particles to extremely large conglomerates.

References

1. Fuhrhop JH, Fritsch D (1986) Acc Chem Res 19:130
2. Ringsdorf H, Scharb B, Venzmer J (1988) Angew Chem Int Ed Engl, 27:113
3. Jayasuriya N, Bosak S, Regen SL (1990) J Am Chem Soc 112:5844—5851
4. Nagawa N, Regen SL (1991) J Am Chem Soc 113:7237
5. Zana R, Yiv S, Kale KM (1980) J Colloid & Interface Sci 77:456
6. Yiv S, Kale KM, Lang J, Zana R (1976) J Phys Chem 80:2651
7. Yiv S, Zana R (1980) J Colloid & Interface Sci 77:449
8. Menger FM, Wrenn S (1974) J Phys Chem 78:1387
9. Kokkinia A, Paleos CM, Dais P (1990) Mol Cryst Liq Cryst 186:239
10. Malliaris A, Binana-Limbele W (1991) Prog Colloid & Interface Sci 84:83
11. Tsiourvas D, Paleos CM, Malliaris A (1992) J Polym Sci, Chem Ed (in press)
12. Malliaris A (1988) Int Revs Phys Chem 7:95
13. Malliaris A (1987) Adv Collod Interface Sci 27:153

Authors' address:

Angelos Malliaris
NRC "Demokritos"
Agia Paraskevi, Athens 153 10, Greece

Monolayers and Langmuir-Blodgett films of nickel(II) tetraaza[14]annulene complexes

F. Bonosi[1], F. Lelj[2], G. Martini[1], G. Ricciardi[2], and M. Romanelli[1]

[1] Dipartimento di Chimica, Universita' di Firenze, Firenze, Italy
[2] Dipartimento di Chimica, Universita' della Basilicata, Potenza, Italy

Abstract: The surface and the aggregation properties of the nickel(II) compounds with the dibenzo- and dinaphtho-tetraazal[14]annulene (TAA) macrocycles have been investigated by their spreading isotherms and electronic spectra. The possibility to form LB films was checked for both compounds. Homogeneous, reproducible LB films were obtained with the dibenzo-TAA derivatives, whereas the high surface viscosity prevented a good multilayer overlapping with the dinaphtho-compound.

Key words: Monolayers — LB films — Ni-TAA complexes

Introduction

In recent years the metal derivatives of tetraaza[14]annulene (TAA) macrocycle 1:

1

$$2\ R_1 = R_2 = CH_3 - (CH_2)_{14} - CO -$$
$$X = benzo$$

$$3\ R_1 = R_2 = CH_3 - (CH_2)_{14} - CO -$$
$$X = naphto$$

have received increasing attention because of their particular semiconducting, conducting and sensor properties [1, 2]. The introduction of alkyl chains in the methylene ring positions gives to 1 amphiphilic properties and makes these compounds suitable for monolayer formation and good candidates for LB film deposition. In this communication, we compare the spreading isotherms and the ability to form LB films of the dibenzo- and dinaphto-derivatives of TAA (compounds 2 and 3, respectively).

Experimental

Samples 1—3 were synthesized [3, 4] according to a modified Goedken and Weiss procedure [5].

The isotherm determination was performed with a KSV LB 5000 apparatus for 2 and with a Lauda FW2 Filmbalance for 3. LB films were obtained with the aid of the KSV LB 5000 apparatus, at a dipping speed of 2 mm/min. The dipping pressure was 25 mN/m for both compounds. The optical spectra were registered with a Perkin-Elmer Lambda 5 UV-VIS spectrometer. For instrumental settings and other experimental conditions see ref. [3].

Results and discussion

Figure 1 shows the spreading isotherms of 2 and 3 over a pure water subphase. Both isotherms were characterized by a slope change at $\pi \sim 10$ mN/m and ~ 15 mN/m with limiting areas A_0 of 48 and 66 Å^2/molecule after the slope changes and of somewhat higher values before the slope changes, for 2 and 3, respectively. The π discontinuities represented phase transitions between different surface arrangements of the macrocycles rather than an actual collapse of the bidimensional structure of the monolayers. This was proved by area relaxation measurements as a function of time at constant π values below and above the isotherms plateau. These latter agreed for a high stability of the monolayers of both compounds and were of the correct shape discussed by Binks [6] to exclude transition to a three-dimensional phase. From simple molecular modeling of the macrocycles in their saddle conformations, 200—280 Å^2 were required for an arrangement of the dibenzo- and dinaphto- TAA rings parallel to the water surface. Thus, edge-on orientation with slightly different tilting angles should be assumed at the interface. However, a slipped stacking overlapping of the macrocycles at the air/water interphase with resultant multilayer formation could not be ruled out. The occurrence of several polar groups in both the macrocycle skeleton and pendant alkyl chains, beside the reported trend of the aryl groups to assume a flat

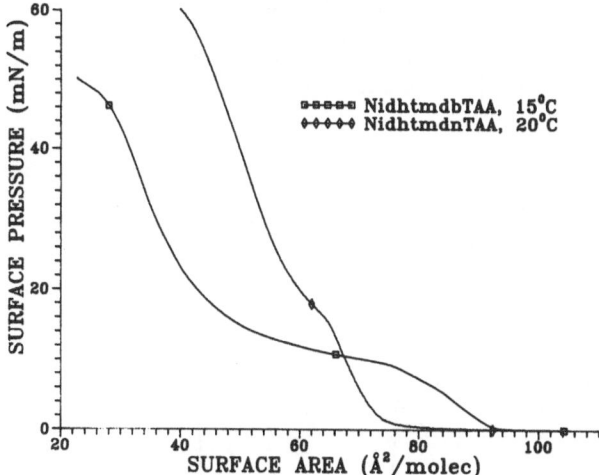

Fig. 1. Spreading isotherms at 288 and 293 K, respectively, of the Nickel (II) compounds with 6,13-bis-hexadecanoyl-5,7,12,14-tetramethyldibenzo[b,i]tetraaza[14]annulene (2) and with 6,13-bis-hexadecanoyl-5,7,12,14-tetramethyldinaphtho[b,i]tetraaza[14]annulene (3)

orientation at the air/water interface [7] agreed in favor of the above arrangement.

The above experimental conditions could suggest a reliable transference of the monolayers to a solid substrate. This was done with good transfer ratios for 2, as shown by a very good linearity between the absorbance at both 600 nm and 395 nm and the number of consecutive layers. The dibenzo-TAA Ni(II) compounds gave therefore good, reproducible LB films. In contrast, with the dinaphto-compounds the resulting LB film-building was very poor, either below or above the transition π value, because the very high monolayer viscosity prevented the overlapping of more than one layer, as further proved by the lack of linearity in the optical spectra.

Figure 2 shows the electronic spectra of 2 and 3 in CHCl$_3$ solution and in the transferred monolayer films. The absorptions in solution were very similar and the typical peaks of TAA [8] were clearly discernible. The Q_α, Q_β, and Soret bands were better resolved in 3 than in 2. A red-shift was observed and caused by the higher electron delocalization in 3. In the transferred films the opposite was verified. Both spectra were obtained with isotropic light at an incidence angle of the light beam with respect to the plate normal, $i = 0^0$. The absorption was significantly wider than in fluid solution, particularly for compound 3, with red-shifts of 10—15 nm of the peaks. In compound 3 the only clear peak

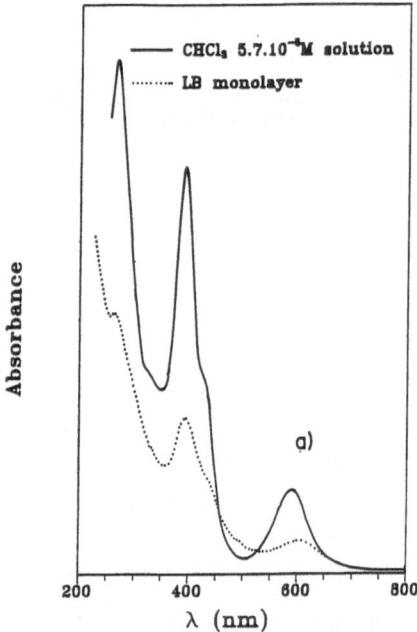

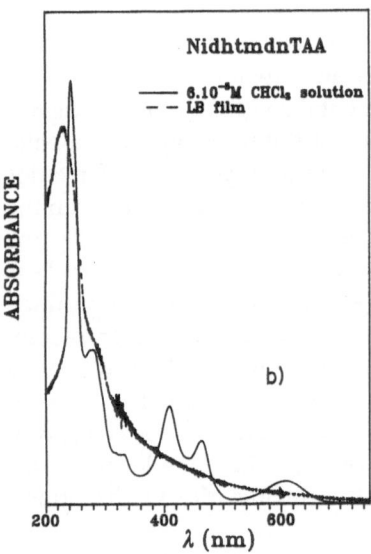

Fig. 2. Electronic spectra of compounds 2 (a) and 3 (b) in CHCl$_3$ solution and in the transferred monolayer films

was that at ~ 230 nm, with a shoulder at ~ 280 nm. Both these absorptions occurred as well-resolved bands in solution, too. No bands in the visible region were resolved in this monolayer and the overall spectrum did not improve with increase of the overlapped layers. The overbroadening of the visible peaks could be due to a very high molecular aggregation, and this was in agreement with the high surface viscosity.

Acknowledgement

Thanks are due to MURST and CNR for financial support.

References

1. Tieke B, Wegmann A (1989) Thin Solid Films 9:109—120
2. Wegmann A, Hunziger M, Tieke B (1989) J Chem Commun 1179—1181
3. Bonosi F, Lelj F, Ricciardi G, Romanelli M, Martini G (1993) Langmuir 8:268—272
4. Bonosi F, Romanelli M, Martini G, Ricciardi G, Lelj F (1993) Thin Solid Films (submitted)
5. Goedken VN, Weiss MC (1980) Inorg Synth 20:115—119
6. Binks BP (1991) Adv Colloid Interface Sci 34:343—431
7. Caminati G, Gabrielli G, Ferroni E (1989) Colloids Surf 41:189—210
8. Rosa R, Ricciardi G, Lelj F, Chizhov V (1992) Chem Phys (in press)

Authors' address:

Prof. Giacomo Martini
Dipartimento di Chimica
Universita' di Firenze
Via G. Capponi 9
50121 Firenze, Italy

Aggregation behavior of an anti-neoplastic etherphospholipid

N. Maurer[1]), O. Glatter[1]), E. Prenner[2]),
and F. Paltauf[2])

[1]) Institute of Physical Chemistry, University Graz
[2]) Institute of Biochemistry and Food Chemistry,
    TU Graz

Key words: Quasi-elastic light scattering — small-angle x-ray scattering — 1-octadecy-2-methyl-sn-glycerophosphocholine

We present a physicochemical characterization of the aggregation behavior of the etherphospholipid 1-octadecyl-2-methyl-sn-glycerophosphocholine (OM-PC) in an aqueous buffer either alone or in combination with diacylphospholipids and cholesterol with dynamic light-scattering and small-angle x-ray scattering.

This compound belongs to a group of synthetic etherphospholipids, 1-alkyl-2-methyl-sn-glycerophosphocholines, which have long been known to act cytostatically on various cancer cell lines in vitro. Several of them, including OMPC, showed promising results in clinical trials with patients suffering from malignant diseases (tumors, leukemia). In combination with cholesterol, vesicles are formed into which other drugs can be incorporated. This offers a further potential in targeting toxic drugs to distinct organs.

The phase behavior of OM-PC was investigated in a temperature range between 2° and 40°C at concentrations between 0.5%—4% (w/v). DLS measurements show that samples prepared at room temperature (RT) always contain a small amount of larger aggregates. Preparation at 40°C renders a purely micellar system. This indicates a phase transition near RT. Size distributions from DLS data give a hydrodynamic radius of 3.7 nm in the maximum of the distribution with a slight polidispersity of 5—10%, as determined from the cumulant fit. This is in good agreement with a diameter of 8 nm from SAXS data. Cooling down to 4°C, the micelles start to grow. The initially clear dispersion becomes turbid, a second jellylike phase has separated. Size distributions show that the system consists of micelles and larger aggregates. While the formation of the low-temperature phase on cooling is a kinetically slow process (hours), the reverse transition is fast.

Authors' address:

Norbert Maurer
Institute of Physical Chemistry
University of Graz
Heinrichstraße 28
A-8010 Graz, Austria

Contribution to the modelization of the surfactant concentration influence on droplet size distributions in oil/water emulsions

B. Mendiboure, C. Dicharry, G. Marion,
G. Morel, J. L. Salager[1]), and J. Lachaise

LTEMPM, Université de Pau, Pau, France
[1]) LFIRP, Universidad de los Andes, Mérida, Venezuela

Abstract: We have worked out a simplified model to forecast the droplet size distribution in emulsions generated by turbulent agitation in a closed vessel. This model is suited only for semidiluted oil/water emulsions stabilized by an ionic surfactant. — In this model, during one step of the calculus, each droplet is submitted to a random transition: it can break into two droplets of half volume, or coalesce with another droplet of same volume, or remain unaltered. — At the instant t, the droplet size distribution is represented by a line vector, the elements of which are the volume percentages of the N possible states for the droplets. During the interval of time dt following the instant t, the random transition is represented by a $[N, N]$ matrix, the elements of which are the transition probabilities in each state. These elements are updated after each step to make allowances for the creation of the oil/water interface and the corresponding surfactant adsorption. The size distribution at the instant $t + dt$ is obtained by multiplying the line vector and the transition matrix. — The transitions result from the competition between the forces driving, breaking and coalescence. The expressions of their probabilities rest essentially on the kinetic theory of gas, the isotropic turbulence theory, and the DLVO theory. — The model has two adjustment parameters: the former linked to the ratio of breaking and coalescence probabilities, the second to the kinetic energy of the droplets in the turbulent medium. A couple of values of these parameters is found to be sufficient to account for the variation of the droplet size distribution in isooctane/water emulsions as a function of the concentration of sodium dodecylbenzene sulfonate.

Key words: Theory of emulsification — lognormal distribution — droplet size distribution surfactant — oil/water emulsion

Introduction

In a first paper [1], we presented a stochastic model to account for the lognormal distribution of the drop sizes of an emulsion generated by a turbulent agitation. In a second paper [2], we showed, as an example, how this model could predict the variations of the droplet size distributions in function of the intensity of mixing. These two papers concerned emulsions generated without surfactant.

In this paper, we improve the model to predict the concentration influence of an ionic surfactant on droplet size distributions in semi-diluted oil/water emulsions.

The model

The emulsion is generated by an isotropic turbulent agitation which produces a wide spectrum of eddies. These eddies induce breakups and coalescences of the droplets, but — and it is a novelty of the model — they can also momentarily keep the size of some droplets.

The drops of the emulsion are ranked in N classes, numbered from 1 to N, between the limits d_{min} and d_{max}. The drops contained in class i are represented by the diameter d_i, geometric mean of the two limits $d_{i\,min}$ and $d_{i\,max}$ of the class.

During a short period of time dt, each drop is submitted to a random transition: it has the probability P_b to break into two droplets of half volume, the probability P_c to coalesce with another droplet of same volume, and the probability P_{iv} to remain unaltered. Of course the three probabilities satisfy the relation:

$$P_b + P_c + P_{iv} = 1 . \tag{1}$$

At the time t, the Markov chain theory allows to represent the droplet size distribution by a line vector, the elements of which are the volume percentages of the N classes, and the limits of which are the instantaneous diameters Min(t) and Max(t) defined by the conditions:

$$P_b = 0 \quad P_c = 1 - P_{iv} \quad \text{for } d < \text{Min}(t) \tag{2}$$

$$P_b = 1 - P_{iv} \quad P_c = 0 \quad \text{for } d > \text{Max}(t) . \tag{3}$$

According to the isotropic turbulence theory, Min(t) is constant and equal to d_{min}, while Max(t) depends on the mass balance of the surfactant between the continuous phase and the surface of the droplets. This balance lays on the surfactant Langmuir isotherm at the oil/water interface; thus, it depends on the area a_0 occupied by a surfactant molecule at this interface and on the ratio k_2/k_1 of the desorption constant by the adsorption constant.

During a time lag dt, the random transitions are represented by a matrix [N, N], the elements of which are the three transition probabilities of all the classes. The distribution at the time $t + dt$ is derived from the product of the matrix [N, N] by the line vector which represents the distribution at the time t.

The initial distribution is located in the two last classes with equal volume percentages. When the random transitions are applied to this distribution, it is possible to follow the evolution of the transient droplet size distribution until its stationary state (Fig. 1).

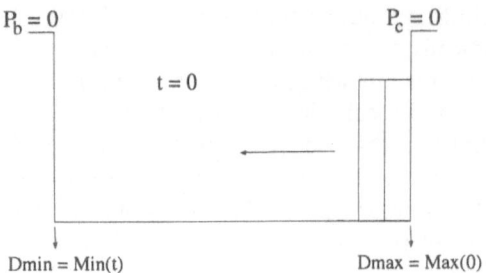

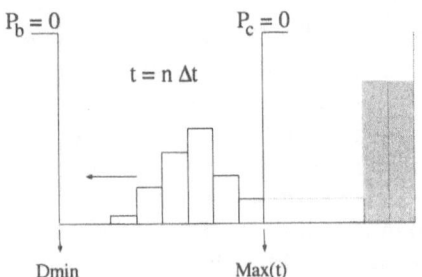

Fig. 1. Building of the initial droplet size distribution and of a transient droplet size distribution of an oil/water emulsion generated by the mechanism of breaking/invariance/coalescence in turbulent medium

Breakup probability P_b is built in, relying on the isotropic turbulence theory and introducing an efficiency factor which depends on the competition between the turbulent forces which tend to break up the droplets and the capillary forces which tend to maintain their cohesion. Inspired by a relation given by [4] for pure liquids, it has the form [3]:

$$P_b = A_1 \varepsilon^{1/3} d_i^{-2/3} dt \, \exp\left[-\frac{\gamma dS}{B\rho_c \varepsilon^{2/3} d_i^{11/3}} \right] . \tag{4}$$

In this expression the first four terms are linked with breakup in turbulent medium [4] and the exponential term is the efficiency factor for breakup. A_1 is a proportionality factor; ε proportional to the energy dissipated per unit mass is written as:

$$\varepsilon = \frac{W^3 D_a^5}{V} , \tag{5}$$

where W is the propeller rotation speed, D_a is the propeller diameter, and V the agitated volume.

In the efficiency factor, γ is the oil/water interfacial tension, dS is the increase of the droplet surface during the time lag dt. B is a proportionality factor and

ρ_c is the continuous phase density. γ varies in function of the surfactant concentration in the continuous phase; this variation is known from the surfactant Gibbs isotherm which also allows to determine the critical micelle concentration (CMC) of the surfactant.

Coalescence probabilities P_c are inspired by the kinetic theory of gases [5–7], an efficiency factor for coalescence being introduced to account for the competition between the turbulent forces which favor coalescence and the electrostatic repulsive forces generated by the ionic surfactant at the surface of the droplets which oppose to it. It has the form [3]:

$$P_c = A_2 \varepsilon^{1/3} d_i^{-2/3} \phi f_{vi} dt \, \exp$$

$$\times \left[-\frac{E_M}{B\rho_c \varepsilon^{2/3} d_{emax}^{11/3} \left[\dfrac{d_i}{d_{emin}} \right]^{-n}} \right] . \tag{6}$$

In this expression, the first six terms are linked with collisions in turbulent medium, and the exponential term is the efficiency factor for coalescence. ε, d_i, dt, B have the same significance as previously. A_2 is a proportionality factor, Φ is the volume fraction of the oil in the emulsion, f_{vi} is the volume percentage of the droplets of the class i.

The numerator of the exponent of the efficiency factor is the electrostatic energy barrier E_M created by the ionic surfactant molecules adsorbed at the droplet surface. It is evaluated from DLVO theory using the electrolyte concentration in the continuous phase, the valence of the ionic molecules which, furthermore, are assumed to be ionized.

The denominator of the exponent of the efficiency factor is the turbulent energy available to produce the coalescence of two droplets of diameters d_i. Its first four terms represent the maximum of energy supplied by eddies; this maximum has been assumed to be the energy of the highest eddy created by the turbulent agitation. Thus, d_{emax} is the size of the highest eddy. The last term of the denominator is an energy transfer coefficient between eddies and droplets. This coefficient is assumed to be a decreasing function of the droplet diameter; it is maximum and equal to unity for the finest droplets so that $d_{emin} = d_{min}$. n must be chosen to ensure the coalescence probability to be active in the classes where are distributed the droplets.

Invariance probabilities P_{iv} have been chosen constant for all the classes, at any time. Its value (0.8) is sufficiently high to ensure dt to be a small enough increment (one millisecond) to maintain their full significations to breakup and coalescence probabilities. It neither affects the stationary distribution, nor the length of the emulsification; it affects only the number of steps of the calculus.

The undimensional step of the calculus is $k_2 dt$. We have taken $k_2 = 50 \text{ s}^{-1}$ [8], which with $dt = 1$ ms gives $k_2 dt = 0.05$.

To reduce the number of the unknown proportionality factors A_1, A_2, B and to satisfy the conditions at the limits of the active interval, we derived from (1):

$$P_b = \frac{P (1 - P_{iv})}{1 + P} \tag{7}$$

$$P_c = \frac{(1 - P_{iv})}{1 + P} \tag{8}$$

with:

$$P = \frac{d_i - \text{Min}(t)}{\text{Max}(t) - d_i} \times \frac{P_b}{P_c} . \tag{9}$$

So, the model has only two adjustment parameters: $A = A_1/A_2$ linked to the ratio of breakup and coalescence probabilities and B linked to the kinetic energy of the droplets (eddies) in the turbulent medium. Furthermore, the limit conditions (2) and (3) are fulfilled.

Experimental verification

To test our model, we have carefully measured the concentration influence of sodium dodecyl benzene sulfonate (SDBS, molecular weight: 348 g) on the emulsification of an isooctane/water emulsion the volume fraction of which was 0.1. The continuous phase was an aqueous NaCl solution, the concentration of which was 1 g/l. Isooctane droplets were first introduced in the aqueous phase with a syringe which created droplets the diameters of which were centered on 70 μm, so that Max(0) could be taken equal to 70 μm.

The turbulent agitation was provided by an Ultra-Turrax T25, used with $W = 13500$ rpm during 360 s to ensure a sufficient homogeneization of the medium. The propeller diameter is $D_a = 1.27$ cm and the agitated volume $V = 100$ cm^3.

The droplet size distributions were determined by laser diffractometry with a Malvern granulometer, model 3601. The geometric mean diameter d_g and the geometric standard deviation σ_g are both decreasing functions of the SDBS concentration (Figs. 2 and 3). A stationary state is achieved for a concentration for which the whole surface of the droplets becomes fully covered with a surfactant monolayer.

We have carefully achieved measurements of interfacial tension and of the SDBS Langmuir isotherm at the isooctane/water interface to determine the surfactant characteristics. We have found:

$$CMC = 8 \times 10^{-4} \text{ mole/l} \quad a_0 = 32.6 \text{ Å}^2$$

$$k_2/k_1 = 2.5 \times 10^{-4} \text{ mole/l}$$

Under the adopted conditions of agitation, we have estimated d_{emin}, d_{emax} and n as follows:

$$d_{min} = d_{emin} = 1.64 \text{ μm} \quad d_{emax} = 10 \text{ μm} \quad n = 2 \ .$$

The theoretical variations of d_g and σ_g derived from our model are reported in Figs. 2 and 3. The best adjustment between theory and experiment is obtained for $A = 2$ (a small enough value to be satisfactory) and $B = 19\,000$ (a value in agreement with data reported in the literature [9]). The precision of the adjustment is of the order of 5% for d_g; it is of the order of 15% for σ_g which is, nevertheless, encouraging in view of the simplicity of the proposed model compared with the complexity of the emulsification phenomena.

Finally, it must be underlined that d_g and σ_g reach limit values for an initial surfactant concentration equal to 2.2 times the CMC; then, the model predicts that the surfactant concentration in the continuous phase is higher or equal to the CMC. This result is in good agreement with experimental observations reported in the literature [10].

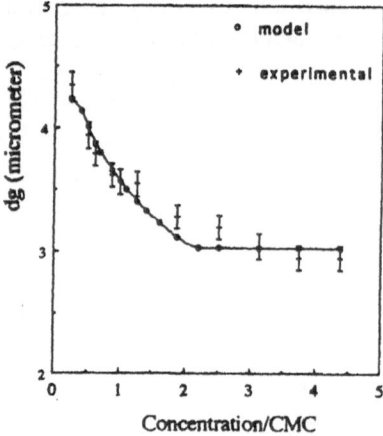

Fig. 2. Experimental and theoretical variations of the geometric mean diameter of the stationary droplet size distribution of an isooctane/water emulsion. The theoretical variations are obtained with $A = 2$ and $B = 19\,000$

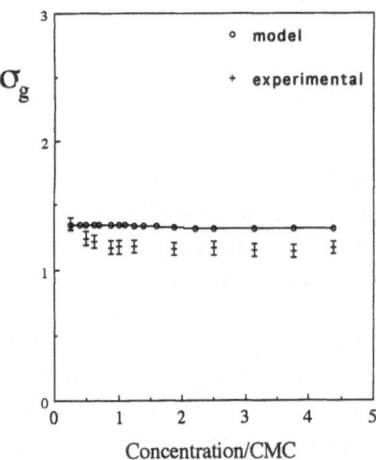

Fig. 3. Experimental and theoretical variations of the geometric standard deviation of the stationary droplet size distribution of an isooctane/water emulsion. The theoretical variations are obtained with $A = 2$ and $B = 19\,000$

Conclusion

The use of the surfactant in our simplified model of emulsification allows to predict the influence of the initial concentration of sodium dodecyl benzene sulfonate on the stationary droplet size distribution of a semidiluted isooctane/water emulsion.

The geometric mean diameter and the geometric standard deviation appear theoretically and experimentally as decreasing functions of this concentration. Limits are reached when the surfactant concentration in the continuous phase is higher than the critical micelle concentration.

Acknowledgement

The authors gratefully acknowledge the Elf Aquitaine Society for financial support and permission to publish this work.

References

1. Mendiboure B, Graciaa A, Lachaise J, Marion G, Salager JL (1989) Progress Colloid Polym Sci 81:274
2. Mendiboure B, Graciaa A, Lachaise J, Marion G, Bourrel M, Salager JL (1991) Progress Colloid Polym Sci 84:338–341
3. Mendiboure B (1992) Thesis, University of Pau, France
4. Coulaloglou CA, Tavlarides LL (1977) Chem Eng Science 32:1289–1297
5. Kuboi R, Komasawa I, Otabe T (1972) J Chem Eng Jpn 5:423–424
6. Delichatsios MA, Probstein RF (1975) J Colloid Interface Sci 51:394–405
7. Coulaloglou CA, Tavlarides LL (1976) AI Ch EJ 22:289–297
8. Van Hunsel S, Joos P (1987) Colloids and Surface 24:139–158
9. Godfrey JC, Obi FIN, Reeve RN (1989) Chemical Engineering Progress 61–69
10. Walstra P (1983) Becher P, (ed) Encyclopedia of Emulsion Technology, Marcel Dekker, New-York, 1:107–114

Authors' address:

Professor J. Lachaise
L.T.E.M.P.M.
Centre Universitaire
de Recherche Scientifique
Avenue de l'Université
64000 PAU, France

trations (water volume fractions ϕ) and temperatures. In order to study the system dynamics, the measurements of the density-density correlation function are performed above and below the percolation threshold, where previous experiments provided evidence of clustering effects. In particular, we study the photon correlation function $F(k, t)$ in a very large time range (50 µs — 10 s). The obtained data are well representative of the slow dynamics in this complex liquid system. In fact, we obtain two different dynamical behaviors below and above percolation respectively. More precisely, below the transition we can represent our data with a narrow distribution of exponential relaxation times; above percolation transition, in addition to this, a long time tail appears in $F(k, t)$. This slow dynamical contribution that persists in time for several orders of magnitude is well represented by the stretched-exponential form of Kohlraush-Williams-Watts (KWW). The overall behavior is consistent with the current mode-mode coupling theories for a dense system. The measured width of the KWW distribution β ($\beta \sim 0.7$) agrees with a recent model developed for percolating hard-spheres.

Authors' address:

Dr. N. Micali
Istituto di Tecniche
Spettroscopiche del CNR
98166 Vill. S. Agata C. P. 55,
Messina, Italy

Long-time behavior of dense microemulsions above and below the percolation threshold: quasi elastic light scattering

D. Lombardo[1]), F. Mallamace[1]) N. Micali[2]),
S. Trusso[2]), and C. Vasi[2])

[1]) Physics Department, Messina University, Italy
[2]) Istituto di Tecniche Spettroscopiche of C. N. R., Messina, Italy

We report the results of quasi-elastic light scattering experiments in dense AOT microemulsions, in the stable phase of water in oil, for different concen-

Small-angle light scattering in microemulsions (spinodal decomposition)

S. H. Chen[1]), D. Lombardo[2]), F. Mallamace[2]),
N. Micali[3]), S. Trusso[3]), and C. Vasi[3])

[1]) Department of Nuclear Engineering, Massachusetts Institute of Technology, Cambridge, U. S. A.
[2]) Dipartimento di Fisica dell'Universita' di Messina, Italy
[3]) Istituto di Tecniche Spettroscopiche del C. N. R., Messina, Italy

Abstract: The phase separation process in a one-phase microemulsions has been examined in an unstable region above its lower cloud point temperature by light-scattering measurements. The three time stages of the spinodal decomposition, i.e., early, intermediate, and late, has been observed. The scaling behavior of the peak in the structure factor, and the scaled structure function, have been studied and compared with the recent theoretical predictions of spinodal decomposition.

Key words: Spinodal decomposition — phase separation — microemulsions

Introduction

In past years considerable interest was devoted to systems which display non-equilibrium statistical behavior. In particular, if a critical mixture is brought from the one-phase region to the two-phase by changing the thermodynamic state variables, then the concentration fluctuation becomes unstable, and phase separation occurs. Phase separation in metastable and unmixing mixtures may proceed either by nucleation or by spinodal decomposition. Many studies have been presented about the kinetics of the dynamics of a phase separating system that is quenched into an unstable two-phase region [1]. However, significant advances have been achieved, by theoretical and experimental studies, since the pioneering study of Cahn and Hillard to characterize the process [2]. Linearized theory and scaling concepts are very useful for analyzing the growth and time evolution of concentration fluctuation in such systems. Experimental studies have been carried out on a large variety of systems, including polymer solutions [3], polymer-polymer mixtures [4], micellar solutions [5], metallic alloys [6], glass-forming mixtures [7] and, in particular, critical binary fluids [1, 8]. It has been observed that the time evolution of phase separation can be divided essentially into three stages, namely: early, intermediate, and late stages, thus characterizing the behavior of phase separation and scattering profiles [1]. The time evolution of spinoidal decomposition is of particular interest due to the ordering processes involved in the phase separation, and many problems are up to now unresolved, as, for example the late-time behavior [1]. The angular distribution of the scattering function has a relevant interest in the intermediate and late stages of spinodal decomposition, because a self-similar structure is formed that reflects the scattering function in the late stage. In simple binary mixtures, spinodal decomposition can be observed only for quench depths of the order of a few mK from the critical temperature. For deeper quenches

spinodal decomposition occurs too fast to be observed. In microemulsions, on the contrary, spinodal decomposition can be observed up to one degree below the critical point [9]. The time evolution of spinodal decomposition being slower in these colloidal systems than in binary liquid mixtures, the various dynamical stages can be observed. The subject of the present work is to report the results of an experimental study on spinodal decomposition, performed by means of time-resolved light-scattering method, in a ternary water-in-oil microemulsion. Our data, showing the three stages of the spinodal decomposition, allow for the evaluation of the power-law relations [10] for the time dependence of both the wave number $k_{m'}$ characterizing the distribution of component fluctuation, and the corresponding intensity I_m given by the scaling concept for the kinetics of phase separation. In addition, the studied structure factor, according to the Furukawa model [11] developed in terms of dynamical scaling concept, suggests the formation of selfsimilar structures.

Results and discussion

The experiment was performed using a time-resolved light-scattering apparatus. In order to reduce multiple scattering effects, we used a fused quarz cell with 1-mm thickness. The sample optical cell was thermostated with a temperature regulation better than $\pm$ 1 mK. The incident He-Ne laser beam was expanded in order to illuminate a wide area of the sample. The scattered light was collected by a lens working as a Fourier-transrom lens, and by this set-up was integrated and averaged over a wide scattering area. The amount of the angular resolution was estimated to be less than 0.1°. The detector is a 500 × 500 pixels CCD sensor, the photocurrent signal was converted to a voltage signal and digitalized by a high-speed frame grabber and then transferred to a computer for data analysis. The data acquisition and storage is quick and the experimental set-up is particularly efficient in following the time evolution of the spinodal decomposition.

The system is a three-component, water-in-oil microemulsion, water-decane-AOT (the surfactant sodium di-2 ethylhexylsulfosuccinate), already studied by means of many different experimental techniques, including light and neutron scattering [12]. In a given range of [water]/[AOT] molar ratios, defined as w, the solution shows a liquid-liquid critical point characterized by a diverging correlation length ξ and a critical slowing down of the order parameter fluctuations [12]. The phase separation occurs in this system on raising the temperature, and above T_c the microemulsion separates in two microemulsions phases with identical droplets, but having different volume fractions ϕ. This latter quantity is the volume fraction of water droplets (volume fraction of water plus surfactant). By fixing the molar ratio to $w = 40.8$, the critical volume fraction is $\phi_c = 0.098$ and the critical temperature is $T_c = 39.860$ °C. Usually, in a pure fluid or in a binary molecular mixture the short range correlation lengths ξ_0 are of the same order of magnitude of a few Å, comparable to the range of the intermolecular potential [12]. In microemulsions ξ_0 can be fairly large (10 − 30 Å); in our system is 13.3 Å [12]. As an ordinary ternary system in a critical microemulsion, ξ diverges like $\xi = \xi_0 \varepsilon^{-\nu}$, where ε is the distance from the critical temperature $\varepsilon = |T - T_c|/T_c$ and ν is a universal exponent. This large value of ξ_0, if compared with binary mixtures, has several experimental implications, such as a larger turbidity, a larger small-angle scattered intensity and a lower surface tension, but the most spectacular effect concerns the possibility to observe the spinodal decomposition by rising the temperature from as far one degree (and even more) from the critical point [9]. For this, we can measure the three kinetic temporal stages.

The spinodal decomposition was obtained in two different ways by thermal quenching or by shear. In this second case, the sample was in the two-phase region, but in order to prevent phase separation, the sample was stirred (the shear flow displaces the coexistence curve towards high temperature; spinodal decomposition appears when the shear is stopped). In the thermal quench the time evolution of the spinodal decomposition was studied with a quick (lower than 0.5 s) increase in the temperature from below to above T_c for the sample put to a temperature T_0 close to T_c (typical values used are $T_c - T_0 \approx 0.05$ K). For both the experimental methods we performed different measurements with temperature jumps $\Delta T = T_f - T_c$ (where T_f is the final temperature) ranging from 0.02 K up to 1 K. As it is well known, spinodal decomposition appears as a ring of scattered light just after the quench of the shear stop; the intensity of the ring increases with time while its radius decreases. For each quench (or shear stop), at time $t = 0$, the scat-

tering function was measured before the phase separation started; these data where subtracted from all subsequent data sets. This scattering function corresponds to the intensity profile $I(k, t = 0)$ and can be fitted with the Ornstein-Zernike expression $I(k, 0) = I_0/(1 + \xi^2 k^2)$, where ξ is the correlation length value for $T = T_0$. Figure 1 shows the evolution of the scattered light profile, after the subtraction of $I(k, 0)$, for the quenched (Fig. 1a) and shaken sample (Fig. 1b); for both, the temperature quench depth is $\Delta T = 0.22$ K.

The Cahn and Hillard theory based on the linearization of a generalized diffusion equation is valid in the early stage of the spinodal decomposition [2]. General features expected from this model for the scattered spectrum are the following: k_m is the wave number where the growth rate has a maximum value in the early stage, and should be the same as the value of $k_m(t)$ for $t = 0$, where $k_m(t)$ (the diameter of the observed ring) is the wave number corresponding to the peak intensity of $S(k, t)$; the value of $k_m(0)$ should coincide with the wave number at a maximum position of scattered intensity; the most dominant fluctuation grows without change in magnitude of its wave number $k_m(0)$. With the evolution of the composition fluctuations the model becomes invalid, the wave number of the most dominant fluctuation changes, decreasing with time as the scattered intensity increases.

The scaling concept [10] for the kinetics of phase separation have been studied in order to characterize the behavior of $k_m(t)$ and $I_m(t)$ (the maximum scattered light intensity of the angular distribution of the scattering profile) and gives the following power-law relations (that hold in the intermediate stage) for the time dependence of $k_m(t)$ (the length scale) and for the intensity:

$$k_m(t) \sim t^{-a} \text{ and } I_m(t) \sim t^b , \qquad (1)$$

with $a = 1/3$ (in the early stage, $a = 0$) and $b > 3a$ [26]. The latter inequality, in the intermediate stage, has been observed ($b = 2.2$) in polymer solutions [3] as predicted by the theory. For k_m, the value $a = 1/3$ is predicted by the droplet colaescence model of Binder and Stauffer [10] and the calculations of Siggia [13]. In Figs. 2a and 2b are reported the time dependence of both $k_m(t)$ and $I_m(t)$ for stirred and quenched samples. From these figures it the early stage can be clearly observed where "the position of the maximum of the ring does not more as the intensity increases" (for the lowest times). This result

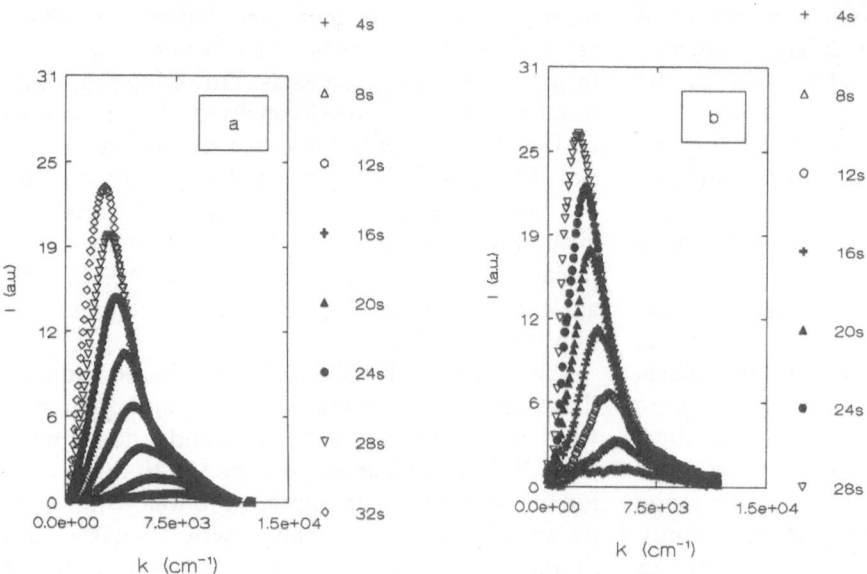

Fig. 1. Evolution of the scattered light profile for the quenched sample (a) and shaken sample (b); for both, the temperature quench depth is $\Delta T = 0.22$ K. The time elapsed from the beginning of phase transition is reported

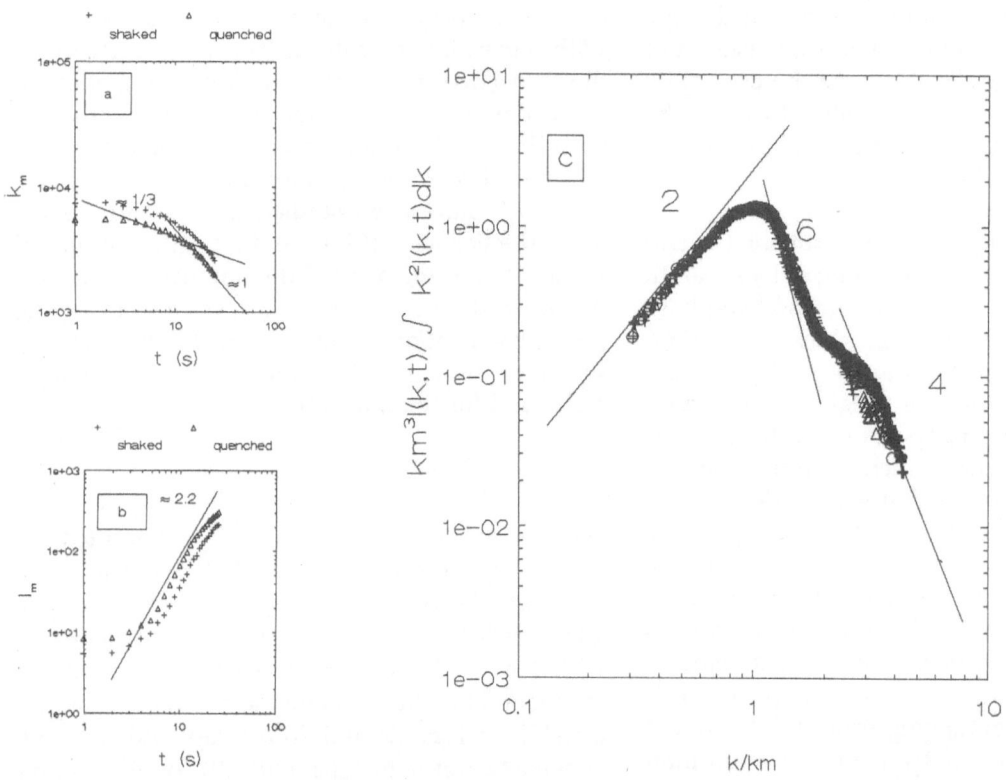

Fig. 2. Double logarithmic plot of the time evolution of the wavevector $k_m(t)$ (a) and of the time evolution of the peak intensity $I_m(t)$ of the mean peak (b); Double logarithmic plot of the scaled structure factor vs the normalized wavenumber. The number denotes the slope of each inserted line

is in agreement with Cahn's theory. As can be observed, in the intermediate stage, the time dependencies of $k_m(t)$ and $I_m(t)$ are well described by the corresponding scaling relations of Eq. (1) with $a \sim 1/3$ and $b \sim 2.2$. For the last step (late stage of the spinodal decomposition) in the time evolution of k_m we observe a significant deviation from the $\sim 1/3$ power dependence; data fitting gives $a \sim 1$. This is presumably due to the effect of the interfacial tension on the spinodal decomposition; microemulsion systems are characterized by a very low value of the interfacial tension. Siggia [13] studied the long-time behavior based on the capillary flow model obtaining $a = 1$.

Of much interest is the result obtained for the variation of the profile of the structure factor as a function of the wavenumber. Using scaling dynamical concepts, Furukawa [11] recently proposed a simple expression, in the region of a self-similar structure, for the structure factor, which can be written as:

$$S(x) \sim \frac{x^\delta}{\left(\dfrac{\gamma}{\delta} + x^{\gamma+\delta}\right)}, \qquad (2)$$

with $x = k/k_m$. Equation (2) suggests the following limiting behavior $S(x) \sim x^\delta$ and $S(x) \sim x^\gamma$ respectively for $x \ll 1$ and $x \gg 1$. In the late stage of phase separation the tail of the structure factor (large scattering wavevector) obeys Porod's law, behaves in fact as k^{-4}, reflecting the formation of a sharp interface ($\gamma = 4$). The scaling of the structure factor suggests the existence of a single-length parameter and the formation of self-similar structures in the phase separating systems. These scaling forms for the structure factor should be related with the k dependence of the dominant peak or the flucuation mode; however, they may be valid for desciling the entire structure in the system. More precisely, from Eq. (2), a k^{-6} tail in the structure function is predicted for $x > 1$. This asymptotic behavior of Furukawa's structure factor appears at relatively low x, say $x \sim 1.5$, and the value $\gamma \sim 6$ has been observed in different systems [4, 8, 14]. The normalized structure factors for the present microemulsion, for some times in the intermediate stage, are shown in Fig. 2c as functions of k/k_m. The normalization of the structure factor, in order to make a close comparison with theoretical predictions, is done by the

use of the form $F(x) = k_m^3 I(k, t)/\int dk\, k^2 I(k, t)$ [8, 15]; the lower and upper bounds of the integral normalizing the scaled function $F(x)$ are $k/k_m = 0.5$ and 1.5, respectively. Several features are obtained from this result and can be summarized in such a way: a) the peak height is ~ 1.4, b) the tail of the main peak is well expressed by a k^{-6} dependence, c) a shoulder peak seems to appear a $k/k_m \sim 2 - 3$, d) the tail at k/k_m obeys Porod's law and gives an indication that the formation of the interface starts in the intermediate stage, e) in $k < 1$ region a k^{-2} dependence seems to hold and can be connected to the interfacial tension [16]. Feature c) has been clearly observed in other experiments, but in our case needs more careful investigation in order to investigate the behavior of the corresponding scaled intensity which can give further information about the kinetics of spinodal decomposition in microemulsion system. Features (a) and (b) are related with Furukawa's predictions and give information that this model, particularly the resulting scaled form for the structure factor, is appropriate to describe the dominant fluctuation mode for spinodal decomposition.

Starting from the observation that a water-in-oil microemulsion gives the possibility to detect the spinodal decomposition as far as one degree (and even more) of the critical temperature, we can conclude that such a system is very attractive for studying the decomposition process and its three stages. We find that the phase separation kinetics in three-component microemulsions is analogous to that of critical binary systems. The present results suggest that k_m and I_m have a time dependence similar to that observed in simple critical mixtures; furthermore, the scaled structure factor obeys the scaling form proposed in terms of scaling dynamical concepts.

References

1. See, for example, Goldburg WI (1981) In: Chen SH, Chu B, Nossal R (eds) Scattering techniques. Plenum Press, New York, p 383; Hashimoto T (1988) In: Phase transition 12:47, and refences cited therein.
2. Cahan JW (1965) J Chem Phys 42:93, Cahan JW, Hillard JE (1958) J Chem Phys 28:258
3. Kuwaraka N, Kubota K (1992) Phys Rev A 45:7385, Lal J, Bansil R (1991) Macromolecules 24:290
4. Bates FS, Wiltzius P (1989) J Chem Phys 91:3258
5. Kuwahara N, Hamano K, Koyama T (1985) Phys Rev A 32:1279

6. Furusaka M, Ishikura Y, Mera M (1985) Phys Rev Lett 54:2611
7. Craievich AF, Sanchez JM, Williams CE (1986) Phys Rev B 34:1278
8. Kubota K, Kuwahara N, Eda H, Sakazume M (1992) Phys Rev A 45:R3377, Wong NC, Knobler CM (1981) Phys Rev A 24:858
9. Roux D (1986) J Phisique 47:733
10. Binder K, Stauffer D (1974) Phys Rev Lett 33:1006, Lebowitz JL, Marro J, Kalos MM (1982) Acta Metall 30:297
11. Furokawa H (1985) Adv Phys 34:703, (1978) Prog Theor Phys 59:1072 (1989) J Phys Soc Jap 58:216
12. Rouch J, Safouane A, Tartaglia P, Chen SH (1989) J Chem Phys 90:3756, Kotlarchyk M, Chen SH, Huang JS (1984) Phys Rev A 29:2054
13. Siggia ED (1979) Phys Rev A 20:595
14. Wiltzius P, Bates FS, Heffner WR (1988) Phys Rev Lett 60:1538, Hashimoto T, Takenaka M, Izumitami T (1989) Polym Commun 30:45, Takenaka M, Izumitami T, Hashimoto T (1990) J Chem Phys 92:4566
15. Chou YC, Goldburg WI (1981) Phys Rev A 23:858
16. Guenoun P, Gastaud R, Perrot F, Beysens D (1987) Phys Rev A 36:4876

Authors' address:

Prof. F. Mallamace
Dipartimento di Fisica
Universita' di Messina
98166 Vill. S. Agata C. P. 55, Messina

Structure of AOT reverse micelles containing hydrophobic or native ribonuclease a

F. Michel[1,2]) and M. P. Pileni[1,2])

[1]) SRSI Lab, bat F74, Université P & M Curie, Paris, France
[2]) CEA, DRECAM/SCM, Gif s/Yvette Cdx France

Abstract: Structural comparison of AOT reverse micelles in the presence and in the absence of native and modified ribonuclease is reported. The intermicellar potential of filled and unfilled micelles determined by SAXS and the percolation threshold are found unchanged.

Introduction

Water-in-oil AOT microemulsions are a dispersion of water droplets in an apolar media. The size of the droplets is ruled by $W = [H2O]/[AOT]$ [1]. The at-tractive interactions between water droplets (AOT-water-isooctane), largely studied at relatively high AOT concentrations [2], favor aggregate formation: dimers, trimers .... If the microemulsion volume fraction Φ is large enough, an aggregate of macro-scopic dimension appears [3].

These microemulsions have the ability to serve as hosts for macromolecules, in particular for en-zymes. Cytochrome c has been reported to strongly perturb the microemulsion: decrease in the micellar radius, stronger attractive interactions between droplets, decrease of the percolatrion threshold [4]. These phenomena have been attributed to the loca-tion of the protein at the interface of the droplet.

In order to investigate the nature of these interac-tions, we have in this present investigation studied the effect of another small enzyme: ribonuclease A. We have made hydrophobic derivatives of ribonuclease A to artificially anchor this enzyme at the interface. We observe that the hydrophobic character of the proteins have little effect on the size, the interaction between droplets and on the percolation process.

Experimental section

AOT and ribonuclease were obtained from Sigma, isooctane from Fluka, and all were used without further purification. The experimental in-vestigations were carried out at a constant W value ($W = 40$) and at a given average ribonuclease number per micelle 4 = [RNAse]/[RM]. [RM] is the reverse micellar concentration and is directly related to the polar volume Φ_W fraction by: $[RM] = 3 \cdot 10^3 \cdot \Phi_W/4 \cdot N \cdot \pi \cdot r_W^3$ where N and r_W are, respective-ly, the Avogadro number and the micellar radius.

Conductivity measurements were done with a Tacussel CD 810 instrument. Small-angle x-ray scat-tering (SAXS) was performed at L. U. R. E. (Orsay) on a D22 difractometer. The wave vector q range under the present investigation is $8 \cdot 10^{-3} \text{ Å}^{-1} < q < 0.11 \text{ Å}^{-1}$.

Modification of ribonuclease

It has been shown previously that reverse micelle can be used to covalently modify enzymes. This method has been chosen to increase the hydrophobic character of ribonuclease. Four dif-ferent reagents were used: myristoyl, stearoyl, cholesterol, and 9-fluorenylmethyl chloroformate, the latter being a chromophore. The chloroformate group reacts with amino groups of the ribonuclease by the following reaction:

RNAse — NH$_2$ + R-O-Cl

→ RNAse-NH-O-R + HCl

Native enzyme is solubilized in a borate buffer (pH = 9.5; 0.1 M) solution and is injected into AOT reverse micelles at fixed water content W = 20. The reagent, previously dissolved in isooctane, is added in the micellar solution. The ratio of the reagent concentration over enzyme concentration is equal to 20. After 15 min, the solution is poured in a 10 — fold volume of cold acetone (−10°C) in order to separate the enzyme from the AOT molecules and from free reagent. This solution is centrifugated at 5000 rpm. The precipitate is redispersed in 100 ml of acetone and again centrifugated. After removing acetone, water is added to the precipitate. The solution is dialyzed at 10°C during 12 h to remove residual borate salt. The remainder is lyophilized. HPLC analysis is performed on a C$_4$ reverse phase column. Fluorenylmethyl ribonuclease is found free of any native enzyme, whereas myristoyl, stearoyl and cholesteryl ribonuclease contain both native and modified enzyme. This is confirmed by electrophoresis.

Structural study of micelles containing native or modified ribonuclease

In the absence of enzyme, electrical conductivity studies have shown that a large and steep increase of the conductivity occurs in w/o microemulsions when the volume fraction Φ of the dispersed phase is increased [5]. These changes have been attributed to the occurrence of a percolation transition. The percolation threshold Φp corresponds to the formation of the first infinite cluster of droplets and is directly related to the interaction between droplets. Figure 1 shows an increase in the conductivity with volume fraction at T = 31°C. Similar conductivity variations are observed for microemulsions containing no, native, and modified ribonuclease. This indicates no changes in the percolation threshold (Table 1), or in the pair potential by the solubilization of ribonuclease and of derivatives.

SAXS is a versatile technique for studying the structure of microemulsions. Reverse micelles interact via short-range repulsive forces (hard sphere type) extended by a strong attractive force. The

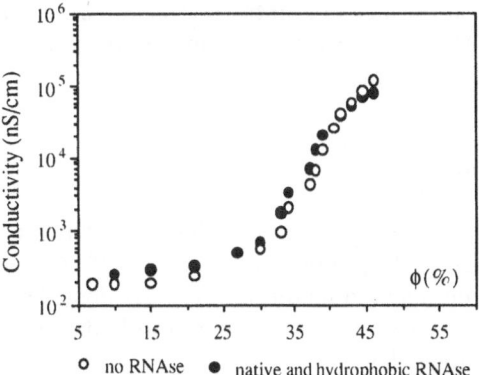

Fig. 1. Conductivity variation with dispersed phase volume fraction. W = 40, T = 31°C

Table 1. Hard sphere diameter σ, sticky parameter τ (Φ = 30% and T = 25°C) and percolation threshold (T = 31°C) of microemulsions at W = 40 containing no ribonuclease, an average of four native ribonuclease and of four modified ribonuclease per micelle

	Ribonuclease		
	No	Native	Hydrophobic
σ(A)	136	146	140
τ	1	1.2	1.2
Φ_p(%)	37	37	38

origin of such attractive forces have been attributed to a combination of i) a relatively long range London-van der Waals force between the droplets, and ii) a short-range attraction which has been ascribed to partial overlap of the surfactant tails during the close approach of the droplets. This potential $V(r)$ can be modeled by an "adhesive sphere potential". Its analytical expression is:

$$V(r)\,/kT = \lim \begin{cases} \infty & if\ r < d \\ ln[12 \cdot \tau \cdot (\sigma\text{-}d)/\sigma] & if\ d < r < \sigma \\ 0 & if\ r > \sigma, \end{cases}$$

where σ is the hard sphere diameter and τ^{-1} the sticky parameter.

A good fit between SAXS experiments and the adhesive sphere model is obtained. Table 1 gives the hard sphere diameter values and sticky parameter obtained from the best fits. We observe that the incorporation of ribonuclease or of ribonuclease derivatives in the microemulsion have no effects on the size nor on the interaction between droplets. This is in good agreement with conductivity measurements.

Conclusion

Water-soluble enzyme, ribonuclease A, has been modified by hydrophobic reagents in AOT reverse micelles: the reaction medium offers the possibility to solubilize hydrophobic (reagent) and hydrophilic (RNAse) molecules.

Structural study (SAXS and percolation) on the hydrophobic enzymes have shown no changes in the form or in the structure of the droplet, as compared to the system with native enzyme. Size and interaction between droplets are the same whether one works with the system containing native ribonuclease or the system containing hydrophobic ribonuclease.

So, the affinity of cytochrome c with the interface of the droplet is due to a different process than that of hydrophobic ribonuclease. With cytochrome c, it concerns strong electrostatic interaction between the opposite charged polar heads of the surfactant and the surface of the enzyme. With hydrophobic ribonuclease, the enzyme is anchored at the interface via hydrophobic molecules bound at the surface of the enzyme. Corresponding investigations, are in progress in our laboratory.

Acknowledgement

C. Nicot and M. Waks and both kindly acknowledged for their fruitful discussions on ribonuclease modification.

References

1. Pileni MP (1989) Structure and reactivity in reverse micelles (ed) Elsevier
2. Safran S, Webman I, Grest G (1985) Phys Rev 32:506
3. Bhattacharya S, Stokes J, Kim M, Huang J (1985) Phys Rev Lett 55:1884
4. Huruguen JP, Pileni MP (1991) Eur Biophys J 19:103
5. Tadros (1991) In: Mittal, Lindman (eds) Surfactant in Solution, Plenum Press NY

Authors' address:

M. P. Pileni
Université P. et M. Curie
Laboratoire S. R. S. II
Bâtiment F (74)
4 place Jussieu
F-75005 Paris, France

Molecular modeling of biomembrane lipids and skeletal polysaccharides

R. H. Mikelsaar

Institute of General and Molecular Pathology, Tartu University, Estonia

Key words: Molecular modeling — biomembrane — lipids — cellulose — hyaluronan

Tartu plastic space-filling atomic models have been used for molecular modeling of biomembrane lipids (lecithin, sphingomyelin, cholesterol, cardiolipin, etc.) and skeletal polysaccharides (cellulose, hyaluronan, chitin, agarose) [1—3]. A new biomembrane lipid bilayer imitation has been built according to which the biomembrane contains trimeric hexagonal-prismatic lipid units, composing dynamic honeycomb-like general structure. The investigation of polysaccharide models showed that cellulose and hyaluronan probably contain antiparallel molecular chains bound by H-bonding network. These chains form sheets and tubular structures.

References

1. Mikelsaar R (1986) Trends Biotechn 4:162—163
2. Mikelsaar R (1987) Mol Cryst Liq Cryst 152:229—257
3. Mikelsaar R, Kusnetsova NYa (1992) In: Kennedy JF, Phillips GO, Williams PA (eds) Lignocellulosics science, technology, development and use. Ellis Horwood, Chichester, pp 479—482

Author's address:

Dr. Raik-Hiio Mikelsaar
Veski Str. 34
Tartu EE 2400, Estonia

Dynamics of adsorption layers at the liquid-fluid interface

R. Miller[1]), G. Loglio[2]), U. Tesei[2],
and A. W. Neumann[3])

[1]) Max-Planck-Institute of Colloid and Surface Science, Berlin, FRG
[2]) Institute of Organic Chemistry, University of Florence, Italy
[3]) Department of Mechanical Engineering, University of Toronto, Canada

Key words: Adsorption kinetics — interfacial relaxations — surfactants — proteins — liquid interfaces

To study the dynamic adsorption behaviour of surfactants and polymers at fluid-fluid interfaces measurements of dynamic interfacial tensions and interfacial dilation rheological investigations are most frequently used. While many studies are performed at the liquid-gas interface only litte work has been done at liquid-liquid interfaces.

The dynamic adsorption behaviour of model surfactants at the aqueous solution — air and aqueous solution — decane interfaces was studied using different experimental techniques: elastic ring method, drop volume technique, and axisymmetric drop shape analysis. The resulting experimental data provide information about the equilibrium adsorption behaviour and the mechanisms controlling the kinetics of adsorption. The results also allow a discussion of the effect of surface active contaminations on the adsorption behaviour of the main surfactant. The importance of the necessary grade of purity of solvents is emphasized.

Authors' address:

Dr. rer. nat. habil. R. Miller
Max-Planck-Institute of
Colloid and Surface Science
Rudower Chaussee 5
D-12489 Berlin-Adlershof, FRG

Periodic viscosity changes at constant shear rate in concentrated kaolin suspensions

H.-J. Mögel, P. Rendtel

Institut für Physikalische Chemie
TU Bergakademie Freiberg, FRG

Key words: Viscosity — kaolin — suspension — aggregation

By the addition of sodium salts to concentrated kaolin suspensions their rheological properties are essentially changed. The non-Newtonian behavior is also influenced by the choice of anions. In a small shear rate range near 10 s^{-1}, we found a periodic time dependence of the appeerent viscosity at constant shear rate and temperature in the suspensions containing the salts NaI, NaSCN, NaBr or NaCl. The added salt solutions are 0.1 molar at pH = 11.5. At other shear rates the time dependence of the viscosity is aperiodic. This effect can be interpreted as a competition between destruction and aggregation of solid kaolin particles.

Authors' address:

Dr. Hans-Jörg Mögel
TU Bergakademie Freiberg
Institut für Physikalische Chemie
Leipziger Straße
D-O-9200 Freiberg, FRG

Monte carlo simulations of langmuir monolayers

H. Stettin[1]), H.-J. Mögel[2])

[1]) Institut für Physikalische Chemie
MLU Halle-Wittenberg, FRG
[2]) Institut für Physikalische Chemie
TU Bergakademie Freiberg, FRG

Key words: Monte Carlo simulations — amphiphilic molecules — Langmuir film, order behavior

Monte Carlo simulations for a liquid supported monolayer of amphiphilic molecules were carried out within a cubic lattice model using two-dimensional periodic boundary conditions. The flexible molecules consist of a hydrophilic head and hydrophobic segments which occupy consecutive lattice sites. Only excluded volume interactions were taken into account. We varied both the head density and the number of segments per molecule. The orientational degree of order obtained for increasing head density is caused by both the excluded volume effect and the constraints of the heads. The influence of the chain length on molecular properties as the mean end-end distance, the mean

Progress in Colloid & Polymer Science　　　　　　　Progr Colloid Polym Sci 93:320 (1993)

end-end projection, the mean height, and the effective layer thickness, as well as the order behavior is calculated. The end-end distance, end-end projection, and effective layer thickness follow a power law in dependence on the number of segments.

Authors' address:

Dr. Heiko Stettin
MLU Halle-Wittenberg
Institut für Physikalische Chemie
Mühlpforte 1
06108 Halle-Germany

Photodimerization of 6-AS in a membrane model system. A kinetic study

M. J. Moreno[1]), M. J. Prieto[2]) and E. C. C. Melo[1])

[1]) Centro de Tecnologia Química e Biológica — IST, Oeiras, Portugal
[2]) Centro de Química Física Molecular, Complexo I — IST, Lisboa, Portugal

Key words: Photoreaction — membrane dynamics — lipophilic probe — fluorescence

Biological membranes are made up of a complex mixture of lipids and proteins; these lipids may vary significantly from each other in their chemical and physical properties. In such mixtures, depending on factors like temperature, pressure, and chemical composition, phase separations leading to domain formation are possible. These phase separations can significantly affect the kinetic and yield of membrane reactions, this being a factor of some specific responses to changes in environmental temperature and composition.

In the present work, the kinetics of the photodimerization of 6-(9 anthroyloxy) stearic acid, 6-AS, in egg lecithin bilayers is characterized by steady-state fluorimetry. The photodimerization rate constant (k_D) is obtained for several temperatures and probe concentrations. The values (e.g., 5.5×10^7 $M^{-1}s^{-1}$ at 20°C 6.8×10^7 $M^{-1}s^{-1}$ at 40°C, for a probe:lipid ratio of 1:1000) are in accordance with the rate constant expected for a diffusion-controlled process in this medium. Furthermore, for probe to phospholipid ratios larger then 1:100, a clear deviation from the model is observed as a result of the non-ideal behavior of the probe-phospholipid solution.

This reaction is a convenient model system for the study of membrane domains, as is well known, for: 1) the mechanism and kinetic, and 2) the dependence of the reaction velocity on probe concentration and probe diffusion coefficient.

Authors' address:

Dr. Eurico Cores Correia de Melo
Centro de Tecnologia Química e Biológica
R. da Quinta Grande, 6
P-2780 Oeiras, Portugal

Phase behavior and rheological properties of alkylglycosides

R. Hofmann[1]), D. Nickel[1]), W. von Rybinski[1]), G. Platz[2]), J. Pölike[2]), and Ch. Thunig[2])

[1]) Henkel KGaA, Düsseldorf, Germany
[2]) Institut für Physikalische Chemie, Universität Bayreuth, FRG

The phase diagrams of different alkylglycosides in water have been determined by measurement of clouding behavior, light scattering, electrical birefringence and rheology. It is shown that alkylglycosides exhibit a phase behavior which is partly different from other nonionic surfactants.

The clouding behavior of alkylglycosides at low concentrations mainly depends on the rate of polymerization and on the addition of electrolyte. For an alkylglycoside with a $C_{8/10}$-alkyl chain a complete miscibility with water has been found above 290 K. Solutions of alkylglycosides with longer alkyl chain length have a more complex phase diagram.

By adding small amounts of long chain alcohols as a third component, lamellar- and L_3-phases appear at low concentrations. This indicates a possible modification of the properties of alkylglycoside/surfactant-systems for the use in a wide range of products.

Authors' address:

G. Platz
Institut für Physikalische Chemie
Universität Bayreuth
Universitätsstraße 30
95447 Bayreuth, FRG

Self-organized criticality of ferritin at a liquid-solid interface

H. Nygren and H. Arwin

Department of Anatomy and Cell Biology,
University of Göteborg and
Department of Physics and Measurement Technology,
Linköping Institute of Technology, Sweden

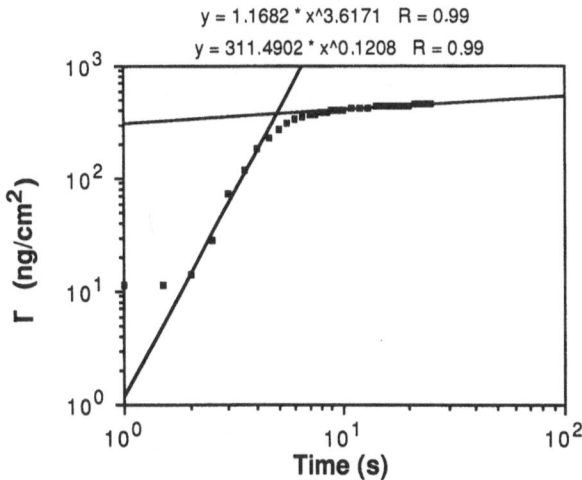

Fig. 1. Adsorption of ferritin to methylized silicon dioxide surfaces measured with off-null ellipsometry as described previously [2]. The bulk concentration of ferritin was 1 mg/ml. Solid lines represent fitted logarithmic functions

Abstract: The adsorption of ferritin at a methylized quartz surface was measured with off-null ellipsometry and transmission electron microscopy (TEM). An initial lag-phase was seen, followed by an accelerating adsorption proportional to $t^{3.6}$. The rate of adsorption decreased below monolayer coverage and a slow continuous binding proportional to $t^{0.12}$ was seen. The adsorbed ferritin molecules were distributed in clusters as seen by TEM. A plot of the velocity time-autocorrelation fuction showed that the decrease of the rate of adsorption was proportional to t^{-3}. Fluctuation of the rate of adsorption was seen during the period of slow adsorption. The results are discussed in relation to recently developed theories of self-organized criticality, a process which can be simulated by using the computer game "game of life".

Key words: Cluster formation, kinetics — liquid-solid interface — protein adsorption — self-organizing systems

Macromolecular reactions at interfaces such as enzyme-substrate interactions, antigen-antibody reactions and protein adsorption have many chemical characteristics, e.g., kinetics and isotherms in common. The initial reaction is often self-cooperative, leading to accelerating rate of adsorption and depletion of reactant in the reaction zone and the reaction becomes mass transport limited [1, 2]. The reaction rate then decreases logarithmically over long periods of time [3, 4]. The present study was undertaken in order to describe the molecular mechanism behind the logarithmic decrease of the reaction rate of macromolecular reactions at interfaces. Adsorption of ferritin at a hydrophobic surface is a suitable model system with the kinetic characteristics described above [1]. The iron nucleus of ferritin makes it possible to use transmission electron microscopy (TEM) and study the spatial distribution of ferritin over time during the process of adsorption.

The adsorption of ferritin at a hydrophobic surface measured with off-null ellipsometry is shown in Fig. 1. An initial lag phase of adsorption is seen, followed by an accelerated binding that can be described by using a power law containing the fractional factor $t^{3.6}$. The rate of adsorption then decreases abruptly and enters a new phase of slow adsorption which may be described by use of a power law containing the fractional factor $t^{0.12}$. An alternative description of the process of ferritin adsorption, the autocorrelation function v(t) [5], is shown in Fig. 2. The decrease of the reaction rate is clearly seen in this plot and can be described by use of a power law containing the factor t^{-3}. A fluctuation in the rate of adsorption after 20 s adsorption time can also be seen in this plot.

The transmission electron micrograph (TEM) in Fig. 3 shows the spatial distribution of ferritin after 10 s of adsorption. The adsorbed molecules are distributed in clusters. Cooperativity of binding at low surface concentrations and a strong decrease of the rate of adsorption below monolayer coverage (Fig. 3) are data which call for a mechanism of desorption. Studies by TEM indicate that critical dissociation of large clusters is the mechanism resulting in self organization of ferritin in small clusters at the surface [6].

A process of self-organized criticality has recently been described theoretically [7] and can be simulated by the computer game "game-of-life" [8]. A simulation based on birth of particles at positions with 1 or 2 neighbors and death of particles at positions with 2—4 neighbors is shown in Fig. 4. The distribution of particles after 87 iterations of these rules (Fig. 4a) is similar to the distribution of ferritin

$$y = 8957.5477 * x\text{^}-3.0029 \quad R = 1.00$$

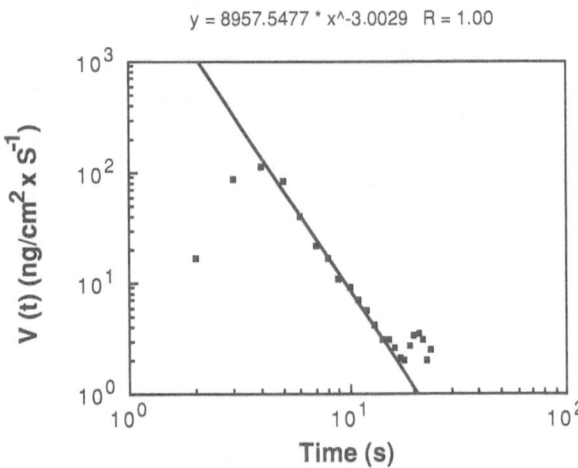

Fig. 2. The velocity time-autocorrelation function [5]. Data from Fig. 1. Solid line represents fitted logarithmic function

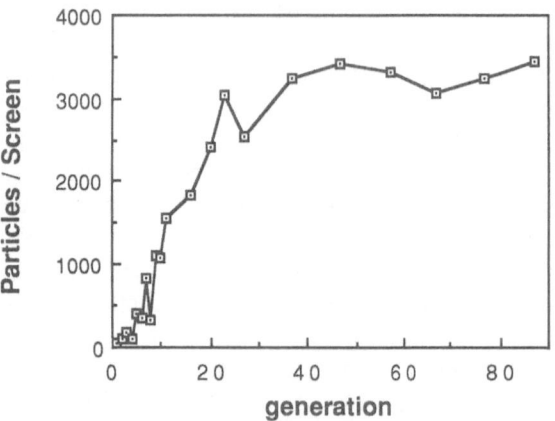

Fig. 3. Electron micrograph of the spatial distribution of adsorbed ferritin after 10 s of adsorption from a bulk concentration of 1 mg/ml

seen in Fig. 3 and the kinetics of binding (Fig. 4b) shows the fluctuations that we also found experimentally (Fig. 2).

The data presented in the present study indicate that novel theories of complex dynamic systems may be useful in the theoretical description of macromolecular reactions at interfaces.

References

1. Nygren H, Stenberg M (1990) Biol Chem 38:67—76
2. Nygren H, Arwin H, Welin-Klintström S (1993) Colloids Surfaces (in press)

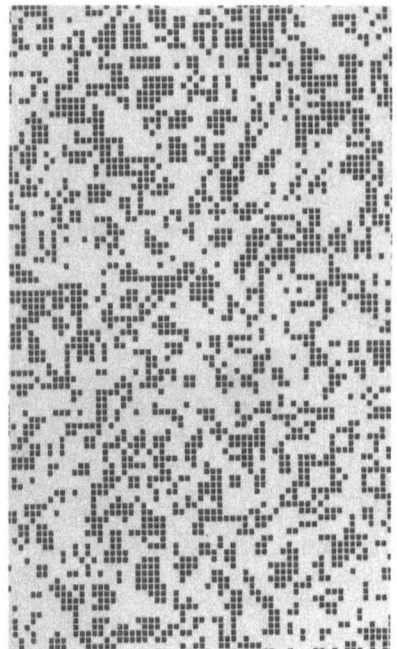

a

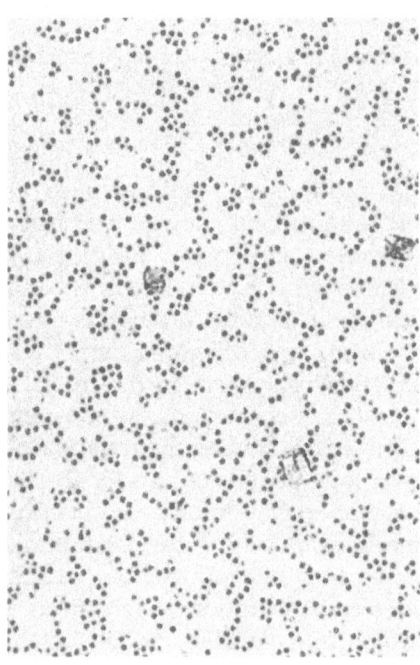

b

Fig. 4. Simulation of self-organized criticality in the game-of-life. Rules: Birth 1, 2; Death 2—4. a) Distribution of particles after 87 iterations from an initial nucleation density of 5 particles per screen. b) The number of particles on the screen (local equilibrium) plotted against the number of iterations

3. Cuypers PA, Willems GM, Kop JM, Corsel JW, Jansen MP, Hermens WT (1987) In: Brash JL, Horbett TA (eds) Proteins at Interfaces ACS Symposium Series 343 pp 208—221
4. Wertén M, Stenberg M, Nygren H (1990) Progr Colloid Polymer Sci 82:349—352
5. Toffoli T, Margalus N (1987) Cellular Automata Machines. The MIT Press Cambridge Mass Chapter 16.6 pp 179—182
6. Nygren H (1992) Progr Colloid Polymer Sci (in press)
7. Bak P, Tang C, Wiesenfeld K (1988) Phys Rev A 38:364—374
8. Bak P, Chen K (1989) Physica D 38:5—12

Authors' address:

Dr. Håkan Nygren
Department of Anatomy and
Cell Biology Division
of Cell Biology Medicinaregatan 5
S-41390 Göteborg, Sweden

Co^{2+} adsorption on hematite: dependence on ionic strength

N. Ogrinc, A. Trkov, and I. Kobal

"J. Stefan" Institute, University of Ljubljana, Slovenia

Abstract: The present work gives a theoretical interpretation of the adsorption of cobalt on spherical colloidal hematite particles suspended in NaNO$_3$ electrolyte, with the main stress on the dependence on ionic strength. The surface complexation model, as one very successful approach used over the past decade in predicting physicochemical phenomena at hydrolyzable oxide-electrolyte interfaces, was chosen. — On the basis of the Gouy-Chapman theory, different models of the electrical triple layer were applied by locating ions of the background electrolyte and cations interacting with surface OH groups at various planes in the layer. It was found that the position of the adsorption curve on the pH axis, its slope, as well as its practical independence of the ionic strength could be satisfactorily explained by a model with an FeOCo$^+$ surface complex located at the "0" plane and all the others on the "β" plane.

Key words: Colloidal suspension — modeling — adsorption — electrical double layer — triple layer model

In a previous study on the adsorption of cobalt on spherical colloidal particles of hematite suspended in NaNO$_3$ as background electrolyte, it was found that there is practically no dependence on ionic strength [1]. The aim of the present work is to give a theoretical interpretation of the observed phenomena on the basis of an electrical triple layer model (TLM) in the surface complexation approach.

In our case the following surface reactions may be expected [2—3]:

$$\mathrm{FeOH_2^+ \rightleftharpoons FeOH + H^+ \; ; \; Ka_1} \tag{1}$$

$$\mathrm{FeOH \rightleftharpoons FeO^- + H^+ \; ; \; Ka_2} \tag{2}$$

$$\mathrm{FeOH + Na^+ \rightleftharpoons FeO^- Na^+ + H^+ \; ; \; K_{Na}} \tag{3}$$

$$\mathrm{FeOH_2^+ NO_3^- \rightleftharpoons FeOH + NO_3^- + H^+ \; ; \; K_{NO_3}} \tag{4}$$

$$\mathrm{FeOH + Co^{2+} \rightleftharpoons FeOCo^+ + H^+ \; ; \; K_{Co,1}} \tag{5}$$

$$\mathrm{2\,FeOH + Co^{2+} \rightleftharpoons (FeO)_2 Co + 2\,H^+ \; ; \; K_{Co,2}} \tag{6}$$

$$\mathrm{FeOH + Co^{2+} + H_2O \rightleftharpoons FeO^- CoOH^+}$$
$$\mathrm{+ \; 2\,H^+ \; ; \; K_{Co,3}} \tag{7}$$

The corresponding equilibrium constants for these reactions were defined according to Leckie's TLM models [4].

Surface potentials ψ_0, ψ_β, and ψ_d were calculated from the surface charge densities σ_0, σ_β, and σ_d and inner- and outer- sphere capacitances C_1 and C_2 following the electroneutrality conditions $\sigma_0 + \sigma_\beta + \sigma_d = 0$. The following situations are the most general:

model 1: background electrolyte ions on the "β" plane while all interacting Co^{2+} ions are on the "0" plane

model 2: protons located on the inner "0" plane while all the other ions (those of the background electrolyte as well as those of the adsorbed cations) are on the outer plane ("β") of the TLM

model 3: the combination of both background electrolyte ions on the "β" plane, while Co^{2+} ions forming FeOCo$^+$ on the "0" plane and the other Co^{2+} ions on the "β" plane.

There are several standard computer programs available for the interpretation of potentiometric titrations and adsorption experiments in the framework of surface complexation model. In our work the SCPLOT was used [5]. The program allows the

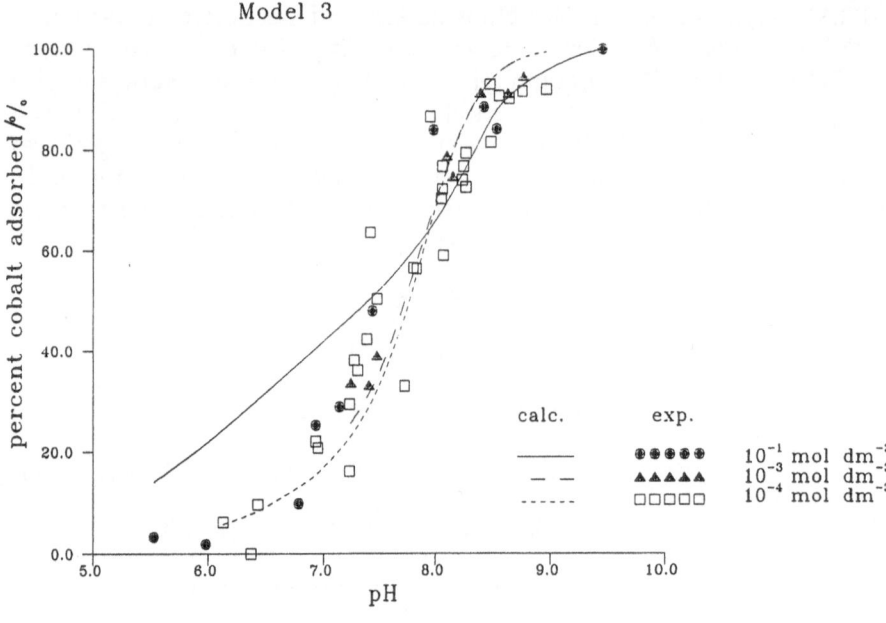

Model 3

Fig. 1. Adsorption curves for 10^{-1}, 10^{-3} and 10^{-4} mol dm^{-3} $NaNO_3$ and $1.7 \cdot 10^{-7}$ mol dm^{-3} initial Co^{2+} concentration at 298 K; the best fitvalues of parameters obtained for titration and adsorption are: $N_s = 8.0 \cdot 10^{-6}$ mol m^{-2}, $Ka_1^{int} = 9.5 \cdot 10^{-9}$ mol dm^{-3}, $Ka_2^{int} = 7.8 \cdot 10^{-9}$ mol dm^{-3}, $K_{Na}^{int} = 1.1 \cdot 10^{-9}$, $K_{NO_3}^{int} = 2 \cdot 10^{-12}$ (mol dm^{-3})^2, $C_1 = 1.9$ F m^{-2}, $C_2 = 1.4$ F m^{-2}, $K_{Co,1}^{int} = 3.4 \cdot 10^{-4}$, $K_{Co,2}^{int} = 2.2 \cdot 10^{-11}$, $K_{Co,3}^{int} = 0$ mol dm^{-3}

optimization of parameters which describe the electrochemical phenomena at the oxide/electrolyte interface: the total concentration of the surface sites (N_s), the inner- and outer-layer capacitance (C_1, C_2) and the equilibrium constants of surface protolysis reactions (Ka_1^{int}, Ka_2^{int}), surface reactions of background electrolyte (K_{Na}^{int}, $K_{NO_3}^{int}$) and surface complexation reactions of adsorbing cations ($K_{Co,1}^{int}$, $K_{Co,2}^{int}$, $K_{Co,3}^{int}$). In addition, surface charge densities, surface potentials and concentrations of surface complexes can be calculated.

Computation started with an initial guess for the fitting values of parameters N_s, Ka_1^{int}, Ka_2^{int}, C_1, C_2, K_{Na}^{int} and $K_{NO_3}^{int}$ and the titration curve (pH versus volume of NaOH added) is calculated and plotted.

Applying best-fit values obtained from titration, and initial guess of $K_{Co,1}^{int}$, $K_{Co,2}^{int}$, $K_{Co,3}^{int}$ values, adsorption curves (percentage of Co^{2+} adsorbed versus pH) were fitted and displayed.

It was found that model 2 gives a pronounced dependence of the adsorption on the ionic strength which does not agree with the experiment. In addition, the slope of the adsorption curve is too low. On the other hand, model 1 shows practically no dependence of the adsorption on ionic strength. Unfortunately, the slope of the adsorption curve is too high and cannot be lowered by any variation of the values of fitting parameters.

Model 3 (Fig. 1) gives a better slope of the theoretical adsorption curve and there is practically

no dependence on the ionic strength either, so we consider this model with Co^{2+} ions forming $FeOCo^+$ at the "0" plane and the other Co^{2+} ions at the "β" plane, as a reasonably acceptable one. A similar situation was found for adsorption of Pb^{2+} and Cd^{2+} ions on geothite [4].

References

1. Kobal I, Hesleitner P and Matijević E (1988) Colloids Surfaces 33:167
2. James RO (1981) In: Anderson MA, Rubin RJ (eds) Adsorption of Inorganics at Solid-Liquid Interfaces. Ann Arbar Science, Miclicgan, pp 219
3. Morel FMM, Yeasted JG, Westall JC (1981) In: Anderson MA, Rubin RJ (eds) Adsorption of Inorganics at Solid-Liquid Interfaces, Ann Arbar Science, Miclicgan, pp 263
4. Hayes KH, Leckie JO (1987) J Colloid Interface Sci 115:564
5. Trkov A, Ogrinc N, Kobal I (1992) accepted for publication in Computer Chem 16:341

Authors' address:

Mrs. Nives Ogrinc
"J. Stefan" Institute
University of Ljubljana
Department of Physical
and Environ. Chemistry
Jamova 39, P. O. Box 100
61111 Ljubljana, Slovenia

Progress in Colloid & Polymer Science   Progr Colloid Polym Sci 93:325 (1993)

Alteration of cellulose thermodynamic properties during activation

I. Osovskaya and G. Poltoratsky

St. Petersburg Technological Institute of Cellulose and Paper Industry, St. Petersburg, Russia

We studied an interaction between a wide range of liquids (26 types) and both natural and wood cellulose of different degrees of regularity accompanied by both the process of cellulose solvation with substances of low molecular weight (water, alcohol) and the phase transformation of the polymer during an action of liquid media (primary amines, concentrated alkaline and acid solutions).

The methods of calorimetry, sorption and densimetry show the main influence of the supermolecular organization, the specific surface of sorbent and physico-chemical properties of the solvent on the measured parameters of the processes which are occurring at the border of phases.

A substantial influence on the glasslike component of cellulose compositions by the process of solvation in non-aqueous solvents of both specific and nonspecific character was discovered.

The small values of thermoeffects of the cellulose solvation in the aproton solvents of both specific and nonspecific character was discovered.

The small values of thermoeffects of cellulose solvation in aproton solvents, and the presence of a substantial swelling prove the nonspecific character of the interaction between these solvents and the pyranous rings of cellulose molecules, which is performed by the van-der-Waals mechanism.

An increase of cellulose sorbing ability for the organic compounds as a result of the additional amorphization of a cellulose structure and the use of organic solvents in sorbtion processes under the determinate balance between water and solvent was shown.

The completed research work allows prognosing the behavior of the cellulose material in the process of its utilization and a proved choice of the concentration of the solutions used.

Author's address:

I. Osovskaya
Technological Institute of Cellulose
and Paper Industry
St. Petersburg, Russia

Structural transition in water-alcohol mixtures at low alcohol concentration: A comparison with the micellization process

M. D'Angelo, A. Fuccello, G. Onori, and A. Santucci

Dipartimento di Fisica — Università di Perugia, Italy

Key words: Micellization — 2-butoxyethanol — IR spectroscopy — surface tension

Several studies related to the properties of alcohol-water mixtures strongly suggest the existence of transitions, in some of these systems, qualitatively similar to the micellization.

In continuation of our previous work [1—4], surface tension and infrared spectroscopy measurements are used here to study 1-propanol- and 2-butoxyethanol-water mixtures and aqueous sodium n-octanoate ($C_8H_{15}O_2Na$), which is known to have a high critical micelle concentration.

In Fig. 1 is plotted the surface tension of 2-butoxyethanol-water mixtures at 3° and 40°C.

Surface tension decreases steeply at low 2-butoxyethanol concentrations and becomes constant at higher concentrations. This phenomenon is similar to that encountered in the formation of micelles. The observed decrease of the critical concentration, c^*, with the temperature (Fig. 2) is typical of surfactants and is consistent with a phase separation model involving a positive standard enthalpy of microaggregates formation.

Figures 3a, 4a and 5a show the infrared spectra, at $T = 30°C$, in the C-H stretching region of 1-propanol (Fig. 3a), 2-butoxyethanol (Fig. 4a) and sodium n-octanoate (Fig. 5a) at three concentrations. From the figures the concentration dependence of the C-H stretching frequencies is evident. In fact, a shift to a lower frequency was observed for the C-H stretching modes of some sodium n-alcanoates upon micelle formation [5].

Figures 3b, 4b and 5b show plots of frequency vs. concentration for the asymmetric methyl stretching absorption [5]. At lowest concentrations, all plots show constant frequency values indicative of the monomeric phase. It follows a range of concentrations where the frequency changes as a function of

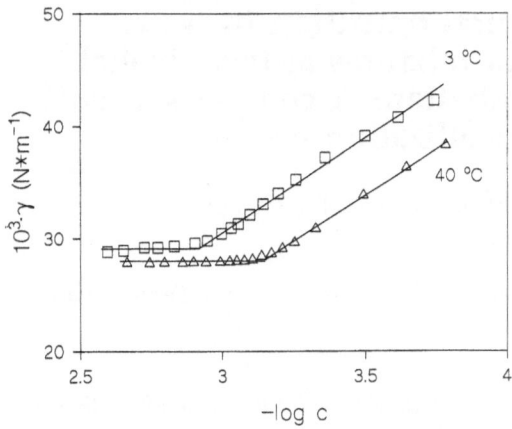

Fig. 1. Surface tension vs. the logarithm of molar concentration of 2-butoxyethanol at 3° and 40°C

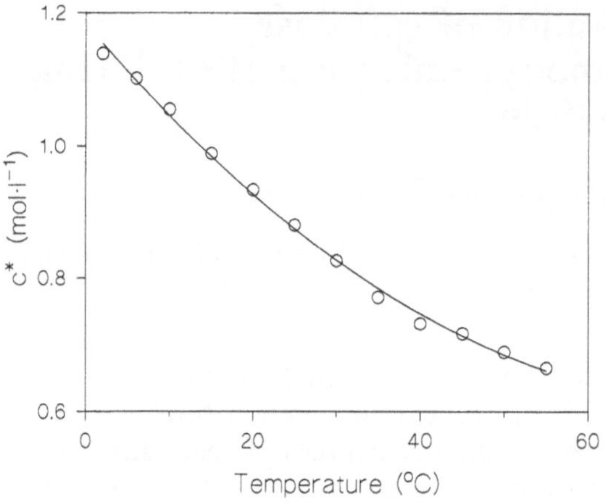

Fig. 2. Critical 2-butoxyethanol concentration vs. temperature. ○: experimental points: — : second order polynomial fitting

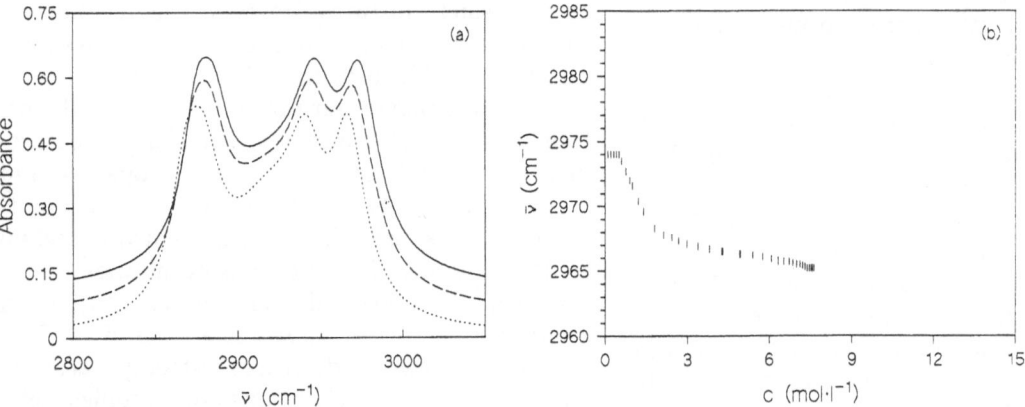

Fig. 3. a) Infrared spectra in the C-H stretching region of water/1-propanol mixtures at three 1-propanol concentrations; (—): 0.8 mol · l⁻¹; (---): 4.9 mol · l⁻¹; (···): 11.0 mol · l⁻¹. b) Concentration dependence of the C-H stretching absorption peaked at highest frequency

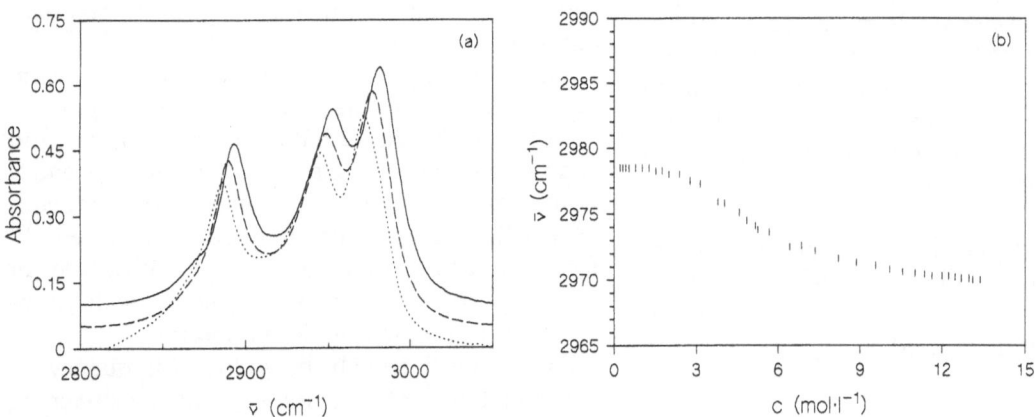

Fig. 4. a) Infrared spectra in the C-H stretching region of water/2-butoxyethanol mixtures at three 2-butoxyethanol concentrations; —: 0.5 mol · l⁻¹, ---: 1.2 mol · l⁻¹; ···: 3.4 mol · l⁻¹. b) Concentration dependence of the C-H stretching absorption peaked at highest frequency

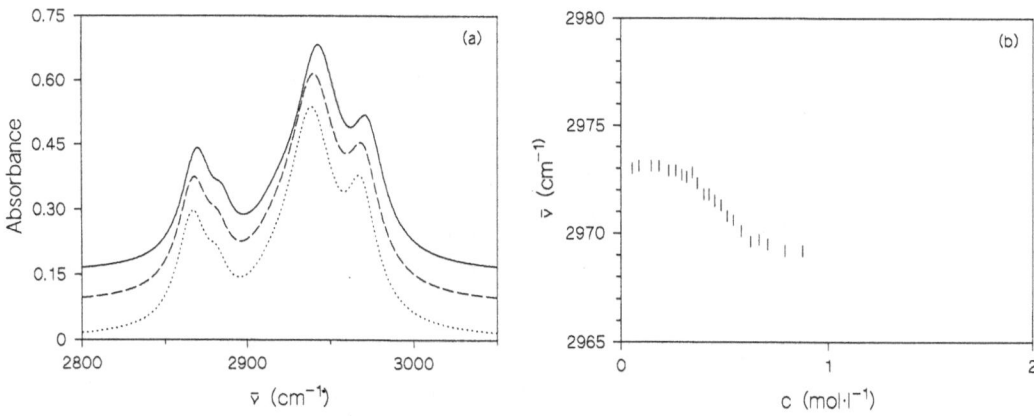

Fig. 5. a) Infrared spectra in the C-H stretching region of water/sodium n-octanoate mixtures at three sodium n-octanoate concentrations; —: 0.5 mol $\cdot$ l^{-1}; ---: 1.2 mol $\cdot$ l^{-1}; $\cdots$: 3.4 mol $\cdot$ l^{-1}. b) Concentration dependence of the C-H stretching absorption peaked at highest frequency

concentration, indicating a transition from the monomeric to the micellar phase. The rate for this phase change is strongly dependent on the system. In the case of 1-propanol a gradual transition is observed toward the values characteristic of the pure alcohol. In the case of 2-butoxyethanol and sodium n-octanoate the transition is more rapid.

The change observed in surface tension and IR spectra for the n-butoxyethanol in water is remarkably similar to that associated with micellization in the case of surfactants and is supporting evidence for microheterogeneities like micelles in this binary system.

References

1. Onori G (1988) J Chem Phys 89:4325—4332
2. Petrillo C, Onori G and Sacchetti F (1989) Mol Phys 67:697—705
3. Onori G (1989) Chem Phys Letters 154:212—216
4. Fioretto D, Marini A, Onori G, Palmieri L, Santucci A, Socino G and Verdini L (1992) Chem Phys Letters 196:583—587
5. Umemura J, Mantsch HH and Cameron DG (1981) J Coll Interface Sci 83:558—568

Authors' address:

Prof. G. Onori
Dipartimento di Fisica
Universita' di Perugia
V. A. Pascoli
I-06100 Perugia (Italy)

Study of the micelles formation in reversed micellar systems by IR spectroscopy

M. D'Angelo, A. Fuccello, G. Onori, and A. Santucci

Dipartimento di Fisica — Universita' di Perugia

Key words: Surfactants — bis(2-ethylhexyl)sodium sulfosuccinate — IR spectroscopy — micelles

Aerosol OT [bis(2-ethylhexyl)sodium sulfosuccinate] molecules aggregate in non polar solvents and form reversed micelles in which large amounts of water can be dissolved. Several features of these systems remain to be solved. Some of these pertain to the influence of water content on the association of the surfactant molecules, and to whether a well-defined critical micelle concentration (CMC) exists in these micellar systems[1]. The present investigation deal with both these problems by using infrared spectroscopy to study the micellar formation process in a AOT/H$_2$O/CCl$_4$ system.

The surfactant concentration dependence of the C-H stretching absorptions of AOT and of the O-H stretching absorptions of H$_2$O have been recorded at concentrations encompassing the CMC and at fixed ratio $W = [H_2O]/[AOT]$ ($0.2 \leq W \leq 10$). Effects detected on the molar extinction coefficients of the bands due to C-H stretchings vibrations (Fig. 1) can be used as indicators of micelle formation. From

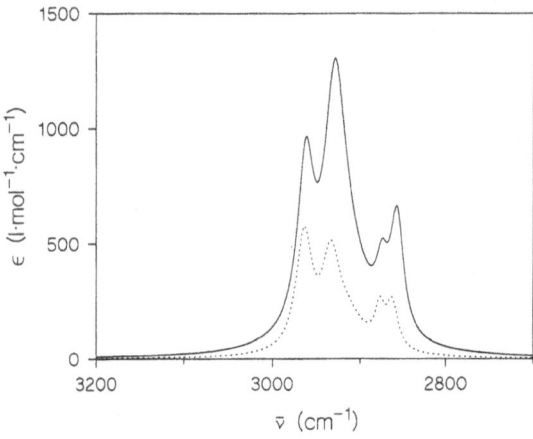

Fig. 1. Molar extinction coefficient vs. wavenumbers for —CH$_3$ and —CH$_2$ stretchings in AOT/CCl$_4$ mixtures. (—): [AOT] = 0.1 · 10^{-3} mol · l^{-1}; (---): [AOT] = 1.2 · 10^{-3} mol · l^{-1}

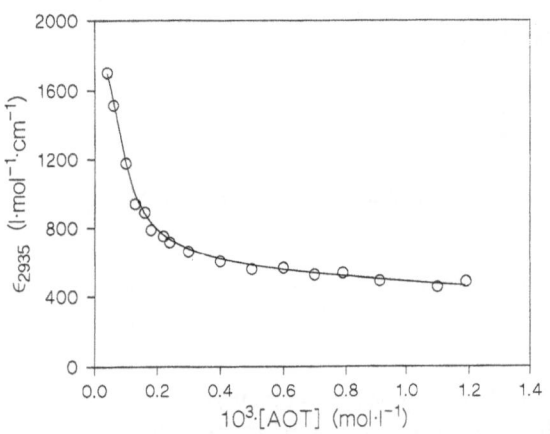

Fig. 2. Molar extinction coefficient at 2935 cm^{-1} as a function of [AOT] in AOT/CCl$_4$ mixtures. (o): experimental points; (—): fitting according to a multiple equilibrium model

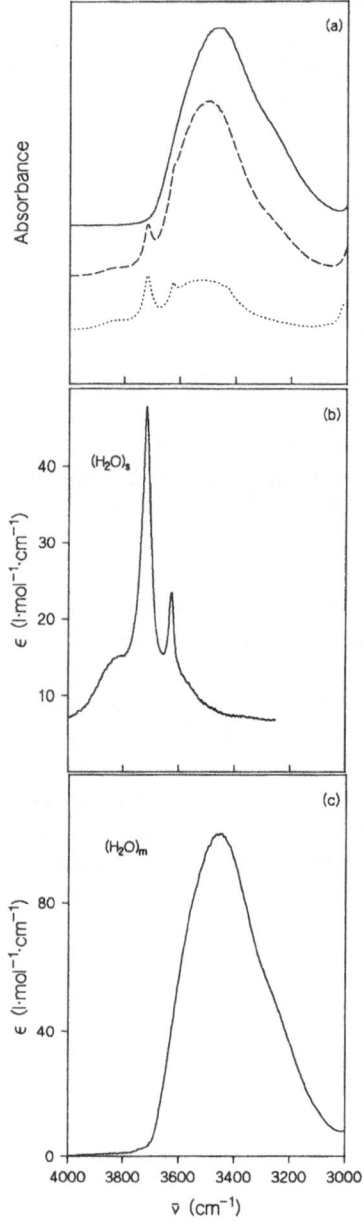

Fig. 3. a) IR stretching absorptions of water in a AOT/H$_2$O/CCl$_4$ system; (—): [AOT] = 0.050 mol · l^{-1}; (---): [AOT] = 2 · 10^{-3} mol · l^{-1}; (...): [AOT] = 0.8 · 10^{-3} mol · l^{-1}. b) Molar extinction coefficient vs. wavenumber for water in the solvent. c) Molar extinction coefficient vs. wavenumber for water inside the micellar core

the data on increasing the surfactant's concentration a noticeable lowering in the molar extinction coefficient measured at the maximum of the antisymmetric and symmetric —CH$_3$ and —CH$_2$ stretching absorptions peaked at 2965, 2935, 2865, and 2878 cm^{-1} is observed.

The ε values (Fig. 2) largely decrease at very low concentrations of surfactant and level off at concentration values ($\sim$ 0.2 · 10^{-3} mol · l^{-1}) comparable with the CMC values reported in the literature for this system [2]. The data can be reproduced equally well by using either a single-equilibrium model monomer $\leftrightarrow$ n-mer (with n = 5 ± 1) or a multiple-equilibrium model [1] (continuous line in Fig. 2). Under these circumstances it is difficult to assign a meaningful CMC.

In the range $0.2 \leq W \leq 10$ examined, the trend of ε vs. [AOT] (Fig. 2) appears to be not dependent on the amount of water initially present in the system. Water in a AOT/H$_2$O/CCl$_4$ system exhibit IR absorptions in the O-H stretching region (see Fig. 3a) that can be attributed to water in the bulk solvent (H$_2$O)$_s$ (Fig. 3b) and to water inside the micellar core (H$_2$O)$_m$ (Fig. 3c). The data show that water is progressively released from the micelle to the solvent on dilution (Fig. 3a and Fig. 4). This

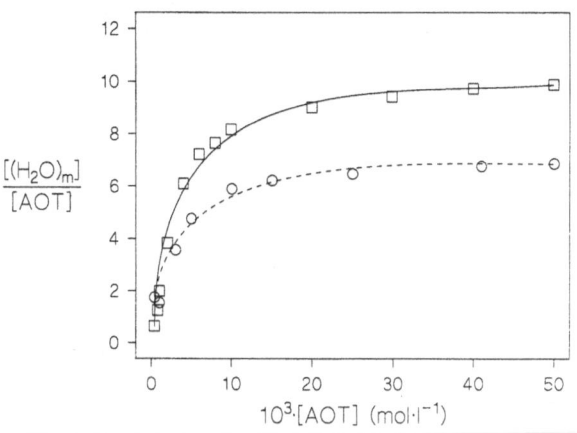

Fig. 4. Mole ratio of water in micellar cores to AOT vs. [AOT]. (□): $W = 10$; (○): $W = 7$. The lines in the figure serves as a guide for the eye

behavior is well explained considering the effect of dilution on the (H$_2$O)$_m$ ↔ (H$_2$O)$_s$ equilibrium. At very low AOT concentrations the water content inside the micellar core appears to be not dependent on the initial value of ratio W (Fig. 4). The result compares well with the findings in Fig. 2 where the trend of ε vs. [AOT] appears to be not dependent on W.

References

1. Eicke HF (1980) Top Current 87:86—145 (and references therein)
2. Chen SH (1986) Ann Rev Phys Chem 37:351—399 (and references therein)

Authors' address:

Prof. G. Onori
Dipartimento di Fisica
Universita' di Perugia
V. A. Pascoli
I-06100 Perugia, Italy

Surface potentials and surface tension of aqueous solutions of trifluorocresol isomers

M. Paluch and M. Węgrzyn

Department of Physical Chemistry and Electrochemistry, Faculty of Chemistry, Jagiellonian University, Kraków, Poland

Key words: Surface potential — adsorbed films

Introduction

At the water/air interface exists the electric potential drop connected with the spontaneous orientation of water molecules. The water dipoles are oriented with their hydrogen atoms in the bulk of water and the oxygen atom towards the gas phase [1]. The consequence of such and orientation is a negative on the air side sign of the surface potential drop of water surface. Introduction of chemical substances with amphipathic structure to the solution causes a change in this potential drop and in the surface tension. The surface potential may decrease or increase. An increase was observed when two or more negative groups are present in the adsorbed molecule and, moreover, if one of them was more hydrophobic than others [2].

The surface properties of such molecules, namely, α, α, α- trifluoro-o, -m and p-cresoles are presented in this paper.

Results and discussion

The surface tension and surface potential of aqueous solutions of trifluorocresol isomers were carried out vs concentration. The surface potential was measured by the flowing jet method [3] and surface tension by the drop weight method [3].

The obtained results showed that trifluorocresols have strong influence on both the surface potential (Fig. 1) and the surface tension (Fig. 2) of free water surface. The adsorption process of trifluorocresol isomer molecules was described by Gibbs' equation, Temkin's isoterm [4], and Helmholtz's equation [5]. The standard free energy of adsorption, the effective dipole moments and the area occupied by the molecule were determined (Table 1).

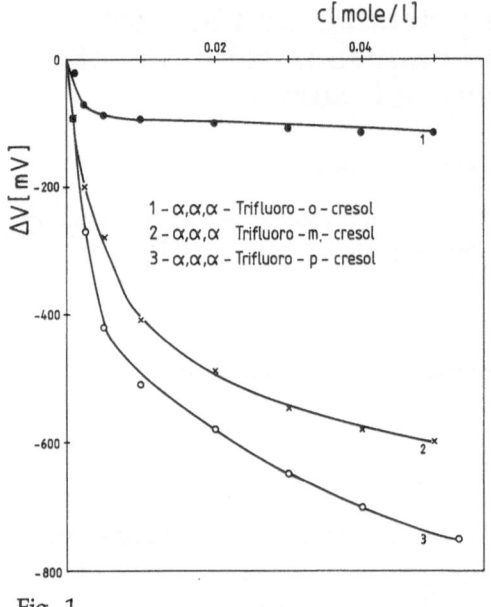

Fig. 1

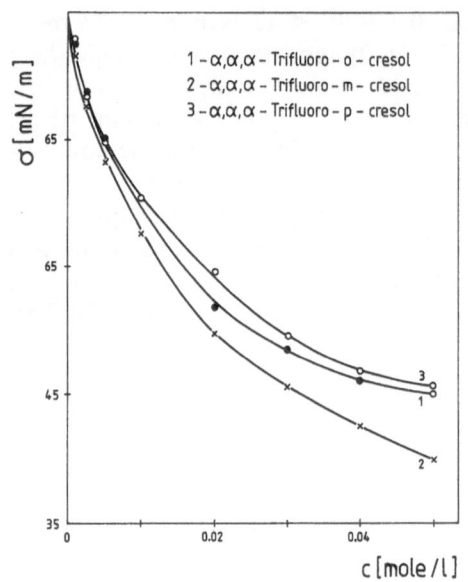

Fig. 2

Table 1. Some parameters of trifluorocresols at water/air interface

Compound	Area occupied on the surface [Å^2]	Effective dipole moment $\bar{\mu}$ [mD]	Standard free energy of adsorp. $\Delta G°$ [Kcal/mol]
a, a, a,-trifluoro-o-cresol	42.5	−142	−4.369
a, a, a,-trifluoro-m-cresol	33.9	−516	−4.483
a, a, a,-trifluoro-p-cresol	35.6	−632	−4.364

References

1. Kamieński B (1959) Electrochim Acta 1:272
2. Paluch M, Filek M (1980) J Colloid Interface Sci 73:282
3. Dynarowicz P, Paluch M (1989) J Colloid Interface Sci 129:379
4. Parsons K (1959) Trudy sovieshcheniya po elektrokhimi, Moscow, p 42
5. Davies JT, Rideal EK (1963) Interfacial phenomena, Acad Press, New York, London, p 70

Authors' address:

Dr. Maria Paluch
Department of Physical Chemistry
and Electrochemistry
Faculty of Chemistry
Jagiellonian University
Karasia 3
30-060 Kraków, Poland

Preparation, characterization and biodistribution of a mixed ^{131}I-Hippuran loaded liposome system

C. Sawas-Dimopoulou[1]), Z. Panagi[1]),
D. Ithakissios[1]), and C. M. Paleos[2])

[1]) Institute of Radioisotopes-Radiodiagnostics
[2]) Institute of Physical Chemistry
    N. C. S. R. Demokritos, Attiki, Greece

Abstract: The preparation, physicochemical characterization and biodistribution of mixed liposomes prepared form dimyristoylphosphatidyl choline-dimyristoylphosphatidylglycerol —NH_4^+ and cholesterol loaded with ^{131}I-Hippuran (IOH) were investigated. The physicochemical behavior of these liposomes was compared to that of

IOH-loaded synthetic vesicles prepared from didodecyl-dimethylammonium bromide (DDAB).

Key words: Mixed liposomes — synthetic vesicles — liposome biodistribution — ^{131}I-Hippuran

Introduction

A prerequisite of efficient drug delivery systems is to remain intact in biological fluids, retaining active ingredients, for periods of time sufficient to reach and interact with target cells [1]. In addition, targeting ability to specific cells coupled with biocompatibility and biodegradability, certainly minimize toxic effects and side-effects [1]. Liposomes and synthetic vesicles of appropriate formulation can, in principle, combine these properties therefore, being, promising candidates for drug delivery (1). The tendency, however, of liposomes to become destabilized is serum renders problematic their wide applicability as drug carriers in vivo. The stability of liposomes in plasma has however been improved by incorporating cholesterol in their membrane structure (1). In addition, liposomes and vesicles stability may be enhanced by appropriate modification of their lipid composition (4) and also by polymerization of these monomeric particles to their polymerized counterparts (5). With this in mind, mixed liposomes derived from DMPC:DMPG — NH_4^+: Cholesterol at a molar ratio of 28:12:6 (µmol) were prepared by employing ^{131}I-Hippuran (IOH) as liposome marker. IOH was used because it is easily detectable and rapidly eliminated from the body. Experiments examined the size, the entrapment efficiency, and the in vivo biodistribution in mice.

Experimental

Liposomes were prepared employing the Bangham film method (6). The dry lipid film was dispersed with 3 ml of saline containing ^{131}I-Hippuran (IOH) at 36.5° — 37°C Vesicles composed of didodecyldimethylammonium bromide and loaded with ^{131}-J-Hippuran were prepared in an analogous manner. Following filtration through a 5 µm Millipore filter IOH-liposomes and IOH-vesicles were submitted to microdialysis or passed through a sephadex G 50 column to eliminate untrapped radioactivity.

The size of liposomes was determined with a Leitz phase contrast optical microscope and its im-

ages were processed with a FG-100 Card of Image Processing INC using Image-Pro Software (Media Cybernetics INC). Mice (Swiss SWR/De) were used for the biodistribution experiments. A γ-counter was used for radioactivity measurements.

Results and discussion

The size of the IOH-liposomes ranged from 0.5 — 4.5 µm with mean diameter 1.8 µm. For DDAB vesicles the size ranged from 0.5 — 4.4 µm and the mean diameter was 1.96 µm.

The radioactivity content of these liposomes measured after microdialysis was 93.84% of the initial dose whereas that of IOH-DDAB-vesicles was only 2.96%. It appears that only the liposomes entrap large volumes of radiopharmaceutical solution.

Leakage was found to be temperature- and medium- dependent for both the liposomes and vesicles. The leakage of IOH after incubation in saline, which generally increases with temperature. Leakage was 14% for the liposomes in 19 h and 36% for the vesicles in 2 h, at 25°C. Because of the low entrapment efficiency of vesicles only the liposomes were used for further experiments.

Incubation studies of liposomes at 36.5°C were also performed for up to 2 h in the presence of plasma or whole blood. Measurements showed that the leakage of radioactivity in plasma was 25% and 28% after 60 and 120 min, respectively. Liposomes retained their initial dimensions. On the other hand, incubation in whole blood for 120 min showed that there was a 23% leakage of radioactivity. These data clearly demonstrate that the liposomes retain a significant amount of radioactivity, apparently due to their in vitro stability in whole human blood and, therefore, they were used in further in vivo experiments.

Tissue distribution of liposomes was determined in mice 5, 30, and 60 min after intravenous injection. The distribution, 30 min after injection is shown in Fig. 1. The results show a very low concentration of radioactivity in liver and spleen and a preferential excretion in urine. It is suggested that liposomes are not recognized by the cells of the R.E.S. It is also shown that biodistribution of free IOH does not differ significantly from that of the liposomes. Thus, in spite of their good in vitro stability in whole blood, they behave in mice like free IOH.

It can therefore be concluded that IOH is either localized at the liposome external interface where it

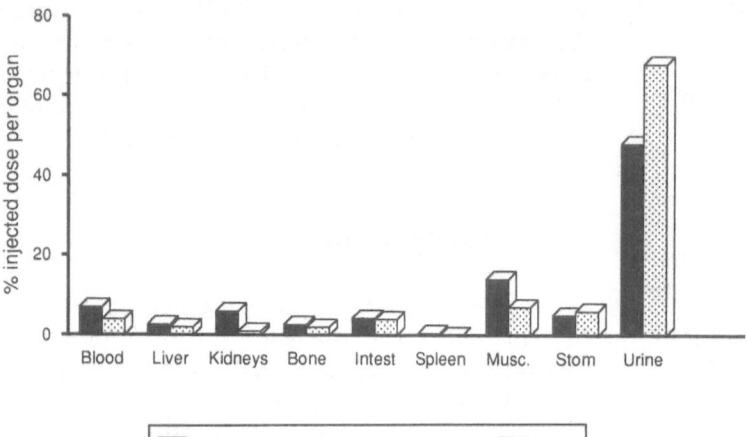

Fig. 1. Biodistribution in mice of IOH liposomes and IOH 30 min after i.v. injection

3. Mayhew E, Rustum XM (1979) Cancer Treat Rep 63:1923
4. Allen TM (1981) Biochim Biophys Acta 640:385
5. Paleos CM (1992) In: Polymerization in Organized Media, Paleos CM (ed), Gordon and Breach Science Publishers, INC, p 283
6. Szoka Jr, Papahadjopoulos D (1980) Ann Rev Biophys Bioeng 9:467

Authors' address:

C. Sawas-Dimopoulou
Institute of Radioisotopes-Radiodiagnostics
N. C. S. R. "Demokritos"
15310, Agia Paraskevi
Attiki, Greece

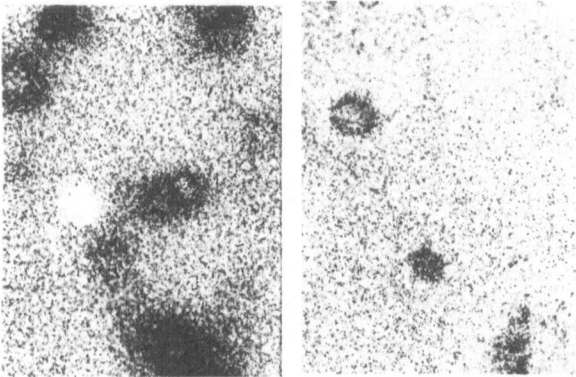

Fig. 2. Autoradiography of the liposomes loaded with ^{131}I-Hippuran. The radioactivity is surrounding the liposomes, with a higher density at the periphery than inside the liposomes. This localization of the radioactivity facilitates of kidney uptake of the radiopharmaceutical. The high ^{131}I-Hippuran affinity for the nephron is well known

can be easily recognized and taken up by the kidneys, or it is liberated from the in vivo destabilized liposomes. Further experiments are under way in order to provide additional information on the mechanism of renal uptake of liposome incorporated IOH.

References

1. Ostro MJ, Cullis PR (1989) American Journal of Hospital Pharmacy 46:1576
2. Weinstein JN (1981) Pure and Appl Chem Vol 53:2241

Concentration dependence of electric light scattering by aqueous dispersions of rodlike virus particles

V. Peikov, S. Stoylov, and I. Petkanchin

Institute of Physical Chemistry Bulgarian Academy of Sciences Sofia, Bulgaria

Aqueous dispersions of charged fd-virus particles are studied electro-optically in concentration region where a Coulomb interaction causes liquid-like ordering (Maier, Schulz, Weber (1988) Macromolecules 22:7087). In the electric light scattering experiments, the interaction is manifested as concentration dependence of the electro-optic effects, low-frequency negative electro-optic effect and decrease of particle rotational diffusion coefficient.

Acknowledgement

Thanks are due to Prof. Weber (Univ. of Konstanz, Germany) for fd-virus particles. Financial support of the Bulgarian National Foundation "Scientific Researches" (No. A 31) is gratefully acknowledged.

Authors' address:

V. Peikov
Institute of Physical Chemistry
Bulgarian Academy of Sciences
Acad. Bonchev Str. Bl. 21
BG-1113 Sofia, Bulgaria

Experimental data are in good accordance with a model of elongated micelles consisting of a core of surfactant chains surrounded by a shell of headgroups while the two surfactants mix homogeneously without any significant enrichment of SDS at the endcaps.

Taking the homogeneous mixing for granted, the shape and surface charge of the micelles were determined for other mixing ratios. A three-axes ellipsoid was chosen as a model for the shape of the micellar core. For the axial ratios a:b:c we found values of $1:(1-2):(1-6)$ depending on the mixing ratio $(1:9 - 9:1)$. We used the Hayter-Penfold model for globular macroion solutions as well as a semi-phenomenological approach according to Farsaci for an attempt to describe the structure factor $S(Q)$ and the micellar interactions.

Authors' address:

Prof. Dr. J. Kalus
Experimentalphysik I
Universität Bayreuth
95440 Bayreuth, FRG

Shape investigation of mixed micelles by small-angle neutron scattering

H. Pilsl[1]), H. Hoffmann[2]), S. Hofmann[2]),
J. Kalus[1]), A. W. Kencono[3]), P. Lindner[4]),
and W. Ulbricht[2])

[1]) Experimentalphysik I, Universität Bayreuth, Germany
[2]) Physikalische Chemie I, Universität Bayreuth, Germany
[3]) National Atomic Energy Agency, Indonesia
[4]) Institut Laue-Langevin, France

Several mixtures of the zwitterionic surfactant tetradecyldimethylaminoxide TDMAO (which normally forms uncharged rodlike micelles) and the anionic sodiumdodecylsulfate SDS (which normally forms charged globular micelles) were studied by small-angle neutron scattering. The aim was to investigate: 1) the mixing behavior of these surfactants, and 2) the influence of surfactant charge on the micellar shape on the other hand. A contrast variation experiment of a 60 mM solution with the mixing ratio TDMAO:SDS = 8:2 was performed. The fact that the SDS chains were protonated or deuterated respectively allowed to get further information about the internal structure of the micelles.

Structural study in AOT reverse micelle containing native and modified α-chymotrypsin

F. Pitré[1,2]) and M. P. Pileni*[1,2])

[1]) Laboratoire S.R.S.I., Université P. et M. Curie, Paris, France
[2]) C.E.N. Saclay, D.R.E.C.A.M., S.C.M., Gif sur Yvette, France

Abstract: Structural studies of reverse micelles containing native and modified α-chymotrypsin are reported. Reverse micelles were used to increase the hydrophobic character of α-chymotrypsin. Three fluorene derivatives were linked to α-chymotrypsin. Determination of the percolation threshold of reverse micellar solution containing native and modified α-chymotrypsin, by conductivity measurements, was performed. Small-angle x-ray scattering was used to determine the intermicellar potential of unfilled and filled reverse micelles. Contrary to the solubilization of a peripheral membrane protein, cytochrome c, in AOT reverse micelles, the incorporation of native and hydrophobic α-chymotrypsin does not induce change of the intermicellar potential.

Key words: Reverse micelle — α-chymotrypsin — acylated α-chymotrypsin — intermicellar potential

Introduction

Reverse micelles are 1 nm to 10 nm radius water droplets dispersed by means of a surfactant in an organic solvent [1]. The ratio of water concentration over surfactant concentration, W = [H$_2$O]/[AOT], determines the size of the reverse micelles [2]. Enzymes can be incorporated into reverse micelles in an active form [3]. Incorporation of enzymes can greatly alter host reverse micelles. Previously, it has been shown that the solubilization of a peripheral membrane protein, cytochrome c, induces large structural modifications with a change in the percolation threshold [4]. Hydrophobic molecules were linked to α-chymotrypsin, in order to create a membrane affinity. Determination of the intermicellar potential by small-angle x-ray scattering of unfilled and filled reverse micelles with α-chymotrypsin chemically modified or not is reported.

Experimental section

AOT was obtained from Sigma, α-chymotrypsin from Fluka, and 9-fluorenyl-methyl chloroformate from Merck. The small-angle x-ray scattering was performed at L.U.R.E. (Orsay, France).

It has been previously shown that reverse micelles can be used to covalently modify enzymes [5]. It has been chosen to use this method to increase the hydrophobic character of α-chymotrypsin. α-chymotrypsin was acylated by 9-fluorenyl-methyl chloroformate. Native α-chymotrypsin is solubilized in a borate buffer (pH = 9.5; 0.1 M) solution and is injected in the AOT reverse micelles. The water content is fixed at w = 20. 9-fluorenylmethyl chloroformate, previously dissolved in isooctane, is added to the micellar solution. The ratio of reagent concentration over enzyme concentration is equal to 20. The acylation reaction is performed at 10 °C. After 20 min, the solution is poured in a 10-fold volume of cold acetone (−10 °C) in order to separate the enzyme from the AOT molecules and from free reagent. This solution is centrifuged at 5000 rpm. The precipitate is redispersed in 100 ml of acetone and centrifugated again. After removing acetone, water is added to the precipitate. The solution is dialyzed at 10 °C, during 12 h to remove the residual borate salt. The remainder is lyophilized. The average number of fluorenylmethyl groups bound to α-chymotrypsin is 3.

Results and discussion

Structural studies were performed for filled reverse micelles containing native or modified α-chymotrypsin close to the maximum solubility of the enzyme.

The maximum solubilization of native α-chymotrypsin increases linearly for different water content, w, with the polar volume fraction. This indicates that the ratio of the maximum concentration of α-chymotrypsin with the micellar concentration (i.e., the average number of enzyme in a reverse micelle) is a constant at fixed w value and increases with w at fixed ϕ_w. A strong increase in the average number of α-chymotrypsin per micelle as a function of w (i.e., the size of the droplets) at fixed ϕ_w is observed. This indicates that α-chymotrypsin needs to be surrounded by water molecules without any contact with surfactant molecules. At w = 40 only one native α-chymotrypsin per micelle can be solubilized in the reverse micelles while three FMOC-α-chymotrypsins can be incorporated on average in a reverse micelle. FMOC-α-chymotrypsin has a higher affinity for reverse micelles than does native α-chymotrypsin.

At fixed temperature, conductivity variation with water volume fraction, ϕ_w, presents three main zones. At small water volume fraction, ϕ_w, the conductivity is slightly higher than that observed in pure oil and it increases linearly with ϕ_w [6]. The increase of the water volume fraction induces an increase in the conductivity by several orders of magnitude. This is the percolation phenomenon. It corresponds to the formation of an infinite cluster. Far above the percolation threshold, the conductivity reaches a limit value. The conductivity measurements of reverse micelles filled by the various α-chymotrypsin derivatives show a similar behavior to that obtained with empty micelles. The percolation onset is not changed by the presence of the enzyme solubilized in the water pool. This indicates no significative changes of the pair potential by solubilization of α-chymotrypsin derivatives.

The average radius of reverse micelles, determined by S.A.X.S., containing native or modified α-chymotrypsin is identical to that of unfilled reverse micelles (Table 1). This indicates that the solubilization of α-chymotrypsin and its derivatives does not perturb large reverse micelles. The adhesives spheres potential is used to determine the intermicellar potential. It is the limit of a square-well potential defined by [7]

Table 1. Values of radius and of sticky parameter at $\phi_w = 20\%$ and $w = 40$

	radius in Å	$1/\tau$
without enzyme	67	0.9
native chymotrypsin	67	0.95
1 modified chymotrypsin per micelle	67	1.1
3 modified chymotrypsin per micelle	69	1

$$U(r)/kT = \lim d \to \sigma F(r) \ ,$$

where:

$$F(r) = +\infty \qquad \text{for } 0 < r < \sigma$$

$$F(r) = -\ln[d/(12\tau(d-\sigma))] \quad \text{for } \sigma < r < d$$

$$F(r) = 0 \qquad \text{for } r > d \ ,$$

where σ and τ^{-1} are respectively the diameter and the stickiness of the particles.

Using the Percus-Yevick (PY) approximation, the explicit and analytical expression of the intermicellar structure factor is obtained [8] by assuming that microemulsions can be considered as a one

component macro-fluid of interacting particles with potential $U(r)$ in the continuous solvent. The S.A.X.S. data are reasonably well fitted for a given sticky parameter, τ^{-1} (Fig. 1). The presence of native α-chymotrpysin and FMOC-α-chymotrypsin does not modify the reverse micelles interaction potential (Table 1). The similar behavior obtained with filled and unfilled reverse micelles is in agreement with the conductivity measurements where no variations of percolation threshold have been noticed. Although the solubilization of FMOC-α-chymotrypsin is increased, no structural changes are seen, in contrast to solubilization of cytochrome c where drastic structural changes are obtained [4].

Conclusion

We have shown that α-chymotrypsin localized in the aqueous core of the reverse micelle does not change the behavior of the reverse micellar system. Although the solubilization of the modified α-chymotrypsin is increased, the strength of the intermicellar potential and the radius of the host reverse micelle are not significatively modified. This increase of the hydrophobicity of the α-chymotrypsin is not sufficient to alter the micellar interface and to produce an intermicellar potential change.

References

1. For a general survey see a) Mittal K, Lindman B (eds) Surfactants in Solution, (Plenum, N.Y.), 1984, and ibid 1987; b) Safran SA, Clark NA (eds) Physics of Complex and Supermolecular Fluids (Wiley, N.Y.), 1987
2. Pileni MP, Zemb T, Petit C (1985) Chem Phys Lett 118:414
3. Pileni MP (ed) (1989) Structure and reactivity in reverse micelles, Elsevier
4. Huruguen JP, Authier M, Greffe JL, Pileni MP (1991) J Phys: Condens Matter 3:865
5. Levashov AV, Kabanov AV, Khmelnitsky YL, Berezin IV, Martinek K (1984) Dokl Akad Nauk SSSR 278:246
6. Eicke HF, Borkovec M, Das-Gupta B (1989) J Phys Chem 93:314
7. Baxter RJ (1968) J Chem Phys 49:2770
8. Regnaut C, Ravey JC (1989) J Chem Phys 91:1211

Authors' address:

Pr Marie-Paule Pileni
Laboratoire S.R.S.I.
Université P. et M. Curie
Bat F BP 52
4, place Jussieu 75005 Paris, France

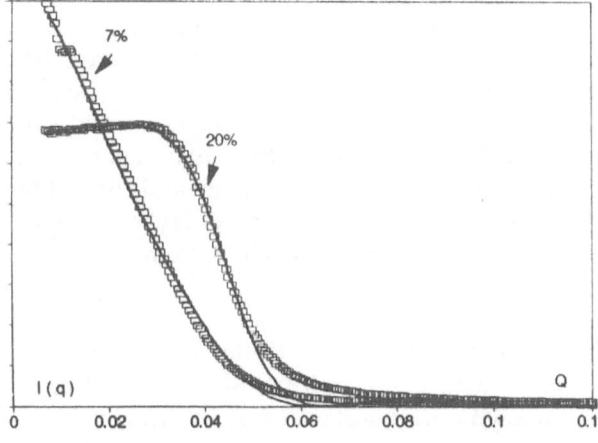

Fig. 1. Experimental (square) and simulated (thick line) x-ray spectra of reverse micelles containing three modified α-chymotrypsin per micelle at two water volume fraction

Progress in Colloid & Polymer Science   Progr Colloid Polym Sci 93:336 (1993)

Electro-optic study of polymer stabilization of colloids

Ts. Radeva and M. Stoimenova

Institute of Physical Chemistry Bulgarian Academy of Sciences Sofia, Bulgaria

The sensitivity of the electro-optic parameters to the stages of surface adsorption of polymers allows the process of colloid stabilization to be followed. Steric stabilization, achieved by adsorption of uncharged polymer, is accompanied by abrupt increase of the relaxation frequency of electric polarizability (Radeva, Stoimenova (1991) Coll & Surf 54:235—244). Electrostatic stabilization, obtained through polyelectrolyte surface recharging, is accompanied by increase in the value of electric polarizability and reduction in its relaxation frequency, the latter being directly related to the type of the adsorbed molecule (Radeva, Stoimenova, JCIS, submitted).

Acknowledgement

Financial support for this project (X-50) of the Bulgarian National Foundation "Scientific Researches" is gratefully acknowledged.

Authors' address:

Ts. Radeva
Institute of Physical Chemistry
Bulgarian Academy of Sciences
Acad. Bonchev Str. Bl. 21
BG-1113 Sofia, Bulgaria

Influence of organic sodium salts on the rheological behavior of aqueous alkaline kaolin suspensions

P. Rendtel and H.-J. Mögel

Institute of Physical Chemistry, TU Bergakademie Freiberg

Key words: Kaolin suspension — rheology — organic substances

In this study the influence of alcohols and sodium salts of organic aliphatic acids with different numbers of carboxyl and hydroxyl groups on the rheological behavior are investigated, whereby the number of carbon atoms changes between two and six. The measurements were carried out with alkaline suspension. The time-dependent behavior and the values of the appearent viscosity are strongly influenced by carboxyl groups. Hydroxyl groups and the number of C-atoms have a secondary priority.

Authors' address:

Petra Rendtel
TU Bergakademie Freiberg
Leipziger Str. 29
Inst. f. Phys. Chemie
09599 Freiberg

Stability and compatibility in blends of silicone and vinylacrylic polymer emulsions

J. Richard, C. Mignaud, and A. Sartre

Rhône-Poulenc-Recherches — Centre de Recherches d'Aubervilliers, Aubervilliers Cedex, France

Special, unexpected properties may generally arise from blends of polymer emulsions; these features depend on the miscibility and stability of the resulting dispersions, and also on the morphology of the films obtained upon drying. In the field of water-repellent paints, blends of silicone and vinylacrylic copolymer emulsions have been previously shown to be a tricky means to associate at high ability to bind mineral pigments and water permeability properties in the same material [1].

In the present work, the stability and compatibility in blends of silicone and vinylacrylic copolymer emulsions are worked out. The system chosen consists of:

— a styrene-butylacrylate (S/BuA) copolymer latex, which is electrostatically stabilized by copolymerized surface carboxyl groups;
— silicone polymer emulsions which are stabilized either by a water soluble polymer such as polyvinylalcohol (PVAL), or a classical anionic emulsifier (SDS).

As far as stability of the dispersion blends is concerned, the key parameters are the nature of the stabilizing system for the silicone emulsion and the pH of the latex. When blending the silicone emulsions which are stabilized by PVAL, with the carboxylated latex at pH values lower than 6, a dense elastic gel is formed. This behavior is attributed to the formation of a hydrogen-bonded interpolymer complex between PVAL and carboxylic groups located at the surface of the latex particles [2].

As regards the morphology of the films, the strong incompatibility of the two polymers is shown; it results in large macroscopic heterogeneities in the dried composite films. Furthermore, semi-quantitative angle-resolved ESCA measurements have evidenced a pronounced diffusion of the flexible, highly mobile silicone species [3] towards the surface of the samples, where a homogeneous top silicone layer of at least a few nm thick is built up. Then, investigations into different ways to increase compatibility of the two phases have been performed. It has been shown that in-plane phase separation at the macroscopic scale can be controlled by two means: lowering the interfacial tension between the two phases upon grafting silane compatibilizers on the vinylacrylic latex particles [4], or decreasing the molecular mobility of the silicone phase upon cross-linking through a reaction with a trifunctional alcaline methylsiliconate. In contrast, diffusion of the silicone chains towards the surface of the composite films can only be significantly restricted by cross-linking the silicone chains.

References

1. a) Pouchol JM (1989) Proceedings: 9th International Conference of the Paint Research Association (Frankfurt, Germany). b) Pouchol JM (1989) Proceedings: Eurocoat Congress 1989 (Nice, France)
2. Staikos G, Bokias G (1991) Makromol Chem 192:2649
3. Schorsch G, Léger L, Cohen-Addad JP (1990) Pour la Science 147:64
4. Brown GR (1989) E. P. Application 327:376

Authors' address:

J. Richard
Rhone-Poulenc-Recherches
Centre de Recherches d'Aubervilliers
52 rue de la Haie Coq
F-93308 Aubervilliers Cedex, France

Solution properties of polysaccharides with immunological activity

T. Fuchs, W. Richtering, and W. Burchard

Institut für Makromolekulare Chemie, Freiburg, Germany

β-1.3/1.6 glucans are known to be immunologically active and some are already used in cancer therapy. The solution behavior of these polysaccharides depends strongly on the solvent. In aqueous solution the macromolecules often form stiff triple helices, whereas in dimethyl sulfoxide the single chains are present. Solutions of the β-1.3/1.6 glucan schizophyllan have been studied by static and dynamic light scattering. The various structures have been characterized in water and dimethyl sulfoxide, respectively. The solution behavior was also investigated in mixtures of water and dimethyl sulfoxide. Large globular aggregates were observed within a certain solvent composition range. These aggregates dissociate irreversibly into triple helices at temperatures above 50°C.

Authors' address:

Walter Richtering
Institut für Makromolekulare Chemie
Stefan-Meier-Straße 31
D-79104 Freiburg, FRG

Spectroscopic characterization of lyotropic phases of perfluoro-polyether derivatives

S. Ristori[1]), G. Gebel[2]), M. Visca[3]), and G. Martini[1])

[1]) Dipartimento di Chimica, Universita' di Firenze, Italy
[2]) Laboratorie de Physico-Chimie Moleculaire, DRFCM/SESAM, Centre d'Etudes Nucleaires de Grenoble, France
[3]) Montefluos spa, Bollate-Milano, Italy

Abstract: Lamellar phases of perfluoropolyethers in binary concentrated, aqueous mixtures have been investigated by small-angle neutron scattering (SANS) and electron paramagnetic resonance (EPR) of TempTMA$^+$ and 5- and 16-doxyl stearic acids. SANS patterns gave sharp maxima which were interpreted on the basis of isotropic and lamellar phases. EPR spectra agreed with the scattering data and gave more details on structural and dynamic features of the investigated systems.

Key words: EPR — perfluorinated compounds — lamellar phases — neutron scattering

Introduction

Perfluoropolyethers (PFPE), with the general chemical composition:

$$CF_3-(O-CF)_n-O-CF_2-COO^-\ M^+$$
$$|$$
$$CF_3$$

are a new class of commercial surfactants, which show most of the typical properties of perfluorinated surfactants, including the formation of micellar solution in water mixtures at low PFPE concentration [1—2] and of liquid crystal lamellar phases at high PFPE concentration [3].

In addition to classical experiments on micellar solutions, PFPE aqueous micellar solutions have also been investigated by electron paramagnetic resonance (EPR) of small and large paramagnetic probes [2, 4]. More recently, scattering techniques have revealed the existence of lamellar phases at high PFPE concentration in water [3]. In this paper, we report on the combined use of scattering and magnetic resonance spectroscopies for the study of lamellar phases of NH$_4$-PFPE in binary water mixtures as compared with the lamellar phases of ammonium perfluooctanoate (NH$_4$-PFO).

Experimental

Materials

The acidic form of PFPE was produced by Montefluos SpA as an intermediate of the hexafluoropropene photo-oxidation. The fraction used in this work (PFPE with equivalent weight, EW, 706, purity 95%) was separated from other EW fractions by iterated vacuum distillation. The ammonium salt was prepared by neutralization of the EW 706 acidic form with a large excess of ammonia water solution.

The following nitroxides (Molecular Probes, Inc., Eugene, OR) were used without further treatment: 4-trimethylamonio-2,2,6,6-tetramethylpiperidine-1-oxyl (Temp-TMA$^+$, 1), 5-, and 16-doxylstearic acid (5-, and 16-DXSA, 2 and 3, respectively). Details on the sample preparation and handling can be found elsewhere [2, 4].

Technique

Small-angle neutron scattering (SANS) experiments were performed on the Paxe spectrometer (Lab. Leob Bruillouin, Saclay, France) under conventional conditions [3]. EPR spectra were recorded with the Bruker 200D spectrometer operating at the X band.

For instrumental settings and experimental conditions, see refs. [2, 4].

Results and discussion

Figure 1 shows the SANS patterns from aqueous mixtures of NH$_4$-PFPE (45% w/w) and NH$_4$-PFO. In both cases a peak series appeared. In the case of NH$_4$-PFO, two peaks were identified. This behavior could only be interpreted on the basis of a biphasic system: an isotropic phase, and a lamellar phase. The second phase agreed for the presence of the higher Q peak, whose position corresponded to that calculated by Boden et al. in oriented samples of Cs-PFO/water lyotropic phases [5]. Accordingly, the scattering pattern observed for NH$_4$-PFPE was also indicative of an ordered lamellar structure. SANS data showed a larger linewidth and no harmonics of the 0,01 Bragg reflection in NH$_4$-PFO, whereas in NH$_4$-PFPE two or three harmonics were detected. This could be attributed to a smaller monodomain size in the perfluorooctanoate liquid crystalline phase.

When nitroxides 1—3 were employed to study the behavior of the NH$_4$-PFO and NH$_4$-PFPE in isotropic solutions, the usual three-line absorptions were observed with lineshapes and ^{14}N coupling constant values dependent of both concentration and temperature. The calculation of the motional properties and of the local ordering was possible on the basis of the computer program written by Schneider and Freed [6].

In lamellar liquid-crystal phases more complex shapes of the EPR absorptions were obtained. Figure 2 shows typical EPR spectra from nitroxides in NH$_4$-PFO and NH$_4$-PFPE water mixtures with

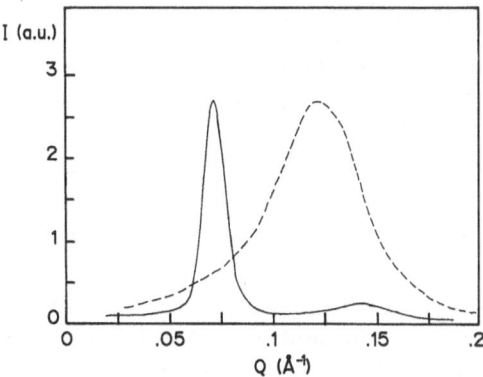

Fig. 1. SANS patterns of binary aqueous mixtures of: (—) ammonium perfluorooctanoate (45% w/w) and (---) ammonium perfluoropolyethers (44.6% w/w)

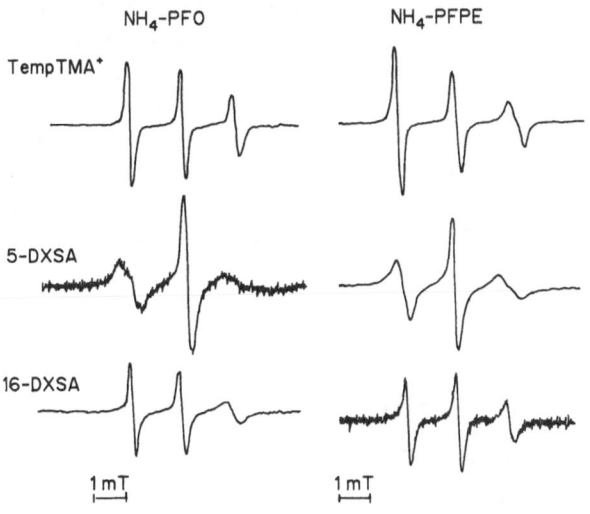

Fig. 2. EPR spectra of TempTMA⁺, 5-, and 16-DXSA in binary water mixtures (~50% w/w) of NH₄-PFO and NH₄-PFPE

weight ratios about 50%. TempTMA⁺ gave invariantly a superposition of two different absorptions due to the probe inserted in different environments with different polarities as it resulted from the different ^{14}N coupling constants of the two signals. Different mobilities also characterized the two environments. Major details on the nature of these spectra were obtained from runs carried out at the Q-band ($\omega \sim 35$ GHz) [7]. A superposition of signals was also observed with 5-DXSA. In this case a clearer separation between two motional domains was apparent and this separation increased with time and with temperature. In contrast, no appreciable spectral overlapping was observed with 16-DXSA, whose shape was typical indeed of fast moving radicals in clearly hydrophobic environment ($\langle A_N \rangle = 1.56$ mT).

A definitive proof of the occurrence of lamellar phases in both systems was achieved from the EPR spectra of mechanically aligned thin films. Figure 3 shows the EPR spectra of 5-DXSA in both NH₄-PFO and NH₄-PFPE 50% w/w films at two different orientation with respect to the magnetic field. Two absorptions again contributed to the overall signals in both orientations, the first one almost independent of the orientation and the second one being strictly orientation dependent. The same probe therefore occupied oriented and unoriented lamellar phases in both samples with a complete orientation dependence of $\langle A_N \rangle$, thus indicating a fairly good order of the oriented phases. This was further proved by the orientation dependence found with nitroxides 1, and 3.

The above experimental observations agreed for a greater facility exhibited by PFPE with respect to

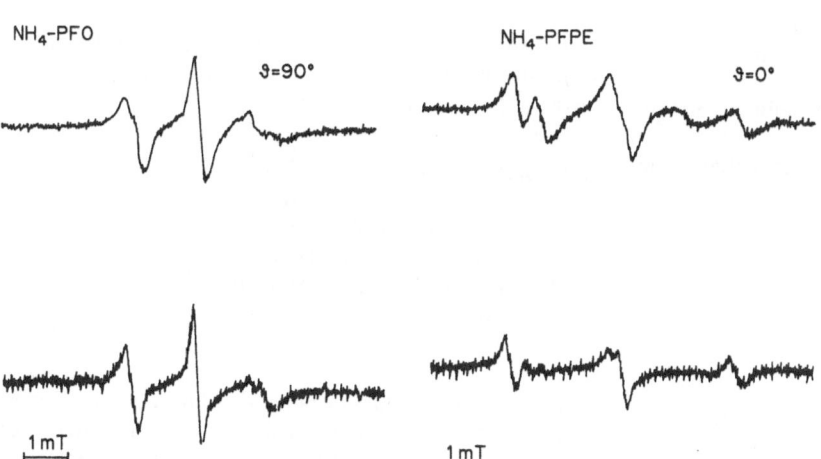

Fig. 3. EPR spectra of 5-DXSA in mechanically aligned NH₄-PFO (left) and NH₄-PFPE (right) aqueous mixtures at 0° and 90° with respect to the magnetic field direction

PFO [8] in forming macroscopically well ordered thin films when deposited on glass surfaces. This was in line with a more flexible interface of the PFPE bilayers in comparison with the PFO bilayers which can be simply inferred from the structures of the two molecules. This suggestion was further verified by NMR. No nematic phase has been detected for the binary system PFPE/water in the concentration range 0.5—70% w/w and from 10 to 80°C, whereas the occurrence of a magnetically oriented phase is well established for the PFO/water system [9]. The difference in the stiffness of the bilayer interface, which is able, in the case of PFPE, to follow the undulations due to thermal motions, might be invoked as the explanation of the different behavior. Consequently, no large anisotropic aggregates, which are thought to be the building units of a nematic phase, could be formed in PFPE systems.

Acknowledgements

Thanks are due to Montefluos spa, MURST and CNR for financial support.

References

1. Chittofrati A, Lenti D, Sanguineti A, Visca M, Gambi CMC, Senatra D, Zhou Z (1989) Progr Colloid Polym Sci 79:218—225
2. Maritni G, Ottaviani MF, Ristori S, Lenti D, Sanguineti A (1990) Colloids Surf 45:177—184
3. Gebel G, Ristori S, Loppinet B, Martini G (1993) J Phys Chem, in press
4. Ristori S, Martini G (1992) Langmuir 8:1937—1942
5. Boden N, Corne SA, Holmes MC, Jackson PH, Parker D, Jolley KW (1986) J Phys (France) 47:2135—2144
6. Schneider DJ, Freed JH (1989) in Biological Magnetic Resonance, Plenum Press, Vol. 8, p 1
7. Ristori S, Ottaviani MF, Visca M, Martini G (to be published)
8. Ristori S, Gebel G, Martini G (1993) Colloids Surf (in press)
9. Weber H, Hoffmann H (1988) Liquid Cryst 3:203—216

Authors' address:

Prof. Giacomo Martini
Dipartimento di Chimica
Universita' di Firenze
Via G. Capponi 9
50121 Firenze, Italy

Surface viscosity of monomolecular films at the air-water interface

F. Ciuchi, A. Relini, R. Ferrando, and R. Rolandi

Department of Physics, University of Genoa, Italy

Interpreting surface shear viscosity in terms of molecular interaction and organization requires viscosity measurements independent of the measuring apparatus. The canal viscosimeter allows such measurements, provided the drag of the subphase is considered. Joly (Le Journal de Physique et le Radium, serie VII T.IX — 1938) furnished a treatment of this physical system which takes into account the subphase drag and introducing an empirical parameter depending on the slit width. We revised Joly's treatment in order to obtain the shear viscosity of different monomolecular films at the air-water interface. The results of our measurements show that the subphase drag parameter depends on the pH of the subphase and on the flow rate of the film.

Authors' address:

F. Ciuchi
Department of Physics
University of Genoa
Via Dodecaneso 33
I-16146 Genoa, Italy

Electrokinetic behavior of polystyrene dispersions in methanol-water mixtures

F. J. Rubio-Hernandez[1]), R. Hidalgo-Alvarez[2]),
F. J. de las Nieves[2]), B. H. Bijsterbosch[3]),
and A. J. van der Linde[3])

[1]) Department of Applied Physics II,
   University of Málaga, Spain
[2]) Department of Applied Physics,
   University of Granada, Spain
[3]) Department of Physical and Colloid Chemistry,
   Wageningen Agricultural University,
   The Netherlands

Abstract: The electrokinetic behavior of negatively charged polystyrene dispersions in methanol-water mixtures has been studied. Data obtained from streaming potential and electrophoresis measurements have been compared. The theory of Levine et al. (J. Colloid Interface Sci., 52 (1975) 136) for systems consisting of narrow capillaries has been used to convert streaming potential data into zeta-potentials. For the electrophoresis data the theory of Dukhin and Semenikhin (Kolloid. Zh., 37 (1975) 1127) has been employed. The electrokinetic behavior of polystyrene latices appears to be extensively influenced by a rather large value of the surface conductance.

Key words: Zeta potential — polystyrene dispersions

Introduction

The classical theory of the electrokinetic phenomena expresses that, for a given system, the electrokinetic potential (ζ-potential) only depends on the properties of the phases in contact, and must be independent of the experimental method employed for its determination. However, several authors [1—3] have found that the ζ-potential as obtained by means of streaming potential or electroosmosis measurements, using the classical Helmholtz-Smoluchowski theory, is smaller than that obtained by means of electrophoretic mobility or sedimentation potential measurements. These discrepancies have usually been explained on the basis of the influence of the surface conductance on the streaming potential or the electroosomotic flow [4, 5], which implies an underestimation of ζ-potentials obtained with these methods. This is the reason why, in most cases, the conversion of streaming potential data into ζ-potentials was carried out by using Smoluchowski's equation, but correcting for the surface conductance effect. However, if the pore radii (a) in a concentrated system are not much bigger than the double layer thickness (κ^{-1}), even accounting for surface conductance in Smoluchowski's equation provides erroneous values for ζ. This case is very usual with low permittivity media, where diffuse layers are very expanded.

Fairly recently, several authors [6—9] have developed new theories that can be applied to the electrokinetic processes that take place in porous plugs with small electrokinetic radii (κa) and/or high ζ-potentials. These theories account for several geometrical structures in terms of, for example, capillary or cell models. They incorporate a corrective factor into the Smoluchowski equation that takes into account the potential distribution inside the porous structure. This correction is the more important the smaller are the pores ($a \ll \kappa^{-1}$).

According to Olivares et al. [9], the theory of Levine et al. for capillary geometry [7] is most ap-

propriate. However, the mathematical process of the theory of Levine et al. is rather involved for interpeting routine measurements of streaming potential, as several authors [2, 9] have pointed out. Trying to simplify the mathematical process, Romm [8] developed a theory in which a semiempirical expression for the corrective factor is obtained. Also, the method of Olivares et al. [9] is mathematically easier, and consists in the introduction of a parameter into the more simple equations of Rice and Whitehead [6] and making use of an iterative process.

Experimental tests on the validity of these theories are very scarce in the literature, especially so when the permittivity of the liquid phase is low. In this case, the double layer thickness is expanded, so that Smoluchowski's condition ($\kappa a \gg 1$) is violated.

Hence, if is very interesting to perform streaming potential measurements on porous plugs as well as electrophoresis measurements on dilute dispersions of the same particles, with the objective of comparing the ζ-potentials obtained by both methods.

Monodisperse spherical polystyrene latices have proved to be very useful model systems for studying various colloidal phenomena. Their versatility has been demonstrated, for example, by Hearn et al. [10]. Consequently, streaming potential measurements of anionic polystyrene plugs and electrophoretic mobility measurements of the same polystyrene particles in dilute dispersions have been made. In this way, we contribute to providing a solid footing for the streaming potential method. This experimental method is very useful in systems such as granulate solids, polymeric films, metallic surfaces, textile fibers and, it is not possible to use electrophoresis [11—14].

Theory

1. Streaming potential

The basic relationship between the streaming potential ($\Delta\phi/\Delta P)_{I=0}$ and the ζ-potential*) has first been formulated by Smoluchowski [15]:

$$\zeta_s = \left(\frac{\Delta\phi}{\Delta P}\right)_{I=0} \frac{\eta\lambda}{\varepsilon_0\varepsilon_r} . \tag{1}$$

---

*) For the meaning of the various symbols, see List of symbols.

The geometry of the plug and the surface conductance are incorporated in Eq. (1) by means of the experimental parameters C and K:

$$\zeta_s = \left(\frac{\Delta\phi}{\Delta P}\right)_{I=0} \frac{\eta CK}{\varepsilon_0\varepsilon_r} \,. \tag{2}$$

When the overlap in the potential distribution in the capillaries is taken into account, a corrective factor S must be introduced in Eq. (2):

$$\zeta = \left(\frac{\Delta\phi}{\Delta P}\right)_{I=0} \frac{\eta CK}{\varepsilon_0\varepsilon_r} \frac{1}{S} \tag{3}$$

S is a function of ζ-potential and electrokinetic radius (κa). Different theories differ in the expression for S:

a) Rice and Whitehead [6] derived:

$$S_{RW} = \frac{1 - \dfrac{2I_1(\kappa a)}{\kappa a I_0(\kappa a)}}{1 - \beta\left[1 - \dfrac{2I_1(\kappa a)}{\kappa a I_0(\kappa a)} - \dfrac{I_1^2(\kappa a)}{I_0^2(\kappa a)}\right]} \,. \tag{4}$$

This theory is limited to values of $\zeta \leqslant kT/e$, but is valid for all values of κa.

b) Levine et al. [7] extended the Rice-Whitehead theory to all values of ζ-potential. Their expression for S is:

$$S_L = \frac{\dfrac{1}{2}(\kappa a)^2(1-G)}{\int_0^{\kappa a} R\cosh\Psi(R)\,dR + \beta\int_0^{\kappa a} R\left(\dfrac{d\Psi(R)}{dR}\right)^2 dR} \,. \tag{5}$$

The inconvenience in using this theory is rooted in its mathematical complexity.

c) Romm [8] has developed a semiempirical corrective factor, valid for all κa and ζ values:

$$S_R = \frac{1 - \dfrac{1}{\omega}(1 - e^{-\omega})}{1 + \dfrac{4HF\zeta\omega}{R^0 T(\kappa a)^2} + \dfrac{\varepsilon\varepsilon_r F_0^2 \omega\zeta^2}{\pi\lambda\eta R^0 T}\left(1 - \dfrac{1}{2\omega}(1 - e^{-2\omega})\right)} \,, \tag{6}$$

where:

$$\omega = \frac{\sinh(\Psi_2/2)}{\Psi_2}\,\Psi_2^{\frac{A}{(0.1+\kappa a)^B}}\,\kappa a\,\tanh\left(\frac{c\Psi_2(\kappa a)^{1/2}}{\sinh(\Psi_2/2)}\right) \,. \tag{7}$$

Romm assigned the values $A = 0.55$, $B = 0.65$, and $C = 0.3$ for these constants.

d) Finally, Olivares et al. [9] have supplied the same corrective factor as Levine et al., but using a less complex mathematical process.

2. Electrophoresis

Electrophoresis consists in the displacement of the charged particles in a colloidal dispersion with respect to the continuous liquid phase. According to the classical Henry theory [16], the electrophoretic mobility is related to the ζ-potential in the following way:

$$\zeta_H = \frac{3}{2}\frac{\eta\mu_0}{\varepsilon_0\varepsilon_r}\frac{1}{f_1(\kappa s)} \,, \tag{8}$$

where $f_1(\kappa s)$ is the consequence of the superposition of the external electric field and the local electric field surrounding the particle.

Dukhin and Semenikhin [28] developed an equation incorporating both the ζ-potential and the diffuse potential ψ_d into the expression for μ_e. In this theory the ionic conduction inside the shear plane is added to the conduction in the mobile part of the diffuse double layer. In other words, the possibility of an anomalous surface conductance is taken into account in DS-theory.

When ζ and ψ_d are both larger than 25 mV the DS-theory supplies the simplified equation:

$$\mu_e = \frac{3}{2}\frac{\varepsilon_0\varepsilon_r\zeta}{6\pi\eta}\left(\frac{1 + Rel\left\{\dfrac{4\ln\cosh(e\zeta/4kT)}{e\zeta/kT}\right\}}{1 + 2Rel}\right) \,, \tag{9}$$

where:

$$Rel = \frac{\lambda_g}{\lambda s} = \frac{\exp(e\psi_d/2kT) + 3m\exp(e\zeta/2kT)}{\kappa a} \,. \tag{10}$$

Materials and methods

Negatively charged monodisperse latexes were prepared according to the method of Furusawa et al. [17]. Essentially, a portion of the styrene monomer, distilled under nitrogen just prior to use, and an aqueous buffer (KHCO$_3$) solution were mixed and saturated with nitrogen in a bottle at 70°C. The buffer was added in order to suppress the formation of hydroxyl groups during the polymerization process. The reaction was subsequently started by adding a nitrogen-saturated $K_2S_2O_8$ solution. The latices were cleaned by ion-exchange over a mixed bed [18]. Surface charge densities σ_0 were obtained by conductometric titration. Table 1 shows the values of σ_0 together with the average particle diameter, as obtained by electron microscopy.

All chemicals were of A. R. quality. All water was purified by reverse osmosis, followed by percolation through charcoal and mixed bed ion exchange resins (Millipore).

Electrophoretic mobilities were obtained with a Zetasizer IIc, by taking the average of six measurements at both stationary levels in a cylindrical cell.

The polystyrene porous plugs required for measuring streaming potentials were obtained by means of centrifugation at 20000 rpm. during 2 h with a Beckman JA-21 centrifuge. The experimental device for streaming potential measurements was initially designed by Van der Put and Bijsterbosch [19] and later refined by Van der Linde and Bijsterbosch [20]. This last device was used in this research. For a detailed description, we refer to reference [20].

The methanol was of analytical-grade, obtained from Carlo Erba, and was used without further purification. The polystyrene latices used here do not dissolve in methanol-water mixtures, as was checked by spectroscopy and conductivity methods.

The dielectric constant values of the mixtures were measured by a Dekameter DK 300 and viscosities by means of a Canon-Fenske viscosimeter. All experiments were performed at 20°C.

Results and discussion

Both latexes studied in this research (L2 and L4) have the same surface charge density (see Table 1), and only differ in their particle size. Therefore, we can compare streaming potential measurements and results obtained by using latex L2 with electrophoretic mobility measurements and results obtained by using latex L4.

In Fig. 1, the electrophoretic mobility of L4 particles is plotted against the volume percentage of methanol in the binary mixtures methanol-water. It can be seen that the μ_e values depend non-linearly upon composition.

Table 1. Particle diameters and surface charge densities of latices

Sample	Diameter (nm)	S. Charge Density (mC/m²)
L2	640 ± 10	−(22.0 ± 0.5)
L4	1080 ± 40	−(21.0 ± 0.5)

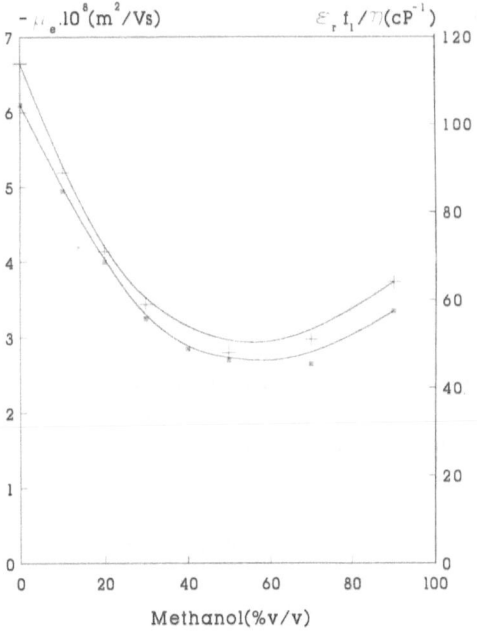

Fig. 1. $\mu_e(.)$ of L4 particles and $\varepsilon_r f_1/\eta (+)$ against methanol percentage (10^{-9} M KBr)

According to Eq. (8), a one-to-one correlation might be expected between the composition dependences of μ_e and $\varepsilon_r f_1/\eta$ if the ζ_H-potential were constant. In order to probe this possibility, the values of $\varepsilon_r f_1/\eta$ of the methanol-water mixtures have been included in Fig. 1. In fact, the composition dependence of $\varepsilon_r f_1/\eta$ follows a similar non-linearity.

Hence, the dependence of μ_e on the methanol content in the aqueous mixtures seems to be mainly due to the variation of the liquid properties. A constant ζ_H-value of −(93 ± 5) mV is obtained for the L4 latex particles (see Fig. 2).

In Fig. 3 streaming potentials of sample L2 are plotted versus the applied pressure difference. The liquid phase consists of different methanol-water

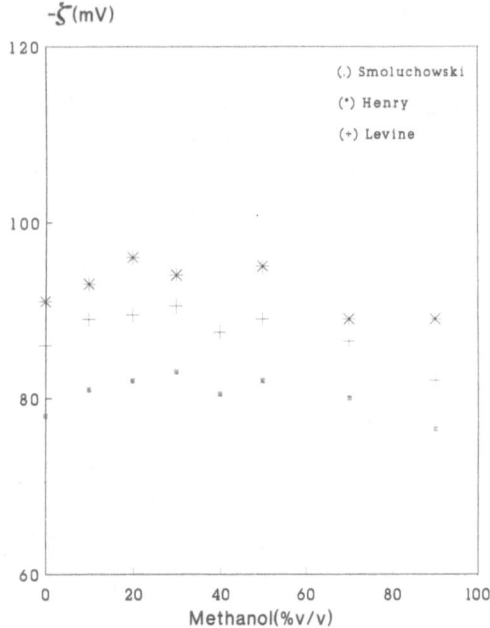

Fig. 2. ζ-potential versus methanol percentage

mixtures (% v/v) and an electrolyte concentration of 10^{-3} mol/dm^3 KBr in all cases, like in the electrophoresis experiments. The dependence is always linear, in agreement with the phenomenological relationship between streaming current, hydrodynamic pressure and streaming potential [21]:

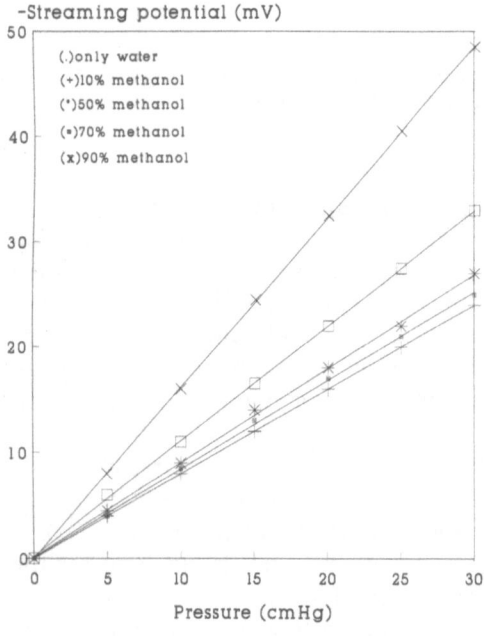

Fig. 3. $\Delta\phi$ vs. ΔP for different methanol content. L2

$$I = L_{21}\Delta P + L_{22}\Delta\phi , \qquad (11)$$

where L_{21} and L_{22} are the phenomenological coefficients that characterize the streaming current and the electric conductance of the porous plug, respectively.

Equation (2) enables to calculate the ζ_s-potential from the slopes $(\Delta\phi/\Delta P)_{I=0}$ in Fig. 3, taking into account the actual conductivity of the plug. In Table 2 the values of the latter are shown. A constant ζ_s-value of $-(80 \pm 4)$ mV is obtained for the L2 particles (see Fig. 2). This constancy of ζ obtained from streaming potential measurements is in qualitative agreement with the electrophoresis results.

As mentioned in the introduction, according to the classical theory of electrokinetic phenomena, the ζ-potential should be independent of the experimental method employed for its determination. However, streaming potential and electrophoresis experiments in this case have provided different ζ-potentials, as can be seen in Fig. 2. These differences are in line with those found by other authors [1—3].

If the electrokinetic radius of the system does not fulfill the condition $\kappa a \gg 1$, the Smoluchowski equation is inappropriate to calculate the ζ-potential. In that case, another theory must be applied.

We have calculated the average pore radius of the porous plug used in our streaming potential experiments, assuming the plug to be equivalent to a bundle of parallel capillaries of equal size. Different equations can be used [22], but Van der Put [22] and Van den Hoven [23] have pointed out that for polystyrene spheres the hydraulic radius of the plug is adequately estimated by:

$$a = (1 - p)\,2s/3p . \qquad (12)$$

The resulting value was 320 ± 15 nm.

The corresponding κa-values, shown in Table 2, vary between 33.4 and 47.8, which means that they are neither so high as to justify the use of the Smoluchowski equation, nor so low as the effects of overlapping of double layers to dominate. Hence, it is expected that the corrective factors (S) are close to unity. Consequently, corrected ζ-potentials will not differ very much from those obtained with the Smoluchowski equation. In Fig. 2 we show the ζ-potential obtained by using S_L (Eq. (5)). ζ_L-potentials are slightly higher than ζ_s-potentials. The other theories studied gave similar corrections. The mean difference of only 5 mV with the electrophoresis results means that Levine's theory is ap-

Table 2. Electric conductance of the plug (K), and electrokinetic radius of L2 porous plug. ($C = 6.7$ cm^{-1})

Methanol (% v/v)	K ($\mu\Omega^{-1}$)	κa
0	144.5	33.4
10	114.5	34.4
20	96.5	35.2
30	80.4	36.0
40	65.5	37.2
50	55.1	38.6
70	42.1	42.2
90	31.5	47.8

propriate to calculate the ζ-potential from streaming potential data.

To study the effect of the ionic strength on the electrokinetic behavior, we have measured streaming potentials (L2) and electrophoretic mobilities (L4) in water and in methanol-water (50% v/v).

The corrective factor S_L in the streaming potential measurements varied from 1 for 10^{-1} M KBr to 0.79 for 10^{-4} M KBr in both cases. In fig. 4 the corresponding ζ-potentials are plotted. Both curves show a plateau at low electrolyte concentration, a known anomalous behavior [24]. This behavior is fully opposite to the variation of the diffuse potential, ψ_d, with the electrolyte concentration, as can

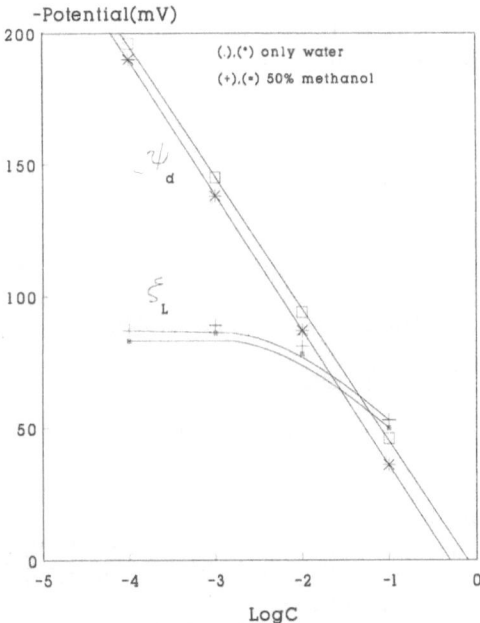

Fig. 4. ζ_L-potential and diffuse potential (ψ_d) versus log[KBr]. L2

be seen in Fig. 4. The diffuse potential has been calculated from the surface charge density by means of the Gouy-Chapman theory of diffuse double layers [21].

Figure 5 shows that a maximum in the ζ-potential is obtained from electrophoretic mobility measurements of latex L4 by using Henry's equation. Various explanations for this maximum have been suggested [25]. One of these presumes the presence of a "hairy layer" on the particle surface [24], the thickness of which varies with the ionic strength. According to this explanation the shear plane

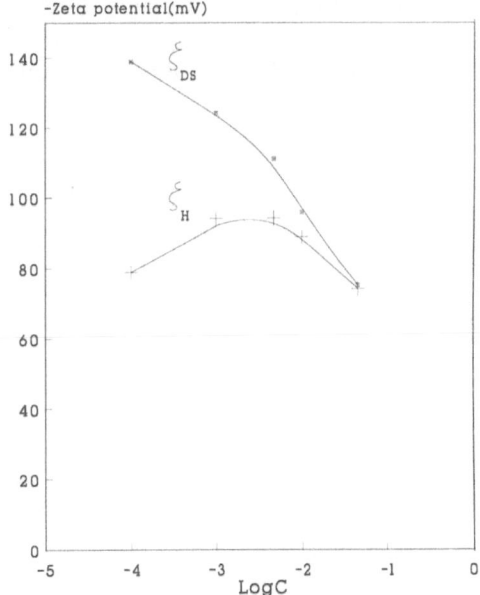

Fig. 5. ζ_H-potential and ζ_{DS}-potential versus log[KBr]

moves away from the surface as the ionic strength decreases. This phenomenon results in two effects. Firstly, it lowers the ζ-potential in the usual way and, secondly, it lowers the electrophoretic mobility by introducing a much greater percentage of ionic conduction in the diffuse layer behind the plane of shear [26]. Recently, also Van der Linde and Bijsterbosch [27] have shown that the maximum in the ζ-$\log C$ curves is due to ignoring the anomalous surface conductance in theories providing ζ-potentials from streaming potential (or electrophoretic mobility) measurements. Henry theory is one of those which explains the maximum in Fig. 5. In the same Fig. 5, we have plotted the ζ-potential obtained from electrophoresis data by using the Dukhin-Semenikhin theory [28]. As can be seen, the maximum disappears, which can be ascribed to the contribution to polarization from all ions being taken into account. This is an indication that the maxima in the ζ-potential of polystyrene as a function of ionic strength can be ascribed to the presence of an anomalous electrical conduction in the solid-liquid interface. Although Levine et al. did not explicitly formulate the contribution of the surface conductance by using the actual conductivity of the plug in the calculations, it is implicitly accounted for, which explains why the maximum is absent in Fig. 4.

The anomalous behavior of polystyrene is more dramatically expressed when the electrokinetic charge density is plotted against the ionic strength. This charge density (σ_{ek}) was calculated from ζ-values by using the Gouy-Chapman theory. In Fig. 6 it can be seen that σ_{ek} decreases when the ionic

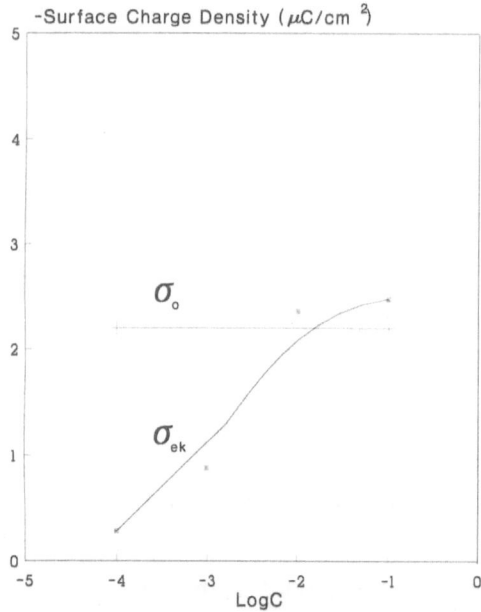

Fig. 6. Electrokinetic charge density versus log[KBr]. L2

strength decreases, while a constancy of σ_{ek} is expected. σ_0, represented by the horizontal straight line in Fig. 6, and σ_{ek} are nearly equal at electrolyte concentrations higher than 10^{-2} M, where the effects of surface conductance and e.d.l. overlap disappear and, therefore, the diffuse and ζ-potentials are the same (see Fig. 4).

List of symbols

a	Pore or capillary radius
C	Cell constant of the plug
e	Proton charge
F	Faraday constant
G	Ratio of the mean electrostatic potential across the capillary to the ζ-potential (ref. [17])
H	Value of the function $H(R)$ defined in ref. [8]
I	Streaming current
I_0, I_1	Zero-order and first-order, respectively, modified Bessel function of the first kind
K	Electrical conductance of the porous plug
k	Boltzmann constant
m	dimensionless ionic drag coefficient
p	Porosity of the plug
R	($= \kappa r$) Dimensionless radial coordinate
r	Radial coordinate
R^0	Gas constant
s	Particle radius
T	Absolute temperature
β	A dimensional parameter defined in reference [6]
$\Delta\phi$	Streaming potential
ΔP	Hydrodynamic pressure
ε_0	Permittivity of vacuum
ε_r	Dielectric constant
Ψ	($= e\psi/kT$) Dimensionless ψ-potential
Ψ_d	Diffuse potential
Ψ_g	($= e\zeta/kT$) Dimensionless ζ-potential
η	Absolute viscosity
κ	Reciprocal double layer thickness
λ	Conductivity of the bulk solution
λ_g	Surface conductance
μ_e	Electrophoretic mobility
ρ	Mass density of the solid phase
ζ	Zeta or electrokinetic potential

References

1. Rastogi RP, Shabh R, Upadhyay BM (1981) J Colloid Interface Sci 83:41
2. Chowdiah P, Wasan DT, Gidaspow D (1983) Colloids & Surfaces 7:291
3. De las Nieves FJ, Rubio-Hernández FJ, Hidalgo-Alvarez R (1988) J Non-Equilib Thermodyn 13:373
4. Overbeek JThG (1952) In: Kruyt HR (ed) Colloid Science, vol 1. Elsevier, Amsterdam
5. Kitahara A, Fujii T, Katano S (1971) Bull Chem Soc Japan 44:3242
6. Rice CL, Whitehead R (1965) J Phys Chem 69:4017
7. Levine S, Marriott JR, Neale G, Epstein N (1975) J Colloid Interface Sci 52:136
8. Romm ES (1979) Colloid J of USSR 41:758
9. Olivares W, Croxton TL, McQuarrie DA (1980) J Phys Chem 84:867

10. Hearn J, Wilkinson MC, Goodall AR (1981) Adv Colloid Interface Sci 14:173
11. Somasundaran P, Kulkarni RD (1973) J Colloid Interface Sci 45:591
12. Wnek WJ, Gidaspow D, Wasan DT (1975) J Colloid Interface Sci 50:609
13. Chowdiah P, Wasan DT, Gidaspow D (1981) AlChE J 27:975
14. Fairhurst D (1990) Polym Mater Sci Eng 62:57
15. Smoluchowski M (1918) Z Phys Chem 93:129
16. Henry DC (1931) Proc Roy Soc (London) Ser A 133:106
17. Furusawa K, Norde W, Lyklema J (1972) Kolloid Z Z Polymere 250:908
18. Van den Hul HJ, Vanderhoff JW (1968) J Colloid Interface Sci 28:336
19. Van der Put AG, Bijsterbosch BH (1981) Acta Polymerica 32:311
20. Van der Linde AJ, Bijsterbosch BH (1989) Colloids & Surfaces 41:345
21. De Groot SR, Mazur P (1969) In: Non-Equilibrium Thermodynamics, North-Holland PC, Amsterdam
22. Van der Put AG (1980) Thesis, Agricultural University, Wageningen
23. Van den Hove ThJJ (1984) Thesis, Agricultural University, Wageningen
24. Hidalgo-Alvarez R, de las Nieves FJ, Van der Linde AJ, Bijsterbosch BH (1986) Colloids & Surfaces 21:259
25. Goff JR, Luner P (1984) J Colloid Interface Sci 99:468
26. Midmore BR, Hunter RJ (1988) J Colloid Interface Sci 122:521
27. Van der Linde AJ, Bijsterbosch BH (1990) Croatica Chim Acta 63:455
28. Semenikhin NM, Dukhin SS (1975) Kolloidn Zh 37:1127

Authors' address:

Dr. F. J. Rubio-Hernández
Escuela Politècnica Universitaria
Department of Applied Physics II
University of Málaga
Campus El Ejido, Málaga 29013, Spain

Langmuir-Blodgett films on substrates with microrelief

V. V. Safronov and E. I. Givargizov

Institute of Crystallography,
Russian Academy of Sciences, Moscow, Russia

Key words: Langmuir-Blodgett films — defects — lateral structure

Physical properties of Langmuir-Blodgett (LB) films apparently depend upon their lateral organization which is formed at a moving meniscus line during LB deposition process. We have shown that it is possible to control the deposition process locally by means of specially designed substrate microrelief.

Fatty acid salt LB film was deposited onto silicon substrates with following etched relief patterns: strips of 5 to 200 µm width and hexagons. The samples were investigated using scanning electron microscope.

Investigated films showed the crack network pattern which appeared to provide information about the local deposition direction. Thus, we were able to see that at relief strips the deposition process has been going parallel to their direction independently of its orientation relative to dipping direction. This effect may be used for controllable deposition of LB films with lateral anysotropy of conductance, optical properties, etc. Another interesting effect is that cracks may terminate at the relief contour. This may help to avoid defect formation at desired sites of substrate. Thus, substrate microrelief may provide a tool for local control of LB film lateral organization.

Authors' address:

V. V. Safronov
Institute of Crystallography
Russian Academy of Sciences
Leninsky pr. 59
117333 Moscow, Russia

Fluorinated copolymer Langmuir-Blodgett films

V. V. Safronov[1], L. D. Budovskaya[2],
V. N. Ivanova[2], and Yu. M. Lvov[1]

[1] Institute of Crystallography, Russian Academy of Sciences, Russia
[2] Institute of Macromolecular Compounds, Russian Academy of Sciences, Sanct-Petersburg, Russia

Key words: Polymeric Langmuir-Blodgett films — copolymers — fluorine

Surface-active polymers are regarded as perspective for practical application Langmuir-Blodgett (LB) materials due to their stability. Meanwhile, a problem of polymeric LB film quality takes place. A possible approach to solution of this problem may be seen in application of copolymers with random distribution of different monomeric units along copolymer chain. This will supply a method for varying polymer chain and, therefore, monolayer properties such as flexibility, balance of hydrophobic/hydrophilic properties, etc.

In this work, we made an attempt to get an idea of how copolymer chain constitution may influence monolayer properties and quality of LB film for the case of copolymer of octafluoroamylmethacrylate with methacrylic acid. We also studied monolayers of a set of fluoroalkylmethacrylates with different fluorocarbon chain length.

Our main results are: monomer molar ratio of the copolymer influences surface pressure of monolayer transition into a state with high viscosity that allows optimization of LB deposition process; LB superlattices containing layers of the copolymer with optimized contents were of perfect order [1]; isotherms of homopolymeric monolayers show phase transition regions.

References

1. Safronov VV, Budovskaya LD, Ivanova VN, Lvov YuM (1992) Biologicheskie Membrany, 9:985—991

Authors' address:

V. V. Safronov
Institute of Crystallography
Russian Academy of Sciences
Leninsky pr. 59
117333 Moscow, Russia

Structure, interfacial tension and solubilization capacity in C_8E_3/water/oil systems

W. Sager

Institut für Physikalische Chemie der Universität Basel, Switzerland

Key words: Phase behavior — microemulsions — C_8E_3 — interfacial tension

The system triethylene glycol monooctylether (C_8E_3)/water/oil is chosen to investigate the formation of surfactant aggregates in Winsor II and W/O microemulsions, because it represents the transition from weakly to strongly structured amphiphilic systems.

The ratio of the rigidity constant to the radius of spontaneous curvature determining the properties of the surfactant layer is calculated on the basis of interfacial tension measurements and the hydrodynamic radii of the dispersed nanophases. The one-phase microemulsion region is characterized by studies of phase diagrams and the structure is confirmed by conductivity, dynamic light scattering and small-angle x-ray scattering. The influence of oil on the solubilization capacity of water can be explained in terms of the dependence of the interfacial tension on approaching the tricritical point and of the critical micelle concentration (CMC) of surfactants in the different hydrocarbons. The area per surfactant molecule at the interface is evaluated from the size dependence of the nanophases on the surfactant-to-water ratio and also from Gibbs adsorption isotherms of the surfactant at the macroscopic interface in Winsor II systems.

Authors' address:

Dr. W. Sager
School of Chemistry
University of Hull
Hull HU6 7RX
North Humberside, U.K.

Preparation of colloidal oxide particles in emulsions

W. Sager[1]), W. Sun[1]), U. D. Schwarz[2]), and H.-F. Eicke[1])

[1]) Institut für Physikalische Chemie der Universität Basel, Switzerland
[2]) Institut für Physik der Universität Basel, Switzerland

Key words: Emulsion precipitation — nano-crystalline materials — colloidal oxide particles

Emulsions as precipitation media for small uniform oxide particles have recently attracted interest in ceramic processing research. The production of reliable and reproducible ceramic components for high technology applications requires a strict control of critical powder characteristics, including chemical homogeneity, small particle size, and a narrow size distribution. In order to precipitate such particles from an aqueous solution, it is necessary to precisely adjust the conditions which allow the controlled formation of seeds and particle growth. Alternatively, heterogeneous systems like emulsions offer the advantage of reducing the "reaction vessel" to the size of the particles [1—3].

We developed a method to prepare nano-sized oxide particles by precipitation in emulsions [4]. It is shown that a suitable combination of surfactants is decisive to obtain spherical uniform particles of such a small size range, since surface active agents are not only necessary in preparing the emulsions, but also in forming spherical particles and stabilizing them against aggregation. Particle precipitation is studied in terms of droplet-droplet exchange and particle growth.

Colloidal particles between 20 and 50 nm are obtained for gel forming oxides, e.g., Fe_2O_3, Y_2O_3, TiO_2, non-gel forming oxides, e.g., ZnO, as well as mixed oxides, e.g., $FeYO_3$. The size dependence of the oxide particles was studied by varying the concentrations of the metal ions and the dispersed phase. In all investigations the particle size, characterized by dynamic light scattering and atomic force microscopy (AFM), was found to be almost independent of the concentrations in the starting emulsions. Contrary to the generally accepted view of each emulsion droplet acting as an isolated micro-reactor, we found within the investigated concentration range the exchange of material between the colliding emulsion droplets to determine the final size of the oxide particles. The comparison between the droplet and oxide particle sizes shows that, depending on the concentration of metal ions in the starting solution, about 2 to 500 droplets have to exchange their contents to obtain oxide particles of about 30 nm.

Characteristics of thin layers prepared from the colloidal oxide particles were investigated by x-ray diffraction and AFM. A hematite layer, heated to 600°C for 30 min, is shown in Fig. 1. The precursor particles have grown to crystallites of 20 to 60 nm. The image also reveals the grain boundary structure between the crystallites, which is typical for ceramics and polycrystalline thin films. The transition from amorphous precursors to the crystalline

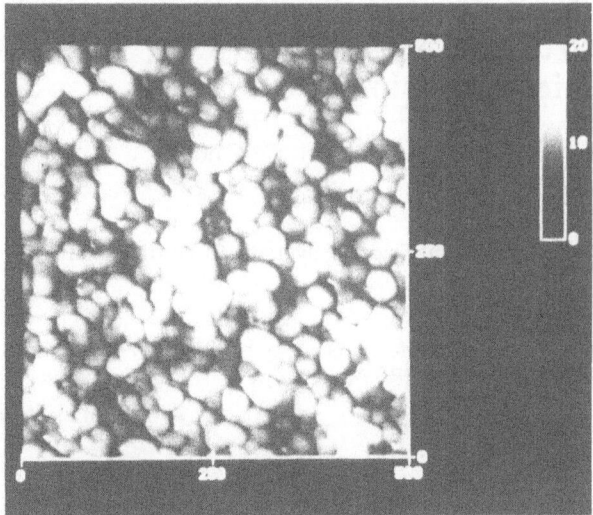

Fig. 1. AFM image (500 × 500 nm²) of a thin Fe_2O_3 layer obtained from a colloidal suspension of ferric hydroxide in decane after heating to 600°C. The maximum height corresponds to 20 nm

state was observed at temperatures much lower than those required for conventional powder routes. The presented preparation method is therefore very attractive for industrial application, since low formation temperatures are required in microelectronic processing routes.

Acknowledgment

The authors are greatly indebted to Dr. C. Schüler, ASEA Brown Boveri AG (ABB), Baden, Switzerland for initiating and continuously supporting this project. We further wish to acknowledge financial support from the Kommission zur Förderung der wissenschaftlichen Forschung, Projekt Nr. 1715.1 (Eidg. Volkswirtschaftsdept.), ABB, and the Swiss National Science Foundation.

References

1. Richardson K, Akinc M (1987) Ceram Intern 13:253
2. Rhine WE, Bowen HK (1988) Ceram Trans 1 (Ceram Powder Sci IIA):119
3. Schmidt H (1991) Ceram Trans 22 (Ceram Powder Sci IV):3
4. Sager W, Eicke H-F, Sun W (1993) (in press Coll Surf)

Authors' address:

Dr. W. Sager
School of Chemistry
University of Hull
Hull HU6 7RX
North Humberside, U.K.

IR study of the water structure in AOT-H$_2$O-CCl$_4$ microemulsions

M. D'Angelo, A. Fuccello, G. Onori, and A. Santucci

Dipartimento di Fisica, Universita' di Perugia, Italy

Key words: Surfactants — bis(2-ethylhexyl)sodium sulfosuccinate — IR spectroscopy — microemulsions

In this paper we report the results of a study on the interaction of water with reversed micelles of bis(2-ethylhexyl)sodiumsulfosuccinate (AOT) in CCl$_4$. This investigation was performed at 30°C as a function of the [H$_2$O]/[AOT] ratio (*W*) by using the IR absorption present in the 3800—3000 cm^{-1} range and due to O-H stretching modes.

Figure 1 shows the IR spectrum of the H$_2$O/AOT/CCl$_4$ system ([AOT] = 0.15 mol · l^{-1}; *W* = 8.5) recorded in the 4000—2500 cm^{-1} range where the absorptions due to the O—H stretchings of H$_2$O and to the C-H stretchings of AOT are present. The contribution to the spectrum of the absorptions due to the vibrational modes of water and AOT are well characterized, because they appear in distinct spectral ranges [1].

Parts a, b, c and d of Fig. 2 show the O-H stretch region for pure water and water in H$_2$O/AOT/CCl$_4$ microemulsions. The spectrum of pure water (Fig. 2a) can be fitted very well in terms of three Gaussian components centered at 3603 ± 6 cm^{-1} (observed for free O-H groups) [2], at 3465 ± 5 cm^{-1} (usually assigned to hydrogen-bonded dimers) [2] and 3330 ± 20 cm^{-1} (related to water in structures with unstrained H bonds) [3]. The IR spectrum of microemulsion water is significantly different than that of pure water, indicating that the water solubilized in the reversed micelles lacks the normal hydrogen-bonded structure present in the bulk water. However, the total peak area, *A*, of the O-H stretching band of water has been found to increase linearly with *W*, as predicted by Beer's law.

In order to quantify the changes in the O-H stretching region, we fitted each spectrum as a sum of three Gaussian components (see parts b, c and d of Fig. 2). The fitted curves are virtually indistinguishable from the measured ones. The parameters characterizing the Gaussian components (peak frequency and bandwidth) are found to depend on the water-to-surfactant ratio, gradually changing with *W* towards the values characteristic of pure water (parts a, and b of Fig. 3). Figure 4 shows the variation of the ratio between the area of any Gaussian component (A_i) to the total peak area (*A*) as a function of *W*. Due to the independence of ε on *W*, it is reasonable to assume that the curves in Fig. 4 are representative of the variations of the different fractions of —OH groups assigned to each Gaussian component. It is seen from the figure that, as expected, the fraction of free —OH groups decreases while that of the linear H bonds increases with *W*. The fraction of peak area attributed to the H-bonded dimers increases initially with *W*, reaching a maximum at *W* ~ 3. The different fractions of peak area changes significantly up to *W* ~ 6; then at higher water contents, gradually approach those of bulk water. This behavior is qualitatively in agreement with data in the literature referring to several physico-chemical properties of water solubilized in reversed AOT micelles [4].

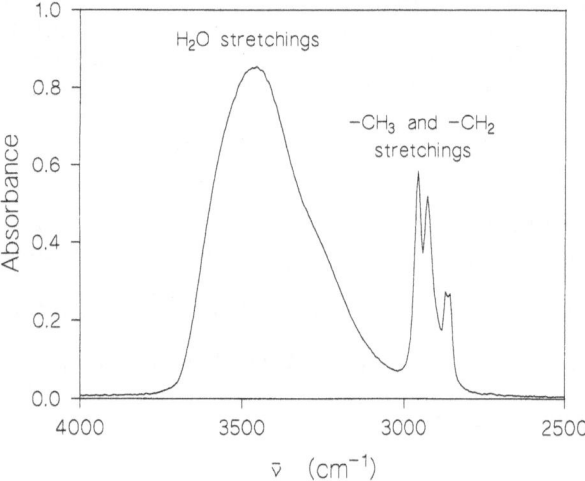

Fig. 1. Infrared spectrum of H$_2$O/AOT/CCl$_4$ system. [AOT] = 0.15 mol · l^{-1}; *W* = 8.5

References

1. Onori G, Ronca M, Santucci A (1991) Progr Colloid Polym Sci 84:88—91
2. Tso TL, Lee EKC (1985) J Phys Chem 89:1612—1618

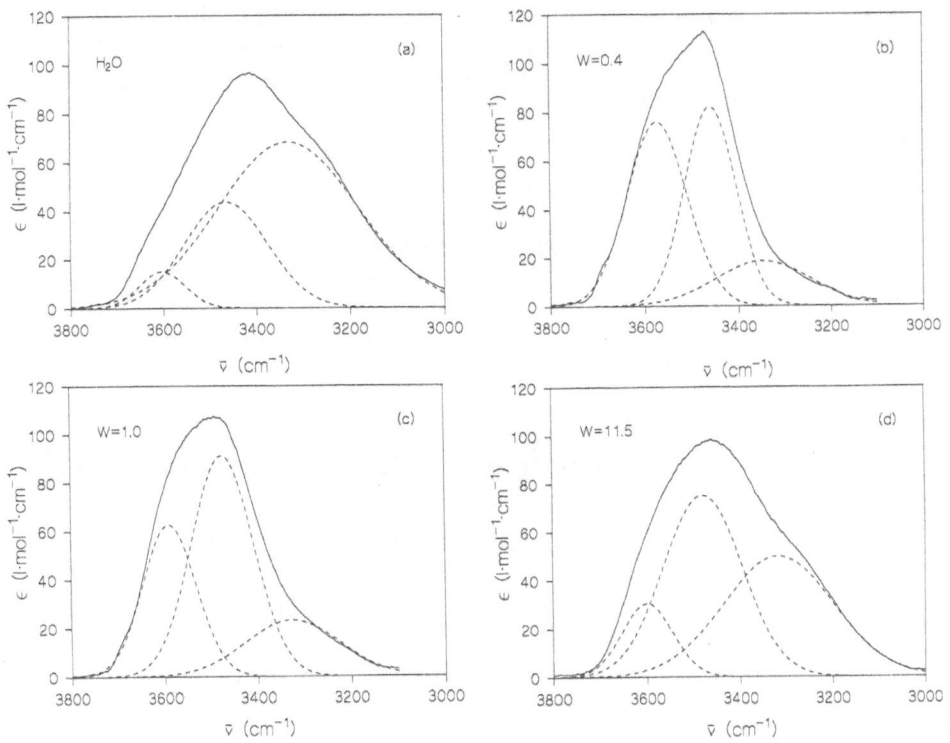

Fig. 2. (—): O-H stretching band for pure water (a) and water in $H_2O/AOT/CCl_4$ system at selected values of W (b), (c), (d); (---): Gaussian components from least squares fitting

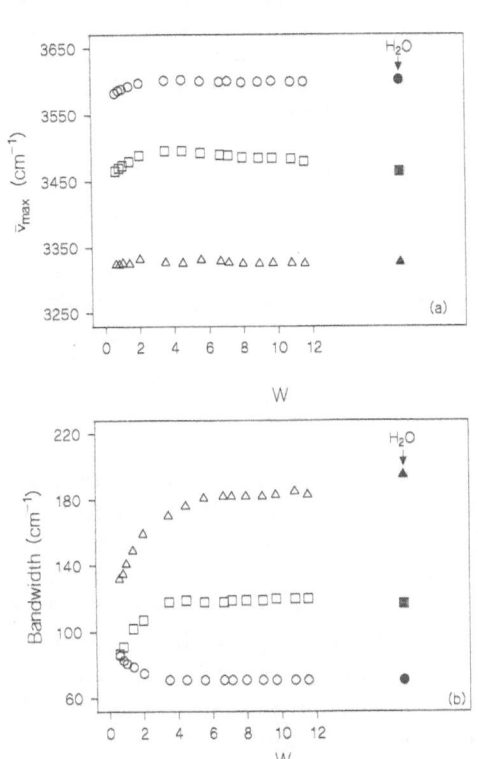

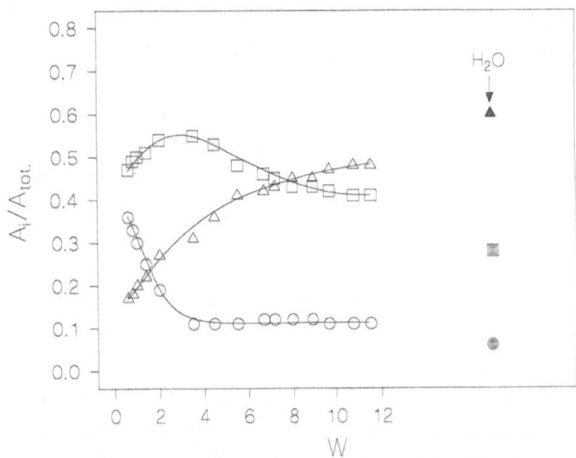

Fig. 4. Ratio of the area of the i-th Gaussian component (A_i) to the total peak area (A) vs. W. Symbols are the same as in Fig. 3. The continuous line serves as a guide to the eyes

Fig. 3. Peak frequency (a) and bandwith (b) of the Gaussian components of the water spectrum in $H_2O/AOT/CCl_4$ system as a function of W. Component centered at ~ 3603 cm^{-1} (○), at ~ 3465 cm^{-1} (□), at ~ 3330 cm^{-1} (△). Solid symbols refer to pure water

3. Walrafen GE (1968) In: Covington HK, Jones P (eds) Hydrogen-Bonded solvent systems. Taylor and Fancis, London
4. Chevalier Y, Zemb T (1990) Rep Prog Phys 53: 279—371

Authors' address:

Prof. G. Onori
Dipartimento di Fisica
Universita' di Perugia
V. A. Pascoli
I-06100 Perugia, Italy

adsorption data information about adsorption isotherms, adsorption mechanisms, and the effect of interfacial active contaminations is available.

Authors' address:

Dr. K.-H. Schano
Labor Automation Berlin
Scharnweberstraße 41
D-12587 Berlin-Köpenick, FRG

Dynamic interfacial tensions measured with an automated drop volume tensiometer

K.-H. Schano[1]), R. Miller[2]), A. Hofmann[3]), and A. Halbig[3])

[1]) Labor Automation Berlin, Berlin-Kopenick
[2]) Max-Planck-Institute of Colloid and Surface Science, Berlin
[3]) LAUDA Dr. R. Wobser GmbH & Co. KG, Lauda-Königshofen

Key words: Interfacial tension — drop volume method — surfactants — liquid interfaces

The automated drop volume method allows to measure dynamic surface and interfacial tensions of pure solvents and surfactant solutions or solutions of other surface active substances, such as polymers, biopolymers, lecithins. Using an automated version, measurements of interfacial tension with high accuracy and reproducibility are possible in large time and temperature intervals.

On the basis of a model of the drop detachment process the hydrodynamic effect at small drop formation times can be taken into account to correct adsorption data in the time interval between 0.5 and 50 s. The method is applied to study the adsorption behavior of different surfactants at the aqueous solution/air, aqueous solution/decane, and water/chloroformic solution interface. From the dynamic

Particle sizing of turbid colloidal systems: a multiple scattering correction

H. Schnablegger and O. Glatter

Institute of Physical Chemistry,
University of Graz, Austria

Abstract: Static light-scattering experiments can be used to determine size distributions of colloidal systems in the range of 50 nm up to several micrometers. Since for light scattering the power of the scattered light is usually high, concentration and/or sample thickness has to be kept low in order to prevent multiple scattering effects. This means that the transmittances of the samples, which are suited for light-scattering investigations are typically above 95% of the incident radiation. We extended the ideas of Hartel for the application on polydisperse colloidal systems with higher concentrations. Thus, it is now possible to evaluate light-scattering data from samples with transmittances of approximately 10% of the incident light. We present simulated data in order to demonstrate the performance of the technique.

Key words: Light scattering — particle sizing — multiple scattering

Introduction

Static light-scattering experiments can be used to determine size distributions of colloidal systems in the range of 50 nm up to several micrometers [1]. Since the intensity of the scattered light is usually high for such large particles, concentration and/or sample thickness has to be kept low in order to prevent multiple scattering effects. The transmittance of a sample is a direct measure for the degree of multiple scattering. As a consequence, the transmittances of the samples which are suited for light-scattering investigations are typically above 95% of the incident radiation. Figure 1 shows the multiple-scattering effect on scattering curves of a monodisperse latex sample with increasing concentrations. It can clearly be seen that the minima of the scattering curves are increasingly smeared out, when more and more multiple scattering comes into play. The goal of this work is to investigate the possibility to extract the size distribution out of scattering functions in the presence of multiply scattered light.

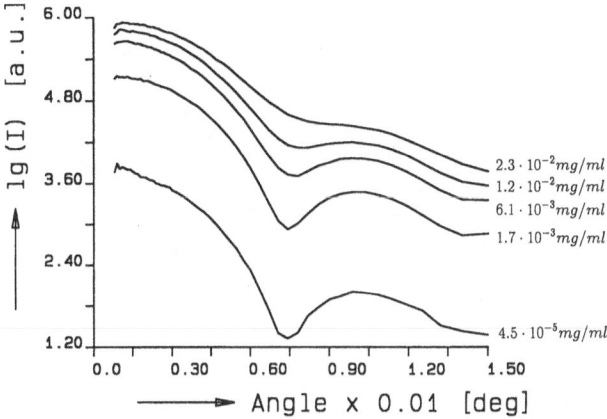

Fig. 1. Experimental static light-scattering functions of monodisperse latex particles of 496 nm diameter and relative refractive index of 1.243 as a function of concentration. Clearly the minima are smeared out when multiple scattering comes into play

The computational method

The main idea for the calculation of the angular distribution of multiply scattered light has been adopted from the work of Hartel [2]. We have extended Hartel's formalism for the application on polydisperse Lorenz-Mie systems, e.g., for the size distribution analysis of turbid colloidal systems of spherical and isotropic scattering objects with higher concentrations and for incident randomly polarized light.

The reconstruction of the size distribution is a nonlinear least squares problem. It is solved by means or the method of linearization [3]. Because of the ill-posedness of the problem the usual regularization techniques [4—6] have been applied additionally.

Results

The performance of the computational procedure has been tested with simulations. A bimodal size distribution of Lorenz-Mie spheres (Fig. 2, full line) and the corresponding theoretical single (triangles) and multiple (crosses) scattering curve with 1% statistical error have been calculated (Fig. 3). In this

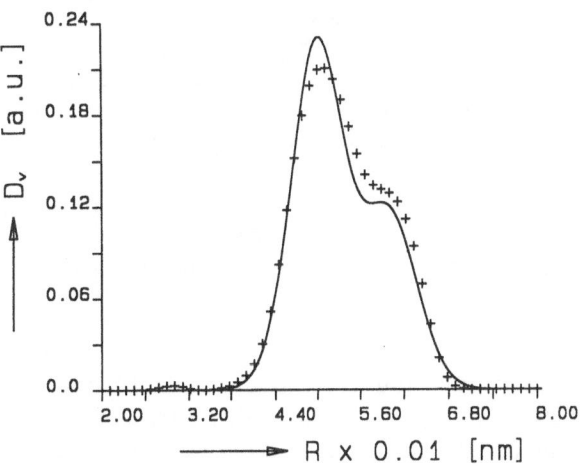

Fig. 2. The reconstruction (pluses) of a simulated volume distribution function (full line) from a scattering curve (see pluses in Fig. 3) with high contributions of multiply scattered light. A beginning lack of resolution can be recognized, indicating the limitations of the method

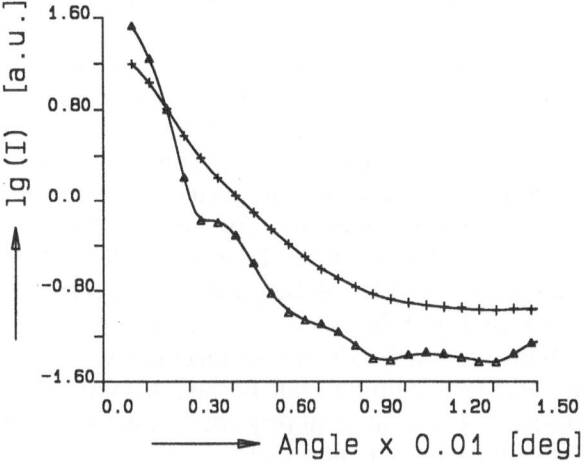

Fig. 3. The simulated multiple scattering function (pluses), the calculated fit (full line through pluses) and the calculated reconstruction (full line through triangles) of the simulated single scattering curve (triangles). The multiple scattering curve has been simulated for the case of 7.02% transmittance of the incident radiation and with 1% statistical noise. Only every third data point is marked with a symbol

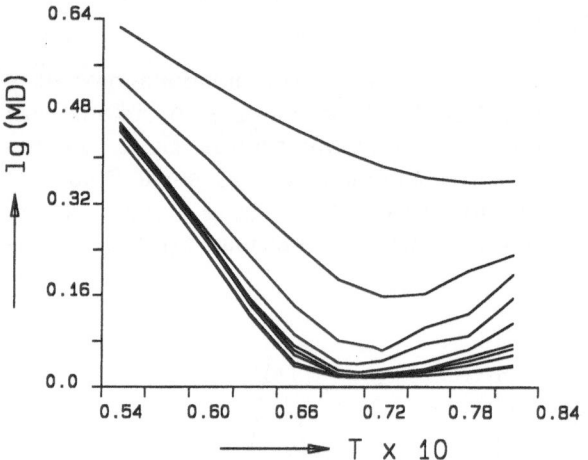

Fig. 4. The mean deviation surface depicted as a function of transmittance (T) and stability parameter. The minimum of this surface indicates the best fitting value of the transmittance of 0.072 ± 0.001

limiting case the theoretical transmittance was 7.02% of the incident light, which corresponds to an aqueous latex suspension (refractive index = 1.598) of concentration = 10^{-3} mg/ml (density = 1 g/ml) and with 22 mm sample thickness. The pluses in Fig. 2 show the beginning lack of resolution. Nevertheless the reconstruction of the single scattering curve (Fig. 3, full line through triangles) is of high accuracy. The excellent fit (Fig. 3, full line through crosses) to the data is a usual phenomenon.

The example above worked well, because the transmittance was known a priori. The determination of the transmittance from the scattering data simultaneously with the size distribution causes instabilities in the nonlinear calculation procedure. Nevertheless, it is possible to calculate the transmittance quasisimulataneously by solving the least squares problem for a set of constant values of the transmittance. In Fig. 4 the mean deviation surface is depicted as a function of transmittance (T) and stability parameter. The minimum of this surface indicates the best fitting value of the transmittance of 0.072 ± 0.001, which is in very good agreement with the simulated value of 0.0702.

Conclusion

We have demonstrated that it is possible, at least theoretically, to calculate size distributions from static light-scattering curves of samples of higher concentration with transmittances of approximately 10%. The transmittance should be known a priori but it can be determined a posteriori within approximately ±1.5%. These findings are restricted to spherical and isotropic colloidal systems with incident randomly polarized radiation.

Verification of these results by experiment is in preparation and will be submitted for publication in Applied Optics.

References

1. Hofer M, Schurz J, Glatter O (1988) J Colloid Interface Sci 122:496—506
2. Hartel W (1940) Licht 10:141—143, 165, 190—191, 214—215, 232—234
3. Bevington PhR (1969) Data Reduction and Error Analysis for the Physical Sciences. McGraw-Hill, New York
4. Lawson CL, Hanson RJ (1974) Solving Least Squares Problems. Prentice-Hall, Englewood Cliffs, N.J.
5. Schnablegger H, Glatter O (1991) Appl Opt 30:4889—4896
6. Schnablegger H, Glatter O (1993) J Colloid Interface Sci 158:228—242

Authors' address:

Prof. Otto Glatter
Institute of Physical Chemistry
University of Graz
Heinrichstraße 28
A-8010 Graz, Austria

Deuterion NMR study on the lyomesophases of nonionic surfactants: quadrupole splittings of hexaethylene glycol dodecyl methyl ether

W. Schnepp, C. Schmidt, and H. Finkelmann

Institut für Makromolekulare Chemie, Universität Freiburg, FRG

Key words: Nonionic surfactants — lyotropic liquid crystals — phase structure — NMR spectroscopy

Recent studies on lyotropic liquid crystalline systems have shown the phase transformations, observed in these systems upon variation of either temperature or concentration, to occur rather gradually, in some cases via narrow regions of intermediate phases. This led us to undertake a detailed study on the system hexaethylene glycol dodecyl methyl ether ($C_{12}E_6C_1$)/water, a typical nonionic surfactant system. Deuterion NMR spectroscopy, which probes the orientation of individual bonds by means of the quadrupole splittings of the nuclei, was used as a tool to investigate the microscopic phase structures. To supplement our previous results on specifically deuterated $C_{12}E_6C_1$ (with deuterions either at the α-alkyl position or at the methoxy end group) [1], here, we present data of the surfactant with a perdeuterated hydrophobic moiety.

Compared with the phase diagrams of the undeuterated [2] and the selectively deuterated systems, which show three mesophases (upon increasing surfactant concentration: hexagonal (H_1), bicontinuous cubic (V_1) and lamellar (L_a)), the phase diagram of the alkyl-perdeuterated surfactant appears to be somewhat different: in particular, the L_a phase is narrower and the V_1 phase has not been confirmed so far.

The analysis of the temperature and concentration dependence of the quadrupole splittings leads to the following results:

i) an almost linear decrease of the splittings with temperature for concentrations at the center of each liquid crystalline phase region;
ii) a much stronger than linear decrease of the splittings with temperature at concentrations close to the phase boundaries of the H_1 phase;
iii) a relatively weak concentration dependence of the splittings at constant temperature within the H_1 phase, but with some irregularities and a tendency of the splittings to decrease towards the phase boundaries;
iv) a ratio of the corresponding quadrupole splittings in the lamellar and in the hexagonal phase considerably smaller than a value of 2, the ratio expected for ideal phase structures (cylindrical micelles in H_1 phase, bilayers in L_a phase).

These results strongly support the model of gradual transformations from nearly perfect to highly defective phase structures going along with concentration- or temperature-induced changes of the hydrophobic/hydrophilic interface curvature.

References

1. Schnepp W, Disch S, Schmidt C (1993) Liquid Crystals 14:843
2. Conroy JP, Hall C, Leng CA, Rendall K, Tiddy GJT, Walsh J, Lindblom G (1990) Prog Colloid Polym Sci 82:253

Authors' address:

Dr. Claudia Schmidt
Institut für Makromolekulare Chemie
der Universität Freiburg
Stefan-Meier-Str. 31
D-79104 Freiburg, FRG

Circular dichroic and calorimetric studies on LDL-subfractions

B. Schuster, R. Prassl, and P. Laggner

Institute of Biophysics and X-Ray Structure Research, Austrian Academy of Sciences, Graz, Austria

Key words: Low-density lipoprotein — subfractions — differential scanning calorimetry — circular dichroism

Low-density lipoprotein (LDL), the primary transport particle for cholesterol in human plasma, is one of the most atherogenic lipoproteins. These complexes with a density of 1.006—1.063 g/ml contain a hydrophobic core of apolar lipids, primarily cholesteryl esters (CE) and triglycerides (TG), surrounded by a polar coat of phospholipids, free cholesterol and apoprotein B-100 (Apo B) [1].

Here, we present a study on well defined LDL-fractions on the basis of their hydrated densities, which are highly homogeneous in their physicochemical behavior. The obtained LDL-subfractions differ in the ratio of lipid to protein as well as the ratio of cholesteryl esters to triglycerides [2].

Near UV circular dichroism (CD) shows different, reversible tertiary structures of Apo B in a temperature range up to 60°C. The spectra at physiological temperature and above are completely different compared to spectra at lower temperatures. The secondary structure of Apo B does not vary essentially below physiological temperature, where the structural conformity for all fractions is high, as detected by far UV CD measurements. At higher temperatures (60°C) the amount of α-helix decreases, whereas it increases for β-sheet depending on the subspecies. Only one of the subfractions differs slightly, there is less β-structure, and the structural stability within the whole temperature range is very high.

Three endothermal transitions were detected by differential scanning calorimetry. The first, reversible one between 21—34°C corresponds to the transition of the CE in the core. The second important core located lipid, TG, influences also the transition behavior. The transition temperature varies between the different subfractions, decreasing (from 28.5 to 25.7°C) with an increase of the TG/CE-ratio; this is in accordance with other investigations [3—5]. The transition enthalpy of the discrete subfractions with a lower TG/CE-ratio are around 430 ± 40 cal/mol CE. Subfractioans with a higher TG/CE-ratio give enthalpies of 680 ± 100 cal/mol CE. It seems that the cholesteryl esters are less constrained in these particles [3]. The ratio of calorimetric to van't Hoff-enthalpy is, in any case, higher than one, so there must be more than two transitions involved. The second transition is also reversible and coincides with a structural change of Apo B and occurs between 40—60°C. The third transition (the temperature range of this transition is between 74 and 82°C) is irreversible and due to disruption of the LDL particle [6]. The ratio of calorimetric to van't Hoff-enthalpy is for both protein transitions more than one. Concerning both protein transitions, we observed a variation in the transition temperature for all investigated LDL-subfractions. Finally, summarizing the enthalpies for the two protein transitions, we obtain a value of 480 ± 80 kcal/mol Apo B for each LDL subfraction.

References

1. Kostner G, Laggner P (1989) In: Fruchart JC, Shepherd J (eds) Human Plasma Lipoproteins. Walter de Gruyter, Berlin pp 23—54
2. Chapman J, Laplaud M, Luc G, Forgez P, Bruckert E, Goulinet S, Lagrange D (1988) J Lipid Res 29:442—458
3. Deckelbaum R, Shipley G, Small D (1977) J Biol Chem 252:744—754
4. Zechner R, Kostner G, Dieplinger H, Degovics G, Laggner P (1984) Chem Phys Lipids 36:111—119
5. Lund-Katz S, Phillips M (1986) Biochemistry 25:1562—1568
6. Walsh M, Atkinson D (1990) J Lipid Res 31: 1051—1062

Authors' address:

Bernhard Schuster, DI
Institute of Biophysics
and X-Ray Structure Research
Austrian Academy of Sciences
Steyrergasse 17/6
A-8010 Graz, Austria

Characterization of submicron-sized drug carrier systems based on solid lipids by synchrotron radiation x-ray diffraction

K. Westesen[1], B. Siekmann[1], and M. H. J. Koch[2]

[1]) Institute of Pharmaceutical Technology, Technical University of Braunschweig, FRG
[2]) European Molecular Biology Laboratory, c/o DESY Deutsches Elektronen Synchrotron, Hamburg, FRG

The physical state of aqueous lipid nanoparticle dispersions prepared by emulsification of molten glycerides using phospholipids, bile salts and poloxamers as emulsifiers [1] was investigated by small-angle and wide-angle synchrotron radiation x-ray diffraction. All carrier systems were in the β-crystalline modification at 20°C. Depending on the matrix constituent, the nanoparticles were either β-crystalline or liquid at body temperature, giving rise to different biopharmaceutical properties. The incorporation of coenzyme Q_{10} did not significantly alter the crystallization tendency of the nanoparticles. Time-resolved x-ray diffraction during temperature scans revealed that the dispersed lipids recrystallize in the α-form after melting and controlled cooling. In contrast, the bulk lipids solidify in the β'-modification and transform rapidly into the β-modification, as demonstrated by simultaneous small- and wide-angle diffraction studies.

References

1. Siekmann B, Westesen K (1992) Pharm Pharmacol Lett 1:123—126

Authors' address:

K. Westesen
Institute of Pharmaceutical Technology
Technical University of Braunschweig
Mendelssohnstr. 1
36106 Braunschweig, FRG

Progress in Colloid & Polymer Science

Progr Colloid Polym Sci 93:357 (1993)

Expansion of stratification domains over the soap film surfaces

D. Langevin[1]) and A. A. Sonin[1,2])

[1]) Laboratoire de Physique,
   Ecole Normale Supérieure, Paris, France
[2]) Institute of Crystallography,
   Russian Acad. Sci., Moscow, Russia

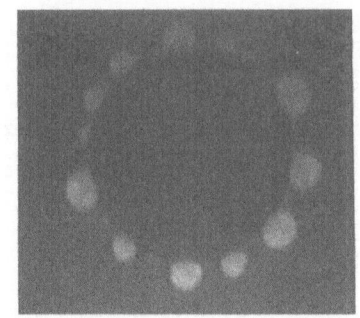

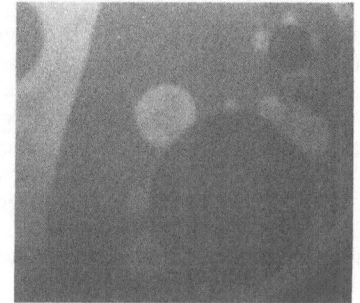

Key words: Stratification — bubble-form instabilities — structural memory

The stratification (layer-by-layer thinning) [1] of the thin circular horizontal soap films formed from micellar 0.1 M/l water solution of DTAB surfactant was studied experimentally by means of the interferometric technique and visual microscopic observations. Several groups of metastable circular stratification domains with different thicknesses were observed. The transfer between the two thickest groups of these domains led to the gradual change in the film thickness h (an analogue to the second order phase transition), while the formation of the two thinnest groups of domains led to the abrupt change in h (an analogue to the first order phase transition). The so-called bubble-form instabilities always appeared at the borders of the thinnest domains (Fig. 1a). It was shown that the stratification domains and bubble instabilities formed the fractal Apollonian nets [2] at the film surface (Fig. 1b). The structural memory effect (the thinner film preservation at the background of the thicker film) was discovered for the black Newton films (Fig. 1c). The following time dependence for a diameter of domains without bubble instabilities (and memory-domains) was observed: $D \sim t^{1/2}$ (the same one, as had been previously described in [3], while for domains with bubble instabilities: $D \sim t$. A theoretical model to explain this behavior, is proposed.

Fig. 1. a) The thinnest stratification domain (black film), expanding over the film surface. The film thickness inside the domain is about 7.5 nm, outside-25 nm. b) The fractal Apollonian net formed by the stratification domains and bubble instabilities. c) The structural memory effect for the black Newton film (shown by the arrow)

References

1. Ivanov IB, Dimitrov DS (1988) In: Ivanov IB (ed) Thin Liquid Films. v 29. Marcel Dekker, New York, pp 379—496
2. Mandelbrot BB (1982) The Fractal Geometry of Nature. W. H. Freeman, San Francisco
3. Kralchevsky PA, Nikolov AD, Wasan DT, Ivanov IB (1990) Langmuir 6:1180—1189

Authors' address:

D. Langevin
Laboratoire de Physique Statistique
Ecole Normale Superieure
24, rue Lhomond
F-75231 Paris Cedex 05, France

Progress in Colloid & Polymer Science

Progr Colloid Polym Sci 93:358 (1993)

Latex aggregation in two dimensions

J. Stankiewicz[1,2], M. A. Cabrerizo Vílchez[1],
R. Hidalgo Alvarez[1], and F. Martínez López[1]

[1] Departamento de Física Aplicada,
Universidad de Granada, Spain
[2] Centro de Física,
Instituto Venezolano de Investigaciones Científicas,
Caracas, Venezuela

Abstract: We studied the aggregation of polystyrene latex in two dimensions. We observed under microscope how salt-induced aggregation of both monomers and clusters' fractal dimension increases with increasing aggregation time. However, it does not depend on the salt concentration in the range studied (0.2—1.0 mol dm^{-3} of NaCl). The question of cluster restructuring is invoked.

Key words: Fractals — polystyrene latex — two-dimensional aggregation

Introduction

Colloidal aggregation is important in many physical, chemical and biological processes. There seem to exist two limiting regimes for it [1]: fast *diffusion* limited cluster aggregation (DLCA) and slow *reaction* limited cluster aggregation (RLCA). Fractal morphology as well as aggregation kinetics are specific to each regime. These two regimes are believed to be universal (independent of the detailed nature of a colloid). Recent papers [2, 3] have shown that, indeed, various systems (colloidal gold, silica and polystyrene) do exhibit such universal behavior.

In DLCA, the fractal dimension D is about 1.45 (1.80) in $d = 2$ ($d = 3$), as obtained by numerical simulations [4, 5]. In RLCA, repulsive forces are not negligible. Much slower rates of aggregation and more compact structures result; $D \cong 1.6$ (2.10) in $d = 2$ ($d = 3$) [6]. The experimentally found fractal dimensions are in good agreement with the numerical results, although there are important unexplained deviations [7, 8].

The number of theoretical predictions and computer simulations for colloid aggregation far exceed the number of experimental results. In $d = 2$, values of D in the range 1.2—1.74 have been obtained experimentally [9—11]. A value of $D = 1.20 \pm 0.15$, found for aggregation of silica particles on an air-water interface [11], is much smaller than values obtained from the DLCA models. This has inspired calculations involving cluster-cluster interactions [12, 13] and long-range interactions between particles [14]. These models yield smaller values of D in $d = 2$; however, there is no clear picture of two-dimensional aggregation thus far.

In this paper, we report results of a two-dimensional aggregation experiment. We have observed, through a phase-contrast microscope, salt-induced aggregation of polystyrene particles on an air-water interface. The fractal dimension has been determined for various electrolyte concentrations. We find that D, obtained in two independent ways, increases as aggregation proceeds. This behavior has been found for all salt concentrations used (0.2—1 mol dm^{-3}).

Experiment

We have studied the behavior of a monodisperse sulfonated polystyrene colloid spread onto a surface of electrolyte solution. The colloid suspensions were made of freshly cleaned polystyrene spheres. The spheres had a diameter of 178 ± 8 nm and their surface charge density was —(17 µC/cm^2) [15]. We used spreading techniques [11, 16] to prepare samples. A small amount of latex in 20% aqueous methanol was carefully dispensed from a microsyringe onto a flat surface of the electrolyte solution. The monomer concentration was about 1×10^7 particles/cm^2. Sodium chloride (NaCl) was added to the Millipore Milli-Q filtered water to induce aggregation. Salt ions cut down the Debye-Hückel screening length, thus lowering the repulsive barrier between charged latex particles [17]. The electrolyte concentration was varied from 0.2 to 1.0 mol dm^{-3} in order to cover diverse regimes of aggregation. To assure DLCA conditions, hydrochloric acid (HCl) was added to a concentration of 1.5 M in two runs.

After evaporation of solvent (usually 10 min) optical observations and photographs were made as aggregation proceeded. We used a phase contrast microscope with a magnification of 400X. Single particles and cluster-cluster aggregation could be seen. The structures formed were digitized from photographs. Equal-mass segments were judged by eye (about five particles per segment). This procedure can introduce systematic errors; however, it does not appreciably affect the scaling relations used to determine the fractal dimension.

Results and discussion

Two fractal dimensions of the aggregates were determined: 1) from the relation $m \sim R_g^{D_1}$, where R_g is the radius of gyration of clusters and m is the cluster mass (number of equal-mass segments counted in the cluster); 2) D_2 is determined from $N(\varepsilon) \sim \varepsilon^{-D_2}$, where $N(\varepsilon)$ is the number of square non-overlapping boxes of side ε needed to cover the *largest clusters observed* (at different aggregation times), within the field of view.

A plot of the logarithm of R_g versus the logarithm of m is shown in Fig. 1 for 79 clusters observed in one of the experimental runs. The slope of the power law, which is followed for more than two decades in mass, gives $1/D_1$. The fractal dimension D_1 as a function of the mean cluster size (mass) $S(t)$ defined by $S(t) = \Sigma m^2 \cdot n_m(t)/\Sigma m \cdot n_m(t)$, where $n_m(t)$ is the number of clusters of mass m at time t, is exhibited in Fig. 2 for various electrolyte concentrations. The data shown is for various stages of aggregation between 10 min and 3 h. For large mean cluster masses, $D_1 \cong 1.48$. However, we believe that D_1 can be larger for larger values of $S(t)$ since it shows a tendency to increase with increasing $S(t)$. A value of $D_1 \cong 1.55 \pm 0.03$ was obtained for a fast aggregation run (1.5 M) after 30 minutes.

Values of D_2 found at various stages of aggregation (between 10 minutes and 17 hours) are shown in Fig. 3. D_2 increases with increasing cluster mass m: from a value of about 1.1 for the smallest clusters ($m \cong 10^2$) to a value of about 1.6 for the largest aggregates observed ($m \cong 3 \times 10^4$). The latter value is in agreement with the value obtained in two-dimensional RLCA simulations [6].

The fractal structures studied are insensitive to electrolyte concentration. This is a surprising result since we had expected to cover different regimes of aggregation by varying salt concentration. Considering only electrostatic interactions, we would expect a variation in the fractal dimension of aggregates formed at different electrolyte concentrations. No such variation takes place however. Furthermore, we note that both D_1 and D_2 increase with increasing aggregation time. It seems that D_2 (and perhaps D_1) tends to a limiting value of about 1.6. To discuss these results we invoke the idea of "restructuring" [19–21] and anisotropic repulsive interactions [11].

We assume that latex particles are trapped by surface tension forces and that they float on a com-

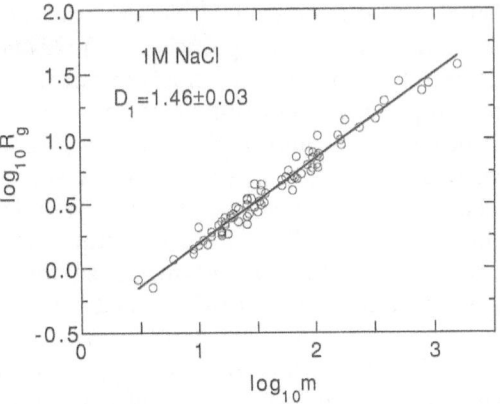

Fig. 1. A plot of radius of gyration R_g versus cluster mass m in log-log scale (m is the number of equal-mass segments counted in the cluster). The inverse of the slope gives the fractal dimension D_1

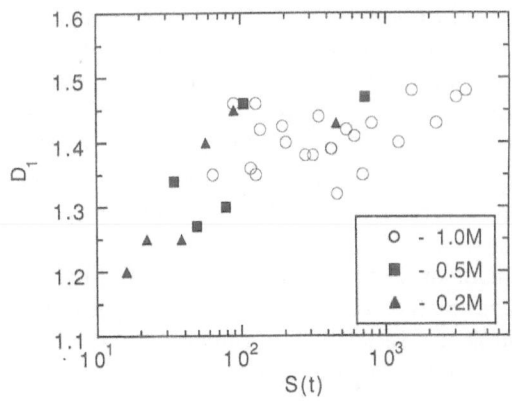

Fig. 2. Variation of the fractal dimension D_1, obtained from the relation between the radius of gyration and the cluster mass, with the mean cluster size $S(t)$ for three different electrolyte concentrations

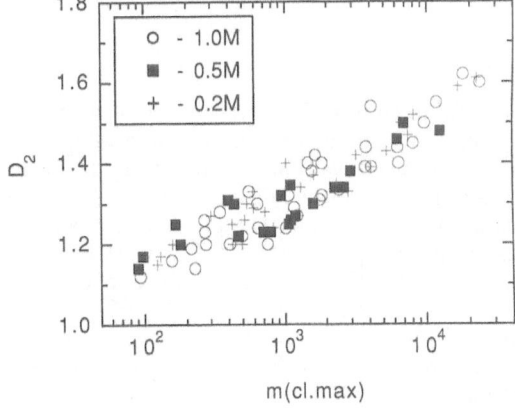

Fig. 3. Variation of the Hausdorff-Besicovitch fractal dimension D_2 with cluster mass for three different electrolyte concentrations

pletely flat air-water interface [18]. Deformations in the water surface around the particles are possible but their effect seems to be unimportant [18]. Anisotropic repulsive interactions favoring end-on approach of clusters would lead to less branching and lower D of clusters as observed at early stages of aggregation [11]. Cluster polarizations may have a similar effect on D as shown by numerical simulations [12, 13]. In our system, these interactions are expected to be insensitive to salt concentration.

The idea of restructuring of fractal aggregates would explain the observed variation of D with m. It is related to not fully irreversible bond formation [19, 20] and consequent rearrangement of particles into more compact structures. Restructuring effects have been reported in gold and silica colloids [19, 21] in $d = 3$. In addition, an appreciable increment of the fractal dimension with time has been found in numerical simulations of a cluster-cluster reversible aggregation model [20]. Interestingly, restructuring would yield aggregates that have about the same fractal dimension as RLCA grown clusters. To our knowledge, restructuring effects had not been observed in two-dimensional systems to date.

In conclusion, our results on two-dimensional colloid aggregation show that anisotropic repulsive interactions and restructuring phenomena could be important in this process. The question of electrolyte dependence and competition between these two mechanisms arises. It should be mentioned that the aggregation kinetics does depend on salt concentration in our experiments [22].

References

1. Lin MY, Lindsay HM, Weitz DA, Ball RC, Klein R, Meakin P (1989) Nature 339:360
2. Lin MY, Lindsay HM, Weitz DA, Ball RC, Klein R, Meakin P (1989) Proc R Soc London Ser A 423:71
3. Lin MY, Lindsay HM, Weitz DA, Ball RC, Klein R, Meakin P (1990) Phys Rev A 41:2005
4. Meakin P (1983) Phys Rev Lett 51:1119
5. Meakin P (1989) In: Avnir D (ed) The Fractal Approach to Heterogeneous Chemistry. John Wiley & Sons Ltd., New York, pp 140—144
6. Meakin P (1988) Phys Rev A 38:4799
7. Carpineti M, Ferri F, Giglio M, Paganini E, Perini U (1990) Phys Rev A 42:7347
8. Zhou Z, Chu B (1991) Physica A 177:93
9. Richetti P, Prost J, Burois P (1984) J Phys Lett 45:1137
10. Skjeltorp T (1987) Phys Rev Lett 58:1444
11. Hurd AJ, Schaefer DW (1985) Phys Rev Lett 54:1043
12. Jullien R (1985) Phys Rev Lett 55:1697
13. Jullien R (1986) J Phys A 19:2129
14. Meakin P, Muthukumar M (1989) J Chem Phys 91:3212
15. Bastos González D (1992) Memoria de Licenciatura, Departamento de Física Aplicada, Universidad de Granada
16. Goodwin JW, Ottewill RH, Parentich A (1980) J Phys Chem 84:1580
17. Lykema J (1987) In Tadros ThF (ed) Solid/Liquid Dispersions. Academic Press, London, p 79
18. Pieranski P (1980) Phys Rev Lett 45:569
19. Aubert C, Cannell DS (1986) Phys Rev Lett 56:738
20. Shih WY, Aksay IA, Kikuchi R (1987) Phys Rev A 36:5015
21. Dimon P, Sinha SK, Weitz DA, Safinya CR, Smith GS, Varday WA, Lindsay HM (1986) Phys Rev Lett 57:595
22. Stankiewicz J, Cabrerizo-Vílchez MA, Hidalgo-Alvarez R (1992) Phys Rev E 47:2663

Acknowedgement

We are grateful to Dr. J. F. Fernández for discussions, encouragement, and comments on the manuscript. All polystyrene latex used was kindly provided by D. Bastos González and D. F. Javier de Nieves. We thank J.Díaz for letting us use his optical equipment and Dr. J. Marro for letting us use his computer facilities. This work was supported in part by the Comisión Interministerial de Ciencia y Tecnología (Spain), project # MAT90-0695-C02-01. One of us (J. S.) acknowledges with thanks financial support from Departamento de Fíisica Aplicada of Universidad de Granada, Junta de Andalucía (Spain), Fundación Gran Mariscal de Ayacucho (Venezuela) and Consejo Nacional de Investigaciones Científicas y Tecnológicas of Venezuela for a sabbatical stay at Universidad de Granada.

Authors' address:

Roque Hidalgo-Alvarez
Biocolloids and Fluid Physics Group
Department of Applied Physics
University of Granada
Campus Fuentenueva
18071 Granada, Spain

Aggregation kinetics in β-FeOOH-potassium oleate system

K. Starchev, V. Peikov, S. P. Stoylov, and I. Petkanchin

Institute of Physical Chemistry,
Bulgarian Academy of Sciences, Sofia, Bulgaria

Electric light scattering measurements of aggregating β-FeOOH particles are reported. The aggregation is induced by adsorption of potassium oleate at the surface of the particles. A power law correlation with the aggregates length and time of aggregation is obtained. The power coefficients are independent of the type of measured average length, which could be related with the self similarity of the aggregates. The fractal dimension determined from these experiments is between 1.7 and 1.9.

Acknowedgement

This work has been supported by the Bulgarian Ministry of Education and Science (Grant No 133/1991).

Authors' address:

V. Peikov
Institute of Physical Chemistry
Bulgarianb Academy of Science
Acad bonchev str. Bl. 21
BG-1113 Sofia, Bulgaria

Electro-optics of semidilute dispersions separation of electric and optic interactions

M. Stoimenova and V. Peikov

Institute of Physical Chemistry,
Bulgarian Academy of Sciences, Sofia, Bulgaria

The application of a new method of treatment of the electro-optic data (Stoimenova, Peikov, JCIS submitted) to the electro-optic study of semidilute dispersions allows the concentration interval in which purely optical interparticle interactions modify the electro-optic responses to be separated and the initial concentrations at which the electric interactions interfere with the electro-optic behaviour to be found. Electro-optic parameters, sensitive to particle electric interactions, are the relaxation frequency of particle electric polarizability, particle hydrodynamic relaxation time, as well as the normalized dependences of electric polarizability on the ionic strength of the medium.

Authors' address:

V. Peikov
Institute of Physical Chemistry
Bulgarian Academy of Sciences
Acad. Bonchev str. Bl. 21
BG-1113 Sofia, Bulgaria

Phase transition and redox change in monolayer films of methylene blue

V. Svetličić[1]), V. Mikac-Dadić[2]), and J. Lugarić[1])

[1]) Rudjer Bošković Institute, Zagreb, Croatia
[2]) Faculty of Printing, University of Zagreb, Croatia

Key words: Phase transition — monolayer film — methylene blue — cyclic voltammetry

Properties of electroactive organic molecules at surfaces can be characterized by purely electrochemical methodology. The redox potential and form of the signal in cyclic voltammetry carry the information on the interaction of a molecule with the substrate (electrode) and between the molecules themselves (supramolecular interactions) for each redox state involved [1—3].

At aqueous solution/electrode interface methylene blue (MB^+) undergoes fast and reversible redox change:

The interaction of the MB$^+$ molecules with the surface of Hg electrode is so strong that the redox signal of the inner monolayer is well separated from the signal of overlaying molecules (Fig. 1) as in the case of underpotential deposition of metallic phases.

The oxidation in the multilayer film (up to 200 equivalent monolayers of LMB) is fast and reversible, and occurs in a very narrow potential range, as it does for metallic phases. The electronic conductivity within the film is due to the formation of a MB$^+$/LMB charge-transfer complex.

The MB$^+$ molecules are flat oriented in the inner monolayer and their lateral interactions are small. The overlaying film grows as a three-dimensional (3D) phase.

NO$_3^-$ induces drastic changes in the organization of the inner monolayer. The phase transition during the redox change in the inner monolayer is analyzed according to BFT model for 2D electrocrystallization. The kinetics of the redox change fits the model for two-dimensional (2D) progressive nucleation and growth.

It is most likely that the presence of NO$_3^-$ induces edge-on orientation of MB$^+$ molecules where lateral interactions are important. This is reflected in the growth of the overlaying film which proceeds layer by layer.

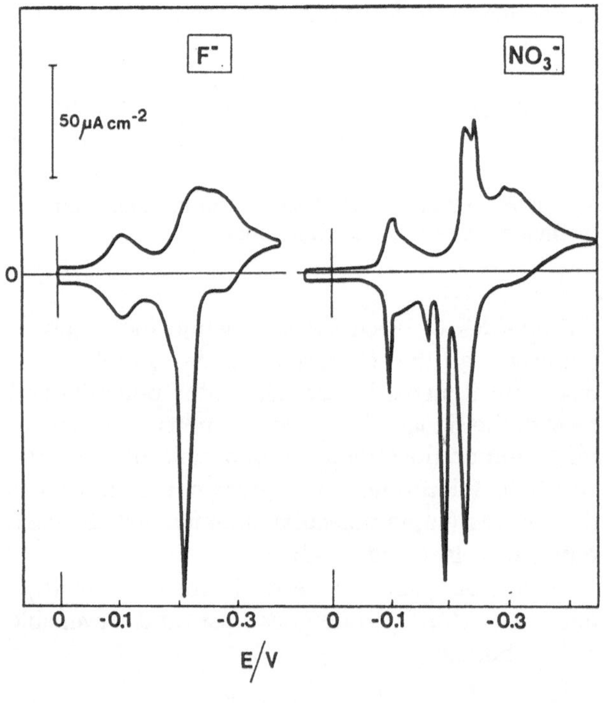

References

1. Soriaga MP (ed) (1988) Electrochemical Surface Science: Molecular Phenomena at Electrode Surfaces. ACS Symposium Series No 378
2. Žutić V, Svetličić V, Clavilier J, Chevalet J (1987) J Electroanal Chem 219:183—195
3. Clavilier J, Svetličić V, Žutić V, Ruščić B, Chevalet J (1988) J Electroanal Chem 250:427—442

Authors' address:

Vesna Svetličić
Rudjer Bošković Institute
POB 1016
41001 Zagreb, Croatia

Fig. 1. Cyclic voltammograms of the entire organic film recorded in 10^{-4} M MB$^+$. $\Gamma_{MB^+} = 3.26 \times 10^{-10}$ mol/cm^2 in 1 M NaF; $\Gamma_{MB^+} = 4.26 \times 10^{-10}$ mol/cm^2 in 1 M KNO$_3$

Self-assembled monolayers of phenothiazines at a Pt-S surface

V. Svetličić[1], J. Clavilier[2], V. Žutić[1],
J. Chevalet[3], and K. El Achi[2]

[1]) Rudjer Bošković Institute, Zagreb, Croatia
[2]) Laboratoire d'Electrochimie Interfaciale du C.N.R.S., Meudon, France
[3]) Laboratoire d'Electrochimie, Université P. et M. Curie, UA 430 C.N.R.S., Paris, France

Key words: Self-assembled monolayers — thionine — phenothiazines — sulphur modified platinum

Spontaneous adsorption of molecules onto a solid surface to form a film consisting of one monolayer of molecules is referred to as the self-assembly of a monolayer. When self-assembled monolayers (SAMs) are composed of reversibly electroactive molecules, the electrochemical techniques may allow a precise determination of surface concentration and an average surface area per molecule. These are commonly determined by integration of voltammetric peaks [1]. The peak width reflects the heterogeneity of surface redox sites when a phase transition is not the limiting step [2].

There is evidence in the literature [3] that the orientation and reactivity of thiophenols chemisorbed on polycrystalline platinum depend dramatically upon their surface concentration. Reversible 2e redox reaction is displayed only by molecules chemisorbed through the sulphur atom when a direct interaction between the aromatic moiety and platinum is blocked. The redox potential was the same as for the unadsorbed molecule. Modification of the surface of polyoriented platinum electrode [4] by sulphur induces a strik-

ing change in the electroactivity of SAMs composed of sulphur containing heteroaromatic molecules such as phenothiazine derivatives. The effect of S-adlayer preparation on the voltammetric behavior of SAMs of thionine is studied by varying S-adatom concentration (Γ_S) at the polyoriented platinum electrode from 0 to the sturation coverage.

Thionine forms electroactive SAMs at sulphur precovered platinum already at submonolayer concentratons of S-adatoms. At bare Pt these molecules are irreversibly adsorbed but are electroinactive [5]. Sudden transition from electroinactive to electroactive thionine monolayer occurs after a critical concentration of S-adatoms is reached. The S-adlayer decreases the interaction of heteroaromatic molecules with platinum so that all molecules become reversibly electroactive [6]:

The reduction potential is 90 mV more negative than the reaction of unadsorbed molecules due to S-S bond formation.

References

1. Soriaga MP, Binamira-Soriaga E, Hubbard AT, Benzinger JB, Pang KWP (1985) Inorg Chem 24:65—73
2. Abruña HD (1988) Coord Chem Rev 86:135—147
3. Mebrahtu T, Berry GM, Bravo BG, Michelhaugh SL, Soriaga MP (1988) Langmuir 4:1147—1151
4. Clavilier J (1980) J Electroanal Chem 107:211—216
5. Clavilier J, Svetličić V (1992) J Electroanal Chem 322:405—409
6. Svetličić V, Clavilier J, Žutić V, Chevalet J (1991) J Electroanal Chem 312:205—218

Authors' address:

Vesna Svetličić
Rudjer Bošković Institute
POB 1016
41001 Zagreb, Croatia

Self-association behavior of DDAB in diluted aqueous solutions

T. Svitova, N. Berezina, E. Zhuravleva, and V. Lobanova

Institut of Physical Chemistry, the Russian Academy of Sciences, Moscow, Russia

The aggregation behavior of the surfactant didodecyl dimethyl ammonium bromide (DDAB) in diluted aqueous solution in concentration range from 10^{-7} to 10^{-2} M was studied using static light scattering, densitometry, freeze-fracture electron microscopy, and tensiometry. It was found that DDAB forms stable, small, unilamellar vesicles with radius ca. 100 Å in diluted aqueous solutions without solvent extraction or sonication, but after consecutive dilution of the solution $2 \cdot 10^{-2}$ M. The phase diagram of DDAB/water system at the temperature range 25—60°C was obtained, and it was found that there are four areas on the phase diagram: 1) molecular solutions; 2) vesicular solutions with small unilamellar vesicles; 3) biphasic region, when two populations of particles coexist — small vesicles and large multilayer ones. 4) Region of random lamellar L_3 phase. Surface and interfacial tension measurements have demonstrated that vesiculation in bulk DDAB solutions results in a rapid decrease of surface and interfacial tension at the solution/air and solution/octane interfaces. It was observed that L_3 phase forms the surfactant phase at the water/octane interface with ultralow interfacial tension about 0.015 mN $\cdot$ m^{-1}.

Authors' address:

T. Svitova
Institute of Physical Chemistry
Russian Academy of Sciences
Leninskii pr. 31
Moscow, Russia

Multicomponent vesicular aggregates from perfluorinated single-chain surfactants

S. Szönyi[1]) and H. J. Watzke[2])

[1]) Laboratoire de Chimie Organique du Fluor,
   Université de Nice, Nice-Sophia Antipolis, France
[2]) Institut für Polymere, ETH-Zürich,
   Zürich, Switzerland

Key words: Self-vesiculation — multicomponent vesicles — perfluoroalkylated amphiphiles — single-chain micellar mixtures

Perfluorinated single-chain amphiphiles have surface active properties distinctly different from hydrocarbon surfactants. A higher critical micellar concentration is generally observed compared to hydrocarbon amphiphiles which shows that one CF$_2$-group equals two methylene groups in alkyl chain. Perfluorinated surfactants are also different in their phase behavior. Their first liquid crystalline phase is a lamellar phase. These properties should inforce the ability to form stable bilayer membranes. However, vesicles are usually produced from double-chain amphiphiles by forced dispersions (e.g., ultrasonication). The high chain melting temperatures of double-chain perfluoro surfactant restrict a wider appreciation of their bilayer forming properties.

Recently, it has been shown that mixtures of oppositely charged single-chain surfactants can form vesicular aggregates under proper conditions [1, 3]. The strong headgroup interactions between these ionic surfactants lead to a strong non-ideal behavior. Mixed micelles may undergo spontaneous reorganization to vesicles at a certain composition when the mixed system cannot form a micelle. We could show that this behavior is also true for single-chain perfluorinated surfactant mixtures when the headgroups are oppositely charged [2].

In this report, we introduce a new route to synthesize a variety of single-chain perfluoro amphiphiles with interactive headgroups. By coupling of suitable headgroups (cationic, anionic, nonionic and zwitterionic) to alkyl oxiran precursors, we can easily obtain the interactive surfactants which showed to be able to form unilamellar vesicles.

Employing electronmicroscopic bare grid techniques (cryoelectronmicroscopy) clear evidence can be obtained about the unilamellar nature of the aggregates down to the smallest diameters ($\approx$ 15 nm) (Fig. 1).

The spontaneously formed vesicles exhibit strong polydispersity depending on the mixed ratios of cationic and anionic components (some mixtures were found to exhibit a bimodal size distribution). The amount of multilamellar vesicles is small. Their appearance in electronmicroscopic investigations is comparable to that of phospholipid liposomes and vesicles prepared from synthetic amphiphiles. Mixtures of fluorocarbon and hydrocarbon amphiphiles can be achieved which also exhibt self-vesiculation to unilamellar vesicles. They provide the means to introduce a variety of functionalized amphiphilic

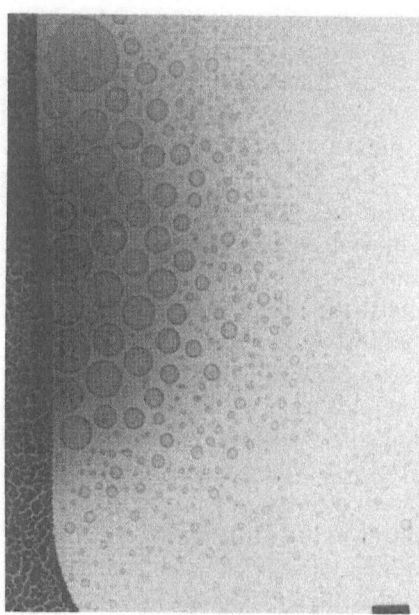

Fig. 1. Electronmicrograph of a cationic-anionic surfactant mixture of single-chain perfluorinated amphiphiles obtained by cryo-electronmicroscopy (bare grid technique). The micrograph shows a polydispersed population of unilamellar vesicles. The aqueous sample was, without further treatment, plunged into liquid propane, transferred to a cold-stage electronmicroscope by cryo-transfer, and inspected. The bar length is 100 nm

systems in closed-shell bilayer structures where the alkylperfluoro components function as membrane stabilizing matrix. Asymmetrical chain length mixtures lead to strong polydispersed size distributions. Occasionally observed high concentration of the vesicular aggregates at the pore border indicates that the vesicles are very stable against fusion. They adopt shape changes due to vesicle-vesicle interactions.

Multicomponent perfluoro single-chain mixtures are a versatile vesiculation system which provides a novel tool to investigate various membrane phenomena. It also provides a rich set of building blocks for vesicular applications.

References

1. a) Kaler EW, Kamalakara Murthy A, Rodriguez BE, Zasadzinski JAN (1989) Science (Washington) 245:1371. b) Kaler EW, Kamalakara Murthy A, Rodriguez BE, Zasadzinski JAN (1992) J Phys Chem
2. Szönyi S, Cambon A, Watzke HJ, Schurtenberger P, Wehrli E (1992) In: Structure and conformation of amphiphilic membranes. Springer proceedings in physics vol 66, Springer-Verlag, Berlin p 198
3. Ambühl M, Bangerter F, Luisi PL, Skrabal P, Watzke HJ (1993) Langmuir 9:36

Authors' address:

Dr. Stephane Szönyi
Laboratoire de Chimie Organique due Fluor
Universite de Nice-Sophia Antipolis
Parc Valrose
F-06108 Nice Cedex 2, France

Neutron-scattering studies of water confined in reversed micelles

E. Bardez[1]), R. Giordano[2]), P. Migliardo[2]), and C. Vasi[3])

[1]) Laboratoire de Chemie Generale, C.N.A.M., Paris, France
[2]) Dipartimento di Fisica, Universita' di Messina, Messina, Italy
[3]) Istituto di Tecniche Spettroscopiche del CNR, Messina, Italy

It is well known that the confinement of water, implying some volume and/or surface interactions of the H_2O molecules with the substrate, into nanoscopic volume (such as reversed micelles) does not allow the H-bond forces to organize typical bulk geometries and structures due to the lowering of the liquid's degrees of freedom.

Neutron-scattering studies [1] allow to extract useful information on the shapes and on the structures (SANS) of droplets, as well as to get a better insight on the diffusive motion of the water protons. The most relevant results of a series of measurements, recently performed at the Loboratoire Leon Brillouin (Saclay) with the PAXE' (SANS) and MIBEMOL (IQENS) spectrometers on the ternary system H_2O/metal-bis(2-ethylhexyl)-sulfosuccinate, metal being Ca^{2+}, Zn^{2+}, Al^{3+}/oil (cyclohexane), can be summarized in the following points, namely:

i) The form factor $F(Q)$ in the micellar growth process depends on the counter-ion, giving rise to spherical reversed micelles in the case of Ca^{2+}, spheres (at low R values, R being the water to surfactant molar ratio) and ellipsoids (for $R > 4$) in the case of Zn^{2+}, spheres (at low R values) and prolate ellipsoids and/or cylinders (for $R > 5$) in the case of Al^{3+}.

ii) The IQENS contribution shows a considerable narrowing of the translational diffusive term, indicating the presence of the Elastic Incoherent Structure Factor (EISF) $A_0(Q)$ typical of the diffusion in restricted geometries. From the extracted $A_0(Q)$, in the case of Ca^{2+} and Zn^{2+} counter-ions, a near spherical form factor of ~3 Å of radius is obtained that corresponds only to the sphere on which the protons of the "interstitial" water can move. Such a kind of water reveals a strong polyrization effect of the O-H stretching band [2] that noticeably differs respect to that of bulk water.

References

1. Migliardo P (1993) Relaxational and Dynamical Properties of Confined H-bonded Liquid systems (to be published in J Phys Cond Matt and references therein)
2. Aliotta F, Migliardo P, Donato DI, Turco-Liveri V, Bardez E, Larrey B (1992) Progr Coll Polym Sci 89: 258—262

Authors' address:

Prof. P. Migliardo
Dipartimento di Fisica
Universita' di Messina
P.O. Box 55
98166 S. Agata (Messina), Italy

Anisotropic light scattering in water polymeric solutions and gels

F. Mallamace[1]), N. Micali[1]), C. Vasi[1]), S. Trusso[1]),
R. Bansil[2], and S. Pajevic[2])

[1]) Istituto di Tecniche Spettroscopiche del C.N.R.,
   Messina, Italy
[2]) Center for Polymer Studies, Department of Physics,
   Boston University, USA

Depolarized Rayleigh-Wing scattering measurements in polymeric aqueous solutions and gels are reported. In particular, the studied system is a polyacrylamide gel, with water as solvent, below and above the sol-gel transition. This transition is obtained by varying, appropriately, the contents of the crosslinking molecules (Bis). All obtained spectra are due to the presence of two distinct dynamical contributions. The large one represents the dynamic of the water molecules hydrogen bond (HB), whereas the narrow one can be connected to the polymeric network. Both these contributions show different behaviors above and below the percolation transition, obtained in the present experiment by changing (at constant temperature) the Bis content in the range 0—2% of the entire quantity of the polymer. The water concentration is 97.5%. Concern the H-bond dynamics, we find that: in the sol region, it behaves in the same way as bulk water, whereas in the gel region we observe a faster dynamics due to the presence of the spanning network. Finally, we show how the presence of this stable structural environment influences the dynamic network of water.

Authors' address:

Dr. C. Vasi
Istituto di Tecniche Spettroscopiche del CNR
98166 Vill. S. Agata
C.P. 55
Messina, Italy

Mechanism of formation of two-dimensional crystals from latex particles on substrata

O. D. Velev[1]), N. D. Denkov[1]),
P. A. Kralchevsky[1]), I. B. Ivanov[1]),
H. Yoshimura[2]), and K. Nagayama[2])

[1]) Laboratory of Thermodynamics
   and Physico-chemical Hydrodynamics,
   University of Sofia
[2]) Protein Array Project, ERATO, JRDC, Tsukuba,
   Japan

The dynamics of two-dimensional ordering of micron-size polystyrene latex spheres on a horizontal glass substrate has been directly observed by means of optical microscopy. It turns out that the ordering starts when the thickness of the water layer containing the particles becomes approximately equal to the particle diameter. By variation of the electrolyte concentration, the charge of the particles and their volume fraction, it is proven that neither the electrostatic repulsion nor the van der Waals attraction between the particles are responsible for the formation of the 2D crystals.

The observations revealed that two main factors are responsible for the ordering. *Attractive capillary forces* (due to menisci formed around the particles) are governing the formation of the nuclei of the ordered phase. The growth of the crystals afterwards proceeds due to the *convective transport* of particles towards the ordered regions. Experiments under controlled crystal growth have been carried out, allowing obtaining of either well ordered monolayers or domains consisting of multilayers (bilayers, trilayers, etc.).

Authors' address:

O. D. Velev
Laboratory of Thermodynamics
Faculty of Chemistry
University of Sofia
J. Boucher Ave. 1
BG-1126 Sofia, Bulgaria

Mercury — a model substance for flotation basic research

K. Volke

KAI e.V. Gruppe Grenzflächenprozesse
Freiberg (Sachs.), Germany

Key words: Mercury — flotation — adsorption — polarography — induction time

1. Introduction

Starting with the research works of Henry (1775—1836), the mercury surface has been a subject of investigation for about 200 years.

In 1957, the Polish electrochemist Pomianowski developed an apparatus by which the model substance mercury could be floated for the first time.

2. Mercury as a model substance

In mineral processing flotation is used only in a few cases for the production of mercury.

Skrylev et al. [1] separated colloidal mercury from waste water by means of aliphatic amines in a flotation machine. Stepanov and Certykovoev [2] flotated mercury by means of mercaptanes and xanthates to separate it from the mercury residues (stupp).

Mercury, therefore, is of great importance as a model substance for the investigation of the flotation microprocesses, although it is seldom flotated. The reason lies in the following advantages over solids, e.g., oxides:

— it is easy to handle in terms of measuring technology, especially in adsorption measurements;
— due to the continuous renewal of its surface (drop electrode) it enables a good reproducibility of the surface processes;
— due to its surface properties marked electrosorption effects are possible.
— influencing the adsorption layer by applying certain potentials is, in principle, possible.
— the surface tensions solid/liquid and solid/gas are readily measurable or known.

A disadvantage is the homogeneous and smooth surface in contrast to solids.

3. Flotation and polarography

Polarography on the one hand and the model substance mercury on the other hand can help in solving the following problems:

— determination of adsorption parameters;
— determination of induction time;
— determination of the concentration of flotation reagents.

The following two collectors were investigated: n-dodecylammonium chloride (DAC) and n-dodecanoic acid (LA). The recovery-pH-curves show a maximum near pK. The flotation activity of DAC is higher than that of LA [3].

The AC-polarographic adsorption investigations show

— extreme values of the parameters interaction coefficient, surface area per molecule, and adsorption equilibrium time at pK;

— the formation of surface-active and flotation-active ion-molecule associates (1:1) at pK;

— that the adsorption parameters, however, are not sufficient for estimating the suitability of a collector [3].

4. Induction time measurements

For the determination of induction time τ_i at the approach of a liquid/gas interface on a mercury drop, an apparatus was developed. Similar investigations were carried out by Usui et al. [4, 5].

In presence of sodium dodecyl sulphate, Kaisheva et al. [5] pointed out that the potential of the electrode has a great influence on the induction time, because the surfactant molecules change the orientation. Own investigations were carried out with LA in a supporting electrolyte (0.01 N KF). τ_i depends on the sinking speed of the liquid surface, but not on the mercury drop size. The addition of LA causes a lowering of the τ_i-values.

References

1. Skrylev LD, Babinec SK, Puric AN (1982) Zh prikl Chim 55:2375
2. Stepanov BA, Certykovoev MV (1982) Cvet Met 1:197
3. Para G, Volke K, Pomianowski A, Pawlikowska-Czubak (1986) Colloid Polym Sci 264:260
4. Usui S, Sasaki H, Hasegawa F (1986) Coll Surf 18:53
5. Kaisheva M, Usui S, Dai Q (1988) Coll Surf 29:147

Authors' address:

Klaus Volke
KAI e.V., Gruppe Grenzflächenprozesse
Chemnitzer Str. 40
D-09599 Freiberg (Sachs.), FRG

Liquid permeability of bidisperse hard-sphere packings

D. M. E. Weesie and A. P. Philipse

Van 't Hoff Laboratory for Physical and Colloid Chemistry, Utrecht University, The Netherlands

Key words: Permeability — colloidal spheres — binary packings

The presented work is a quantitative study of flow in porous media, consisting of dense, random, macroscopically homogeneous packings of well-characterized, uncharged, bidisperse colloidal silica spheres.

The liquid permeability of these packings has been studied as a function of time, pressure, and the volume ratio of small to large spheres.

A semi-empirical (Kozeny-Carman) model predicts the permeation results fairly well. For monodisperse packings the liquid permeability agrees with exact calculations of Zick and Homsy for Stokes flow in an fcc-array of monodisperse spheres. We also expect agreement between the model and calculations for Stokes flow in binary sphere arrays.

Authors' address:

Dominique M. E. Weesie
Van 't Hoff Laboratory
for Physical and Colloid Chemistry
Utrecht University
Padualaan 8
NL-3584 CH Utrecht, The Netherlands

Characterization of drug-loaded human low-density lipoproteins

K. Westesen[1], A. Gerke[1], M. H. J. Koch[2], D. Svergun[2], M. Masquelier[3], S. Vitols[3], and C. Peterson[3]

[1] Institute of Pharmaceutical Technology, Technical University of Braunschweig, FRG
[2] European Molecular Biology Laboratory, c/o DESY Deutsches Elektronen Synchrotron, Hamburg, FRG
[3] Department of Clinical Pharmacology, Karolinska Hospital, Stockholm, Sweden

Physicochemical properties of isolated human and LDLs loaded with various anticancer drugs were compared at different temperatures. The results of chemical analysis are compatible with those found in the literature. While synchrotron radiation small-angle x-ray scattering (SRSAXS) gives particle sizes comparable to those in the literature, photon correlation spectroscopy (PCS) gives much smaller sizes. Analysis of SRSAXS curves indicates that the particle shape of LDLs deviates from spherical symmetry. The existence of a change of structural characteristics between 20° and 38°C was confirmed. ^{1}H NMR spectra indicated that the triglyceride molecules have some motional freedom, even in the unperturbed systems at 20°C. A pronounced pertubation of the lipid core at 20°C was only found for LDL heavily loaded with WB4291. In contrast, loading with Vincristin or AD32 did not result in a detectable perturbation of the structured core. ^{19}F NMR spectroscopy data suggested two chemical environments for AD32 molecules and their predominant location in the core.

Authors' address:

K. Westesen
Institute of Pharmaceutical Technology
Technical University of Braunschweig
Mendelssohnstr. 1
36106 Braunschweig, FRG

Characterization of phospholipid stabilized oil-in-water emulsions with special regard to parenteral emulsions

K. Westesen[1]), M. H. J. Koch[2]), and D. Svergun[2])

[1]) Institute of Pharmaceutical Technology,
   Technical University of Braunschweig,
   Braunschweig, FRG
[2]) European Molecular Biology Laboratory,
   c/o DESY Deutsches Elektronen Synchrotron,
   Hamburg, FRG

A concentration series of small unilamellar vesicles (SUVs) and a selection of i.v. emulsions were studied by synchrotron radiation small-angle x-ray scattering (SAXS). Between 0.25 and 3% phospholipids, the scattering of the SUV concentration series was proportional to the concentration of single bilayers. Most of the SAXS experiments were performed on centrifugation infranatants of the i.v. emulsions which still contained large numbers of small vesicles, whereas the electron density contrast was substantially improved due to the separa-

tion of the large emulsion droplets. The scattering curves of the infranatants exhibit the characteristic scattering of single phospholipid bilayers. In contrast to quantitative ^{31}P NMR experiments using paramagnetic shift reagents, SAXS data allow to detect the bilayer structures and to estimate their concentration independently of the amount of charged surface active components. The contents of bilayer structures found in the centrifugation infranatants differ for various clinically used and model i.v. emulsions.

Authors' address:

K. Westesen
Institute of Pharmaceutical Technology
Technical University of Braunschweig
Mendelssohnstr. 1
36106 Braunschweig, FRG

Photochemically induced phase transitions in lyotropic liquid crystalline surfactant systems

T. Wolff, B. Klaußner, and D. Nees

Physikalische Chemie, Universität Siegen

The phase transition temperatures from nematic or from hexagonal lyotropic liquid crystalline phases to the isotropic micellar phase (T_{NI} and T_{HI}, respectively) in surfactant — water — mixtures is sensitive to the addition of small amounts of certain aromatic compounds which cause an increase or decrease of transition temperatures by several degrees. Other aromatic compounds do not show this effect. In some cases, it is possible to photochemically transform compounds of the former class to compounds belonging to the latter. Consequently, isothermal phase transitions can be induced by exposing the sample to suitable light.

We have studied these effects in cationic micellar systems of cetyltrimethylammonium bromide (CTAB) [1—3] as well as in anionic systems of potassium myristate (KMy). As an example of a photoisomerizable solubilizate, we used 4-hydroxystilbene (4HS) which undergoes trans → cis isomerization upon irradiation. Transition temperatures T_{NI} (in the CTAB-system) and T_{HI} (in the KMy-system) were found to increase in the presence of solubilized trans-4HS. Photochemical isomerization leads to a further increase in both systems. A photochemically induced isothermal phase transition can thus be induced when systems in the isotropic phase are irradiated at temperatures close to T_{NI} or T_{HI}, respectively.

In the described, as well as in previous experiments, it turned out that longish isomers (such as trans-4HS) always induce transition temperatures exceeding those induced by more isometric isomers (such as cis-4HS). A study of a series of hydrocarbon solubilizates — differing in anisometry only — confirms this influence of solubilizate shapes: transition temperatures decrease gradually when 1,4-dicyclohexylbenzene (rod-like shape) as a solubilizate is exchanged by equimolar amounts of bicyclohexyl, trans-decaline, cyclooctane, cis-decaline, and adamantane (spherical shape).

References

1. Wolff T, Klaußner B, von Bünau G (1990) Progr Colloid Polym Sci 83:176—180
2. Wolff T, Seim D, Klaußner B (1991) Liq Cryst 9:839—847

Authors' address:

Prof. Dr. Thomas Wolff
Universität Siegen
Physikalische Chemie
57068 Siegen, FRG

Enzyme-induced percolation of w/o microemulsions

A. Xenakis[1], V. Papadimitriou[1], and P. Lianos[2]

[1] National Hellenic Research Foundation, Inst. Biological Research & Biotechnology, Athens, Greece
[2] Physics Section, School of Engineering, University of Patras, Greece

Abstract: Water-in-oil microemulsions based on 2-bis-ethylhexylsulfosuccinate sodium salt (AOT) containing chymotrypsin or trypsin have been studied by conductivity and luminescence decay measurements. In all cases, percolation processes occurred. When the enzymes were added, the percolation threshold shifted towards higher water content values. This shift depends on the enzyme concentration. The luminescence decay profiles of $Ru(bpy)_3^{2+}$ in the presence of quencher also depended on the enzyme presence and its concentration. A "percolation" model is proposed in order to discuss our results.

Key words: Microemulsions — reverse micelles — percolation — chymotrypsin — trypsin — AOT

Introduction

Water-in-oil microemulsions, sometimes called reverse micelles, are fine dispersions of water in a non-polar organic solvent stabilized by surfactant molecules. When the anionic surfactant AOT is used, the reverse micelles are spherical droplets surrounded by a monolayer of AOT molecules [1]. The size of the reverse micelles is directly related to the water content and can be expressed by the hydration ratio $w_0 = [H_2O]/[AOT]$.

Over recent years, it has been well established that many enzymes can be incorporated in the water droplets of a (w/o) microemulsion without loss of activity [2, 3]. Most studies have been directed at investigating changes in protein conformation and enzymatic activity [4].

In the present work, we have studied the behavior of w/o microemulsions containing two well studied proteolytic enzymes, α-chymotrypsin and trypsin [5, 6]. Electric conductivity and time-resolved luminescence spectroscopy were used in order to investigate the effect that these enzymes may have on the structure of the AOT reverse micelles [7]. For the latter method, we used $Ru(bpy)_3^{2+}$ as lumophore and $Fe(CN)_3^{3-}$ as quencher [8].

Materials and methods

Materials: α-chymotrypsin from bovine pancreas type II (Sigma), trypsin from bovine pancreas type III (Merck), as well as 2-bis-ethylhexylsulfosuccinate sodium salt (AOT) (Sigma), tris (2,2'-bipyridine) ruthenium dichloride hexahydrate, $Ru(bpy)_3^{2+}$ (GFS Chemicals) and potassium hexacyanoferrate (III) $K_3Fe(CN)_6$ (Merck) were used as supplied. All other chemicals were of the highest available purity and double distilled water was used throughout this study.

Preparation of microemulsions. A solution of 0.1 M AOT in isooctane was prepared, the water content of which was periodically checked by Karl Fischer titrations. Reverse micelles were formed by the addition of the appropriate amount of a concentrated solution of α-chymotrypsin or trypsin in 50 mM Tris/HCl pH 9 buffer sodium. Solubilization was achieved by gentle shaking within a few seconds. The desired value of w_0 was adjusted by the addition of a supplementary amount of buffer solution.

Conductivity measurements: The conductivity of the microemulsions was measured with a Metrohm 644 conductometer using a thermostated microcell. The cell constant, c, was equal to 0.97 cm^{-1}.

Luminescence decay measurements: Nanosecond decay profiles were recorded with the photon counting technique using a specially constructed hydrogen flash, ORTEG and Schlumberger-Enertec electronics and a Nucleus Multichannel scaler card with an IBM-PC. A Melles-Griot interference filter was used for excitation (450 nm) and a cutoff filter (600 nm) for emission. All samples were deoxygenated by the freeze-pump-thaw method. The decay profiles were recorded in 500 channels at 5 ns per channel and were analyzed by least-square fits using the distribution of the residuals and the autocorrelation relation function of the residuals as fitting criterion. The concentration of the lumophore was maintained at 10^{-5} M. All measurements were performed in thermostated cells at 35°C. The decay time of free $Ru(bpy)_3^{2+}$ then ranged around 530 ns. The quencher was $Fe(CN)_6^{3-}$ at a constant concentration equal to 5×10^{-4} M.

Results and discussion

Conductivity

The conductivity of AOT based w/o microemulsions was measured in the presence of various a-chymotrypsin or trypsin concentrations at constant temperature. Figure 1A shows the variation of the conductivity as a function of w_0, for different α-chymotrpysin concentrations [E]. A sharp increase of the conductivity appears at high w_0 values. This abrupt change is due to the increased probability of the reverse micelles to form large clusters allowing the exchange of the conducting species (surfactant counter-ions, buffer ions). This is attributed to the appearance of a percolation procedure of the distinct droplets [9, 10]. The percolation phenomenon is observed for all enzyme concentrations. However, the percolation threshold is shifted towards higher w_0 values when the α-chymotrpysin concentration increases (Fig. 1B). It is interesting to note the quite linear relationship between w_0 and [E]. On the other hand, the conductivity values decrease by increasing the enzyme concentration at constant w_0. The behavior was exactly the same when trypsin was added at various concentrations in the AOT microemulsions.

The presence of enzyme, either a-chymotrypsin or trypsin, seems to affect the mobility of the various conducting species. This could be explained by a possible localization of the buffer ions on the vicinity of specific enzyme sites, limiting the ex-

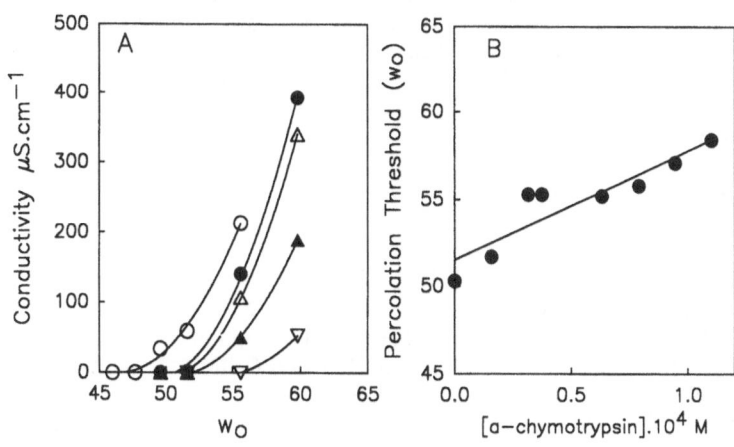

Fig. 1. A) Variation of the conductivity of AOT microemulsion systems containing α-chymotrypsin vs. w_0, for different enzyme concentrations: ○: 0.158 · 10^{-4} M; ●: 0.315 · 10^{-4} M; △: 0.375 · 10^{-4} M; ▲: 0.630 · 10^{-4} M; ▽: 1.10 · 10^{-4} M. B) Variation of the percolation threshold as a function of enzyme concentration

change between the droplets, and leading to a conductivity decrease. Such an assumption is justified by the luminescence data presented below.

Luminescence quenching data

The luminescence decay profiles of 10^{-5} M Ru(bpy)$_3^{2+}$ in the presence of 5×10^{-4} M Fe(CN)$_6^{3-}$ were analyzed with a percolation model given by the following equations [11].

$$I(t) = I_0 \exp(-k_0 t) \exp(-C_1 t^f + C_2 t^{2f}) \qquad (1)$$

$$K(t) = f C_1 t^{1-f} - 2 f C_2 t^{1-2f} , \quad (s^{-1}) \qquad (2)$$

where k_0, C_1 and C_2 are constants; f is a non-integer positive exponent of geometrical nature [11]. Smaller f means that the reaction domain is more restricted. f, C_1 and C_2 are calculated through a fitting procedure using Eq. (1) and they are used to calculate the time-dependent reaction rate $K(t)$ by means of Eq. (2). An idea of how $K(t)$ evolves with time can be given by tabulating the value of $K(t)$ at the beginning of time (K_1), at the end (K_L) and at the average over the total time range (K_{av}) [11]. Table 1 shows f and K values calculated for $w_0 = 25$, i.e. below the electric percolation threshold. f values show that for $[E] > 0.145 \times 10^{-4}$ M the

Table 1. Values of f, K_1, K_L, and K_{av} for the system with $w_0 = 25$ obtained by analyzing luminescence decay profiles of 10^{-5} M Ru(bpy)$_3^{2+}$ in the presence of 5×10^{-4} M Fe(CN)$_6^{3-}$ according to Eq. (1)

Enzyme concentration (10^{-4} M)	f	K_1 (10^6 s^{-1})	K_L	K_{av} (10^6 s^{-1})
0	0.38	54	0	1.2
0.145	0.38	54	0	1.2
0.483	0.27	72	0	1.0
0.870	0.23	73	0	1.0
1.200	0.18	56	0	0.7

reaction (quenching) becomes highly restricted. This is also reflected in the values of K_{av} which decrease with increasing enzyme concentration, as well as the recorded luminescence intensity (not shown) which is lower in the presence of enzyme. K_L values are all zero, in accordance with the fact that this system is below the electric percolation threshold. K_1 values are invariable within experimental error, or, at the limit, they increase. This implies a possible decrease of the number of micelles in the presence of enzyme, which is in accordance with the values of f. The above results indicate that the mobility of the ionic species in a water-in-oil microemulsion decreases in the presence of enzyme.

In conclusion, the presence of enzyme in w/o microemulsions affects the electric percolation threshold and induces a progressive decrease of the electric conductivity. The electric charge carriers may be immobilized within pockets (sub-compartments) formed with the help of the cosolubilized enzyme.

References

1. Luisi PI, Straub B (eds) (1984) Reverse Micelles. Plenum Press, London
2. Luisi PL, Giomini M, Pileni MP, Robinson BH (1988) Biochim Biophys Acta 947:209
3. Martinek K, Levashov AV, Klyachko Y, Berezin IV (1986) Eur J Biochem 155:453
4. Cazianis CT, Xenakis A, Evangelopoulos AE (1987) Biochem Biophys Res Commun 148:1151
5. Walde P, Peng Q, Fadnavis NW, Battistel E, Luisi PL (1988) Eur J Biochem 173:401
6. Barbaric S, Luisi PL (1981) J Am Chem Soc 103:4239
7. Huruguen JP, Authier M, Greffe JL, Pileni MP (1991) Langmuir 7:243
8. Lianos P, Zana R, Lang J, Cazabat AM (1986) In: Mittal KL, Bothorel P (eds) Surfactants in Solution, Vol 6. Plenum, New York, p 1365
9. Lagues M, Ober R, Taupin C (1978) J Phys Lett 39:L—487
10. Van Dijk MA (1985) Phys Rev Lett 55:1003
11. Lianos P, Modes S, Staikos G, Brown W (1992) Langmuir 8:1054

Authors' address:

A. Xenakis
National Hellenic Research Foundation
Institute for Biological Research & Biotechnology
48, Vas. Constantinou Ave.
GR-11635 Athens, Greece

Progress in Colloid & Polymer Science

Progr Colloid Polym Sci 93:373 (1993)

Effect of alcohols on the structure of AOT reverse micelles with respect to different enzyme activity

A. Xenakis[1]), H. Stamatis[1]), A. Malliaris[2]), and F. N. Kolisis[3])

[1]) Inst. of Biological Research & Biotechnology, National Hellenic Research Foundation, Athens, Greece
[2]) N.R.C. "Democritos", Athens, Greece
[3]) Chemical Engineering Dept., National Technical University of Athens, Greece

Abstract: The activity of lipases from *Rhizopus delemar* and *Penicillium simplicissimum* has been studied in esterification reactions of various aliphatic alcohols with fatty acids in microemulsions containing AOT in isooctane. The effect on the enzyme activity of the alcohol chain length and their structure has been examined. *P. simplicissimum* showed higher reaction rates in the esterification of long chain secondary alcohols, primary alcohols showed lower rates, while the tertiary ones exhibited the slowest rates. *R. delemar* lipase showed a preference for the esterification of short chain primary alcohols, while the secondary alcohols had a low rate and the tertiary ones could not be catalyzed. The effect of the alcoholic substrates on the reverse micellar structure as well as the localization of the enzyme, were examined by fluorescence quenching measurements and spectroscopical studies.

Key words: Lipases — selectivity — esterifications — microemulsions — reverse micelles

Introduction

The use of lipases for the transformation of water insoluble substrates is becoming increasingly recognized [1]. In addition to hydrolysis of triglycerides, lipases can be used in esterification and transesterification reactions in low-water content media [2]. This catalytic process is heterogeneous and is favored in w/o microemulsions, which are dispersions of oil and water stabilized by surfactant molecules [2]. In these water containing microdroplets enzyme molecules can be entrapped thus avoiding direct contact with the unfavorable organic medium while retaining their catalytic ability [3]. Microemulsions have been employed for lipase-catalyzed hydrolysis of triglycerides [4], transesterification [5, 6], and glyceride synthetic reactions [7, 8].

In the present study, w/o microemulsions were used to host the esterification of various alcohols by fatty acids, catalyzed by *Penicillium simplicissimum* and *Rhizopus delemar* lipases. *P. simplicissimum* is an extracellular enzyme which shows a high stabiality in water-immiscible organic solvents and it is nonspecific in the hydrolysis of mono-, di-, and triolein [9], while *R. delemar* lipase is 1,3 regio specific in hydrolytic reactions.

Materials and methods

Lipase from *P. simplicissimum* was supplied by Dr. U. Menge, GBF, and was purified by the method proposed by Sztajer et al. [9], while the one from *R. delemar* (Fluka, Basel, Switzerland) was purified according to [10]. Bis-(2-ethylhexyl)sulfosuccinate sodium salt (AOT) (99%, Sigma, USA), the fluorescent probe $Ru(bpy)_3Cl_2 \cdot 6\,H_2O$ (G. F. Smith Chemical Co.) and the quencher $K_3Fe(CN)_6$ (Merck) were used as supplied. All alcohols, organic solvents, fatty acids and assay reagents used were of the highest purity. Doubly distilled water was used throughout this study. The esterification reaction was carried out as described previously [11].

Fluorescence quenching measurements were performed using a Perkin-Elmer 650-40 fluorometer. All solutions were measured without previous degassing, in equilibrium with atmospheric oxygen. The excitation wavelength used was 450 nm, at which no absorption was detected due to any of the constituents of the solution and the emission wavelength was 620 nm. In small confinements, e.g., microemulsions [12], under certain conditions a linear relationship exists between $\ln(F_0/F_i)$ and $[Q]$, where F_i and F_0 are the fluorescence intensities in the presence and in the absence of quencher, respectively, and $[Q]$ is the concentration of the added quencher. The slope of the $\ln(F_0/F_i)$ vs. $[Q]$ plot represents the reciprocal of the concentration of the reverse micelles, assuming that the sizes are monodispersed.

Results and discussion

In order to elucidate the role of the structure of the alcohols on lipase activity the esterification of primary, secondary and tertiary alcohols with varying chain lengths catalyzed by *P. simplicissimum* and *R. delemar* lipases was investigated.

Table 1 shows the measured values of the initial velocities for the different esterification reactions catalyzed by the two lipases. The effect on *P. simplicissimum* lipase activity was examined following the rate of the esterification of pentanol-1, pentanol-2 and 2-methylbutan-2-ol with oleic acid. As shown in Table 1, the degree of esterification of the secondary alcohol, pentanol-2, is much higher than that for the primary pentanol-1, with the tertiary alcohol 2-methylbutan-2-ol the reaction was very slow. A different specificity was observed in the esterification of the same alcohols by *R. delemar* lipase, i.e., the reaction rate for pentanol-1 was three times higher than for pentanol-2. It is interesting to note that *R. delemar* lipase cannot catalyze the esterification of tertiary alcohols. The preference of *P. Simplicissimum* lipase for the secondary alcohols in this type of reaction was also confirmed in studies of esterification of cyclohexanol and n-hexanol with lauric acid.

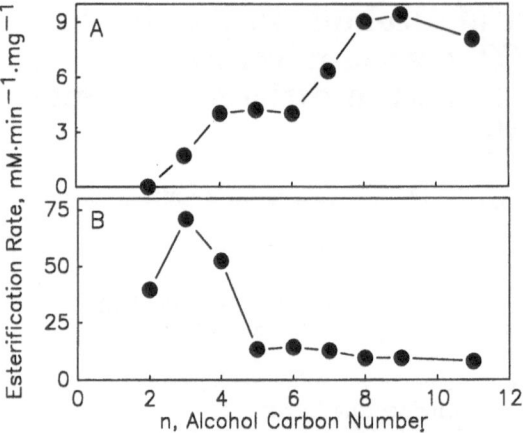

Fig. 1. Effect of primary alcohol chain length on the esterification of 50 mM lauric acid catalyzed by A) *P. simplicissimum*, and B) *R. delemar* lipase in microemulsions. The alcohol concentration was 150 mM

Table 1. Rates of esterification of various alcohols with oleic acid

Alcohols[a]	Initial velocity, mM min^{-1} mg^{-1}	
	P. simplicissimum	*R. delemar*
Pentanol-1	9.7	17.6
Pentanol-2	43.7	5.8
2-Methyl-butanol-2	1.7	—
Hexanol	7.6	15.6
Cyclohexanol	11.2	—
(—)Methanol	75.0	—
Nerol	58.0	—
Geraniol	58.0	—
Cholesterol	4.0	—

[a] Esterification of 50 mM oleic acid with 120 mM of each alcohol.

The role of the chain length (C = 2 to 11) of various primary alcohols on lipase activity was examined by studying their esterification with lauric acid catalyzed by the two lipases. As shown in Fig. 1A, the activity of *P. simplicissimum* lipase increases with increasing chain length of the primary alcohol. *R. delemar* lipase shows a maximum esterification rate for alcohols with C = 3 and a profound decrease with further increasing of the chain length

(Fig. 1B). Similar results have been reported for *R. delemar* lipase in other microemulsions [13], while in other reaction media the findings were different [14, 15]. Our observation that secondary and tertiary alcohols are less efficiently catalyzed by *R. delemar* lipase, is in agreement with the work of Tsujisaka et al. [15] with the same enzyme but in a different system. It should be emphasized that *P. simplicissimum* lipase catalyzes the esterification of secondary alcohols (Table 1) with high efficiency, contrary to the catalytic esterification of lipases reported so far.

The particular behavior of *P. simplicissimum* lipase as well as the differences observed in the selectivity of *R. delemar* lipase with respect to the alcohol chain length, can be attributed to various factors such as the effect of structure and size of the reverse micelles, the availability of the substrates, and the location of the enzyme in the microemulsion.

In order to elucidate modifications of the reverse micellar structure upon adding the substrates possibly responsible for the observed differences of the enzymatic activity, we have employed the fluorescence quenching technique. A series of microemulsions having the same composition, but each containing the same amount of different alcohols, was tested. Plots of $\ln(F_0/F_i)$ vs. [Q] (Fig. 2), were linear, indicating the existence of distinct confinements in all cases. The micellar concentration measured from the slopes of these curves was found to be of the order of 5×10^{-4} M, and to be

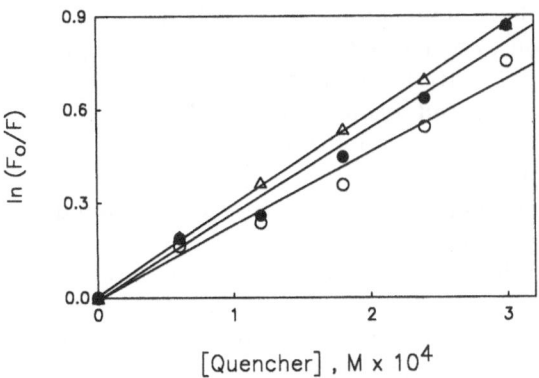

Fig. 2. Typical examples of Stern-Volmer plots obtained in 0.1 M AOT/isooctane microemulsions upon addition of 30 mM of (−)menthol (○), hexanol (●) and cyclohexanol (△). Concentration of fluorophor $[Ru(bpy)^{2+}] = 2 \cdot 10^{-5}$ M

independent of the type of alcohol used. Using simple geometrical considerations, and assuming monodispersity of the droplets, the radii of the reverse micelles were calculated to be ca. 20 Å, a value in good agreement with the literature [16, 17]. It seems therefore that within the validity of the static fluorescence quenching technique, the presence of the alcohol molecules does not significantly disturb the structure of the reverse micelles. Dynamic light scattering measurements on similar AOT microemulsions with the same alcohols have also led to the same conclusion [13].

The fluorescence spectra of these enzymes due to the tryptophan residues [18] showed differences depending on the type of lipase used and on the water content. Namely, in the case of *R. delemar* lipase the emission maximum shifts from 336 nm to 346 nm when w_0 is increased from 6 to 30, while in a water solution the maximum is at 349 nm. On the other hand, the emission maximum of *P. simplicissimum* lipase does not change upon varying w_0, and remains around 335 nm.

The observed fluorescence shift is related to the variation of the polarity in the dispersed water phase [19], due to the increase of w_0. This behavior of the two lipases could be attributed to their different site of localization in the reverse micelles. *R. delemar* lipase seems to reside in the water core of the microdroplets, thus being more sensitive to polarity variations, while *P. simplicissimum* lipase is probably located near the microinterface. It is evident that if the enzyme molecule is close to the interface, it will better catalyze the apolar substrates,

whereas if it is located in the water core, the polar substrates will be favored (Fig. 1). Moreover, the different sites of localization of the two lipases can possibly be explained in terms of their hydrophobicity. Indeed, *P. simplicissimum* lipase has been reported to exhibit a high ratio of apolar amino acids and a high stability in apolar solvents [9], suggesting a pronounced hydrophobic character. On the other hand, *R. delemar* lipase can be considered as relatively hydrophilic, since it has been reported as glucosylated protein [20].

In conclusion, our results indicate that the observed differences in enzyme specificities can be attributed to the different sites of solubilization of the enzymes in these w/o microemulsions. In addition the different esterification rates are related to the partitioning of the enzymes in the various sites of the reaction medium.

Acknowledgement

The authors wish to thank Dr. U. Menge from GBF for his generous gift of *P. simplicissimum* lipase. This work has been supported by EEC in the frame of BRIDGE.

References

1. John VT, Abraham G (1990) In: Dordick J (ed) Biocatalysis for industry. Plenum, New York
2. Zaks A, Klibanov AM (1985) Proc Natl Acad Sci USA 82:3192
3. Luisi PL, Straub B (eds) (1984) Reverse Micelles Plenum Press, London
4. Xenakis A, Valis TP, Kolisis FN (1989) Progr Colloid Polym Sci 79:88
5. Bello M, Thomas D, Legoy MD (1987) Biochem Biophys Res Commun 146:361
6. Holmberg K, Osterberg E (1987) Progr Colloid Polym Sci 74:98
7. Hayes DG, Gulari E (1991) Biotechnol Bioeng 38:507
8. Fletcher PDI, Freedman RB, Robinson BH, Rees GD, Schomacker R (1987) Biochim Biophys Acta 912:278
9. Sztajer H, Lunsdorf H, Erdmann H, Menge U, Schmid R (1992) Biochim Biophys Acta 1124:253
10. Valis TP, Xenakis A, Kolisis FN (1992) Biocatalysis 6:267
11. Stamatis H, Xenakis A, Provelegiou M, Kolisis FN (1993) Biotechnol Bioeng 42:103
12. Turro NJ, Yekta A (1978) J Am Chem Soc 100:5951
13. Hayes DG, Gulari E (1990) Biotechnol Bioeng 35:793
14. Morita S, Narita H, Matoba T, Kito M (1984) J Am Oil Chem Soc 61:1571
15. Okumura S, Iwai M, Tsujisaka Y (1979) Biochim Biophys Acta 575:156
16. Kotlarchyk M, Chen SH, Huang JS (1982) J Phys Chem 86:3273

17. Zulauf M, Eicke HF (1979) J Phys Chem 83:480
18. Weber G (1960) Biochem J 75:345
19. Lakowicz JR, Weber G (1973) Biochemistry 12:4171
20. Haas MJ, Allen J, Berka TR (1991) Gene 109:107

Authors' address:

A. Xenakis
Inst. of Biolog. Research & Biotechnology
National Hellenic Research Foundation
48, Vas. Constantinou Avc.
116 35 Athen, Greece

Small-angle x-ray scattering studies on the polydispersity of iron micelles in ferritin

P. Zipper[1]), M. Kriechbaum[2]),
and H. Durchschlag[3])

[1]) Institute of Physical Chemistry, University of Graz, Austria
[2]) Department of Physics, Princeton University, USA
[3]) Institute of Biophysics and Physical Biochemistry, University of Regensburg, FRG

Key words: Ferritin — polydispersity of iron micelles — small-angle x-ray scattering — contrast variation — computer simulations

Ferritin, the major iron-storage protein in mammals, consists of an approximately spherical protein shell from 24 subunits, surrounding a core of microcrystalline ferric oxide hydrate. The ferric micelle contains a variable number of iron atoms (up to 4500). Both full ferritin and iron-free apoferritin have already been a subject of previous small-angle investigations using x-rays or neutrons (e.g., [1—4]).

The present study is primarily concerned with small-angle x-ray scattering (SAXS) studies on various intermediate states of horse-spleen ferritin differing in their iron content [5]. These intermediates were produced by partial removal of iron from full ferritin (by treatment with $Na_2S_2O_4$ and EDTA) or by incorporation of iron into apoferritin (by treatment with $(NH_4)_2Fe(SO_4)_2$). The iron content of the intermediates was checked spectrophotometrically. SAXS measurements were per-

formed in dilute buffer or in 66% (w/v) sucrose solution to match out the scattering of the protein moiety (contrast variation). The scattering data were processed using program ITP [6], yielding distance distribution functions, $p(r)$, and (mean) radii of gyration, R_G.

Experiments in buffer revealed pronounced changes of $p(r)$ functions (Fig. 1) and of R_G values upon release or uptake of iron. Less pronounced changes were derived from the contrast variation experiments. The data obtained in sucrose solution were also analyzed in terms of size distribution functions [7] of the iron micelles. The samples investigated exhibited a bimodal polydisperse distri-

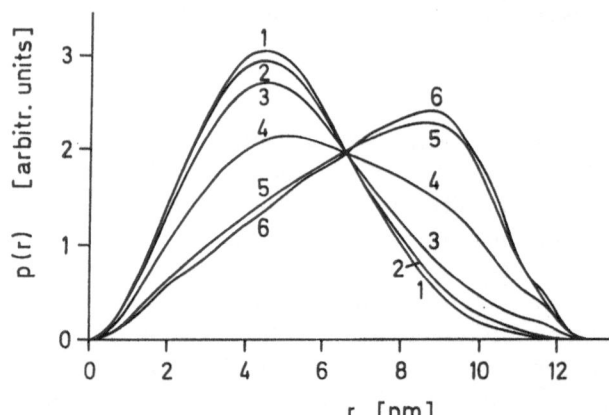

Fig. 1. Distance distribution functions $p(r)$ derived from SAXS measurements of native and treated ferritin (samples 1—5, Fe content 20, 18, 10, 4.5, 0.5% (w/w), R_G = 3.73, 3.82, 4.08, 4.67, 5.24 nm) and of apoferritin (sample 6, R_G = 5.33 nm) in dilute buffer. The functions are normalized to the same area and correspond to particles of 12.5 ± 0.5 nm diameter. The changes in the shape of the functions and the shift of the maximum from 4.5 nm (for native ferritin) to 8.7 nm (for apoferritin) are caused by the alterations of the electron density distribution upon release of iron

bution of micellar size. The type of distribution was retained during the release of iron, but its features were found to vary in a characteristic fashion with the iron content (Fig. 2). This behavior was interpreted to reflect the existence of two types of micelles differing not only in size (mean diameters in native ferritin ≈ 1.8 and 7.2 nm, respectively), but presumably also in the rate of iron release.

Model calculations based on the assumption of three kinds of coexisting particles (two ferritin species differing in micellar size and apoferritin)

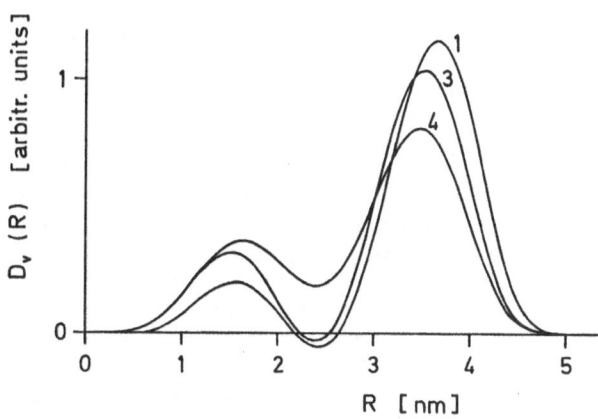

Fig. 2. Size distribution functions derived from SAXS measurements of native and treated ferritin (samples 1, 3, 4; cf. Fig. 1) in sucrose solution. $D_v(R)$ represents the distribution of volumes of spherical particles whose size is expressed by their radius R

Position-resolved x-ray scattering of moldings from semicrystalline polymers

P. Zipper[1]), A. Jánosi[1]), E. Wrentschur[1]),
P. M. Abuja[2]), and C. Knabl[1])

[1]) Institute of Physical Chemistry, University of Graz, Austria
[2]) Present address: Institute of Biophysics and X-Ray Structure Research, Austrian Academy of Sciences, Graz, Austria

Key words: Semicrystalline polymers — polypropylene — structure — texture — position-resolved x-ray scattering

were found to simulate the experimentally observed changes in the $p(r)$ functions upon iron release very well, whereas models assuming a synchronous degradation of uniformly sized micelles turned out to be incompatible with the experimental results. The findings indicate that iron is simultaneously released from large and small micelles, whereby large micelles are intermediately transformed into small ones.

References

1. Bielig HJ, Kratky O, Steiner H, Wawra H (1963) Monatsh Chem 94:989—991
2. Fischbach FA, Anderegg JW (1965) J Mol Biol 14:458—473
3. Bielig HJ, Kratky O, Rohns G, Wawra H (1966) Biochim Biophys Acta 112:110—118
4. Stuhrmann HB, Haas J, Ibel K, Koch MHJ, Crichton RR (1976) J Mol Biol 100:399—413
5. Kriechbaum M (1986) Thesis, Univ Graz, Austria
6. Glatter O (1977) J Appl Cryst 10:415—421
7. Glatter O (1980) J Appl Cryst 13:7-11

Authors' address:

Prof. Dr. Peter Zipper
Institut für Physikalische Chemie
Universität Graz
Heinrichstrasse 28
A-8010 Graz, Austria

The cross-section of plastic parts extruded or injection-molded from polypropylene or other semicrystalline polymers is not homogeneous but exhibits a sandwich-like architecture. Details of this layered structure do not only depend on the processing conditions to a high degree, but may also vary considerably within the moldings with increasing distance from the gate. The elucidation of the morphology and texture in the moldings is of importance because of the relations between processing conditions, structure, and end use properties. As previously shown, position-resolved x-ray scattering is a potent tool for the characterization of the parts in terms of structure and texture [1—3].

In our experimental approach the cross-section of plastic parts is scanned with a fine x-ray beam aligned parallel to their surface and the intensity of scattering is measured as a function of both the scattering angle and the position in the cross-section. The line-shaped x-ray beam (cross-section about 60 μm × 2 mm at the sample) is obtained by means of a Kratky camera; the narrow width of the beam renders a high resolution of position in the direction of interest (viz. along the normal to the surface). For scanning, the sample is moved in steps of 10 or 20 μm through the beam. The same setup is used both for small-angle and wide-angle experiments.

The analysis of intensity maps (Fig. 1) obtained from wide-angle measurements on cross-sections using a linear position-sensitive detector delivers detailed information on the layered structure, the distribution of different polymer modifications, the size of crystallites as well as on the state of orienta-

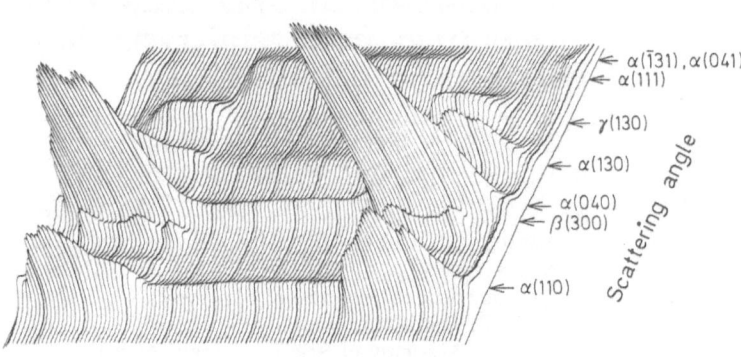

Distance from surface

Fig. 1. Wide-angle x-ray intensity map (intensity as a function of scattering angle, plotted vs. the distance from surface) of a sample cut from an injection-molded polypropylene plate (thickness of cross-section $\approx$ 2.1 mm, scanned in steps of 20 μm, angular range from 12.8 to 22.5°). The intensity map of the cross-section displays reflections due to different crystalline modifications: α, β, γ; the pertaining indices are shown. The pronounced differences in intensity between different regions in the cross-section reveal a much higher degree of orientation in the surface layers than in the core

tion of crystallites and non-crystalline material. This wealth of information may serve as the basis for a better understanding of mechanical properties [4] of the plastic parts.

Acknowledgement

The work ws supported by the "Fonds zur Förderung der wissenschaftlichen Forschung" (project P7446).

References

1. Fleischmann E, Zipper P, Jánosi A, Geymayer W, Koppelmann J, Schurz J (1989) Polym Eng Sci 29:835—843
2. Zipper P, Jánosi A, Wrentschur E (1990) Österr Kunststoff Z 21:54—59
3. Zipper P, Jánosi A, Wrentschur E, Abuja PM (1991) J Appl Cryst 24:702—708
4. Fleischmann E (1990) Thesis, Univ Leoben, Austria

Authors' address:

Prof. Dr. Peter Zipper
Institut für Physikalische Chemie
Universität Graz
Heinrichstrasse 28
A-8010 Graz, Austria

Low gravity: a new research tool

H. J. Sprenger

INTOSPACE, Hannover, Germany

The weightlessness of space, called "microgravity" (10^{-6} g) or "low gravity", is available on flights of the Space Shuttle, unmanned re-entry capsules, and sounding rocket flights. For very short times (seconds) the state of low gravity can be achieved also on earth (drop towers) or parabolic flights of an airplane.

The most important advantages of microgravity are: strongly reduced thermal convection, the absence of buoyancy and sedimentation, and the possibility of containerless processing. These unique features allow for the quantification of gravity-related phenomena and the investigation of the influence of weak forces on physico-chemical process otherwise hidden by gravity.

The main role of microgravity can be sonsidered to be an important additional tool for terrestrial fundamental and applied research.

The influence of gravity is most important in systems with at least one fluid phase.

Typical examples are:

— Density differences in multi-phase systems lead to flows and demixing.
— Concentration and density changes at advancing solidification fronts cause convection induces flows and disturb transient diffusion patterns.
  Aggregation of particles in suspension by weak (van der Waals) forces are limited in size because of convection and sedimentation.
— Crystal growth of materials with homogeneous distribution dopants suffers from temperature and concentration fluctuations.
— Heterogeneous nucleation by crucible and container walls impedes the achievement of metastable or amorphous state.
— Measurements of thermophysical parties of fluids are strongly falsified by convective flows.

Main advantages of microgravity therefore can be supposed for the investigation of

— crystallization and solidification (phase transitions)
— aggregation phenomena (multi-phase fluids)
— capillary forces (surfaces and interfaces).

Application of microgravity for the study of aggregation phenomena in colloid systems

The main objectives for research in colloid systems are:

— Study of van der Waals forces.
— Application of statistical mechanics to kinetic phenomena and rheology.
— Growth of large fractals.
— Light scattering and neutron scattering for determination of particle distributions.
— Rheological measurements.
— Colloidal phenomena in biological systems (cellular and molecular level).
— Numerical modeling of experimental results.

The results of such studies are important for:

— Improvements in ceramic processing.
— Preparation of metal-matrix composites.
— Processing and products of food industry.
— Pharmaceutical and cosmetic products.
— Rheological processes.
— Behaviour of aerosols.
— Understanding of biological processes.

One of the parameters in the experimental investigation of colloid systems which has not received much attention so far is the utilization of microgravity. A microgravity environment is ideal for studies on the prevention, respectively the investigation of particle agglomeration, since convection and sedimentation are considered as strongly dominant effects. There is little knowledge on combined effects of different interaction forces, which are more or less dominant in terrestrial systems and are dependent on volume fraction, wettability, and type of solution phase. Hence microgravity offers the unique opportunity to eliminate gravity effects and to establish the significance of other forces.

In this way, the reduction of gravity constraints would allow to study the basic influence of weak forces leading to collisions and aggregations in dispersions, the possibility of maintaining a stable dispersion in metallic melts, the growth of fractal aggregates and crystals from a solution/suspension. If these forces are found to be of major importance, relevant microgravity experiments will allow to quantify necessary gravity corrections in models.

We have already started a concerted action aimed at elaborating the main open scientific questions and to give the appropriate ansers about how to solve them and how to introduce the achieved knowledge into applications by industry. *Such an approach would consider microgravity as an additional research tool (like vacuum or accelerators) and should only be taken into account if the terrestrial work, elaborated in the significance and usefulness of microgravity experimentation.* With respect to this argumentation, this approach will be multidisciplinary, i.e., it will also encompass related research areas in which colloidal phenomena can play an important role. The envisaged range covers selected fields of materials science and biology, such as sol-gel processing, rheological phenomena in composite casting, suspension polymerization, micelle formation and cell aggregation phenomena.

Author's address:

Heinz J. Sprenger
INTOSPACE
Hannover, FRG

Electrophoretic light scattering: a coherent detection method using amplitude weighted phase statistics

A. Sobotta, K. Schätzel, W. Weise, and S. Egelhaaf

Institut für Angewandte Physik, Kiel, FRG

Key words: Light scattering — electrophoresis — phase analysis

We present a new method to measure small electrophoretic mobilities of colloidal dispersions using the phase of scattered laser light and computing the Amplitude Weighted Phase Structure Function (AWPS). Processing phases instead of frequencies allow a better separation of electrophoretic and distributing diffusive motion than do conventional techniques [1]. Our AWPS-arrangement with non-coherent detection works fine for particles with diameters larger than 80 nm [2]. Using smaller particles requires higher concentrations of the dispersions to get — for a given maximum laser power — a sufficient signal on the detectors. In the case of many particles (more than 10 000 in the measurement volume) the coherent detection of the scattered light yields a much better signal-to-noise ratio than does the noncoherent [3]. In our experiment the phase information is obtained by means of a 4-detector Mach Zehnder interferometer with polarization coding using polarization-dependent beam splitters in combination with quarter- and half-wave-plates. The 4-detector setup suppresses fluctuations of the beam amplitudes.

Mobility measurements of latex particles with diameters of more than 80 nm show a very good agreement with our noncoherent setup.

References

1. Schätzel K, Merz J (1984) J Chem Phys 81:2482
2. Schätzel K, Weise W, Sobotta A, Drewel M (1991) J Colloid Int Sci 143:287
3. Drain LE (1972) J Phys D: Appl Phys 5:481

Authors' address:

Andreas Sobotta
Institut für Angewandte Physik
Leibnizstr. 11
24098 Kiel, FRG

disturbed by electroosmosis. Both effect add up to give the measured (apparent) particle mobility. In order to find undisturbed particle mobilities, we use a theory of electroosmotic velocity profiles of the fluid for the case of a long cylindrical sample cell and a sinusoidal electric field [1]. The electroosmosmotic mobility is strongly frequency dependent in the low frequency range (we use up to 30 Hz), while the electrophoretic particle mobility is not. Therefore, measured (apparent) mobilities are frequency dependent due to the electroosmotic part and a phase shift between the dielectric field and the measured velocity is introduced. We measure this phase shift and use it to correct apparent mobilities for electroosmosis.

Our measurements are performed by light scattering (real fringe laser Doppler) and phase processing of the detected signal [2, 3]. This technique does not imply the restrictions concerning field amplitude and frequency that limited conventional spectral analysis methods. Therefore, the field strengths may be as low as 1 V/cm or less at 20 Hz for mobilities of order of magnitude 10^{-8} m^2/Vs.

References

1. Schätzel K, Weise W, Sobotta A, Drewel M (1991) J Colloid Int Sci 143:287
2. Schätzel K, Merz J (1984) J Chem Phys 81:2482
3. Miller J, Schätzel K, Vincent B (1991) J Colloid Int Sci 143:532

A highly sensitive method to measure electrophoretic and electroosmotic mobilities in real time

W. Weise, K. Schätzel, and A. Sobotta

Institut für Angewandte Physik, Kiel

Key words: Light scattering — electrophoresis — electroosmosis — phase analysis

Measurements of electrophoretic mobilities of particles suspended in some fluid are always

Authors' address:

Wieland Weise
Institut für Angewandte Physik
Leibnizstr. 11
24098 Kiel, FRG

Author Index

Subject Index